ELSEVIER

APPLIED MYCOLOGY AND BIOTECHNOLOGY

VOLUME 5

GENES AND GENOMICS

APPLIED MYCOLOGY AND BIOTECHNOLOGY

VOLUME 5

GENES AND GENOMICS

Edited by

Dilip K. Arora
Centre of Advanced Study in Botany
Banaras Hindu University
Varanasi, India

Randy M. Berka
Novozymes Biotech, Inc.,
1445 Drew Avenue,
Davis, CA 95616-4880, USA

ELSEVIER
2005

Amsterdam – Boston – Heidelberg – London – New York – Oxford
Paris – San Diego – San Francisco – Singapore – Sydney – Tokyo

ELSEVIER B.V.
Radarweg 29
P.O. Box 211, 1000 AE Amsterdam
The Netherlands

ELSEVIER Inc.
525 B Street, Suite 1900
San Diego, CA 92101-4495
USA

ELSEVIER Ltd.
The Boulevard, Langford Lane
Kidlington, Oxford OX5 1GB
UK

ELSEVIER Ltd.
84 Theobalds Road
London WC1X 8RR
UK

First edition 2005

Library of Congress Cataloging in Publication Data
A catalog record is available from the Library of Congress.

British Library Cataloguing in Publication Data
A catalogue record is available from the British Library.

ISBN: 0-444-51808-8

Transferred to digital print 2007
Printed and bound by CPI Antony Rowe, Eastbourne

Editors

Dilip K. Arora
Centre of Advanced Study in Botany
Banaras Hindu University
Varanasi, 221005 India
E-mail: aroradilip@yahoo.co.in

Randy M. Berka
Novozymes Biotech, Inc.
1445 Drew Avenue
Davis, CA 95616-4880, USA
E-mail: Rambo@novozymesbiotech.com

Editorial Board

*Present affiliation: National Bureau of Agriculturally Important Microorganisms, Kusmaur (Post: Kaithauli), Post Bag No. – 06, Mau Nath Bhanjan, Uttar Pradesh 275 101, INDIA

Contents

Contributors

Luciano Angelo   Faculdade de Ciências Farmacêuticas de Ribeirão Preto, Universidade de São Paulo, São Paulo, Brazil.

Guus Bakkeren   Agriculture and Agri-food Canada, Pacific Agri-food Research Center, 4200 Highway 94, Summerland, B.C. Canada VOZ1Z0.

Randy M. Berka   Novozymes Biotech, Inc., 1445 Drew Avenue, Davis, California 95616-4880, USA.

Elena V. Bashkirova   Novozymes Biotech, Inc., 1445 Drew Avenue, Davis, California 95616-4880, USA.

Frederick J. Bowring   School of Biological Sciences, Flinders University, PO Box 2100, Adelaide, South Australia 5001.

David E.A. Catcheside   School of Biological Sciences, Flinders University, PO Box 2100, Adelaide, South Australia 5001.

Dan Cullen   USDA Forest Products Laboratory, Madison, Wisconsin 53705, USA.

Hernan D. Folco	Wellcome Trust Centre for Cell Biology, University of Edinburgh, Scotland, United Kingdom.
Scott E. Gold	Department of Plant Pathology, University of Georgia, Athens, GA 30602-7274, USA.
Gustavo H. Goldman	Faculdade de Ciências Farmacêuticas de Ribeirão Preto, Universidade de São Paulo, São Paulo, Brazil.
Maria Helena de S. Goldman	Faculdade de Ciências Farmacêuticas de Ribeirão Preto, Universidade de São Paulo, São Paulo, Brazil.
Anke Grünler	Institute for Microbiology, Friedrich-Schiller-University Jena, Neugasse 24, D-07743 Jena, Germany.
Ian Grainge	Section of Molecular Genetics and Microbiology, University of Texas at Austin, Austin, TX 78712, USA.
Steven D. Harris	Plant Science Initiative and Department of Plant Pathology, University of Nebraska, N234 Beadle Center, Lincoln, NE 68588-0660, USA.
Harm J Hektor	Biomaoe Technology, Nijenbough 4, 9747 AG Groningen, The Netherlands.
J. Stephen Horton	Department of Biological Sciences, Science and Engineering Center, Union College, Schenectady, NY 12308-2311, USA.
Makkuni Jayaram	Section of Molecular Genetics and Microbiology, University of Texas at Austin, Austin, TX 78712, USA.

Steven J. Klosterman Department of Plant Pathology, University of Gorgia, Athens GA 30602-7274, USA.

Luis F. Larrondo Departamento de Genética Molecular y Microbiología, Facultad de Ciencias Biológicas, Pontificia Universidad Católica de Chile, Santiago, Chile and Instituto Milenio de Biología Fundamental y Aplicada, Santiago, Chile.

André Rodrigues Lopes Faculdade de Ciências Farmacêuticas de Ribeirão Preto and Faculdade de Filosofia, Ciências e Letras de Ribeirão Preto, Universidade de São Paulo; Av. do Cafe S/N, CEP 14040-903, Ribeirão Preto, São Paulo, Brazil.

Giuseppe Macino Dipartimento di Biotecnologie Cellulari e Ematologia, Universita' di Roma "La Sapienza", Viale Regina Elena 324, 00161 Rome, Italy.

Everaldo dos Reis Marques Faculdade de Ciências Farmacêuticas de Ribeirão Preto, Universidade de São Paulo, São Paulo, Brazil.

Xiang Jia Min Centre for Structural and Functional Genomics, Concordia University, 7141 Sherbrooke Street West, Montreal, QC H4B 1R6, Canada.

David Moore School of Biological Sciences, The University of Manchester, 1.800 Stopford Building, Manchester M13 9PT, United Kingdom.

Helena Nevalainen Department of Biological Sciences, Macquarie University, Sydney, NSW 2109, Australia.

Maria Garcia-Pedrajas	Department of Plant Pathology, University of Georgia, Athens, GA 30602-7274, USA.
Annette Pickford	Dipartimento di Biotecnologie Cellulari e Ematologia, Universita' di Roma "La Sapienza", Viale Regina Elena 324, 00161 Rome, Italy.
Tiina Pakula	VTT Biotechnology, P.O. Box 1500, FIN-02044 VTT, Finland.
Merja Penttila	Department of Biological Sciences, Macquarie University, Sydney, NSW 2109, Australia.
Stefanie Pöggeler	Lehrstuhl für Allgemeine und Molekulare Botanik, Ruhr-Universität Bochum, Universitätsstr. 150, D-44780 Bochum, Germany.
Geoffrey D. Robson	School of Biological Sciences, The University of Manchester, 1800 Stopford Building, Manchester M 139PT, United Kingdom.
Alberto L. Rosa	Instituto de Investigación Médica "Mercedes y Martín Ferreyra", INIMEC-CONICET, Córdoba, Argentina.
Michael W. Rey	Novozymes Biotech, Inc., 1445 Drew Avenue, Davis, California 95616-4880, USA.
Rick Rink	Biomaoe Technology, Nijenbough 4, 9747 AG Groningen,The Netherlands.
Karin Scholtmeijer	BiOMaDe Technology, Nijenborgh 4, 9747AG Groningen, The Netherlands.
Christine Schimek	Institute for Microbiology, Friedrich Schiller-University Jena, Neugasse 24, D-07743 Jena, Germany.
Camile P. Semighini	Plant Science Initiative and Department of Plant Pathology, University of Nebraska, N234 Beadle Center, Lincoln, NE 68588-0660, USA.
Esteban D. Temporini	Division of Plant Pathology and Microbiology, Department of Plant Sciences, University of Arizona, Tucson, AZ, USA.

Velentino Te'O   Department of Biological Sciences, Macquarie University, Sydney, NSW 2109, Australia.

Rafael Vicuña   Departamento de Genetica Molecular y Microbiologia, Facultad de Ciencias Biologicas, Pontificia Universidad Catolica de Chile, Santiago, Chile and Instituto Milenio de Biologia Fundamental y Aplicada, Santiago, Chile.

Kersti Voigt   Institute for Microbiology, Friedrich Schiller-University Jena, Neugasse 24, D-07743, Jena, Germany.

Yuri Voziyanov   Section of Molecular Genetics and Microbiology, University of Texas at Austin, Austin, TX 78712, USA.

Conor Walsh   School of Biological Sciences, The University of Manchester, 1800 Stopford Building, Manchester M 139PT, United Kingdom.

Johannes Wöstemeyer   Institute for Microbiology, Chair of General Microbiology and Microbial Genetics and Fungal Reference Centre, Friedrich-Schiller-University Jena, Neugasse 24, D-07743 Jena, Germany.

Han AB Wösten   Deptt. of Microbiology, Institute of Biomembranes, University of Utrecht, Padualaan 8, 3584 CH Utrecht The Netherland.

P. Jane Yeadon   School of Biological Sciences, Flinders University, PO Box 2100, Adelaide, South Australia 5001.

Peijun Zhang   Centre de Génomique Fonctionnelle de Sherbrooke, Faculté de Medicine, Université de Sherbrooke, 3001, 12e Avenue Nord, Sherbrooke, QCJ1H5N4, Canada.

Preface

The fungal kingdom represents an extremely diverse group of organisms, and members of this group are found in virtually every ecosystem on Earth. The most familiar species to us are those fungi that are important pathogens of plants and animals and those that are used commercially for production of enzymes, biochemicals, and antibiotics. Evolutionary studies suggest that the first filamentous ascomycetes diverged from yeasts approximately one billion years ago, which is shortly after the fungi diverged from plants and animals. This raises a number of interesting questions about the evolution of fungal genomes and the functional equivalence of genes in fungi and higher eukaryotes.

It was just five years ago that the community of fungal biologists bemoaned the fact that numerous bacterial genomes as well as yeast, mouse, and human genomes were being completed, but the study of genomes from filamentous fungi was lagging far behind. Thankfully, the pace of fungal genome sequencing has increased dramatically since then. As of this writing no less than eight fungal genomes have been completed as high quality drafts, and several of these are in various stage of the finishing and annotation processes. In addition, genome sequencing projects have been initiated for approximately 20 other fungal species. The scientific community is attempting to choose the most informative organisms to sequence from the more than 1.5 million fungal species that inhabit our planet. Their decisions are justifiably based on which fungi present serious threats to human health, serve as important model organisms in biomedical research, exhibit unique or diverse natural capabilities, play critical roles in natural environmental processes, and provide a wide range of evolutionary comparisons at key branch points in the 1 billion years spanned by the fungal evolutionary tree.

Not surprisingly, the genome of *Neurospora crassa* was the first to be completed. This organism is arguably the most thoroughly characterized fungal species from the standpoint of its genetics, biochemistry, and cell biology. The 40-megabase genome of *N. crassa* contains about 10,000 protein-coding genes. This is more than twice the number identified in the fission yeast *Schizosaccharomyces pombe*. Analysis of the gene models produced some unexpected insights into the biology of *N. crassa* including the identification of genes associated with photobiology, secondary metabolism, and important differences in Ca^{2+} signaling as compared with plants and animals. Additional clues were obtained about a genetic phenomenon that is unique to fungi called repeat-induced point mutation (RIP). Apparently RIP has had a profound impact on genome evolution, greatly slowing the creation of new genes via duplication, producing a genome with an unusually low proportion of closely related genes.

Publication of the *Neurospora* genome was followed by one which described the sequencing and analysis of the *Phanerochaete chrysosporium* genome, a white rot basidiomycete fungus which has the capacity to completely metabolize cellulose and lignin, two of the most abundant carbon polymers on Earth. The *P. chrysosporium* genome revealed numerous genes encoding secreted oxidases, peroxidases and hydrolytic enzymes that cooperate in lignocellulose degradation, a pivotal process in the global carbon cycle. These observations may provide a framework for further development of industrial bioprocesses for biomass utilization, organopollutant degradation and fiber bleaching.

Fungal genome sequence data continues to be collected and analyzed at several institutions, including the Broad Institute (United States), the Joint Genome Institute (United States), the Computational Biology Research Center (Japan), the Sanger Institute (United Kingdom), The Institute for Genomic Research (United States), Gene Alliance (Germany), and MWG Biotech (Germany). However, the majority of fungal genome research is currently being orchestrated by the Fungal Genomics Initiative (FGI) which represents a partnership between the Broad Institute and the scientific community including a number of academic and industrial groups. Several U.S. government agencies (Department of Energy, USDA, NHGRI, and NSF) have recognized the need and benefits of these projects with generous funding. Despite the recent surge in the sequencing of genomes from filamentous fungi, post-genomic experimentation is still in a relative state of infancy compared to the advanced state-of-the-art for bacteria and higher eukaryotes. The benefits expected from fungal genome analysis are now just a small trickle like the source of a mighty river. Nevertheless, significant contributions are being made by fungal biologists who are applying the tools of bioinformatics and functional genomics to uncover new information about the cellular and biochemical processes that have evolved in this diverse group of organisms.

This volume of Applied Mycology and Biotechnology is a continuation of the series entitled "Fungal Genomics" (see volume 3 and 4). As with previous books of this series, we are cognizant of the hurdles that must be overcome to develop a comprehensive volume on fungal genomics because of the range and complexity of the information emanating from current studies in the fungal biology community. Nevertheless, we have attempted to assemble chapters that characterize the diversity of scientific approaches and thought processes to provide readers with a reference point from which to embark for future investigations. This issue features descriptions of experimental work directed toward understanding and extracting information from fungal genomes, including recombination, introns, pathogenesis, DNA microarrays, gene regulation, development and morphogenesis and gene silencing.

We are grateful to the members of the editorial board for their help in assembling this volume as well as the authors who generously contributed chapters. We also thank Ms. Hetty Verhagen, Rathbone Clare and Ana-Bela sa Dias of Elsevier Life Sciences for their technical assistance, and Devendra Fuloria of National Bureau of Agriculturally Important Microorganisms (NBAIM) for his valuable help in preparing the camera ready copy of this volume.

Dilip K. Arora
Randy M. Berka

**Applied Mycology
and Biotechnology**
An International Series
Volume 5. Genes and Genomics
© 2005 Published by Elsevier B.V.

ELSEVIER

Recombination in Filamentous Fungi

Frederick J. Bowring, P. Jane Yeadon and David E. A. Catcheside
School of Biological Sciences, Flinders University, PO Box 2100, Adelaide, South Australia
5001 (david.catcheside@flinders.edu.au).

Although their contribution to our early understanding of meiotic recombination is unequalled, in the latter part of the last century filamentous fungi fell behind their more genetically tractable single-celled cousin, *Saccharomyces cerevisiae* in this area. However, an increase in ease of DNA manipulation resulting from recent genome projects, coupled with the unique biology of filamentous fungi has resulted in something of a revival of their utility in unraveling the ways in which genetic information is shuffled during meiosis. In this review we consider features of meiotic recombination whose elucidation has depended upon a substantial contribution from filamentous fungi. We then focus on the more recent contributions this group of fungi has made to the understanding of how genetic recombination is achieved, emphasizing how the unique biology of filamentous fungi has been exploited to address questions not easily answered using other model systems.

1. INTRODUCTION

During the last century the single-celled fungus, *Saccharomyces cerevisiae* provided the greatest volume of data to guide our understanding of the mechanism of genetic recombination. Nevertheless, data from experiments with the filamentous fungi have contributed key insights into the mechanics of recombination from the earliest phases of the analysis of this process and continue to do so to this day. Early work on the genetic systems of the filamentous ascomycete, *Neurospora crassa* by Shear and Dodge (1927) and by Lindegren (reviewed in Lindegren 1973) led Beadle and Tatum to choose *N. crassa* for their experiments, which demonstrated for the first time that genes carry the genetic information for the production of specific enzymes (Beadle and Tatum 1941). This laid the foundation for the use of this fungus as a model for investigation of fundamental molecular processes, including recombination, in eukaryotes. *Sordaria macrospora*, another filamentous ascomycete, has even traveled into space and shown that meiosis and sporulation proceed normally in the absence of gravity (Hahn and Hock 1999).

The Fungal Genome Initiative of the Broad Institute (http://www.broad.mit.edu/ annotation/fungi/fgi/index.html) comprises a range of collaborative projects, with

Corresponding author: David E. A. Catcheside

each of several groups working on a specific fungus, and has led to rapid progress in characterization of the genomes of several filamentous fungi. The lead project is sequencing (Galagan et al. 2003) and annotation (Borkovich et al. 2004) of the *N. crassa* genome. The *Magnaporthe grisea* (rice blast) genome project is also well advanced, draft sequences are available for *Aspergillus nidulans, Fusarium graminearum* and *Ustilago maydis* and genome sequencing for several other filamentous fungi is underway. The availability of sequence data for complete genomes, coupled with development of efficient gene manipulation tools has spawned a renaissance in the use of filamentous fungi to address fundamental biological questions - a position lost during the 1950's when the focus of attention on uncovering the fundamental processes of life turned largely to prokaryotic systems because of the advantages they then offered in tractability for genetic analysis. The genetic systems of the ascomycete filamentous fungi make them peculiarly suitable for probing the intricacies of recombination during meiosis, the reductional division in eukaryote life cycles.

In this review, we will summarize the historical background to the study of recombination, highlighting the involvement of filamentous fungi. We will evaluate several of the models that purport to explain recombination, with reference to data from filamentous fungi and emphasize recent insights from filamentous fungi that might not have been available using other model organisms.

2. HISTORICAL BACKGROUND

2.1 Meiosis and Genes

During meiosis, homologous chromosomes first pair then segregate to opposite poles. Thus, the first meiotic division yields two daughter cells in which the number of chromosomes is halved. Sister chromatid separation during a subsequent mitosis-like division yields four haploid progeny. In the case of plants and animals, these haploid cells form gametes, each requiring fusion with another gamete to establish the next generation. In haploid organisms, the meiotic products are the new generation. Some filamentous fungi, including *N. crassa, Sordaria fimicola* and *Ascobolus immersus*, experience an additional mitotic division before meiotic products are packaged, yielding eight haploid spores in which each chromosome is a double-stranded derivative of one of the eight single strands of the bivalent in the premeiotic diploid.

Separation of homologs in meiosis was first documented in the grasshopper *Brachystola magna* (Sutton 1902) providing an explanation for the phenotypic ratios seen in F2 progeny of true-breeding pea varieties (Mendel 1865) The observation that chromosomes and Mendelian factors, or genes, appear to be inherited in the same way prompted the first suggestion that genes are located on chromosomes (Wilson 1902; Sutton 1903).

2.2 Linkage and Crossing Over

Bateson, Saunders and Punnet (1905) first showed that some sweet pea genes were frequently inherited together but on occasions, segregated independently. Work with the vinegar fly *Drosophila melanogaster* yielded the first experimental evidence that genes are carried by chromosomes as the gene for a white eye mutant was inherited with the X-chromosome (Morgan 1910). As more genes on the *Drosophila* X-chromosome were discovered, varying degrees of linkage between pairs was observed, with each pair separated by a distance assumed (Morgan 1911) to reflect the frequency with which chromosomal exchanges, or crossovers (Morgan and Cattell 1912) occurred between them. The first linkage map of gene loci on chromosomes based on the relative frequency of crossovers was made by Sturtevant (1913) who placed six genes on the map of the *Drosophila* X-chromosome. The predicted correlation between the frequency of chiasmata (Janssens 1909) visible connections between homologous chromosomes during meiosis, and the frequency of crossing over, was demonstrated in *Zea mays* (Creighton and McClintock 1931). In most organisms, normal meiotic chromosome segregation was found to be dependent on the occurrence of at least one such crossover event per bivalent (Baker et al. 1976) and linkage maps were considered to reflect the physical distance between loci governing the probability of an exchange event between them (Haldane 1919).

2.3 Gene Conversion

Recombination was observed not only between genes but also within them. Small numbers of wild type progeny were obtained from crosses between *Drosophila lozenge* mutant males and females carrying different *lozenge* alleles (Oliver 1940). Since this apparent reversion correlated with crossing over nearby, it was assumed to be a result of reciprocal exchange between alleles.

Work with fungi, in which all products of a single meiosis are retained in each ascus, provided the first indication that recombination was not always reciprocal. Mutations in *S. cerevisiae* sometimes segregated abnormally at meiosis, giving an ascus with spores in which three were wild type and one mutant or three mutant and one wild type, a phenomenon called "gene conversion" (Lindegren 1953). Initially thought to be due to a rare and peculiar chromosomal aberration, gene conversion was soon found to be widespread in Ascomycetes.

Mary Mitchell's study of the *pdx* locus of *N. crassa* (Mitchell 1955) demonstrated that wild type progeny arising from crosses between two pyridoxine-requiring mutant strains were not due to abnormal segregation of entire chromosomes. Segregation of genetic markers flanking the *pdx* locus was normal, so the gene conversion effect was limited to a short chromosomal region. Since back-crossing showed that asci with wild type progeny did not hold the predicted reciprocal double mutants, this allelic recombination could not be due to crossing over. However, in asci showing conversion, the frequency of crossing over between *pdx* and a closely linked marker, *co*, was much increased. Similar results were obtained at other loci in *Neurospora* (Case and Giles 1958a, 1958b; Stadler 1958; Murray 1960) and *A. nidulans* (Pritchard 1955). Studies of spore color mutations demonstrated conversion in *S. fimicola* (Olive 1959) and *A. immersus* (Lissouba et al. 1962) and showed that, as in *Neurospora*, the occurrence of conversion resulted in an increase in crossing over nearby (Kitani et al. 1962).

2.4 Post-meiotic Segregation

Interestingly, Olive (1959) found not only 6:2 and 2:6 ratios of spore color in *Sordaria*, but also 5:3 and 3:5 ratios. Also in *Sordaria*, an additional class of abnormal asci was found (Kitani et al. 1962) with aberrant 4:4 segregation of spore color, which differed in two spore pairs that would normally be genetically indistinguishable (sister spores from the post-meiotic mitosis). Since genes flanking the spore color locus showed identical segregation in sister spores, the aberrant allele segregation was not due to abnormal chromosomal behavior. The spore color mutation must have segregated in the mitotic division after meiosis, showing that the chromosome resulting from recombination was a heteroduplex (hDNA) a DNA molecule in which the two strands are not fully complementary. Post-meiotic segregation (PMS) is also seen in *S. cerevisiae*, but at a much lower frequency (Fogel et al. 1979). Aberrant 4:4 asci are even rarer in yeast than the frequency of 5:3 segregation would predict (Fogel et al. 1979). but are seen in crosses defective for some DNA mismatch repair proteins (Williamson et al. 1985). Aberrant 4:4 segregation is also rare in *Neurospora*, but more frequent than in yeast, and has been detected at the *pan-2* locus (Case and Giles 1964).

3. MODELS OF MEIOTIC RECOMBINATION

The demonstration that DNA is the genetic material (Avery et al. 1944) and deduction of its structure (Watson and Crick 1953) enabled the first plausible molecular models for meiotic recombination (Whitehouse 1963; Holliday 1964). In these models, a requirement for specific base pairing between strands derived from non-sister chromatids forced the points of exchange between homologous chromosomes to be in the correct register. Correction of mismatched bases within the resultant hDNA provided an explanation for gene conversion. The manner in which chromatids are broken and heteroduplex formed differs in detail between models, all of which drew heavily on data from yeast and filamentous fungi.

Holliday (1964) proposed that strands of the same polarity are broken at equivalent places in non-sister chromatids and, following a reciprocal exchange of pairing with the intact homologous strands, the free ends are rejoined in reciprocal combination forming heteroduplex on each homolog (Fig. 1). Additional symmetric heteroduplex DNA can be formed by branch migration of the resultant junction between the duplexes (Fig.1 C) by winding sequence off one duplex and on to the other (Meselson 1972). Resolution of what has become known as a Holliday junction, by breakage and rejoining of pairs of DNA strands, can occur in two ways (Fig. 1D) one leading to a crossover (Fig. 1F) and the other to restoration of the parental combination of flanking markers (Fig. 1E).

Sigal and Alberts (1972) showed that the relationship between the strands of the two duplexes at a Holliday junction is such that the two types of resolution can occur with equal ease. In each case, repair of mismatches in heteroduplex may result in conversion of markers within this region. As heteroduplex tracts will be symmetrically arranged on the two chromosomes following resolution of the junction (Fig. 1E and 1F) the Holliday model predicts that conversion is equally likely to occur on either chromosome involved in the event. Although detection of events that are best explained by asymmetric distribution of heteroduplex in yeast (Fogel and Mortimer 1970) and *Neurospora* (Stadler and Towe 1971) made amendment of the Holliday model a necessity, the Holliday junction has remained a feature of each successive recombination model. Meselson and Radding (1975) suggested that the break initiating recombination occurs in one strand of one DNA duplex only (Fig. 2). followed by extension of the resultant 3' end by DNA synthesis, displacing the strand 5' of the break (Fig. 2B). It is then proposed that the displaced strand invades the homologous duplex, induces a

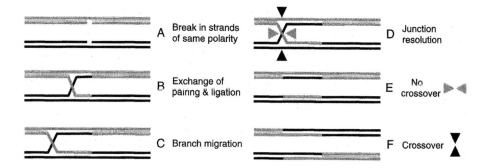

Fig. 1: Holliday's model for meiotic recombination (adapted from Holliday 1964). Breaks in single strands of like polarity in homologous chromosomes (A) is followed by exchange of pairing. This establishes symmetric hDNA and a Holliday junction linking the two DNA duplexes (B). The Holliday junction can migrate, extending the symmetric heteroduplex to the left or right of the initiating breaks (Meselson 1972) (C). Depending on how the junction is resolved by strand breakage and reciprocal rejoining (D), the event leads to conversion only (E) or both crossover and conversion (F).

break in the strand of like polarity (Fig. 2C) and is ligated to form a Holliday junction (Fig. 2D). Although migration of the junction can lead to heteroduplex on both the initiating and invaded duplex, synthesis extending only the initiating strand ensures that the heteroduplex formed in the initial invasion is asymmetric, occurring primarily on the invaded chromosome (Fig. 2F). Demonstration that the chromosome on which recombination is initiated is the more likely to experience conversion (Gutz 1971; Catcheside and Angel 1974; Nicolas et al. 1989) rapidly made this model untenable.

Into the 21st century, most additional work on the elucidation of the mechanisms of recombination has used *S. cerevisiae* as the experimental organism, reflecting the ease of genetic manipulation and early development of molecular tools in this species. Yeast has both haploid and diploid phases, has high levels of meiotic recombination at some recombination hotspots, and, because of the phenomenon of mating type switching, isogenic strains can readily be generated. In addition, growth conditions can be manipulated to ensure that most cells enter meiosis at the same time.

Synchronous meiosis has been used to demonstrate that meiosis-specific DNA double strand breaks (DSBs) occur close to the *ARG4* promoter of yeast (Sun et al. 1989) where recombination is known to be initiated (Nicolas et al. 1989). Removal of the initiation site eliminates the DSB (Sun et al. 1989). DSBs at other yeast recombination initiation sites (Cao et al. 1990; Fan et al. 1995) also show a correlation with recombination, as meiotic recombination cannot be detected in mutant strains that do not make DSBs (Alani et al. 1994; Cao et al. 1990). DSBs result in no substantial double strand gap in the DNA but are processed to give long, 3'-overhanging single-stranded tails (Sun et al. 1989, 1991). Joint molecules, isolated from meiotic yeast cells (Collins and Newlon 1994; Schwacha and Kleckner 1994, 1995), form mainly between homologs but rarely between sister chromatids and are present at times and positions that suggest they are likely to be recombination intermediates (Schwacha and Kleckner 1994). Their structure suggests (Schwacha and Kleckner 1995) that they contain double Holliday junctions although Collins and Newlon (1994) whose methodology did not use a cross-linking agent, found joint molecules to contain only a single Holliday junction.

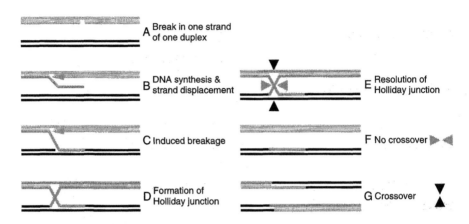

Fig. 2: The Meselson and Radding (1975) model for meiotic recombination. DNA is broken in one strand of one duplex (A). The 3' end is extended by DNA synthesis using the initiating chromosome as template, displacing the old strand of like polarity (B) which interacts with the homologous chromosome and induces a break (C), which is ligated to form a heteroduplex region and a Holliday junction (D). The junction is resolved (E) to give conversion with (G) or without (F) a crossover.

In the double-strand break-repair (DSBR) model (Szostak et al. 1983; Sun et al. 1991) a break in both strands of one homolog (Fig. 3 A) is processed by resection of the 5' ends to generate single strands ending 3' (Fig. 3 B). Invasion of the homologous duplex by a 3' end (Fig. 3 C) leads to the formation of a D loop due to DNA synthesis, providing a template for repair synthesis of the second 3' end (Fig. 3 D). Ligation of the free ends produces paired Holliday junctions (Fig. 3 E). Branch migration can give symmetric heteroduplex and resolution of the two Holliday junctions in the same or opposite sense yields crossover (Fig. 3 G) or non-crossover (Fig. 3 F) outcomes associated with conversion resulting from correction of mismatched bases in heteroduplex DNA. Unlike the Meselson and Radding model, the DSBR model explains the higher probability of conversion of the initiating chromosome. However, like earlier models, it also suggests that half of the chromosomes with a conversion event should also have a crossover near the converted site, although most mitotic (Pâques and Haber 1999) and some meiotic conversions (Bowring and Catcheside 1996) yield few associated crossovers.

Synthesis-dependent strand annealing (SDSA) variants of the DSBR model (Nassif et al. 1994; Ferguson and Holloman 1996; Pâques et al. 1998) account for the predominance of conversion of the initiating chromosome and allow for a low association of conversion with crossing over. In the simplest of SDSA models (Fig. 4) following extension by repair synthesis, the invading strand is re-annealed with the initiating duplex (Fig. 4 F) and no Holliday junction is formed, a circumstance that leads only to conversion of the initiating chromosome and no crossover. Strong support for SDSA models comes from the finding that replication errors occurring

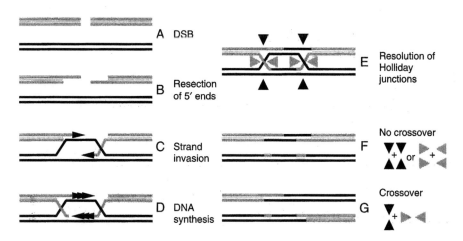

Fig. 3: The DSBR model for meiotic recombination (Szostak et al. 1983 and modified by Sun et al. 1991). Recombination is initiated by a break in both strands of the initiating duplex (A). The 5' ends are resected (B) and one of the 3' ends invades the homologous duplex, where it is extended by DNA polymerase using the homolog as template (C). The D loop formed by displacement of a strand of the invaded chromosome provides a template for repair of the second strand of the initiating chromosome (D). The ligation of ends forms two Holliday junctions, each of which can be resolved in either of two ways (E), leading to non-crossover (F) or crossover chromosomes (G).

during repair synthesis usually appear only on the initiating chromosome (Strathern et al. 1995; Pâques et al. 1998).

The SDSA model (Fig. 4) has been elaborated to allow crossovers and explain how the newly replicated DNA can be retrieved from the template with which it will otherwise be progressively entwined. Pâques et al. (1998) proposed that return of the newly synthesized DNA to the initiating duplex could be achieved by migration of a short replication bubble (Fig. 5 D$_1$) or by migration of a half Holliday junction behind the extending 3' end (Fig. 5 D$_2$). To account for crossovers, it was proposed that the displaced strand of the invaded chromosome may become involved in pairing with the second 3' end of the initiating chromosome (Fig. 5 E) leading to an intermediate that can be processed to form double Holliday junctions as in the DSBR model. The degree of association between conversion and crossovers may thus be a function of which of these two pathways is favored.

An alternative explanation for a paucity of nearby crossovers in chromosomes that have experienced conversion is control of the manner of scission of Holliday junctions. Sister chromatid gene conversion, a frequent mode of DSB repair in somatic cells, rarely yields reciprocal exchange (Johnson and Jasin 2000) suggesting symmetric scission of the resultant paired Holliday junctions. The RecQ helicase in humans, in combination with topoisomerase IIIα, is known to resolve double Holliday junctions *in vitro* such that flanking sequences are not recombined (Wu and

8

Hickson 2003). Similarly biased junction resolution in some meiotic cells could account for variability in the proportion of conversion events observed to have a nearby crossover.

4. GENES THAT REGULATE RECOMBINATION IN SPECIFIC LOCATIONS
4.1 *ccf* and *cv* Loci in *Ascobolus*

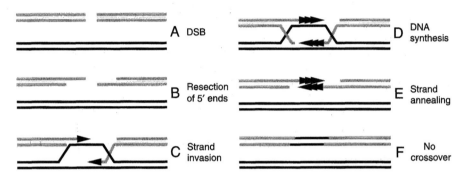

Fig. 4: A simple SDSA model for meiotic recombination (Pâques and Haber 1999). This suggested mechanism proceeds as the DSBR model (Fig. 3) for steps A-D. Then, instead of ligation leading to paired Holliday junctions, the newly synthesized DNA ends are unraveled from the template to anneal one with the other (E). Remaining gaps are closed and ligated (F). The initiating chromosome only is converted and there is no crossing over.

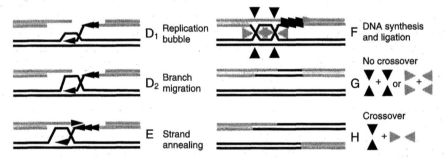

Fig. 5: A modified SDSA model to account for crossing over in some of the progeny that experience conversion (Pâques and Haber 1999). This suggested mechanism proceeds as the DSBR (Fig. 3) and simple SDSA (Fig. 4) models for steps A-C. Extension of a 3′ end by DNA synthesis following invasion of the homologous duplex leads to it becoming wound around the template in the helical turns of a DNA duplex. Pâques and Haber (1999) envisaged two ways in which this entanglement might be avoided, by helicases maintaining a short replication bubble (D_1) or by branch migration of a half Holliday junction (D_2). Instead of annealing of the newly synthesized strands, as in Fig. 4 E, strand annealing might sometimes occur between the displaced strand of the invaded duplex and the second 3′ end of the initiating chromosome (E) leading to the formation of Holliday junctions (F), which can be resolved with or without a crossover (G, H).

Various genes influencing gene conversion in specific locations of the genome have been discovered in *Ascobolus*. These conversion control factors, or *ccf* loci, fall into three classes. The first class contains only *ccf-2*, which influences recombination at the adjacent *wI* locus (Lamb and Helmi 1978; Helmi and Lamb 1983). The *ccf-2* gene has three alleles, *ccf-2(P)*, *ccf-2(91)* and *ccf-2(K)*, and crosses homozygous for *ccf-2(P)* have the highest conversion frequency in *wI*. Alleles *ccf-2(91)* and *ccf-2(K)* appear to act as partially dominant reducers of recombination, as crosses of either to *ccf-2(P)* yield conversion frequencies less than half the sum of the two homozygotes (Lamb and Helmi 1978). The *ccf-3* and *ccf-4* alleles, which act on specific but remote loci, represent the second class (Helmi and Lamb 1983). The *ccf-3E* allele is dominant to *ccf-3e*, acting to increase conversion at the *wI* locus. Dominance of the *ccf-4R* allele depends on the alleles of *ccf-2* and *ccf-3* present in the cross such that, in the presence of *ccf-3E*, *ccf-4R* reduces recombination in *wI* only when *ccf-2(91)* is absent (Helmi and Lamb 1983).

The final class suppresses recombination at a nearby locus only when heterozygous, and includes *ccf-5* alleles which, when heterozygous, reduce recombination at the *w9* locus (Howell and Lamb 1984; Lamb and Shabbir 2002) and *ccf-1* alleles, with a similar effect at *w62* (Emerson and Yu-Sun 1967). Similarly, when heterozygous, alleles of *cv1*, *cv2*, *cv4* and *cv6* each reduce recombination at the *b1*, *b2*, *b4* and *b6* loci respectively (Girard and Rossignol 1974). The *b4* and *b6* loci are linked and, interestingly, heterozygosity for *cv4* and *cv6* not only reduces conversion in *b4* and *b6* but also reduces crossing over between them (Girard and Rossignol 1974).

4.2 Fine Control Loci in *Neurospora*
4.2.1 *cog*

The *cog* locus appears similar to *ccf-2*, with *cog⁺* analogous to *ccf-2(P) cog*, located centromere-distal of the *his-3* locus (Bowring and Catcheside 1991; Yeadon and Catcheside 1995a, 1998) influences allelic recombination within *his-3* and crossing over in the chromosomal segments surrounding the gene (Angel et al. 1970). Two *cog* phenotypes, high (*cog⁺*) and low (*cog*) frequency recombination, have been described (Angel et al. 1970), with the chromosome that bears *cog⁺* almost exclusively the one experiencing conversion (Catcheside and Angel 1974; Yeadon and Catcheside 1998; Yeadon et al. 2001). The presence of *cog⁺* increases allelic recombination frequency six-fold and crossovers between *his-3* and the centromere-distal gene, *ad-3*, four-fold when compared to similar crosses where *cog* is homozygous (Catcheside and Angel 1974).

In crosses heterozygous for TM429, a translocation that separates the proximal end of *his-3* from *cog⁺*, conversion of alleles proximal to TM429 requires that *cog⁺* be on the intact chromosome (Catcheside and Angel 1974). Thus, recombination is initiated at *cog⁺* and can cross the translocation heterology only if it travels along the unbroken chromosome.

The conclusion that the chromosome on which conversion is initiated is usually the recipient of information is supported by data from both budding and fission yeasts. Where the activity of a recombinator has been altered, such as the *M26* mutation that generates a recombination hotspot at the *ade6* locus in *Schizosaccharomyces pombe* (Gutz 1971) or the promoter deletion that removes hotspot activity in *cis* at *ARG4* of *S. cerevisiae* (Nicolas et al. 1989), the chromosome with the more active hotspot is predominantly the one that experiences conversion.

Interestingly, *cog* alleles are co-dominant in effect on both intergenic and allelic recombination in the *his-3* region (Yeadon et al. 2004 b). Crossing over between *lys-4*, centromere-proximal to *his-3*, and *ad-3* occurs in 20% of progeny, a genetic interval of 20 centiMorgans (cM), when *cog⁺* is homozygous. When *cog⁺* is heterozygous, the same interval is 9.4 cM and only 2 cM if *cog* is homozygous. Similarly, the *his-3* to *ad-3* interval varies depending on the *cog* alleles in the cross, measuring 15.5 cM (*cog⁺/cog⁺*), 7.5 cM (*cog⁺/cog*) and 1.4 cM (*cog/cog*). In addition, in heteroallelic crosses, two copies of *cog⁺* result in approximately twice the average frequency of allelic recombination as that of a single copy. Thus, the two *cog* alleles appear to operate independently of one another to initiate recombination events.

Co-dominance of *cog* alleles may suggest a difference in mechanism between *cog* and *ccf-2*. However, sequence heterology close to *cog* reduces recombination slightly (Yeadon et al. 2004 a, and discussed in section 7.2). so the apparent partial dominance of the low frequency *ccf-2* alleles may be due to variation in allelic sequences. Perhaps the level of heterology is greater at *ccf-2* than at *cog* or it may be that the recombination machinery is more sensitive to sequence mismatch in *Ascobolus* than in *Neurospora*.

4.2.2 *ss*

Recombination between *nit-2* alleles depends on the wild-type origin of the strains in the cross, with crosses of homozygous origin yielding higher allelic recombination frequencies than those where the two strains are from different backgrounds. This heterozygous suppression of recombination, analogous to the *ccf-1*, *ccf-5* and *cv* elements found in *Ascobolus*, is due to a genetic element called *ss* (Catcheside 1981) and closely linked to *nit-2*. Recombination is suppressed two to twenty fold depending upon the *ss* alleles involved.

4.2.3 *rec* Genes

Like *ccf-4* in *Ascobolus*, each of the three known *Neurospora rec* genes, *rec-1*, *rec-2* and *rec-3*, exerts its influence in a specific region of the genome (Fig. 6). Each dominant *rec⁺* allele is thought to produce a diffusible product (Catcheside 1977) and reduces recombination in its target regions by about an order of magnitude. The *rec-1* gene acts between *his-1* alleles on linkage group V (Jessop and Catcheside 1965) and *nit-2* alleles on LG I (Catcheside 1970) but not at ten other loci tested (Catcheside and Austin 1969). The *rec-2* locus, on LG V (Smith 1968) is known to affect recombination in intervals that include *his-3* to *ad-3* (Angel et al. 1970; Catcheside and Corcoran 1973) *sn* to *arg-3* (Catcheside and Corcoran 1973). both on LG I, and *his-5* to *pyr-3* (Smith 1966) on LG IV.

In the presence of *rec-2⁺*, recombination between *his-3* alleles is reduced 30-fold in crosses containing *cog⁺* and four-fold in crosses homozygous *cog*, to the same low level (Angel et al. 1970). In addition, recombination events that occur in the presence of *rec-2⁺* appear to be initiated at the 5′ end of *his-3* and not at *cog* (Catcheside and Angel 1974; Yeadon and Catcheside 1998). It seems likely that the allelic recombination frequency attributable to each *cog* allele reflects the frequency with which recombination is initiated there and that the *rec-2⁺* product prevents initiation at either *cog* allele (Catcheside and Angel 1974).

The *rec-3* gene product regulates recombination in the *am* locus on LG V (Catcheside 1966) and *his-2* on LG I (Catcheside and Austin 1971) and has three known alleles, *rec-3*, *rec-3ᴸ* and *rec-3⁺*. The *rec-3⁺* allele is equally dominant to both *rec-3* and *rec-3ᴸ* at the *his-2* locus but at *am*, the effect of *rec-3ᴸ* on recombination is intermediate between that of *rec-3* and *rec-3⁺* (Catcheside 1974). Thus, like *ccf-4*, dominance of *rec-3ᴸ* may depend on the initiator with which it interacts.

Although, like *ccf-4*, the dominant allele of each of the *Neurospora rec* genes reduces recombination in its target region, it is possible that *rec-2* is more like *ccf-3*, the dominant allele of which increases recombination at the Ascobolus *w1* locus. How can this be? Unpaired genes are not expressed during meiosis in *Neurospora*, a phenomenon known as meiotic suppression of unpaired DNA (MSUD; Shiu et al. 2001). Part of the *rec-2* region sequence is absent from the equivalent region in *rec-2⁺* strains, which in turn contains sequence not present in *rec-2* (F. J. Bowring and D. E. A. Catcheside, unpublished). Thus, if it is *rec-2*, rather than *rec-2⁺*, that produces a functional protein, and this protein is required for recombination in *his-3*, MSUD may prevent expression except in *rec-2* homozygotes. One could therefore imagine a mutant form of *rec-2* that would be a recessive reducer of recombination in *his-3*. In opposition to this proposal, *rec-2* strains transformed with DNA from the *rec-2⁺* region show reduced recombination, although the reduction is much less than is observed with indigenous *rec-2⁺* (F. J. Bowring and D. E. A. Catcheside, unpublished). Also, it seems unlikely that alleles of all three *rec* genes reflect insertion/deletion events, nor can MSUD explain the locus-dependent dominance of *rec-3ᴸ*. So, even if MSUD determines the apparent dominance of *rec-2⁺*, the *rec-3⁺* gene must produce a product whose mechanism may resemble *ccf-4* in *Ascobolus*.

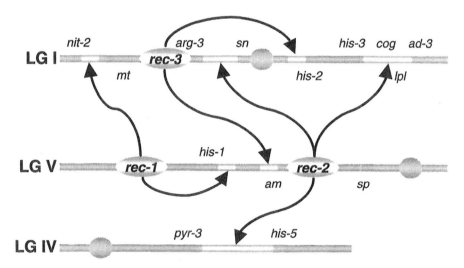

Fig. 6: A partial linkage map of *Neurospora*, showing the locations of the known *rec* genes. The arrows indicate the regions in which each *rec* gene influences recombination. The discs represent the centromeres. Only part of each chromosome is shown.

5. POLARITY

Lissouba and Rizet (1960). working with a series of allelic spore color mutants in *A. immersus*, found a linear relationship between conversion frequency of each mutant site and the position of the mutation in the locus, the first observation of a conversion

12

gradient. Polarity of conversion was also observed in *A. nidulans* (Siddiqi 1962) and in *Neurospora* (Murray 1963; Stadler and Towe 1963) Fogel and Mortimer (1970) showed that conversion at *ARG4* of yeast involves a tract of DNA sequence, rather than a single site. If a site at the low end of the conversion gradient is converted, other sites towards the high end of the gradient are likely to have been converted along with it. Case and Giles (1958 a) generated a genetic map of the *Neurospora pan-2* locus, positioning 37 mutant alleles across the gene. Consistent with Fogel and Mortimer's (1970) observation, alleles in the proximal end of *pan-2* experience recombination more frequently than those at the distal end.

Polarity of gene conversion suggests that recombination starts in a particular place and proceeds in a linear fashion. In *Neurospora* crosses homozygous for a translocation that moved part of LG IR including the *me-6* gene into LG V between *sp* and *inos*, inverting the region with respect to the centromere (Murray 1968). the direction of polarity in *me-6* was the same as in structurally normal homozygotes, showing that polarity is a property of the inverted region itself. In addition, translocation of *me-2* from LG IV to LG VI did not change the pattern of polarity in *me-2*. Murray (1968) concluded that polarity is dependent on sequences near *me-2*, within which recombination is presumably initiated.

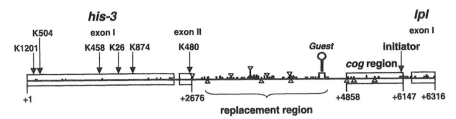

Fig. 7: A physical map of the *his-3* region in *Neurospora*. Sequence heterology in the *his-3* to *cog* interval is shown in St Lawrence 74A/Lindegren Y8743 diploids. The centromere is to the left of the Figure, which is to scale. Coding sequences and the *cog* region (Yeadon and Catcheside 1995 a) are indicated by open boxes enclosing the line that represents the sequence alignment. The position, relative to the first nucleotide of the *his-3* coding sequence (Legerton and Yanofsky 1985). is shown below the sequence line. Sequence polymorphism is indicated by vertical bars, the height of which represents the number of mismatches, from one to seven, in each 10 bp section of sequence. Insertion/deletions are shown as triangles. If a triangle is above the sequence line, the St Lawrence sequence is longer; if below, the Lindegren sequence is longer. The hairpin symbol indicates the position of the *Guest* element (Yeadon and Catcheside 1995) in the St Lawrence sequence. The location of each mutant site (K1201, K458, K26, K874 and K480) and of the putative recombination initiation site is indicated by a short arrow. The bracket indicates the sequence replaced in strains made by targeted transfection.

5.1 Polarity of Recombination Initiated at *cog* in *Neurospora*

Use of molecular techniques such as polymerase chain reaction (PCR) and DNA sequencing have added to knowledge of polarity of recombination at *his-3* of *Neurospora*. Identification of the altered sequences responsible for *his-3* mutant alleles K1201, K504, K458, K874, K26 and K480 (Yeadon and Catcheside 1999; Yeadon et al. 2001; Rasmussen et al. 2002; Yeadon et al. 2002) yields a physical map (Fig. 7) with sites in the same order as in the genetic map (Angel et al. 1970) showing that polarity reflects relative physical position along a chromosome.

Although there are multiple strain specific differences in the *cog*-region (Yeadon and Catcheside 1995 a) sequences, the *cog* variants in the Lindegren *A*, Emerson *A* and St Lawrence 74*A* strains all appear to have the same low frequency recombination, or *cog*, phenotype. In *cog^La*, the only naturally occurring *cog^+* allele known (Yeadon and Catcheside 1995b, 1999) and first found in Lindegren Y8743 (Angel et al. 1970) a ten bp sequence including two single nucleotide polymorphisms (SNPs) is required for the high frequency recombination phenotype (Yeadon and Catcheside 1998). Using targeted transplacement, *Neurospora* strains have been constructed in which 1.8 kb of native sequence between *his-3* and *cog* (Fig. 7) has been replaced with foreign sequences (Yeadon et al. 2002; Rasmussen et al. 2002) including the human *Herpes simplex thymidine kinase* (*TK*) gene. All transplacement strains are both *cog^+* and *rec-2*, to ensure that local recombination occurs at high frequency. Each of the manipulated strains is also a histidine mutant, either K26 or K480, to enable selection for His^+ progeny and allow determination of allelic recombination frequency. Analysis of His^+ progeny of *TK* heterozygotes indicates that recombination is initiated on the distal side of the replacement region (Yeadon et al. 2001; Fig. 7), 1.2 kb proximal of the SNPs that are necessary for high frequency recombination. In addition, a peak in conversion close to these SNPs (Yeadon and Catcheside 1998) suggests that the initiator may be located here (Fig. 7), ~3.4 kb from *his-3* (Yeadon et al. 2004 a).

In the strains made by transplacement, the distance between *cog* and the *his-3* K480 mutant site varies from 1.7 kb to nearly 6 kb. The frequency of recombinant His^+ progeny yielded by crosses heteroallelic K26/K480 increases exponentially as the distance from *cog* declines (Yeadon et al. 2002), suggesting that recombination events terminate in a stochastic fashion, one possible explanation for polarity. Analysis of these data suggested that, in *cog^+* homozygotes, recombination is initiated at *cog* in at least 17% of meioses, that most conversion tracts are very short and few extend more than 14 kb. Since a similar exponential relationship between distance separating markers and the chance of co-conversion has also been found at the *rosy* locus of *Drosophila* (Hilliker et al. 1994) and at *ade6* of *S. pombe* (Grimm et al. 1994) the distance that recombination events progress along a chromosome may be a stochastic process in most organisms.

6. THE EFFECT OF GENETIC BACKGROUND ON RECOMBINATION
6.1 At *Neurospora his-3* and *nit-2* Loci

Although recombination frequency in the *his-3* region depends upon which alleles of *cog* and *rec-2* are present, it seems that these are not the only factors involved. His^+ frequency, for a single pair of mutant *his-3* alleles, and the frequency of crossing over, in both the *lys-4* to *ad-3* and the *his-3* to *ad-3* intervals, varies over a two-fold range in crosses with identical *cog* and *rec-2* genotype (Yeadon et al. 2004 a) Similarly, allelic recombination at the *nit-2* locus is not solely determined by the *ss* and *rec-1* alleles present but also varies about 1.5 fold with parental provenance (Catcheside 1970). Analysis of recombination frequencies from crosses between strains with the same or similar genetic backgrounds suggests that genes with small effects on recombination are highly polymorphic in our laboratory strains.

6.2 Within the *A* and *B* Mating Type Factors in *Schizophyllum commune*

Mating type compatibility in *S. commune* is determined by two unlinked factors, *A* and *B*, both of which must be heterozygous for mating to occur (Simchen 1967). The *A* factor is determined by two linked loci, *a* and *β* (Raper et al. 1958; Raper et al. 1960). The frequency of recombination between the *a* and *β* loci varies with genetic background. A single unlinked *rec* locus has a major effect, with the dominant *rec⁺* allele reducing recombination approximately four-fold, from 15% to 4% (Simchen 1967). In addition, as at the *his-3* locus of *Neurospora*, there appear to be other genes with small effects, resulting in a spread of recombination frequencies from 1% to 8% in crosses including *rec⁺* and 8% to 25% in the absence of *rec⁺*.

Recombination in the *B* factor is similarly regulated, but the two systems appear to be independent. Each regulatory system also seems to be differently affected by temperature, suggesting that the gene products involved have different temperature optima (Stamberg 1968).

7. SEQUENCE HETEROLOGY AND RECOMBINATION

In order to analyze recombination events, it is necessary for the chromosomes involved in the event to differ in some detectable way. However, heterozygosity usually alters the events being studied, sometimes markedly. Indeed, in *S. cerevisiae*, heterology has profound effects on both gene conversion and crossing over.

7.1 The *b2* Locus of *A. immersus*

Despite the effect on recombination events, heterozygosity has been used to reveal details of recombination intermediates at the *b2* locus of *A. immersus* (Nicolas and Rossignol 1989 and references therein) hDNA is most often initiated at the end of the gene closer to the 1F1 mutation, with events extending towards A4 at the other end, but in 20% of cases appears to be initiated elsewhere. Asymmetric hDNA is most common close to the initiator, and declines in frequency towards the other end of the gene, while symmetric hDNA increases in the same direction. Heterozygosity appears to have no effect on the spread of asymmetric hDNA but blocks extension of symmetric hDNA. Close to the initiator, mismatches in *b2* reduce both kinds of hDNA, suggesting that heterology interferes with initiation, perhaps at the strand invasion step (Nicolas and Rossignol 1989). In addition, crossing over is increased by large heterologies and by mismatches upstream of the heterology, and tends to occur between the initiator and the heterology, suggesting that such heterology blocks the migration of Holliday junctions and forces their resolution (Nicolas and Rossignol 1989).

The nature of mutations, inferred from their mode of generation, has also illuminated details of mismatch repair (MMR) in *Ascobolus* (Nicolas and Rossignol 1989). In some cases, the direction of MMR is determined by the type of mismatch, as single-base insertion/deletions are repaired to match the shorter sequence. Large heterologies, in contrast, are repaired to match the invading strand in hDNA. Both frame shifts and large heterologies do not show PMS and must be efficiently repaired, and both can impose their direction of repair on nearby mismatches. A mutation that usually experiences a high frequency of PMS will be converted more often, with a reduction in PMS, if it is close to a frame shift mutation, and will be repaired in the same direction as the insertion/deletion. A large heterology influences the direction of repair of both high and low PMS mutations. Influence over MMR direction is a local effect, suggesting that MMR tracts are much shorter than the total length of hDNA.

7.2 The *his-3* Region of *Neurospora*

Between *his-3* and *cog*, different wild type sequences diverge (Fig. 7) by up to 3.4% (Yeadon and Catcheside 1999). Analysis of recombinant His⁺ progeny of crosses heterozygous for these silent markers showed that 40% of conversion tracts are discontinuous, with patches of sequence from each parent (Yeadon and Catcheside 1998). A similar analysis of a TM429 heterozygote showed that conversion can occur on both sides of the translocation breakpoint (Yeadon et al. 2001) supportive of the genetic data of Catcheside and Angel (1974). In combination with the finding of interrupted conversion tracts in normal sequence chromosomes (Yeadon and Catcheside 1998) conversion on both sides of TM429 suggests a template-switching variant of the SDSA model (Pâques and Haber 1999; Yeadon et al. 2001) for repair of a double strand break close to *cog* (Fig. 8). DSB repair initiated on the normal-sequence chromosome results in invasion of the chromosome carrying the translocation and use of this chromosome as

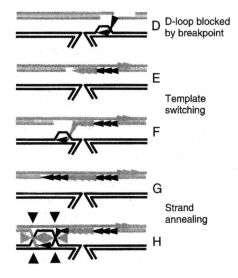

Fig. 8: Template-switching variant of the SDSA model, showing how recombination can cross a chromosomal break in a translocation heterozygote. Strand breakage, resection of the 5' ends and the initial invasion of the homolog for repair synthesis proceed as in the DSBR (Fig. 3 A-C) and simple SDSA (Fig. 4 A-C). models. However, the D-loop cannot migrate past the structural discontinuity (D). If repair synthesis is free to alternate between the invaded and initiating chromosome (E and F), the event can cross the translocation breakpoint. Strand annealing may occur to the initiating chromosome (G) or to a D loop on the invaded chromosome (H), in which case double Holliday junctions result. Conversion without a crossover can result from annealing of the newly synthesized DNA to the initiating chromosome (G) or by symmetric resolution of the Holliday junctions. Crossovers, which may have associated conversion events arising from mismatch repair of heteroduplex DNA, result from asymmetric resolution of Holliday junctions, as shown in Fig. 5.

template to copy sequence close to *cog*. Sequence copying then returns to the initiating chromosome, or its sister, until the translocation heterology is passed, after which the TM429 chromosome can once again be the copy template. Demonstration that in *msh-2* mutants, presumably defective in mismatch repair, conversion tracts are still discontinuous (described in more detail in section 9.2) provides additional support for template switching during recombinational repair. Like other SDSA models, resolution of either single (Ferguson and Holloman 1996) or double Holliday junctions (Pâques et al. 1998) may yield a crossover.

The presence of naturally occurring mismatches close to *cog* reduces recombination in *his-3* slightly, to 70% of the homologous level (Yeadon *et al.* 2004 a) perhaps by interfering with initiation as suggested by Nicolas and Rossignol (1989) for the *b2* gene of *Ascobolus*. Mismatches within *his-3* reduce allelic recombination three-fold (Yeadon et al. 2004 a) perhaps by decreasing the chance of resolution of a recombination intermediate in this location. In diploids heterozygous for *TK* inserted between *cog* and *his-3*, a recombination event must cross a 2.2 kb region in which the homologous chromosomes are completely heterologous in order to reach *his-3*. This necessity has no effect on allelic recombination, so such heterology does not increase the chance that recombination events will terminate. Perhaps, at this locus of *Neurospora*, the stage of recombination after the recombination event is established but prior to resolution involves only DNA synthesis and thus extension of asymmetric hDNA, also unaffected by heterology in *Ascobolus* (Nicolas and Rossignol 1989).

Although heterozygosity for *TK* does not affect conversion on the downstream side of the heterology, such large heterologies do not stimulate crossing over as they appear to do in *Ascobolus* (Nicolas and Rossignol 1989). Indeed, heterozygosity for sequences inserted between *his-3* and *cog* halves the genetic interval between *his-3* and *ad-3* (P. J. Yeadon and D. E. A. Catcheside, unpublished). Since *his-3* to *cog* is usually 25-30% of the *his-3* to *ad-3* interval (Angel et al. 1970; Catcheside and Angel 1974) the heterology seems to prevent crossing over in the heterologous region and may also reduce exchanges in nearby homologous sequences.

8. RECENT CONTRIBUTIONS OF FILAMENTOUS FUNGI TO AN UNDERSTANDING OF MEIOTIC RECOMBINATION

Although in the final quarter of last century the relative contribution of filamentous fungi toward an understanding of meiotic recombination dipped slightly, there has been a resurgence in their importance with the advent of effective DNA manipulation methods. It has become possible to use advantageous traits of filamentous fungi to address aspects of meiosis and meiotic recombination that were difficult or intractable with other systems.

Recent work in *Neurospora* on polarity and the effects of genetic background and sequence heterology on recombination have been described above (sections 5.2, 6.1 and 7.2 respectively).

8.1 The Relationship Between Gene Conversion and Crossing Over

Smith (1965) proposed three criteria that he considered would allow a map of alleles to be oriented with respect to markers flanking the mapped gene. Criterion I, which was considered to have the highest predictive value, assumed that recombination within a gene was due to events in which crossing over was a frequent outcome, and relied upon the configuration of flanking markers amongst intragenic recombinants from crosses of the general configuration shown below (Fig. 9).

Prototrophic intragenic recombinants, wild-type at both m1 and m2 positions in gene A, fall into two non-parental classes with respect to flanking markers, either Pd or pD. If the size of the pD class exceeded that of the Pd class, Smith (1965) suggested that m1 must be proximal since the former was thought to arise by a single crossover in region II whereas the latter required crossovers in regions I, II and III and seemed likely to be less frequent. Conversely, if Pd was larger than pD, then m2 was considered proximal. Although this approach predicted the correct orientation of the *Neurospora his-3* gene (Bowring and Catcheside 1991; Yeadon and Catcheside 1995a, 1999) this may have been fortuitous, since an unambiguous orientation of the *Neurospora am* gene was not possible by this method despite three independent studies.

Fincham (1967). Smyth (1973) and Rambosek and Kinsey (1983) each produced fine-structure maps of the *am* gene and although the linear order of alleles was consistent, with the am^6 allele at one end of each map, the predicted orientation differed between studies. Fincham found am^6 to be the most distal allele, Smyth placed it most proximal and, almost as if it wished to curry favor with neither Fincham nor Smyth, the extensive data set of Rambosek and Kinsey did not suggest an orientation. The *de facto* standard adopted was that of am^6 as the most proximal allele, as determined by Smyth (1973) but this was subsequently shown to be incorrect (Bowring and Catcheside 1995) casting doubt on the validity of the Smith (1965) criterion I for the orientation of genes.

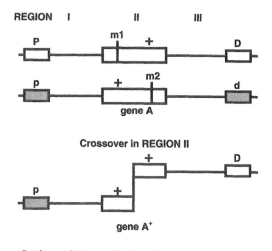

Fig. 9: m1 and m2 are mutant alleles of gene A, flanked by P/p and D/d, alleles of genetic markers proximal and distal to the centromere respectively. Where m1 is closest to P the pD class is expected to be in excess among A^+ recombinants since only a single crossover in region II is required (adapted from Smith, 1965).

In formulating criterion I, Smith ignored the contribution of gene conversion to the formation of gene A^+ progeny but Smyth (1970) recognized that the majority of Am^+ progeny would arise by gene conversion. However, since a large proportion of conversion events were, and still are, thought to have associated crossovers, Smyth considered that Am^+ progeny with a non-parental association of flanking markers (the "R" class of recombinant progeny) arose from a single recombination event, conversion with an adjacent associated crossover. Smyth also assumed that crossovers in the flanking intervals, independent of the event at *am*, had a negligible effect on the size of the R class.

Fincham (1974) proposed that the majority of the R class could arise from gene conversion with a separately initiated crossover nearby but outside of the mutant sites. More importantly, Fincham considered that a bias for these crossovers to occur on one side of the gene was responsible for the inflation in size of one of the classes of R prototrophs. When Fincham formulated his hypothesis the data of Rambosek and Kinsey (1983) were not available. However, the conflict between studies (Smyth 1973; Fincham 1967; Rambosek and Kinsey 1983) with respect to orientation could be reconciled if it was assumed that genetic distance to markers flanking *am* makes a significant contribution to the disparity in the numbers in the flanking marker exchange class.

Though both Smyth and Fincham used the *sp* locus as proximal marker, *his-1* and *inl* were used respectively as distal marker *his-1* is 3cM from *am* and *inl* 8.8 cM away (Fincham 1967). In each case, the application of criterion I put the high conversion allele, am^6, at the same end as the marker most distant from *am*. Now, if m1 is converted to + more frequently than m2 and region I is larger than region III (Smyth 1973) we might expect pD to be in excess of Pd since more crossing over is expected in the larger interval (region I). We would thus conclude that m1 is proximal. Conversely, if m1 is converted to + more frequently than m2 and region III is larger than region I (Fincham 1967) Pd would be the larger class and we would conclude that m1 is distal. An associated prediction is that as the lengths of regions I and III approach equality then so should the numbers in the recombinant classes pD and Pd, as indeed was observed by Rambosek and Kinsey (1983), using flanking markers almost equidistant from *am*.

Fincham (1974) had already considered that a disparity in distance to flanking markers might contribute to a difference in the number of pD and Pd Am+ recombinants but rejected this because it required that events at *am* promote exchanges nearby without participating in them. This concept was, and remains, at odds with a central tenet of the majority of recombination models - gene conversion with and without flanking marker exchange are considered to be alternative outcomes of the basic recombination event.

Bowring and Catcheside (1996) repeated one of the Smyth (1973) crosses in order to assess the extent to which events in *am* promote the occurrence of distinct crossovers nearby. In addition to classical phenotypic markers *sp* and *his-1*, *am* was also flanked by molecular markers 8 kilobases (kb) proximal and 5 kb distal. Of 44 crossovers between the phenotypic markers, only 14 were within the 13 kb region bounded by the molecular markers, sufficiently close to *am* to be considered associated, and in the majority of cases the crossover occurred some distance from *am*. Only 7% of conversion events at *am* had an associated crossover. This suggested that at *am* at least, gene conversion with and without crossing over may not be alternative outcomes of the basic recombination event. However, low frequency recombinators such as that near *am* receive little attention so it is not clear whether this lack of association is a unique or is a general phenomenon.

The recombination frequency at *am* is modest, but work in recent years has tended to concentrate on loci with a high recombination frequency since more recombinants can be collected for a comparable amount of work. For example, in a screen of over six thousand octads (F.J. Bowring and D.R. Stadler, unpublished) only seven with a recombination event in *am* were identified, none of which had an associated crossover. Conversely, in the *his-3* region of *Neurospora*, recombination can be much more frequent than at *am*, and depends on the alleles of *cog* and *rec-2* in the cross. As described above, the distance between flanking genetic markers *lys-4* and *ad-3* can be as low as 2cM (in the absence of *cog*[+]) or as high as 20cM (when *cog*[+] is homozygous). Crosses heteroallelic K1201/K874 may yield from 0.01% (in the presence of *rec-2*[+]) to 1% (*cog*[+] homozygous) His[+] spores (Yeadon et al. 2004 b). In addition, an octad with a recombination event at *his-3* was isolated in a screen of 150 octads (P. J. Yeadon, F. J. Bowring and D. E. A. Catcheside, unpublished). So it is clear that, like yeast hotspots, *cog*[+] is a high frequency recombinator. As at yeast loci, chromosomes that have experienced recombination at *his-3* are highly likely also to have a local crossover. Indeed, the flanking markers *lys-4* and *ad-3* are recombined in 40-45% of recombinant His[+] progeny (Yeadon et al. 2004 b; P. J. Yeadon, F. J. Bowring and D. E. A. Catcheside, unpublished).

We previously considered that, as both recombination frequency and flanking marker exchange are similar at *his-3* and at yeast hotspots, but flanking marker exchange is rare at *am* where recombination is infrequent, the apparent association between conversion and crossing over might be due to the relative frequency of the two types of event at high frequency hotspots rather than to a mechanistic relationship. However, the frequency of flanking marker exchange is the same in His[+] progeny from all crosses heteroallelic K1201/K874, regardless of the alleles of *cog* and *rec-2* in the cross (Yeadon et al. 2004 b). and is thus unaffected by whether recombination is initiated at a high frequency or not. It now seems more likely that the contribution of SDSA and DSBR pathways of recombinational repair and/or the degree of bias in junction resolution may differ between loci even in a single organism. Since our recent screen of 150 octads also detected conversion at *lys-4* and *cot-1* loci (P. J. Yeadon, F. J. Bowring and D. E. A. Catcheside, unpublished). it may be feasible to identify loci at which conversion is frequent but crossing over is uncommon, allowing additional investigation of the relationship between gene conversion and crossing over.

8.2 Initiation of Recombination

It has been known for some time that meiotic recombination in *S. cerevisiae* is initiated by a DSB (Nicolas et al. 1989). As well as suffering a number of meiotic defects, *S. cerevisiae* SPO11 mutants are deficient in both DSBs and in meiotic recombination (Klapholz et al. 1985). Exposure of SPO11 mutants to ionizing radiation partially rescues the phenotype, suggesting that the resulting lesions can substitute for DSBs generated by the Spo11 protein. It was thus concluded that Spo11p is responsible for generating meiotic DSBs in this yeast, although *in vitro* catalytic activity has not been demonstrated (Keeney, Giroux and Kleckner 1997).

Recent work using *Coprinus cinereus* (Celerin et al. 2000). and *S. macrospora* (Storlazzi et al. 2003). suggests that DSBs are also responsible for initiation of meiotic recombination in these fungi. *Spo11* mutants of both species display a similar spectrum of meiotic defects to *S. cerevisiae* SPO11 mutants, including a failure of homologs to synapse, aberrant segregation of chromosomes and reduced spore production and viability. As in *S. cerevisiae*, exposure of meiotic tissue to ionizing radiation partially rescues the phenotype in both species. While this and other evidence (Dernburg et al. 1998; Romanienko and Camerini-Otero 2000) indicate that *spo11* catalyzed DSBs are likely to be involved in initiation of meiotic recombination in all organisms, evidence suggests that not all recombination events require *spo11* generated DSBs for initiation. For example, there is a 24 cM region of *S. cerevisiae* chromosome III that experiences few DSBs (reviewed in Pâques and Haber 1999). In a Sordaria SPO11 null mutant, although meiotic recombination is reduced about ten-fold across a large interval spanning the chromosome 4 centromere, residual crossing over remains (Storlazzi et al. 2003).

In *Neurospora spo11* mutants, there may be a variable effect on recombination depending upon which part of the genome is examined (F. J. Bowring, P J. Yeadon, R. G. Stainer and D. E. A. Catcheside, unpublished). There appears to be a 10-fold reduction in crossing over in an interval on LG VII in a *Neurospora spo11* mutant, but in the *his-3* region on LG I there is no apparent reduction in the frequency of either gene conversion or crossing over following meiosis in three different *spo11* mutants. Since none of the three alleles is a deletion, for each was generated using repeat-induced point mutation (Selker 1990). there is a slim chance that all are leaky. However, as in *Coprinus* and *Sordaria spo11* mutants, meiosis in homozygotes of all three *Neurospora* mutants is severely disrupted. Moreover, one allele contains a premature stop codon preceding three of the five conserved Spo11 motifs (Bergerat et al. 1997) yielding a predicted peptide lacking two invariant residues that appear to be essential for recombination in *S. cerevisiae* (Diaz et al. 2002).

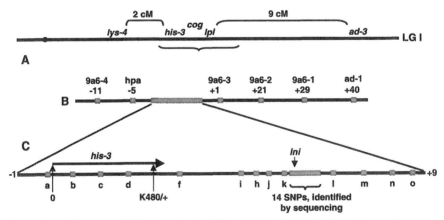

Fig. 10: Part of *Neurospora* LG IR, showing the SNPs to be identified. (A) shows the genetic map; the small disc to the left of *lys-4* represents the centromere. The bracketed region is expanded in the middle view (B). The positions (-11 to +40; B and C) indicate the location with respect to approximate distance from the *his-3* start codon, in kb. The names given to each SNP are shown above the overview and below the expansion (C) SNPs "e" and "g" are not shown, as these could not be detected by the technology used by the Australian Genome Research Facility (AGRF). "h" and "i" are switched in position (C) as the original "h" also could not be detected by the AGRF. "*ini*" (C) is the putative *cog* initiation site.

8.3 Use of Genetic and Silent Sequence Markers to Provide a Detailed Description of Recombination in the *his-3* Region of *Neurospora*

Neurospora laboratory strains are highly polymorphic, both genetically and at the sequence level (Yeadon and Catcheside 1999). We have identified many positions at which the LG I sequence differs in St Lawrence 74A and Lindegren Y8743 wild type strains. We have chosen some of these markers, spanning more than 50 kb of LG I (Fig. 10) for analysis of the outcome of recombination events initiated at *cog*. Each selected SNP can be identified using technology designed by the Australian Genome Research Facility or, in the case of the 14 SNPs close to the *cog* initiation site (Fig. 10) in a single sequencing reaction.

We have isolated 150 octads from a cross in which not only are all the above SNPs segregating, but also the genetic markers *lys-4*, *his-3* and *ad-3* (Fig. 10) and the unlinked markers *cot-1* and *am*. We have observed four instances of gene conversion, two at *cot-1* and a single event at each of the *his-3* and *lys-4* loci. Completion of the SNP-testing may determine the frequency with which conversion is initiated at *cog*⁺ and will provide a detailed picture of unselected recombination events in this region.

9. THE IMPACT OF GENOMICS

Genome projects allow comparison of genes involved in homologous recombination and it is becoming evident that whilst the suite of genes involved in yeasts and in multicellular eukaryotes share orthologs (Borkovich et al. 2004; Young et al. 2004) each has unique members, suggesting the process is likely to differ in detail if not in principle across the diversity of eukaryotes. Moreover, sequence based homology infers functional homology but does not demonstrate it. Since recombination genes in filamentous fungi show greater similarity to those of animals and plants than do those of yeasts, studies of filamentous fungi are likely to play an important future role in revealing the details of recombination in higher eukaryotes.

9.1 Demonstrating Functional Homology in a Multicellular Eukaryote

The *Ustilago maydis brh2* gene, an ortholog of mammalian *BRCA2*, has a role in DNA repair, recombination, and ultimately in genome stability (Kojic et al. 2002). Human BRCA2 interacts with human DSS1 in a yeast 2-hybrid system (Marston et al. 1999) and since BRCA2 has a role in DNA repair and recombination it was thought that if this was an evolutionarily conserved interaction, human DSS1 should also have a role in the maintenance of genome stability. *S. cerevisiae* and *S. pombe* both lack a BRCA2 ortholog but each has a DSS1 ortholog. Mutation of the DSS1 ortholog in these yeasts led to modest growth retardation but little evidence of an effect on recombinational repair. In search of another candidate model system, Kojic et al. (2003) examined the *U. maydis* DSS1 ortholog. They demonstrated a DSS1/brh2 interaction and a role for DSS1 in recombinational repair. Since the DSS1/brh2 interaction is present in both humans and *U. maydis*, Kojic et al. (2003) concluded that the role of DSS1 in recombinational repair is probably also conserved, a conclusion that may not have been reached if investigation had been confined to yeasts.

9.2 Mismatch Repair Genes in *Neurospora*

The mismatch repair (MMR) pathway recognizes unpaired nucleotides and pairing between non-complementary nucleotides. During recombination, allelic differences, point mutations, insertions and deletions yield hDNA with mismatched bases (Modrich 1991; Strand et al. 1993) correction of which by MMR may generate novel sequence combinations.

MutS, MutL and MutH are the main proteins involved in MMR in bacteria. MMR is initiated when MutSp binds to a mismatch (Su et al. 1988) and MutLp binds to the MutS-mismatch complex, increasing the stability of the complex (Grilley et al. 1989; Mankovich et al. 1989) The binding of MutL activates MutH endonuclease (Grafstrom and Hoess 1987) which cleaves the unmethylated strand at the GATC site (Au et al. 1992) thus targeting repair to the newly synthesized strand.

Six *MutS* (*MSH1-MSH6*), and four *MutL* (*PMS1* and *MLH1-MLH3*) homologs have been identified in *S. cerevisiae* (reviewed in Pâques and Haber 1999; Borts et al. 2000). A heterodimer of MSH proteins recognizes the mismatch and an MLH protein heterodimer binds to the mismatch/MSH complex. The MSH and MLH proteins involved in each dimer depend on the nature of the mismatch (Fig. 11). All Mshp heterodimers involved in MMR include Msh2p, whereas, only two of the three MMR-related Mlhp heterodimers involved include Pms1p.

Mutations in *MSH2* and *PMS1* in *S. cerevisiae* increase PMS, due to lack of correction of mismatches in hDNA, and also have a mutator phenotype, with genome-wide microsatellite instability and up to a 1000-fold increase in spontaneous mutation (Strand et al. 1993). In addition, the viability of spores from MMR-deficient diploids is reduced compared to that of spores from wild type crosses, to 60-80% in *S. cerevisiae* (Williamson et al. 1985) and 86% in *S. pombe* (Rudolph et al. 1999). Also in *S. pombe*, a *msh2Δ* (deletion) mutant exhibits abnormal meiotic chromosome structures, with linear elements frequently aggregated, a rare occurrence in the wild type (Rudolph et al. 1999).

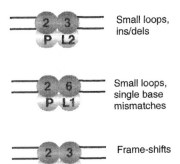

Small loops, ins/dels

Small loops, single base mismatches

Frame-shifts

Fig. 11: Different MSH and MLH heterodimers recognise different types of mismatch (adapted from Borts et al. (2000). The darker discs represent MSH and the lighter discs MLH proteins, where for example "2" means Msh2p and "L2" Mlh2p.

As in *S. cerevisiae*, *Neurospora* has six *MutS* and four *MutL* homologs, of which *msh-2* and *pms1* are easily identified (Borkovich et al. 2004). Strains defective in *msh-2* and *pms1* have been generated (L. Y. Koh and D. E. A. Catcheside unpublished) by repeat-induced point mutation (RIP; Selker 1990). Although neither gene is deleted, both predicted protein sequences have multiple amino acid substitutions. Since the start codon is altered in the RIPed *msh-2* sequence and a stop codon replaces residue 38, we can be certain *msh-2^{RIP}* is a null mutant. The first stop codon in the *pms1^{RIP}* mutant replaces residue 184, so it is also unlikely that the Pms1 protein is fully functional.

Both *msh-2^{RIP}* and *pms1^{RIP}* appear to have a mutator function, as the mitotic reversion rate of the *his-3* K458 allele is increased 20 fold in *msh-2* and *pms1* mutants compared to wild type strains (L. Y. Koh and D. E. A. Catcheside unpublished). Moreover, while wild type *Neurospora* crosses yield 85-95% viable spores, *msh-2^{RIP}* and *pms1^{RIP}* homozygotes yield 20-30% colorless spores and only 60-70% of spores are viable. Allelic recombination at *his-3* in *msh-2^{RIP}* homozygotes is three-fold more frequent than in wild type diploids and conversion tracts in recombinant progeny are at least as patchy as in progeny of wild type crosses (L. Y. Koh and D. E. A. Catcheside unpublished). The *pms1^{RIP}* homozygotes show no obvious recombination phenotype, although recombination in *his-3* may be slightly increased. The double homozygote has the same phenotype as the *msh-2^{RIP}* homozygote, suggesting that, as in *S. cerevisiae*, Msh-2p and Pms1p operate in the same pathway.

If Msh-2p is necessary for MMR in *Neurospora*, as is its homolog in *S. cerevisiae* (reviewed in Borts et al. 2000) retention of interruptions to conversion tracts in *msh-2* knockouts provides support for the template-switching hypothesis of recombination (Yeadon et al. 2001). The increase in recombination in *his-3* may reflect a high frequency of restoration repair within the gene in wild type crosses. Alternatively, since the crosses analyzed include mismatches both within *his-3* and close to *cog*, lack of Msh-2p may reduce or negate any effect of sequence heterology on initiation and/or resolution of recombination (Yeadon et al. 2004 a).

9.3 Meiotic Insights from *Coprinus*

Zolan and associates have taken advantage of the synchrony of meiosis in *C. cinereus* to examine chromosome behavior and the role of certain recombination genes during this period. As well as providing the first evidence of programmed cell death during meiosis in a non-animal, Celerin and co-workers (2000) demonstrated that synapsis of homologs does not always follow pairing of homologous regions.

Li et al. (1999) used fluorescence *in situ* hybridization (FISH) of chromosome 8- and 13-specific probes to examine pairing between homologs during meiosis. They demonstrated that pairing occurs rapidly following nuclear fusion and that a substantial amount of pairing is present by the time that nucleoli fuse. In an examination of the effect on meiosis of disruption of the *mre11* homolog, *rad11*, Gerecke and Zolan (2000) found that while the homologous chromosomes of *rad11* mutants failed to synapse, some pairing between homologs remained. This pairing, less frequent than in the wild-type but at higher levels than in the corresponding mutants of *S. cerevisiae*, confirmed that synapsis is not an inevitable outcome of homolog pairing.

In order to avoid confusion, a definition of what we intend to convey by the terms "presynaptic pairing" and "synapsis" is appropriate here. By "synapsis" we mean the alignment of homologs along their entire length, as occurs during pachynema. When using "pairing" or "presynaptic pairing" we mean the proximity of homologous regions of chromosomes, which occurs before synapsis and is usually only evident when particular chromosomal regions are marked in some way (e.g. Li et al. 1999).

For some time, two classes of organism have been recognized according to whether *spo11* is required for synapsis of homologs during pachynema. Homologs fail to synapse when *spo11* is disrupted in the first class of organism, which initially comprised only *S. cerevisiae*, but now includes *C. cinereus* (Celerin et al. 2000) mouse (Romanienko and Camerini-Otero 2000) *S. macrospora* (Storlazzi et al. 2003) and *N. crassa* (F. J. Bowring, P. J. Yeadon, R. G. Stainer and D. E. A. Catcheside, unpublished). In the second class, which currently has two members, *D. melanogaster* (McKim and Hayashi-Hagihara 1998) and *Caenorhabditis elegans* (Dernburg et al. 1998), synapsis of homologs occurs irrespective of *spo11* function. However, the work of Celerin et al. (2000) indicates that this classification system may need further refinement.

Mutation of the *C. cinereus spo11* ortholog disrupts synapsis of homologs and formation of synaptonemal complex but presynaptic pairing of homologs persists at about 70% of wild-type levels. Conversely, in another class I organism, *S. cerevisiae*, both synapsis and pairing are severely reduced in *spo11* mutants (Weiner and Kleckner 1994). To our knowledge, there are no data regarding the effect of *spo11* disruption on presynaptic pairing in mouse. However, some *SPO11*-independent homolog recognition has been inferred for *S. macrospora* (Storlazzi et al. 2003) and as indicated above, homologous recombination in one region of LG I is not reduced in *Neurospora spo11* mutants despite an apparent absence of synapsis. By definition, some form of interaction between homologs must occur during homologous recombination. Therefore, assuming that the *Neurospora spo11* mutants are null, it is possible that presynaptic pairing occurs independently of *spo11* function in this organism also.

9.4 An Integrated Approach to the Study of Meiosis and Recombination in *Sordaria*

We have already considered some of the recent findings of Zickler and co-workers who have used a combined cytological-genetic approach to exploit some of the unique features of *S. macrospora*. *Sordaria* chromosomes are sufficiently large for examination under the light microscope and ascus length, nucleolar shape/position and nuclear volume alter during meiosis, providing staging information independent of chromosome behavior. Their work has provided new insights into the influence of double-strand breaks on chromosome behavior throughout meiosis.

van Heemst et al. (1999) tagged Spo76p with green fluorescent protein. GFP-tagged Spo76p exhibited the same localization pattern as the untagged protein suggesting that the tag does not interfere with function. Moreover, the tagged gene complemented a *SPO76* mutation to the same extent as the gene alone. Spo76p was found to localize to chromosomes from karyogamy through leptonema, zygonema and pachynema, and the labeled protein has been used as a marker to highlight chromosomes during meiosis and mitosis (van Heemst et al. 1999; Storlazzi et al. 2003). Apparently, the *Sordaria SPO76* gene has a role in the maintenance of cohesion between sister chromatids during meiosis. An absence of functional Spo76 caused an absence of chiasmata, and a loss of crossing over was predicted.

Storlazzi et al. (2003) demonstrated that, as expected, an absence of *spo11* catalyzed DSBs leads to chromosome non-disjunction at meiosis I, but that mis-segregation also occurs during meiosis II. They suggested that this may be due to an absence of localized disruption of sister cohesion, normally occurring at chiasmata. Second division mis-segregation also seems to occur in *Neurospora spo11* mutants where chromatids can be seen streaked across the ascus at Interphase II (Fig. 12).

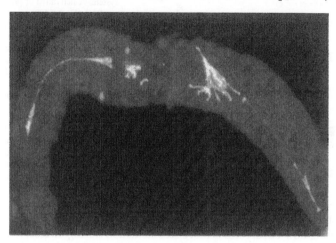

Fig. 12: Z-projection of Interphase II chromosomes in a *spo-11^RIP1/RIP1* homozygote (F. J. Bowring, P. J. Yeadon, R. G. Stainer and D. E. A. Catcheside, unpublished).

10. PUTTING RECOMBINATION TO WORK

Evolution of genes proceeds by mutation and selection, with meiotic recombination enabling eukaryotes to sort advantageous mutations from the majority that are deleterious. A variety of *in vitro* DNA mutation and shuffling systems have been developed for rapid generation of novel gene variants, which can then be screened *in vivo* for improvements in chosen directions. For a recent review, see Neylon (2004). For example, mutations can be generated by errors in reassembling gene fragments by PCR and shuffled by iteration of fragmentation and reassembly (Stemmer 1993). Such systems have been used successfully for directed evolution of a variety of proteins with improvements to thermostability, protein-ligand binding and substrate selectivity.

Discovery that meiotic recombination in *Neurospora* tolerates substantial heterology in recombining sequences (Yeadon and Catcheside 1995 a; Yeadon et al. 2004 a) has stimulated development of a directed evolution system in which both shuffling and screening are achieved *in vivo* (Catcheside et al. 2003). Vectors have been constructed (Fig. 13) that allow the target DNA to be inserted between the *cog* recombination hotspot and mutant *his-3* alleles on LG I (Rasmussen et al. 2002) permitting pairs of heterologous genes from different species to be recombined. Recombinants are enriched by selecting His+ progeny (Fig. 13) and the resultant novel genes, expressed in a eukaryotic system, can be screened directly for advantageous variants. By duplicating the target gene in one of the parents, it is

26

possible to use RIP (Selker 1990). to generate novel variants of heterologous genes in *Neurospora* (E. Cambareri and W. D. Stuart, unpublished). RIP and shuffling can be achieved in a single meiotic cycle, in principle permitting directed evolution of single heterologous genes *in vivo* in *Neurospora*.

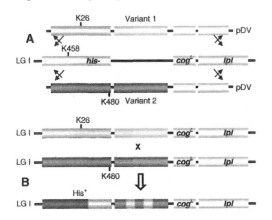

Fig. 13: The Flinders gene diversification system. A derivative of pDV2 (*his-3* K26) or pDV3 (*his-3* K480) carrying a variant of a foreign gene is transfected, by electroporation, into conidia of *his-3* K458 recipients of either mating type (A). Since conidia are multinucleate and the K458 allele complements either K26 or K480, transformants do not need histidine to grow. Pure cultures are isolated in which every nucleus has the foreign gene inserted between *his-3* and *cog*. A mating pair, one K26 and the other K480 and each with a different variant of the foreign gene, yields histidine independent progeny, which may also have experienced recombination within the foreign gene (B).

11. CONCLUSION

Establishing how eukaryotes achieve recombination during meiosis remains one of the more challenging problems for molecular biology. This is partly because of the genetic "uncertainty principle" (Paques and Haber, 1999) sequence alterations that serve as recombination reporters are likely to influence the very process they are designed to betray. It is also becoming clear that recombination pathways are probably not the same in all organisms. Recombination in filamentous fungi appears to show greater similarity to the process in higher eukaryotes than it does to recombination in yeasts. Since the genomes of several filamentous fungi have been sequenced, increasing the ease of their genetic and molecular manipulation, it is likely they will play a substantial role in a full understanding of meiotic recombination in multicellular organisms. At the very least, they are likely to help fill in some of the critical gaps in our understanding of recombination.

Acknowledgements: This work was supported by grants from the Australian Research Council and the Flinders University Research Budget.

REFERENCES

Alani E, Reenan RAG and Kolodner RD (1994). Interaction between mismatch repair and genetic recombination in *Saccharomyces cerevisiae*. Genetics 137:19-39.

Angel T, Austin B and Catcheside DG (1970). Regulation of recombination at the *his-3* locus in *Neurospora crassa*. Aust J Biol Sci 23:1229-1240.

Au KG, Welsh K and Modrich P (1992). Initiation of methyl-directed mismatch repair. J Biol Chem 267:12142-12148.

Avery AT, CM MacLeod and McCarty M (1944). Studies on the chemical nature of the substance inducing transformation of pneumococcal types. Induction of transformation by a desoxyribonucleic acid fraction isolated from pneumococcus Type III. J Exp Med 79:137-158.

Baker BS, AT Carpenter, MS Esposito, RE Esposito and Sandler L (1976). The genetic control of meiosis. Ann Rev Genet 10:53-134.

Bateson W, Saunders ER and Punnett RC (1905). Experimental studies in the physiology of heredity. Rep Evol Comm Roy Soc 2:1-55 and 80-89.

Beadle G and Tatum E (1941). Genetic control of biochemical reactions in *Neurospora*. Proc Natl Acad Sci USA 27:499-506.

Bergerat A,dc Massy B, Gadelle D, Varoutas PC, Nicolas A, and Forterre P (1997). An atypical topoisomerase II from Archaea with implications for meiotic recombination. Nature 386:414-417.

Borkovich KA, Alex LA, Yarden O, Freitag M, Turner GE, Read ND, Seiler S, Bell-Pedersen D, Paietta J, Plesofsky, M Plamann, U Schulte, G Mannhaupt, FE Nargang, A Radford, C Selitrennikoff, JE Galagan N, Dunlap JC, Loros JJ, Catcheside DEA, Inoue H, Aramayo R, Polymenis M, Selker EU, Sachs MS, Marzluf GA, Paulsen I, Davis R, Ebbole DJ, Zelter A, Kalkman E, O'Rourke R, Bowring FJ, Yeadon PJ, Ishii C, Suzuki K, Sakai W and Pratt R (2004). Lessons from the genome sequence of *Neurospora crassa*: Tracing the path from genomic blueprint to multicellular organism. Microbiol Mol Biol Rev 68:1-108.

Borts RH, Chambers SR and Abdullah MFF (2000). The many faces of mismatch repair in meiosis. Mutation Res 451:129-150.

Bowring FJ and Catcheside DEA (1991). The initiation site for recombination *cog* is at the 3' end of the *his-3* gene in *Neurospora crassa*. Mol Gen Genet 229:273-277.

Bowring FJ and Catcheside DEA (1995). The orientation of gene maps by recombination of flanking markers for the *am* locus of *Neurospora crassa*. Curr Genet 29:27-33.

Bowring FJ, Catcheside DEA (1996). Gene conversion alone accounts for more than 90% of recombination events at the am locus of *Neurospora crassa*. Genetics 143:129 136.

Cao L, Alani E and Kleckner N (1990). A pathway for generation and processing of double-strand breaks during meiotic recombination in *S. cerevisiae*. Cell 61:1089-1101.

Case ME and Giles NH (1958 a). Recombination mechanisms at the *pan-2* locus in *Neurospora crassa*. Cold Spring Harbor Symp Quant Biol 23:119-135.

Case ME and Giles NH (1958 b). Evidence from tetrad analysis for both normal and aberrant recombination between allelic mutants in *Neurospora crassa*. Proc Natl Acad Sci USA 44:378-390.

Case ME and Giles NH (1964). Allelic recombination in *Neurospora*: tetrad analysis of a three-point cross within the *pan-2* locus. Genetics 49:529-540.

Catcheside DEA (1970) Control of recombination within the *nitrate-2* locus of *Neurospora crassa*: an unlinked dominant gene which reduces prototroph yield. Aust J Biol Sci 23:855-865.

Catcheside DEA (1981) Genes in *Neurospora* that suppress recombination when they are heterozygous. Genetics 98:55-76.

Catcheside DEA, Rasmussen JP, Yeadon PJ, Bowring FJ, Cambareri E, Kato E, Gabe G and Stuart WD (2003). Diversification of heterologous genes *in vivo* in *Neurospora*. Appl Microbiol Biotech 62:544-549.

Catcheside DG (1966). A second gene controlling allelic recombination in *Neurospora crassa*. Aust J Biol Sci 23:855-865.

Catcheside DG (1974). Fungal Genetics. Ann Rev Genet 8:279-300.

Catcheside DG (1977). The Genetics of Recombination. London, Edward Arnold.

Catcheside DG and Angel T (1974) A *histidine-3* mutant, in *Neurospora crassa*, due to an interchange. Aust J Biol Sci 27:219-229.

Catcheside DG and Austin B (1969). The control of allelic recombination at histidine loci in *Neurospora crassa*. Aust J Biol Sci 56:685-690.

Catcheside DG and Austin B (1971). Common regulation of recombination at the *amination-1* and *histidine-2* loci in *Neurospora crassa*. Aust J Biol Sci 24:107-115.

Catcheside DG and Corcoran D (1973). Control of non-allelic recombination in *Neurospora crassa*. Aust J Biol Sci 26:1337-1353.

Celerin M, Merino ST, Stone JE, Menzie AM and Zolan ME (2000). Multiple roles of Spo11 in meiotic chromosome behavior. EMBO J 19:2739-2750.

Collins I and Newlon CS (1994). Meiosis-specific formation of joint DNA molecules containing sequences from homologous chromosomes. Cell 76:65-75.

Creighton HB and McClintock B (1931). A correlation of cytological and genetical crossing-over in *Zea mays*. Proc Natl Acad Sci USA 17:492-497.

28

Emerson S and Yu-Sun CCC (1967). Gene conversion in the Pasadena strain of *Ascobolus immersus*. Genetics 55:39-47.

Dernburg AF, McDonald K, Moulder G, Barstead R, Dresser M and Villeneuve AM (1998). Meiotic recombination in *C. elegans* initiates by a conserved mechanism and is dispensable for homologous chromosome synapsis. Cell 94:387-398.

Diaz RL, Alcid AD, Berger JM and Keeney S (2002). Identification of residues in yeast Spo11p critical for meiotic DNA double-strand break formation. Mol Cell Biol 22:1106-1115.

Fan Q, Xu F and Petes T (1995). Meiosis-specific double-strand breaks at the *HIS4* recombination hotspot in the yeast *Saccharomyces cerevisiae*: control in cis and trans. Mol Cell Biol 15:1679-1688.

Ferguson DO and WK Holloman (1996). Recombinational repair of gaps in DNA is asymmetric in *Ustilago maydis* and can be explained by a migrating D-loop model. Proc Natl Acad Sci USA 93:5419-5424.

Fincham JRS (1967). Recombination within the *am* gene of *Neurospora crassa*. Genet Res Camb 9:49-62.

Fincham JRS (1974). Negative interference and the use of flanking markers in fine-structure mapping in fungi. Heredity 33:116-121.

Fogel S and Mortimer RK (1970). Fidelity of gene conversion in yeast. Mol Gen Genet 122:165-182.

Fogel S, Mortimer RK, Lusnak K and Tavares F (1979). Meiotic gene conversion: a signal of the basic recombination event in yeast. Cold Spring Harbor Symp Quant Biol 43:1325-1341.

Galagan JE, Calvo SE, Borkovich A, Selker EU, Read ND, Jaffe D, FitzHugh W, Ma LJ, Smirnov S, Purcell S, Rehman B, Elkins T, Engels R, Wang S, Nielsen CB, Butler J, Endrizzi M, Qui D, Lanakiev P, Bell-Pedersen D, Nelson MA, Werner-Washburne M, Seltrenikoff CP, Kinsey JA, Braun EL, Zelter A, Schulte U, Kothe GO, Jedd G, Mewes W, Staben C, Marcotte E, Greenberg D, Roy A, Foley K, Naylor J, Stange-Thomann N, Barrett R, Gnerre S, Kanal M, Kamvysselis M, Mauceli E, Bielke C, Rudd S, Frishman D, Krystofova S, Rasmussen C, Metzenberg RL, Perkins DD, Kroken S, Cogoni C, Machino G, Catcheside D, Li W, Pratt RJ, Osmani SA, DeSouza CPC, Glass L, Orbach MJ, Berglund AJ, Voelker R, Yarden O, Plamman M, Seller S, Dunlap J, Radford A, Aramayo R, Natvig DO, Alex LA, Mannhaupt G, Ebbole DJ, Freitag M, Paulsen I, Sachs MS, Lander ES, Nusbaum C and Birren B (2003). The genome sequence of the filamentous fungus *Neurospora crassa*. Nature 422:859-868.

Gerecke EE and Zolan ME (2000). An *mre11* mutant of *Coprinus cinereus* has defects in meiotic chromosome pairing, condensation and synapsis. Genetics 154:1125-1139.

Girard J and Rossignol J-L (1974). The suppression of gene conversion and crossing over in *Ascobolus immersus*: evidence for modfiers acting in the heterozygous state. Genetics 76:221-243.

Grafstrom RH and Hoess RH (1987) Cloning of *mutH* and identification of the gene product. Gene 22:245-253.

Grilley M, Griffith J and Modrich P (1989). Bidirectional excision in methyl-directed mismatch repair. J Biol Chem 268:11830-11837.

Grimm C, Bahler J and Kohli J (1994). *M26* recombinational hotspot and physical conversion tract analysis in the *ade6* gene of *Schizosaccharomyces pombe*. Genetics 135:41-51.

Gutz H (1971). Site specific induction of gene conversion in *Schizosaccharomyces pombe*. Genetics 69:317-337.

Hahn A and Hock B (1999). Chromosome mechanics of fungi under speceflight conditions – tetrad analyses of two-factor crosses between spore color mutants of *Sordaria macrospora*. FASEB 13:s149-S156.

Haldane JBS (1919). The combination of linkage values, and the calculation of distances between the loci of linked factors. J Genet 8:299-309.

Helmi S and Lamb BC (1983). The interactions of three widely separated loci controlling comversion properties of *w* locus in *Ascobolus immersus*.Genetics 104: 23-40.

Hilliker AJ, Harauz G, Reaume AG, Gray M, Clark SH and Chovnick A (1994). Meiotic conversion tract length distribution within the *rosy* locus of *Drosophila melanogaster*. Genetics 137:1019-1026.

Holliday R (1964) A mechanism for gene conversion in fungi. Genet Res Camb 5:282-304

Howell WM and Lamb BC (1984). Two locally acting genetic controls of gene conversion, *ccf-5* and *ccf-6*, in *Ascobolus immersus*. Genet Res 43:107-121.

Janssens FA (1909). Spermatogénèse dans les Batraciens. V. La théorie de la chiasmatypie. Nouvelles interprétation des cinèses de maturation. Cellule 25:387-411.

Jessop AP and Catcheside DG (1965). Interallelic recombination at the *his-1* locus in *Neurospora crassa* and its genetic control. Heredity 20:237-256.

Johnson RD and M Jasin (2000). Sister chromatid gene conversion is a prominent double-strand break repair pathway in mammalian cells. EMBO J 19:3396-3407.

Keeney S, Giroux CN and Kleckner N (1997). Meiosis-specific DNA double-strand breaks are catalyzed by Spo11, a member of a widely conserved protein family. Cell 88:375-384.

Kitani Y, Olive LS and El-Ani AS (1962). Genetics of *Sordaria fimicola*. V. Aberrant segregation at the *g* locus. Am J Bot 49:697-706.

Klapholz S, Waddel CS and Esposito RE (1985). The role of the SPO11 gene in meiotic recombination in yeast. Genetics 110:187-216.

Kojic M, Kostrub CF, Buchman AR. and Holloman WK (2002). BRCA2 homolog required for proficiency in DNA repair, recombination, and genome stability in *Ustilago maydis*. Mol Cell 10:683-691.

Kojic M, Yang HJ, Kostrub CF, Pavletich NP and Holloman WK (2003). The BRCA2-interacting protein DSS1 is vital for DNA repair, recombination, and genome stability in *Ustilago maydis*. Mol Cell 12:1043-1049.

Lamb BC and Helmi S (1978). A new type of genetic control of gene conversion, from *Ascobolus immersus*. Genet Res 32:67-78.

Lamb BC and Shabbir G (2002). The control of gene conversion properties and corresponding-site interference: the effects of *conversion control factor 5* on confversion at locus *w9* in *Ascobolus immersus*. Hereditas 137:41-51.

Legerton TL and Yanofsky C (1985). Cloning and characterisation of the multifunctional *his-3* gene of *Neurospora crassa*. Gene 39:129-140.

Li L, Gerecke EE and Zolan ME (1999). Homolog pairing and meiotic progression in *Coprinus cinereus*. Chromosoma 108:384-392.

Lindegren CC (1953). Gene conversion in *Saccharomyces*. J Genet 51:625-637.

Lindegren CC (1973). Reminiscences of B. O. Dodge and the beginnings of *Neurospora* genetics. Neurospora Newsl 20:13-14.

Lissouba P, Mousseau J, Rizet G, Rossignol JL (1962) Fine structure of genes in the ascomycete *Ascobolus immersus*. Adv Genet 11:343-380

Lissouba P and Rizet G (1960). Sur l'existence d'une génétique polarisée ne subissant que des échanges non réciproques. C.r hebh Séanc Acad Sci Paris 250:3408-3410.

Mankovich JA, McIntyre CA and Walker GC (1989). Nucleotide sequence of the *Salmonella typhimurium mutL* gene required for mismatch repair: Homology of *mutL* to *HexB* of *Streptococcus pneumoniae* and to *PMS1* in the yeast *Saccharomyces cerevisiae*. J Bacteriol 171:5325-5331.

Marston NJ, Richards WJ, Hughes D, Bertwistle D, Marshall CJ and Ashworth A (1999). Interaction between the product of the breast cancer susceptibility gene BRCA2 and DSS1, a protein functionally conserved from yeast to mammals. Mol Cell Biol 19:4633-4622.

McKim KS and Hayashi-Hagihara A (1998). *mei-W68* in *Drosophila melanogaster* encodes a Spo11 homolog: evidence that the mechanism for initiating recombination is conserved. Genes Dev 12:2932-2942.

Mendel G (1865). Versuche über Pflanzenhybriden. Verh Naturf-Ver Brünn 4:3-47. In "Classic Papers in Genetics" (1959) ed. JA Peters, Prentice-Hall, NY, USA.

Meselson MS (1972). Formation of hybrid DNA by rotary diffusion during genetic recombination. J Mol Biol 71:795-798.

Meselson MS and CM Radding (1975). A general model for genetic recombination. Proc Natl Acad Sci USA 72:358-361.

Mitchell MB (1955). Aberrant recombination of pyridoxine mutants of *Neurospora*. Proc Natl Acad Sci USA 41:671-684.

Modrich P (1991). Mechanisms and biological effects of mismatch repair. Ann Rev Genet 25:229-253.

Morgan TH (1910). Sex limited inheritance in *Drosophila*. Science 32:120-122.

Morgan TH (1911). An attempt to analyse the constitution of the chromosomes on the basis of sex limited inheritance in *Drosophila*. J Exp Zool 11:365-414 .

Morgan TH and Cattell E (1912). Data for the study of sex-linked inheritance in *Drosophila*. J Exp Zool 13:79-101.

Murray NE (1960). Complementation and recombination between *methionine-2* alleles in *Neurospora crassa*. Heredity 15:207-217.

30

Murray NE (1963). Polarized recombination and fine structure within the *me-2* gene of *Neurospora crassa*. Genetics 48:1163-1183.

Murray NE (1968). Polarized intragenic recombination in chromosome rearrangements of *Neurospora*. Genetics 58:181-191.

Nassif N, Penney J, Pal S, Engels WR and Gloor GB (1994). Efficient copying of non-homologous sequences from ectopic sites via P-element induced gap repair. Mol Cell Biol 14:1613–1625.

Neylon C (2004). Chemical and biochemical strategies for the randomization of protein encoding DNA sequences: library construction methods for directed evolution. Nucl Acids Res 32:1448-1459.

Nicolas A and Rossignol J-L (1989). Intermediates in homologous recombination revealed by marker effects in *Ascobolus*. Genome 31:528-535.

Nicolas A, Treco D, Schultes NP and Szostak JW (1989). Identification of an initiation site for meiotic gene conversion in the yeast *Saccharomyces cerevisiae*. Nature 338:35-39.

Olive LS (1959). Aberrant tetrads in *Sordaria fimicola*. Proc Natl Acad Sci USA 45:727-732.

Oliver CP (1940). A reversion to wild type associated with crossing over in *Drosophila melanogaster*. Proc Natl Acad Sci USA 26:452-454.

Pâques F and Haber JE (1999). Multiple pathways of recombination induced by double strand breaks in *Saccharomyces cerevisiae*. Microbiol Mol Biol Rev 63:349-404.

Pâques F, WY Leung and JE Haber (1998) Expansions and contractions in a tandem repeat induced by double-strand break repair. Mol Cell Biol 18:2045-2054.

Pritchard RH (1955). The linear arrangement of a series of alleles of *Aspergillus nidulans*. Heredity 9:343-371.

Rambosek. JA and Kinsey JA (1983). Fine structure mapping of the *am* (GDH) locus of *Neurospora*. Genetics 105:293-307.

Raper JR, Baxter MG and Middleton RB (1958). The genetic structure of the incompatibility factors in *Schizophyllum commune*. Proc Natl Acad Sci USA 44:889-900.

Raper JR, Baxter MG and Ellingboe AH (1960). The genetic structure of the incompatibility factors in *Schizophyllum commune*. The *A* factor. Proc Natl Acad Sci USA 46:833-842.

Rasmussen JP, Bowring FJ, Yeadon PJ and Catcheside DEA (2002). Targeting vectors for gene diversification by meiotic recombination in *Neurospora crassa*. Plasmid 47:18-25.

Romanienko PJ and Camerini-Otero RD (2000). The mouse *Spo11* gene is required for meiotic chromosome synapsis. Mol Cell 6:975-987.

Rudolph C, Kunz C, Parisi S, Lehmann E, Hartsuiker E, Fartmann B, Kraemer W, Kohli J and Fleck O (1999). The *msh2* gene of *Schizosaccharomyces pombe* is involved in mismatch repair, mating type switching and meiotic chromosome organisation. Mol Cell Biol 19:241-250.

Selker EU (1990). Premeiotic instability in *Neurospora*. Ann Rev Genet 24:579-613.

Shear CL and Dodge BO (1927). Life histories and heterothallism of the red bread-mold fungi of the *Monilia sitophila* group. J Agr Res 34:1019-1042.

Sigal N and Alberts B (1972). Genetic recombination: the nature of a crossed strand-exchange between two homologous DNA molecules. J Mol Biol 71:789-93.

Schwacha A and Kleckner N (1994). Identification of joint molecules that form frequently between homologs but rarely between sister chromatids during yeast meiosis. Cell 76:51-63.

Schwacha A and Kleckner N (1995). Identification of double Holliday junctions as intermediates in meiotic recombination. Cell 83:1-20.

Shiu PK, Raju NB, Zickler D and Metzenberg RL (2001). Meiotic silencing by unpaired DNA. Cell 107:905-916.

Siddiqi OH (1962). The fine structure of the *paba-1* region of *Aspergillus nidulans*. Genet Res Camb 3:69-89.

Simchen G (1967). Genetic control of recombination and the incompatibility system in *Schizophyllum commune*. Genet Res Camb 9:195-210.

Smith BR (1965). Interallelic recombination at the *his-5* locus in *Neurospora crassa*. Heredity 20:257-276.

Smith BR (1966). Genetic controls of recombination. Heredity 21:481-498.

Smyth DR (1970). Genetic control of recombination in the *amination-1* region of *Neurospora crassa*. PhD Thesis, Australian National University.

Smyth DR (1973). A new map of the *amination-1* locus of *Neurospora crassa* and the effect of the *recombination-3* gene. Aust J Biol Sci 26:1355-1370.

Stadler DR (1958). Gene conversion of cysteine mutants in *Neurospora*. Genetics 44:647-655.

Stadler DR and Towe AM (1963). Recombination of allelic cysteine mutants in *Neurospora*. Genetics 48:1323-1344.

Stadler DR and AM Towe (1971). Evidence for meiotic recombination in *Ascobolus* involving only one member of a tetrad. Genetics 68:401-413.

Stamberg J (1968). Two independent gene systems controlling recombination in *Schizophyllum commune*. Mol Gen Genet 102:221-228.

Stemmer WPC (1993). DNA shuffling by random fragmentation and reassembly: *In vitro* recombination for molecular evolution. Proc Natl Acad Sci USA 91:10747-10751.

Storlazzi A, Tessé S, Gargano S, James F, Kleckner N and Zickler D (2003), Meiotic double-strand breaks at the interface of chromosome movement, chromosome remodeling and reductional division. Genes Dev 17:2675-2687.

Strand M, Prolla TA, Liskay RM and Petes TD (1993). Destabilization of tracts of simple repetitive DNA in yeast by mutations affecting DNA mismatch repair. Nature 365:274-276.

Strathern JN, Shafer BK and McGill CB (1995). DNA synthesis errors associated with double-strand-break repair. Genetics 140:965–972.

Sturtevant AH (1913). The linear arrangement of six sex-linked factors in *Drosophila*, as shown by their mode of association. J Exp Zool 14:43-59.

Su SS, Lahue RS, Au KG and Modrich P (1988). Mispair specificity of methyl-directed mismatch correction *in vitro*. J Biol Chem 263:6829-6835.

Sun H, Treco D and Szostak JW (1989). Double strand breaks at an initiation site for meiotic gene conversion. Nature 338:87-90.

Sun H, Treco D and Szostak JW (1991). Extensive 3'-overhanging, single-stranded DNA associated with the meiosis-specific double-strand breaks at the *ARG4* recombination initiation site. Cell 64:1155-1161.

Sutton WS (1902) On the morphology of the chromosome group in *Brachystola magna*. Biol Bull Mar Biol Lab, Woods Hole 4:24-39

Sutton WS (1903). The chromosomes in heredity. Biol Bull Mar Biol Lab, Woods Hole 4:231-248.

Szostak JW, Orr-Weaver TL, Rothstein RJ and Stahl FW (1983). The double strand break repair model for recombination. Cell 33:25-35.

van Heemst D, Käfer E, John T, Heyting C, van Aalderen Mand Zickler D (2001). BimD/SPO76 is at the interface of cell cycle progression, chromosome morphogenesis, and recombination. Proc Natl Acad Sci USA 98:6267-6272.

Watson JD and Crick FHC (1953). A structure for deoxyribose nucleic acid. Nature 171:737-738.

Weiner BM and Kleckner N (1994). Chromosome pairing via multiple interstitial interactions before and during meiosis in yeast. Cell 77:977-991.

Whitehouse HLK (1963). A theory of crossing-over by means of hybrid deoxyribonucleic acid. Nature 199:1034-1040.

Williamson MS, Game JC and Fogel S (1985). Meiotic gene conversion mutants in *Saccharomyces cerevisiae*. I. Isolation and characterization of *pms1-1* and *pms1-2*. Genetics 110:609-646.

Wilson EB (1902). Mendel's principles of heredity and the maturation of the germ-cells. Science 16:991-993.

Wu L and Hickson ID (2003). The Bloom's syndrome helicase suppresses crossing over during homologous recombination. Nature 426:870-874.

Yeadon PJ, Bowring FJ and Catcheside DEA (2004a). Sequence heterology and gene conversion at *his-3* of *Neurospora crassa*. Curr Genet 45:289-301.

Yeadon PJ, Bowring FJ and Catcheside DEA (2004b). Alleles of the hotspot *cog* are co-dominant in effect on recombination in the *his-3* region of *Neurospora*. Genetics 167:1143-1153.

Yeadon PJ and Catcheside DEA (1995a). The chromosomal region which includes the recombinator *cog* in *Neurospora crassa* is highly polymorphic. Curr Genet 28:155-163.

Yeadon PJ and Catcheside DEA (1995b). *Guest*: a 98 bp inverted repeat transposable element in *Neurospora crassa*. Mol Gen Genet 247:105-109.

Yeadon PJ and Catcheside DEA (1998). Long, interrupted conversion tracts initiated by *cog* in *Neurospora crassa*. Genetics 148:113-122.

Yeadon PJ and Catcheside DEA (1999). Polymorphism around *cog* extends into adjacent structural genes. Curr Genet 35:631-637.

32

Yeadon PJ, L. Y. Koh, Bowring FJ, Rasmussen JP and Catcheside DEA (2002). Recombination at *his-3* in *Neurospora* declines exponentially with distance from the initiator, *cog*. Genetics 162:747-753.

Yeadon PJ, Rasmussen JP and Catcheside DEA (2001). Recombination events in *Neurospora crassa* may cross a translocation breakpoint by a template-switching mechanism. Genetics 159:571-579.

Young JA, Hyppa RW and GR Smith (2004). Conserved and nonconserved proteins for meiotic DNA breakage and repair in yeasts. Genetics 167:593-605.

**Applied Mycology
and Biotechnology**
An International Series
Volume 5. Genes and Genomics
© 2005 Elsevier B.V. All rights reserved

2

ELSEVIER

EST Data Mining and Applications in Fungal Genomics

Peijun Zhang[1] and Xiang Jia Min[2]
[1]Centre de génomique fonctionnelle de Sherbrooke, Faculté de medicine, Université de Sherbrooke, 3001, 12e Avenue Nord, Sherbrooke, QC J1H 5N4, Canada (peijun.zhang@usherbrooke.ca); [2]Centre for Structural and Functional Genomics, Concordia University, 7141 Sherbrooke Street West, Montreal, QC H4B 1R6, Canada (jack@gene.concordia.ca).

EST (expressed sequence tag) technology has long been used for gene discovery. As more and more EST data have become publicly available, the usage of ESTs has expanded to other areas, such as *in silico* genetic marker discovery, *in silico* gene discovery, construction of gene models, alternative splicing prediction, genome annotation, expression profiling, and comparative genomics. In comparison with whole genome sequencing, EST technology is simpler and less costly, especially in the case of large genomes. Moreover, since ESTs represent "the expressed parts" of genomes, they are more immediately informative about the transcriptomes. On the other hand, ESTs are not suitable for the studies related to "the control parts" of genomes, such as promoters and transcription enhancing/inhibiting elements. In addition, information for rarely expressed genes is also difficult to mine from EST data. EST data mining requires bioinformatics resources such as databases, data retrieving tools and analysis algorithms. Bioinformatics tools are also required to deal with EST errors and contaminations. Many of these tools are freely available to the academic community. In this chapter, EST data resources, tools used for processing and mining EST data, and applications of ESTs in genomics, particularly, fungal genomics, are reviewed.

1. INTRODUCTION

The availability of whole genome sequences and expressed sequence tags (ESTs) has accelerated our understanding of living things. The former only provides encoded information that needs to be deciphered, i.e., to identify genes, promoters, and other functional elements, while the latter provides us with a partial dictionary that can illustrate the gene expression patterns in different tissues, at different developmental

---

Corresponding author: Peijun Zhang

stages, and in response to different environmental stimuli.

ESTs are single-pass sequencing reads from either the 3' or 5' end of cDNAs, with limited or sequence quality is relatively poor. The scales of EST projects are much more flexible than no editing. ESTs differ from genomic sequences in three ways. 1) obtaining ESTs is relative inexpensive: 2) ESTs represent actively expressed genes at the time of tissue sampling; 3) whole genome sequencing, thus it can be handled by most laboratories. Since the advent of EST technology in the early 1970's (Putney et al. 1983) EST data have been accumulating exponentially. At the time of writing, there are over 21 millions EST entries in the NCBI dbEST database alone (Release 060404).

Initially, ESTs were used as an approach for gene discovery (Adams et al. 1991). As more complete genome sequences become available, one might wonder whether EST sequencing is still important. The answer would be yes. This is justified by the following facts. 1) Whole genome sequencing is still a costly undertaking and cannot be taken as a routine experimental procedure for the foreseeable future. 2) Some large genomes, such as wheat, are polyploidy and are subject to chromosomal duplication events. Sequencing such genomes is difficult and inefficient. 3) ESTs are representative of transcriptomes and as such reveal details otherwise invisible in genomic sequences: expression profiles, alternative splicing, RNA editing, etc. Therefore, ESTs are not only alternatives to full genome sequences, but also essential elements for functional genomics.

Exploiting EST data is not trivial and cannot be done without bioinformatics tools. First, databases are required to store and organize EST data, secondly, data retrieving tools are required to retrieve needed EST entries from EST databases, and finally, analysis tools are required to mine EST data. All these require modern computational techniques and algorithms.

In this chapter, we review the current status of EST technology with an emphasis on data mining tools and applications. In section 2, the background knowledge of ESTs and EST technology are briefly reviewed. In section 3, EST-related data sources and tools are described. While a description of the algorithms and detailed usages of these tools is beyond the scope of this review, we will cover the various applications of these tools and hardware requirements for installing them. In section 4, the applications of EST data mining are discussed. In the last section, the common EST sequence errors and the strategies to deal with such errors are discussed. Our discussion will focus on fungal ESTs except in cases where no reports related to fungi are available.

2. EST TECHNOLOGY

Since the term EST was first used by Adams et al. (1991). EST technology has advanced rapidly. Nowadays, most EST projects depend heavily on high-throughput technology and advanced software. However, the basic procedures for obtaining ESTs remain unchanged, though modifications may be seen regularly. The procedures include RNA extraction and mRNA purification, cDNA synthesis and cDNA library construction, and DNA sequencing and sequence processing.

2.1 Construction of cDNA Libraries

cDNA libraries used for fungal EST projects are not different from cDNA libraries used for other purposes. cDNA libraries usually are constructed from mRNAs isolated from a target fungal species at different developmental stages and under different culture conditions (Nugent et al. 2004). Often total RNA is isolated by routine protocols such as the Trizol method (Nugent et al. 2004). Poly(A)$^+$ RNA is isolated from total RNAs such as using Oligotex mRNA Spin-Column (Qiagen, Mississauga, ON, Canada) and is used for synthesis of cDNAs that are then cloned into a vector such as using the Superscript Plasmid System (Invitrogen Life Technologies) (Nugent et al. 2004) or lambda ZAPII vector (Stratagene, USA) (Felipe et al. 2003) for amplification. The details for the protocol of cDNA library construction may vary for products from different sources and the protocol provided by the manufacturers should be followed.

2.2 cDNA Library Normalization

Expression levels vary vastly for different genes in an organism under a give condition and also between developmental stages. A normalization procedure is required if the library is to be used for gene discovery. This procedure increases the frequency of clones derived from rare mRNA species and reduces the frequency of clones derived from abundant mRNA species. Therefore normalization increases the probability of selecting unique clones for EST sequencing (Soares et al. 1994) A procedure based on re-association kinetics for normalization was described in detail by Soares et al. (1994) and has been successfully applied for *Bombyx mori* ESTs (Mita et al. 2003).

2.3 Generation of Subtractive cDNA Libraries

Suppression subtractive hybridization (SSH) is a new and highly effective method for constructing subtracted cDNA libraries (Diatchenko et al. 1996, Diatchenko et al. 1999). It combines normalization and subtraction in a single procedure. The normalization step equalizes the abundance of cDNAs within the target population and the subtraction step excludes the common sequences between target and driver populations. SSH dramatically enriches the low-abundance cDNAs synthesized from differentially expressed mRNAs. This method has been used to construct a differentially-expressed gene subtracted cDNA library from two cell lines with different metastatic phenotypes (Liang et al. 2004), to identify symbiosis-regulated genes in *Eucalyptus globules-Pisolithus tinctorius* ectomycorrhiza (Vioblet et al. 2001) and differentially expressed genes in *Agaricus bisporus* (Morales and Thurston 2003)

Ray et al. (2004) recently developed a Negative Subtraction Hybridization (NSH) method and applied it in *A. nidulans*. With NSH, a plasmid library made from cDNAs prepared from cells grown under a given physiological condition is screened with cDNAs prepared from another physiological condition as probes. Plasmids with inserts that failed to hybridize to cDNA probes predominantly represent condition-specific or rare transcripts. When applied to *A. nidulans*, NSH was a cost-effective method for isolating novel full-length cDNAs of differentially expressed transcripts and rare transcripts.

2.4 DNA Sequencing and Sequence Quality Control

Plasmid DNA from each target clone is usually isolated and purified by routine protocols such as alkaline lyses. Nucleotide sequences are determined using commercial kits such as ABI Prism BioDye Terminator (Applied Biosystems, http://www.appliedbiosystems.com/index.cfm), which is based on the dideoxy chain-termination method (Sanger et al. 1977) and DNA sequencers such as ABI 377 and ABI 3700.

Base callers read sequencing results (trace files) produced by sequencers and convert them into nucleotide sequences of four bases (AGCT) and N (for unsolved base call). At the same time, base callers also produce quality values for called bases. Many EST sequences in the public domain submitted before base callers were available do not have quality strictly controlled. Nowadays, all EST and genome sequencing projects are heavily relying on base callers. The Phred software is the first generation of base caller with function for assigning a quality value for each base (Ewing and Green 1998; Ewing et al. 1998). TraceTuner is the "next generation" of base calling software developed by Paracel (http://www.paracel.com/products/tracetuner.php), Which achieves significant improvement of base calling accuracy by utilizing optimizations for processing dye blobs and a mobility shift correction method. More recently, a new generation program with improved base calling accuracy, called AutoEditor, has been developed and applied in recent genome sequencing projects (Gajer et al. 2004). However, its application to EST sequence quality control remains to be explored.

3. EST DATA SOURCES AND DATA MINING TOOLS

EST data, as well as genome sequence data, are among the most important resources for genomics researchers. Like other DNA sequence data, publicly accessible EST data are deposited in several primary public databases, such as GenBank and EMBL. Small-scaled species-specific databases also exist. These small databases can be very useful to researchers who work in related areas. In addition to data sources, data retrieval and search tools are required to fetch required EST data from EST databases, and data management and analysis tools are required to manage and mine EST data. In this section, we discuss some important EST data resources and tools.

3.1 EST Data Sources

As mentioned above, most EST data are stored in publicly accessible databases. Most of these databases supply similarity and keyword search tools through the corresponding websites. The web-based data retrieval/search tools are convenient but are only useful for retrieving small batches of data of interest. In order to retrieve large batches of data, or to pipe the retrieved data directly to other data mining tools, users should download the databases and install them locally, or use other internet-based data retrieve tools, such as netblast and SeqHound (see below). Our intention is not to cover all EST data sources available publicly. Instead, we discuss fungus-related EST data sources only.

3.1.1 dbEST

dbEST (Boguski et al. 1993) is a subset of the GenBank database. It is one of the largest EST depositories in the world. At the time of writing (Release 060404), this database contains more than 21 million entries. Recent dbEST information is available at http://www.ncbi.nlm.nih.gov/dbEST/dbEST_summary.html.

Like other data in GenBank, EST data can be accessed through Entrez. There are two ways to retrieve EST data through Entrez. The first is a query model (http://www.ncbi.nim.nih.gov/dbEST), where users submit keyword containing queries. The second is a batch model (http://www.ncbi.nim.nih.gov/ entrez/ batchentrez.cgi) where users retrieve entries based on lists of GI or accession numbers. dbEST can be searched by using the NCBI BLAST user interface (http://www.ncbi.nim.nih.gov/BLAST). To narrow the search space, the EST database, or a subset of the EST database can be selected. dbEST can also be searched through a locally installed netblast tool (see below). dbEST datasets are also downloadable from the NCBI ftp server (ftp://ftp.ncbi.nih.gov/repository/dbEST/) in FASTA format.

3.1.2 EMBLEST

The EMBLEST database contains essentially the same entries as NCBI dbEST, since EMBL and NCBI exchange data daily. EMBLEST can be accessed through the SRS database query tool at EMBL-EBI (http://srs.ebi.ac.uk) and mirror sites such as NIAS (http://www.dna.affrc.go.jp/srs6/). SRS allows users to retrieve batch entries by various categories such as taxonomy and organism. In NIAS, nucleotide sequence data are divided into EST (EMBLEST) and non-EST (EMBLOTHER) sub-databases, thus simplifying the retrieval of batch EST data. The following example shows how to retrieve an EST dataset for a given organism (in this case baker's yeast) from the NIAS site.

1) Click the 'Databanks' button to go to the DATABANKS page.
2) Click the 'EMBLEST' link to go to EST database query page.
3) Click the 'Organism' under name or 'org' under short name.
4) Click the 'List Value' with default keyword (*) to see all entries. To see all entries for an organism, input the name of the target organism.
5) To see all EST entries of baker's yeast, click the 'baker's yeast' link.
6) Click the 'Save' button.
7) Change options if necessary and click the 'Save' button.

To retrieve EST datasets for all fungi, select 'Taxon' instead of 'Organism' and use keyword 'fungi'.

3.1.3 COGEME phytopathogenic fungi and oomycete EST database

The EST data of COGEME phytopathogenic fungi and oomycete EST database were generated though the COGEME project (Consortium for the functional genomics of microbial eukaryotes) (http://cogeme.ex.ac.uk) (Soanes et al. 2002). To date, the database contains 36,022 entries from 14 fungal genomes (*Blumeria graminis*, *Botryotinia fuckeliana*, *Cladosporium fulvum*, *Collectotrichum trifolii*, *Cryphonectria parasitica*, *Fusarium sporotrichioides*, *Gibberella zeae*, *Leptosphaeria maculans*, *Magnaporthe grisea*, *Mycosphaerella graminicola*, *Phytophthora infestans*, *Phytoththora sojae*, *Sclerotinia scelerotiorum*, and *Verticillium dahliae*). The user interfaces for keyword query search and BLAST search are available at the web site. Sequence data can be downloaded from the same web site.

3.1.4 Fungal EST data at the University of Oklahoma's Advanced Center for Genome Technology

At the University of Oklahoma's Advanced Center for Genome Technology (http://www.genome.ou.edu), 7 fungal EST databases are available. They are *Aspergillus nidulans* (mixed vegetative and 24 hr asexual development culture, 12,485 sequences), *Aspergillus flavus* (vegetative mycelia, 1,253 sequences), *Neurospora crassa* (evening cDNA library, 9,148 sequences), *Neurospora crassa* (morning cDNA library, 10,871 sequences), *Neurospora crassa* perathecia (wild type, fruiting bodies, about 3,000 reactions), and *Fusarium sporotrichioides* (7,495 sequences). Keyword searching and sequence downloads are available from the website.

3.1.5 *Aspergillus oryzae* EST Database

The *Aspergillus oryzae* EST Database (http://www.nrib.go.jp/ken/EST/db) contains 21,735 sequences (7,714 contigs after clustering) from 9 different cultivation conditions, including liquid cultures with glucose, with glucose at 37°C, with maltose, at alkaline pH, and without carbon source, solid cultures of soy bean, wheat, and rice, and germination stage. The website provides BLAST similarity search and keyword search functions.

3.1.6 *Dictyostelium discoideum* cDNA Database

The *Dictyostelium* cDNA database (Dicty_cDB, http://www.csm.biol.tsukuba.ac.jp/cDNAproject.html) contains nearly 15,000 ESTs generated from various stages of the organism, especially slug (multicellular) and vegetative (unicellular) stages (Itoh et al. 2001). The EST and assembled contig sequences can be downloaded from the web site. The web site also provides several search interfaces to retrieve cDNA annotation information.

3.1.7 *Aspergillus niger* EST Database

The *A. niger* EST database contains 12,819 ESTs generated by the Fungal Genomics Project at Concordia University (Adrian Tsang, Reginald Storms, and Gregory Butler, unpublished results). Those ESTs are cleaned by LUCY (Chou and Holmes 2001) and assembled by Phrap into 5,202 unigenes. These ESTs and unigenes are functionally annotated using BLASTX against GenBank non-redundant protein database with TargetIdentifier/Annotator (see Section 3.2.3.20) and can be accessed at https://fungalgenomics.concordia.ca/public/annotation/Aspergillus_niger/summary.php.

The EST sequences can be downloaded from https://fungalgenomics.concordia. ca/fungi/Anig.php.

3.2 Data Retrieval, Search, Management, and Mining Tools
3.2.1 Data retrieving tools
3.2.1.1 SRS

SRS (sequence retrieval system, Lion Bioscience AG) is an integrated database system which handles many public biological databases and tools. There are more than 50 SRS servers around the world (http://srs.ebi.ac.uk). Different servers may use different versions of SRS. Databases and tools available may also vary. Web-based SRS allows users to perform BLAST and keyword search and batch dataset retrieval. A Unix-based command-line interface, Getz, is also available (ftp://ftp.edi.ac.uk/pub/software).

3.2.1.2 SeqHound

SeqHound (http://www.bluesprint.org/seqhound) (Michalickova et al. 2002) is an integrated biological sequence database and sequence retrieving platform which allows users to retrieve sequences and other related data through local installation or server/client model. The server/client model is very useful to users without the capacity to install the whole system. The client-side API (application programming interface) is available in Perl, C/C++, Visual Basic, and Java. We have integrated Perl client-side API into several locally developed bioinformatics tools (Zhang, unpublished data). The performance of SeqHound is highly satisfactory.

SeqHound holds all NCBI Entrez databases, 3-D structure databases, and Gene Ontology databases. API data retrieving functions include several categories such as sequence fetch, structure fetch, genome iterators, Gene Ontology, RPS-BLAST, functional annotation, protein neighbors, and taxonomy. An identifier conversion function is also available.

The latest release of the SeqHound code is available from sourceforge.net (http://sourceforge.net/project/showfiles.php?group_id=1791&package_id=39608) under GNU General Public License.

3.2.2 Database search tools

A similarity search, where using a query sequence to find records that are identical or similar to the query, is probably the most frequently used bioinformatics operation. Similarity searching is extensively used in various bioinformatics applications such as gene identification, genome annotation, gene indexing, and comparative genomics. In this section we discuss some popular similarity search-related tools with emphasis on their general functionality and availability.

3.2.2.1 BLAST

The BLAST database search is one of the basic bioinformatics operations. It is regularly used for simple similarity search and high-throughput bioinformatics projects. For simple similarity search with single queries or a small batch of queries, web-based BLAST is a convenient tool. The web-based BLAST is available at most biological sequence database web sites. However, the web-based BLAST is not suitable to large bioinformatics projects. Locally installed BLAST packages are required in such situations.

The most popular packages available publicly are NCBI BLAST (blastall), Washington University WU-BLAST (wu-blastall), and NCBI netblast (blastl3). Both blastall and wu-blastall require locally installed databases. BLAST is useful for in-house generated sequence data or downloaded public datasets. Installation of large public databases, such as the whole GenBank, requires a large storage device. For laboratories in which high-end computer hardware is not available, netblast is a good alternative. Contrary to blastall and wu-blastall, netblast (network-client BLAST) does not require locally installed databases. It searches remote databases (NCBI GenBank) over the internet.

NCBI blastall is freely available from NCBI ftp server (ftp://ftp.ncbi.nlm.nih.gov/blast/executables/). Netblast is freely available from ftp://ftp.ncbi.nih.gov/blast/blastl3/CURRENT/. The old, unsupported version of WU-BLAST is available from http://blast.wustl.edu/blast/executables/. A license is required for the latest version, which can be obtained from http://blast.wustl.edu/licensing.

3.2.2.2 NHGRI::Blastall

The Perl module NHGRI::Blastall (freely available at http://genome.nhgri.nih.gov/blastall/Blastall.shtml) is used to integrate all the three BLAST programs mentioned above into other programs. In addition, NHGRI::Blastall parses BLAST results.

3.2.2.3 Zerg and Zerg::Report

Zerg and Zerg::Report (http://bioinfo.iq.usq.br/zerg/) are Perl modules for parsing BLAST reports. Zerg scans BLAST reports for elements and returns their codes and values. Zerg::Report returns a nested hash data structure. Both modules are freely available from the above web site.

3.2.2.4 Boulder Perl modules

Boulder Perl modules, developed by Lincoln D. Stein (lstein@cshl.org), include several biological database parsers. Among them, Boulder::GenBank, Boulder::Unigene, and Boulder::BLAST are useful for EST data mining. Boulder Perl modules can be downloaded freely from CPAN (www.cpan.org).

Boulder::GenBank is a Perl module to fetch and parse GenBank records from NCBI GenBank (through NetEntrez), local GenBank (through Yank), or local flat file. Similar to Boulder::GenBank, Boulder::Unigene parses UniGene flat files. However, Boulder::Unigene does not fetch records from the UniGene database. Boulder::Blast is a BLAST output report parser parsing both NCBI and WU BLAST reports.

3.2.2.5 GENE2EST

GENE2EST (http://woody.embl-heidelberg.de/gene2est) (Gemund et al. 2001) is a web-based tool to search ESTs using genomic sequences as queries. GENE2EST takes a genomic sequence and masks repeats and queries EST databases. GENE2EST generates reports in the BLAST report format, alignment output, and graphic output for gene structures and other information.

3.2.3 EST cleaning, clustering, assembling, annotation and management tools
3.2.3.1 LUCY

LUCY (Chou and Holmes 2001) is an EST cleaning package which trims vector splicing sites, vector sequences, and poly-A/T tails and extracts high quality regions. LUCY requires base-calling quality data as input. LUCY is freely available for academic users from http://www.tigr.org/softlab. The newer version LUCY2 (Li and Chou 2004) is available at www.complex.iastate.edu.

3.2.3.2 Phrap

Phrap is a program of a sequencing lab package including Phrap, Phred, Consed, and Cross_Match. Phrap is developed for genome sequencing projects. It has been used for EST assembling. Phrap is fast, generates accurate consensus sequences, and produces consensus quality estimates. Phrap handles very large projects. Phrap is freely available for academic usage by getting license directly from the authors (http://www.phrap.org/consed/consed.html#howToGet).

3.2.3.3 CAP3

Like Phrap, CAP3 (Huang and Madan 1999) was developed for assembling shotgun sequences. It is now widely used for EST assembly and integrated into several EST tools (see below). In comparison to Phrap, CAP3 generates shorter contigs with fewer errors. Using human EST data, Liang et al (2000) found that CAP3 out-performed Phrap and TIGR Assembler (see below). CAP3 is available upon request from the author (Xiaoqiu Huang, huang@mtu.edu). In addition, several web-based CAP3 user interfaces are available, including http://pbil.univ-lyon.fr/cap3.php, http://fenice.tigem.it/ bioprg/ interfaces/cap3.html, and http://deepc2.zool.iastate.edu/aat/cap/cap/html.

3.2.3.4 stackPACK

StackPACK (http://www.sanbi.ac.za/relnotes.html) (Christoffels et al. 2001) is an EST project package developed at the South African National Bioinformatics Institute, University of Western Cape. StackPACK can run on Unix and Linux platforms. StackPACK takes data in GenBank and FASTA formats, and Phred quality data. Poor quality sequences are masked with either RepeatMasker or Cross_Match. StackPACK uses d2_cluster to perform clustering, and consensus sequences generated with Phrap and CRAW. StackPACK can be run in either configurable command line model or web-based model. The package is freely available from the above web site.

3.2.3.5 TIGR Assembler

The TIGR Assembler (http://www.tigr.org/software/assembler/) (Huson et al. 2001) is a sequence fragment assembly program which builds contigs from small sequence reads, such as EST data. The program works under RedHat Linux, Solaris and Digital Linux 4.0 and 5.x, and probably any other Linux systems. TIGR Assembler can be downloaded freely from the above web site.

3.2.3.6 EST-ferret

The EST-ferret (http://legr.liv.ac.uk/EST-ferret/) (Wengzhou Li, unpublished data) is a Perl-based package for large-scale EST data projects. It integrates several programs/databases such as Phred, CAP3, NCBI BLAST, and Gene Ontology. The EST-ferret runs on Linux systems. The package can be downloaded from the web site by registered users.

3.2.3.7 PHOREST

PHOREST (http://www.biol.lu.se/phorest/) is a web-based tool for the comparative analysis of EST data. It can perform EST clustering, uniset generation, gene discovery and annotation, comparison between data of two EST projects, and BLAST similarity searches. The package is freely available from the above web site.

3.2.3.8 CLOBB

CLOBB, or cluster on the basis of BLAST similarity (http://www.nematodes. org/CLOBB) (Parkinson et al. 2002) is a highly portable EST cluster program. It requires only NCBI BLAST and Perl. CLOBB can generate clusters from raw EST data (in FASTA format) or from combination of previous build and EST data. CLOBB does not generate consensus sequences. CLOBB is suitable for small to medium datasets. CLOBB is freely available upon request from the authors.

3.2.3.9 FELINES

FELINES, or Finding and Examining Lots of Intron 'N' Exon Structures (http://www.genome.ou.edu/informatics.html) (Drabenstot et al. 2003) is a package including 5 applications developed in Perl. The package provides utilities to automate intron and exon database construction and analysis from EST and genomic sequence data.

The package is divided into three layers. The alignment layer (wiscrs.pl) pairs and aligns EST sequences to their homologous genomic sequences. The script employs NCBI BLAST (blastall) to pair EST and genomic sequences. E-value, the cutoff of which can be defined by users, is used to evaluate the homology of the pairs. Next, alignments are generated by Spidey, a tool for aligning mRNAs to genomic sequences (Wheelan et al. 2001). The extraction layer (gumbie.pl) extracts the intron and exon regions from the Spidey alignment files produced by the alignment layer and parses the resulting datasets into the respective database. The parsing filter criteria are fully customizable in an option file. The analysis layer consists of three applications (icat.pl, findmner.pl, and cattracts.pl). The icat.pl performs two tasks. First, it filters imported intron databases against the defined filter criteria to ensure that the imported intron datasets are comparable to those generated by FELINES. Secondly, icat.pl extracts user defined conserved intron motifs. The findmners.pl identifies all user-defined fixed-length sequences and reports their statistics. The cattracts.pl searches multiple consensus elements simultaneously. The FELINES package is available at the website mentioned above.

3.2.3.10 ESTWeb

ESTWeb (http://bioinfo.iq.usp.br/estweb/) (Paquola et al. 2003) is a package for processing and storage of large-scale EST data using the web as the user interface. The package is not an online tool, local installation is required. The package requires installation of several applications (such as Apache, PostgreSQL, NCBI BLAST, Megablast, Phred, phd2fasta, cross-match, Phrap) and Perl modules (such as GD, CGI, DBI, DBI::Pg, XML::Dumper, and Digest::MD5). The package is freely available from the web site above.

3.2.3.11 PipeOnline

PipeOnline (http://stress-genomics.org, Ayoubi et al. 2002) is an automated EST processing package which deals with large collections of trace files or sequence files in FASTA format. PipeOnline screens and removes vector sequences, assembles redundant input files and generates unigene EST data sets and consensus sequences, and performs functional annotation. The package is free for academic use. Local installation of the package requires several third-party software programs, including Phred, Cross-Match, Phrap, NCBI BLAST, MySQL, and Apache. PipeOnline run on Unix systems and requires Perl.

3.2.3.12 ESTprep

ESTprep (Scheetz et al. 2003) is a highly configurable EST processing package. It takes trace files as input and outputs EST sequences and reports those which meet configurable quality criteria. ESTprep was developed in C and is freely available from http://genome.uiowa.edu/pubsoft/software.html.

3.2.3.13 ESTAP

ESTAP (Mao et al. 2003) is an automated EST processing package which performs cleaning (low quality ends, vectors, adaptors, poly-A/T, chimera, and contaminated sequence from the cloning vector, the host, or user defined organisms), similarity search, and assembling. A web interface is also included in the package to manage projects and for viewing original and cleaned data and analysis results. ESTAP is available from http://www.vbi.vt.edu/~estap.

3.2.3.14 EST-PAGE

EST-PAGE (Matukumalli et al. 2004) is written in Perl employing BioPerl, Phred, PostgreSQL or MySQL. The package reads trace files, cleans EST sequences (vectors, *E. coli* contamination), submits to dbEST, and performs clustering and annotation. The source code is available upon request from the authors.

3.2.3.15 ESTAnnotator

ESTAnnotator (Hotz-Wagenblatt et al. 2003) is a tool for the high throughput annotation of ESTs by automatically running a collection of bioinformatics applications. It performs a quality check and masks repeats, vector parts and low quality sequences. Then it performs database searching and EST clustering. Finally, the outputs of each individual tool are gathered and the relevant results presented in a descriptive summary. ESTAnnotator is available at http://genome.dkfz-heidelberg.de.

3.2.3.16 TIGR Gene Indices clustering tools

TIGR Gene Indices clustering tools (TGICL) is a software pipeline package for automating analysis of large EST and mRNA databases in which the sequences are first clustered based on pairwise sequence similarity, and then assembled by individual clusters to produce longer and more complete consensus sequences (Pertea et al. 2003). The clustering is performed by a modified version of NCBI's MegaBLAST, and the resulting clusters are then assembled using CAP3. The system can run on multi-CPU architecture including SMP and PVM. The package can be downloaded from http://www.tigr.org/tdb/tgi/software/.

3.2.3.17 ESTScan

ESTScan (http://www.ch.embnet.org/software/ESTScan.html, Iseli et al. 1999) is a coding region identification and translation tool. A HMM (Hidden Markov Model) - based algorithm is integrated into the program which can efficiently correct frame shifts resulting from EST sequence errors. Command-line ESTScan is written in Perl wrapping the HMM in C. Command-line ESTScan takes about 20 command-line parameters. A web-based interface is also available at the URL above. The package can be downloaded from http://iubio.bio.indiana.edu: 7780/ perl/ custom/ index. cgi? Dir =/public/ molbio/ dna/ analysis/ESTScan.

3.2.3.18 DIANA-EST

DNA Intelligent Analysis for ESTs (DIANA-EST) (Hatzigeorgiou et al. 2001) is an artificial neural networks-based program to identify coding regions with ESTs. DIANA-EST takes care of frame shift EST errors. The program is available upon request from the author.

3.2.3.19 ATGpr/ATGpr_sim

ATGpr is a program for predicting whether a cDNA contains a translation initiation site or not (Salamov et al. 1998). The algorithm in the program used the linear discrimination method including the following sequence characteristics: 1) positional triplet weight matrix around an ATG, 2) the length of an open reading frame (ORF), 3) 5' UTR-ORF hexanucleotide difference, 4) signal peptide characteristic, 5) presence of another upstream in frame ATG, and (6) upstream cytosine nucleotide frequency (Salamov et al. 1998). Nadershahi et al. (2004) compared the performance of several programs including ATGpr, NetStart, and ESTScan and found that ATGpr had relatively high sensitivity, specificity, and overall accuracy in identifying translation start sites and rejecting incomplete sequences. The program is available at http://www.hri.co.jp/atgpr. ATGpr_sim is an improved version of ATGpr by integrating the similarity information of a query sequence with a hit in BLASTX (Nishikawa et al. 2000). ATGpr_sim program is available at http://www.hri/co.jp/ atgpr/ ATGpr_sim.html. These two programs can only process one sequence per submission.

3.2.3.20 TargetIdentifier/Annotator

TargetIdentifier and Annotator were implemented for identifying full length EST clones and functionally annotating EST derived sequences (Xiang Jia Min, Gregory Butler, Reginald Storms and Adrian Tsang, unpublished results). The algorithm utilizes the frame predicted by BLASTX and examines the presence of stop codons in the 5' untranslated region and in-frame start codon in the query sequence as well as uses the alignment parameters of the high score pair in the BLASTX for predicting if an EST is full-length or not. The programs need the EST sequences in FASTA and the pre-computed BLASTX output as input. The tools are accessible at https:// fungalgenome. concordia.ca/tools/TargetIdentifier.html.

4. APPLICATIONS IN FUNGAL GENOMICS AND COMPARATIVE GENOMICS
4.1 Gene Discovery and Genome Annotation

The primary purpose for most EST projects is to generate ESTs that can be used for gene discovery in a target organism. In the previous section, the tools used for EST assembling and EST data mining were described. In this section we briefly summarize the progress of EST application in gene discovery and genome annotation.

4.1.1 Gene discovery

Like most other EST projects, fungal EST projects, particularly for fungal species whose genomes have not been sequenced, are aimed at gene discovery in the organism. The genes can be further explored to identify candidates of industrial or pharmaceutical interest or to build a gene catalog (gene index). The gene catalog can be used to make microarray for studying transcriptomes at different developmental stages, under different culture conditions, or upon interactions between pathogenic fungi and their hosts. For example, Sacadura and Saville (2003) generated 2,871 ESTs from *Ustilago maydis* and assembled them into 1,293 unique sequences. A subset of identified genes was used to investigate the expression of those genes during teliospore germination. Nugent et al. (2004) generated 7,455 ESTs that were assembled into 3,074 unique sequences, from a forced diploid culture of *U. maydis* growing as filaments and examined the expression of a number of genes using Northern hybridization. Felipe et al. (2003) identified 2,160 genes assembled from 3,938 ESTs from the dimorphic and pathogenic fungus *Paracoccidioides brasiliensis* and characterized the transcriptome. Similar projects have been reported in other fungal species including *Gibberella zeae* (Trail et al. 2003) *Verticillium dahliae* (Neumann and Dobinson 2003). *Paxillus involutus* (Johansson et al. 2004) *Conidiobolus coronatus* (Freimoser et al. 2003) and *Aspergillus nidulans* (Sims et al. 2004).

Though the details may vary slightly in different gene discovery projects, the general procedures are similar, that is, they are mainly dependent on similarity search based approaches (Lindlof 2003). Usually BLAST or a similar program is used to search nucleotide or protein databases in which entries have been functionally annotated (see details in Section 3.1.2 for the tools). The databases commonly used for gene discovery include, the GenBank non-redundant protein and nucleotide databases (http://www.ncbi.nlm.nih.gov/), the UniProt/Swiss-Prot (http://www.ebi.ac.uk/ swissprot/ index.html) protein database, and the COG (clusters of orthologous groups) database (Tatusov et al. 2003). The integrated protein domain/motif database InterPro (Mulder et al. 2003) is often searched for finding functional domains in ESTs and EST-assembled sequences using InterProScan (Zdobnov and Apweiler 2001). For functional categorization of ESTs, the Gene Ontology Annotation (GOA) database (Camon et al. 2004). is often used. For example, Felipe et al. (2003) used BLASTX to search against the GenBank nr, COG, and GO databases, and InterProScan to search against the InterPro database for functional annotation of *Paracoccidioides brasiliensis* ESTs. BLAST programs have also been used for transcriptome comparisons with other fungal species (Felipe et al. 2003).

For gene discovery purposes, both in-house generated and in the public domain ESTs can be used. For example, Faria-Campos et al. (2003) mined the EST databases from eight microorganisms in the quest for new proteins by searching the ESTs against proteins in the COG database. They found 4,093 ESTs from those eight organisms that are homologous to COG genes and these are candidates for further protein characterization. In addition to the similarity based search tools described in Section 3, other tools which may be used in the gene discovery process are also available. Several of such tools are described as follows.

InterProScan is a tool that scans given protein sequences against the integrated protein signature database InterPro (Zdobnov and Apweiler 2001). The InterPro database includes PROSITE, Pfam, PRINTS, ProDom, SMART, and TIGRFAMs member databases (Mulder et al. 2003). The database is available via web server at http://www.ebi.ac.uk/interpro and anonymous FTP at ftp://ftp.ebi.ac.uk/pub/ databases/ interpro. InterProScan program is available from the EBI ftp server at ftp://ftp.ebi.ac.uk/ pub/software/unix/iprscan/. The public web interface is at http://www.ebi.ac.uk/interpro/scan.html and email submission server as interproscan@ebi.ac.uk.

SignalP consists of two different predictors based on neural network and hidden

Markov model algorithms for predicting the presence and location of signal peptide cleavage sites in amino acid sequences from different organisms including Gram-positive prokaryotes, Gram-negative prokaryotes, and eukaryotes (Nielsen et al. 1997, Emanuelsson and von Heijne 2001). The current version SignalP 3.0 server with improved prediction accuracy is available at http://www.cbs.dtu.dk/services/SignalP/. This tool is particularly useful for EST projects aiming at finding genes encoding secreted proteins. For example, Harcus et al (2004) analyzed EST sequences from the nematode *Nippostrongylus brasiliensis* for putative amino-terminal secretory signals using SignalP. They found that secreted proteins may be undergoing accelerated evolution, either due to relaxed functional constraints, or in response to stronger selective pressure from host immunity.

Gene Ontology (GO) is widely accepted for description of genes and their products in an organism. There are a good number of tools available for GO related analysis at http://www.geneontology.org/GO.tools.html. Goblet is a software package which performs annotation based on GO terms for anonymous cDNA or protein sequences (Hennig et al. 2003). The Goblet server is accessible at http://goblet.molgen.mpg.de.

UITagCreator is a tool for creating a large set of synthetic tissue identification tags that are used for determining tissue sources of ESTs within pooled cDNA libraries (Gavin et al. 2002). The source code and installation instructions, along with detection software usable in concert with created tag sets, are freely available at http://genome.uiowa.edu/pubsoft/software.html.

ESTs from fungal-infected plant tissues are composed of a mixture of plants and fungal sequences. Separating plant and fungal sequences in EST pools requires advanced algorithms. Hsiang and Goodwin (2003) reported that standalone TBLASTX with a matching database comprised of a single plant and a single fungal genome appears to be a faster and more accurate method than BLASTX searches of the GenBank non-redundant database to distinguish fungal and plant sequences in mixed EST collections. Maor et al. (2003) developed an algorithm and PF-IND software that utilize the differences in codon usage bias to discriminate between plant and fungal sequences. The software includes five pairs of fungi and their host plants. The software can distinguish between homologous fungal and plant genes and also helps identify the correct reading frame of ESTs. PF-IND is an internet application composed of a database, a JAVA applet main analysis program and active server pages. The application and the database are stored and run on the Microsoft IIS5 server at Tel-Aviv University (http://www2.tau.ac.il/lifesci/plantsci/as/geneSort_new.asp, Maor et al. 2003).

4.1.2 Genome annotation

ESTs including complete and partial cDNAs are expressed genes and hence are extremely useful for confirmation of genes predicted by a gene finding program such as GenScan (http://genes.mit.edu/GENSCAN.html, Burge and Karlin 1997) and Glimmer (http://www.tigr.org/software/glimmer/, Delcher et al. 1999) for a sequenced genome and for finding new genes in a genome. By aligning the transcribed sequences to a genomic sequence, the gene structure including introns and exons can be resolved. Several popular programs are available for aligning ESTs to genomic sequences, including EST_GENOME, SIM4, DDS/GAP2, GeneSequer, Spidey, BLAT, MGAlignIt, and PASA.

EST_GENOME, also called est2genome, is a software tool to align a set of spliced nucleotide sequences (ESTs, cDNAs or mRNAs) to an unspliced genomic DNA sequence, inserting introns of arbitrary length when needed (Mott 1997). The program outputs a list of exons and introns it has found, in the format of a list of matching segments. This format is easy to parse into other software. The program also indicates the gene's predicted direction of transcription based on the splice site information. The full sequence alignment can also be printed. The details of the algorithm and usage were described by Mott (1997) and at http://www.hgmp.mrc.ac.uk/Software/EMBOSS/Apps/est2genome.html. The original est_genome version is available from: ftp://ftp.sanger.ac.uk/pub/pmr/est_genome.4.tar.Z. The online accessible tool is at http://bioweb.pasteur.fr/seqanal/interfaces/est2genome.html.

SIM4 is a similarity-based tool for aligning expressed DNA sequences (EST, cDNA, mRNA) with a genomic sequence, developed by Florea et al. (1998). A user guide, details of description for the program, source code and installation procedures are available at http://globin.cse.psu.edu/globin/html/docs/sim4.html. The program is available online at http://pbil.univ-lyon1.fr/sim4.php for single sequences and at http://sky.bsd.uchicago.edu/batch_sim4.htm for a batch of sequences. SIM4 has been used for aligning ESTs to the genome of the filamentous fungus *Neurospora crass* to validate predicted genes from that genome (Galagan et al. 2003).

DDS/GAP2 is one of two sets of programs in the Analysis and Annotation Tool (AAT) developed by Huang et al. (1997). DDS/GAP2 is used to compare a genomic DNA query sequence with a cDNA database. The other set of programs, DPS/NAP is for comparing the DNA query sequence with a protein database. Each set contains a fast database search program and a rigorous alignment program. The database search program quickly identifies regions of the query sequence that are similar to a database sequence. Then the alignment program constructs an optimal alignment for each region and the database sequence. The alignment program also reports the coordinates of exons in the query sequence. These tools are available online at http://genome.cs.mtu.edu/aat/aat.html.

GeneSeqer is tool for aligning ESTs to a genomic sequence. It can align not only ESTs and cDNAs from same species as the genomic DNA but also ESTs from fairly divergent species that still share a common gene space (Usuka et al. 2000). Its performance has been demonstrated for *Arabidopsis* genome annotation (Brendel et al. 2004). The source code is available for download at http://bioinformatics.iastate.edu/bioinformatics2go/gs/download.html.

Spidey was implemented by Wheelan et al. (2001). for aligning spliced sequences to genomic sequences using local alignment algorithms and heuristics. It is slightly more accurate than SIM4 and est2genome and could be used for cross-species alignment, such as alignment of mouse mRNAs to human genomic sequences (Wheelan et al.

2001). Spidey is available as a standalone program and as a web service, both are accessible at http://www.ncbi.nlm.nih.gob/spidey/.

BLAT is a BLAST-like alignment tool developed by Kent (2002) for rapid mRNA/DNA and cross-species protein alignment. It is similar in many ways to BLAST, but it effectively unsplices mRNA onto the genome sequence, and it is much more fast and accurate. BLAT is available as a client/server version and a standalone program. The client/server version is especially suitable for interactive applications, and is available via a web interface at http://genome.ucsc.edu. The source code and executables of the software can be downloaded at http://www.soe.ucsc.edu/~kent.

MGAlign is a novel, rapid, memory efficient and practical method for aligning mRNA/EST sequences to their cognate genome sequences (Lee et al. 2003). MGAlignIt is a freely available web service, based on MGAlign. The web service is located at http://origin.bic.nus.edu.sg/mgalign/mgalignit.html. The MGAlign software is available in binary form for several platforms including Microsoft Windows, Sun Solaris, Linux, Mac OS X and SGI IRIX from the above site.

The Program to Assemble Spliced Alignment (PASA) tool was implemented by Haas et al. (2003). PASA uses a novel algorithm to assemble clusters of overlapping transcript alignments (ESTs and full-length cDNAs) into maximal alignment assemblies, and thereby to incorporate all available transcript data including splicing variations into gene structure annotation. This tool has been successfully applied to automate updating of Arabidopsis gene annotations (Haas et al. 2003). The PASA pipeline package including PASA as well as the pipeline to generate transcript alignments, to compare alignment assemblies to existing gene model annotations, to update gene structure annotations based on transcript alignments, and to automatically model new genes based on full-length cDNA containing alignment assemblies is available at http://www.tigr.org/software/.

Genome annotation and computational identification of complete gene models in eukaryote genomes requires multiple sources of evidence including *ab initio* gene prediction, splice site prediction and EST/cDNA alignment to a genome (Allen et al. 2004). The Combiner is a program designed by Allen et al. (2004) to construct gene models by using combined evidence generated from EST/cDNA alignment and *ab initio* gene finders. The program achieves by taking advantage of the many successful methods for computational gene prediction and can provide substantial improvements in accuracy over any individual gene prediction program. Three different algorithms for combining evidence in the Combiner were implemented and tested on 1,783 confirmed genes in *Arabidopsis thaliana*. The results show that the Combiner can produce dramatic improvements in sensitivity and specificity in gene finding (Allen et al. 2004). The software is available at http://www.tigr.org/software/combiner/.

The generic genome browser (GGB) (http://www.gmod.org/) is a good system for displaying the EST/cDNA alignment to a genome (Stein et al. 2002). GGB displays genomic annotations as interactive web pages using the Bio::Graphics module of the Bioperl package. It supports scrolling, zooming, gene ID or full-text search of annotations, and third party annotation. It is equally suitable to genomes in early stages of sequencing, "rough draft" genomes, and finished genomes, and is highly configurable (http://www.gmod.org/ggb/gbrowse.shtml). Many model organism databases such as the *Saccharomyces* Genome Database (SGD) (http://www.yeastgenome.org/cgi-bin/SGD/gbrowse/yeast) and the WormBase (http://www.wormbase.org/db/seq/gbrowse/wormbase) use GGB for genome annotation. The dictyBase also uses the GGB to display genes predicted from *Dictyostelium discoideum* genome and cDNA/EST alignment to the genome (http://dictybase.org/).

4.2 *In Silico* DNA Polymorphism and Genetic Marker Mining

DNA polymorphisms and genetic markers have broad applications in genetic and genomic research such as linkage mapping, gene tagging, genetic variation, and evolution. Traditional methods for identification of polymorphisms are labor-intensive and costly. EST-based polymorphism identification is an effective and cheap alternative. This *in silico* approach has been successfully used to mine several genetic markers such as SSR- and SNP-markers.

4.2.1 SSR

Microsatellites or simple sequence repeats (SSRs) are stretches containing repeats of short units of nucleotides. SSR-markers have been broadly used in plant genetic research. Generally, EST-based SSR-marker mining includes the following steps: 1) EST data collection and pre-processing, 2) SSR-containing EST hunting, 3) clustering, and 4) primer design. The resulting primers are subject to experimental confirmation.

EST sequences can be downloaded from several public databases such as NCBI dbEST and EMBL (see Section 2). As we know, EST sequences are error-prone and may contain contaminations. If 'raw' EST sequences are used, poor quality sequences and contaminations have to be corrected or removed first in a pre-processing stage. Poly-A or poly-T tails also need to be removed. An alternative is to use gene indices (see Section 4.6) as a starting data source. Since EST sequences in gene indices have been strictly filtered, no further pre-processing is required. SSR-containing EST hunting is usually done by using dedicated tools, such as Sputnik (http://espressosoftware .com/pages/sputnik.jsp), SSRIT (http://arsgenome.cornell. edu/cgi-bin/rice/ ssrtool. pl, Kantety et al. 2002) and MISA (http://pgrc.ipk-gatersleben.de/misa/, Thiel et al. 2003). MISA can detect both perfect and compound SSRs. The clustering step is used to remove redundant sequences and to generate consensus sequences. stackPACK2 (Miller et al. 1999) can be used for the clustering process (Kantety et al. 2002; Thiel et al. 2003). The primer designing process can also be automated by using special tools such as PRIMER3 (Rozen and Skaletsky 2000).

Several reports have shown that this approach is a feasible alternative to traditional ones. For instance, Kantety et al. (2002) analyzed 260,000 ESTs from grass plants. They found that the frequency of SSR-containing ESTs varied from 1.5% in maize to 4.7% in rice. After clustering, 3,599 consensus and singleton sequences were obtained. Thiel et al. (2003) identified 1,856 SSR-containing ESTs from 24,595 barley ESTs. Three hundred and eleven primer pairs were experimentally tested and 76 SSR-markers were integrated into a barley genetic consensus map. In *Medicago truncatula*, 4,384 SSR-containing ESTs were identified from over 147,000 ESTs. Four hundred and six of 616 primer pairs designed produced SSR bands in all genotypes tested. About 70% of those markers produced polymorphism in alfalfa, *M. truncatula*, and other annual medics (Eujayl et al. 2004).

Though most published reports are related to plants, at least one report related to fungi has been published. van Zijll de Jong et al. (2003) investigated EST-based SSR markers for grass endophytes *Neotyphodium coenophialum* and *N. lolii*. They found that 9.7% of *N. coenophialum* ESTs and 6.3% of *N. lolii* ESTs contain SSR loci. Fifty primer pairs for *N. coenophialum* and 57 for *N. lolii* were tested with Neotyphodium and Epichloë species. A high proportion of these primer pairs produced amplified bands and most of them revealed genetic variation.

In comparison to SSR markers identified by other means, the EST-based SSR markers tend to be more representative. The possible reason is that the EST-based SSRs are located in expressed genes, which are well spread on chromosomes, such as in euchromatic regions. This was demonstrated by Areshchenkova and Ganal (2002) in their work on isolation of SSR markers for tomato from either genomic libraries or ESTs. While all markers obtained from genomic libraries were mapped in centromeric regions, eight of 11 EST-based markers mapped in euchromatic regions.

4.2.2 SNP

Single nucleotide polymorphisms (SNPs) are probably the most frequent DNA sequence variation. SNPs are very useful resources in genetic and genomic studies in animals, plants, and fungi. SNPs can be discovered by using two major approaches. One of them is based on wet lab techniques such as oligonucleotide hybridization, ligation, primer extension, and sequencing of PCR products, which is inefficient and costly. Another, a more efficient alternative, is an *in silico* approach using genomic or EST sequences. The EST-based approach is discussed here.

The general procedures for the EST-based SNP mining are similar to those for the EST-based SSR mining. However, two aspects should be treated specifically: exclusion of low-score candidates and discrimination of paralogues. Basically, SNP mining is to find allelic heterogeneity. Since ESTs are not high quality sequences, not all heterogeneities detected are true SNPs. An evaluation step should be added to the process to filter out candidates that may result from sequence errors. Similarly, gene duplication (paralogues) can also generate heterogeneities. Discrimination of paralogues should be taken into account. In addition, raw EST sequences should be more rigorously cleaned in a pre-processing step.

Several SNP mining algorithms have been published. Most of them use trace data instead of sequence data since new filter criteria can be applied to trace data. Three algorithms are reviewed below.

4.2.2.1 SNPpipline

Single nucleotide polymorphism pipeline (SNPpipeline) (Buetow et al. 1999) is a

SNP mining package using EST sequence and trace data. SNPpipeline employs Phred, Phrap, DEMIGLACE, PHYLIP, and the CHLC database. Phred is used to re-call sequences from trace files. Low quality sequences are trimmed. The cleaned sequences are assembled and aligned with Phrap (no clustering is required since Unigene clusters have been used as starting data). The assemblies are refined with PHYLIP based on a distance matrix of substitutions observed between all possible pairs of sequences within each assembly. DEMIGLACE is used to identify candidate SNPs based on sequence alignment data, base calling scores, and distance data produced above. DEMIGLACE filters out the following possible candidates: 1) those whose neighboring sequence quality scores drop by 40% or more, 2) those whose peak amplitude is below the fifteenth percentile of all base calls for that nucleotide type, 3) those in regions with a high number of disagreements with the consensus, 4) those whose base call with an alternative call whose peak area is 25% or more of that of the principle call, and 5) those occurring in only one read direction.

The identified candidates are then stored in the CHLC database for primer designing. The package also includes a graphical interface for visual evaluation of the results.

Starting with 8,262 Unigene human EST sets (10 or more trace files are available for each set), more than 3,000 SNPs have been identified. Validation of 192 candidates showed that 82% of them identify a variation in a sample of ten Centre d'Etudes Polymorphism Human (CEPH) individuals (Buetow et al. 1999).

4.2.2.2 POLYBAYES

POLYBAYES (Marth et al. 1999) uses only ESTs for which trace files are available. All participating EST sequences have to be re-called to obtain base quality data. The algorithm used in POLYBAYES differs from that in SNPpipeline in two ways. First, POLYBAYES uses genomic sequence as a reference (anchor) sequence. Secondly, POLYBAYES uses statistics (Bayesian inference) to discriminate paralogues and to filter out low-confidence candidates. The algorithm used in PLOYBAYES is summarized as follows: 1) masking known repeat sequences in the human genomic sequences with RepeatMasker, 2) similarity search against NCBI dbEST with WU-BLAST (p-value cutoff 10^{-50}), 3) re-calling sequences from trace files with Phred, 4) clustering, 5) pairwise aligning each cluster member to the reference sequence with Cross_Match, 6) aligning ESTs using a procedure called "sequencing padding", 7) identification and removal of paralogues through a statistical approach (Bayesian interference), taking sequence quality into account, 8) identification of SNPs using an approach similar to that used above.

4.2.2.3 SNiPpER

SNiPpER (Kota et al. 2003) has been designed to mine SNPs from EST data without corresponding trace files. Obviously, an alternative scoring system is required since no sequence quality data (from trace data) are available. The scoring algorithm is based on the fact that true SNPs occur repeatedly while sequence noise occurs more randomly. Therefore, any observed polymorphism can be assigned a score reflecting the likelihood of the polymorphism not occurring randomly. Two factors contribute a score: the frequency of the same polymorphism occurring in a given position and the frequency of similar polymorphism patterns occurring in the neighborhood. A high score indicates that the polymorphism is true. Obviously, the algorithm can only detect frequent polymorphisms. Detection of rare minor polymorphisms is difficult since no extensive redundancy of ESTs with minor polymorphisms is available.

Using the algorithm, the authors mined barley (*Hordeum vulgare* L.) SNPs. From 271,630 ESTs representing 23 varieties, 3,069 SNP candidates were identified. From a set of 63 SNPs, 54 (86%) were validated using a direct sequencing approach and 28 of them were mapped to distinct loci within the barley genome.

4.3 *In Silico* Gene Discovery

The primary usage of early EST projects was for gene discovery and it still remains true for present EST projects (see above). Intensive wet-lab procedures are involved in the traditional EST-based gene discovery approach (see section 4.1.1). The *in silico* approach differs from the traditional one in that no wet-lab procedures are used. Another difference is that ESTs are used as queries to search against DNA or protein databases in the traditional EST-based gene discovery approach while protein or DNA sequences are used as queries to search EST databases in *in silico* approach. The *in silico* approach takes advantage of the availability of large datasets of ESTs from wide resources and advanced bioinformatics tools and algorithms.

Early *in silico* EST-based gene discovery projects have employed simple algorithms, such as BLAST searches (Liu et al. 1998) while recent efforts use more complicated algorithms in which various bioinformatics resources and tools may be combined.

Schultz et al. (2000) developed an algorithm coupling a BLAST search with a domain identifying protocol. Firstly, BLAST similarity searches of 1.6 million human ESTs with proteins of 50 families of extracellular proteins and 50 families involved in intracellular signaling cascades from *Homo sapiens, Drosophila melanogastger, Caenorhabditis elegans, Arabidopsis thaliana* and *Saccharomyces cerevisiae* were conducted. The results were filtered by removing hits identical to human proteins. Other hits (E-value cutoff: 1.0) were clustered. The consensus sequences were used for reciprocal searches of a non-redundant protein database. The hits were manually checked to ensure their qualities. Using this algorithm, they identified 1,197 novel human signaling genes.

Wittenberger et al. (2001) developed an algorithm to discover genes by searching protein sequences against EST database using two BLAST steps. Firstly, 189 human GPCR (G-protein coupled receptor) proteins were searched against 3.6 million ESTs with TBLASTN. About 1,500 distinct ESTs (E-value < 10) obtained in the first step were then searched against a protein database with BLASTX. ESTs with high similarity hits were collected and ESTs whose hits identical to known GPCRs were manually removed from the collection. The remaining ESTs might represent new GPCRs. Further analysis of 47 mammalian ESTs allowed 14 to be assigned as candidates for new putative GPCRs.

Lynn et al. (2003) used a different algorithm to discover genes involved in the TLR

(Toll-like receptor) pathway in chicken. Their algorithm includes 1) clustering and assembling 330,338 chicken ESTs with stackPackv2.2 and Phrap, 2) generating consensus sequences with Craw and translating them to proteins using ESTScan, 3) searching human proteins involved in the TLR pathway and AMPs (antimicrobial proteins) against the translated proteins with BLASTP, 4) comparing the resulting hits to the same proteins from other species. Four new chicken TLR members in full sequence and 6 in partial sequence were identified in the work.

Brown et al. (2003) found 43 new MFS (major facilitator super-family) of transporters using a semi-automated approach similar to that used by Lynn et al. above. Firstly, members of a core family of MFS proteins were collected by searching a seed sequence each from 28 MFS families with PSI-BLAST (position-specific iterative BLAST). The core family members were then searched against dbEST (E-value cutoff: 1×10^{-5}). The collected ESTs were cleaned and previously characterized ESTs were removed from the collection. Next, the ESTs in the collection were assembled with CAP3. The contigs and singletons produced represent new gene candidates. Lastly, the candidates were analyzed through nonautomated protocols including removal of ESTs matched to characterized proteins and nonhuman genes, transmembrane topology prediction, motif analysis, and mapping the ESTs to human genome sequences.

Torto et al. (2003) developed an algorithm called PexFinder to identify extracellular effector proteins from the plant pathogen *Phytophthora*. The algorithm can be easily adapted to other similar projects. The algorithm used the following steps: 1) translating 5' ESTs from the first ATG codon, 2) identifying signal peptides using SignalP 2.0, a program identifying signal peptides and their cleave sites, 3) querying original EST database with signal-containing ESTs and removing redundant sequences, 4) searching the nonredundant EST (containing extracellular protein signals) against GenBank nonredundant database for annotation. From 2,147 ESTs, the program identified 261 ESTs corresponding to 142 nonredundant *Pex* (Phytophthora extracellular protein) cDNAs, 78 of which were novel. Experimental evaluation found two new necrosis-inducing cDNAs, *crn1* and *crn2*.

4.4 Alternative Splicing Prediction

Alternative splicing (AS) refers to the biological process in which more than one mRNA (isoforms) is generated from the transcript of a same gene through different combinations of exons and introns. It is an important mechanism to increase the diversity of gene products (Graveley 2001). It is estimated by large-scale genomics studies that 30-60% of mammalian genes are alternatively spliced (Mironov and Pevzner 1999; Brett et al. 2000; Kan et al. 2001; Modrek et al. 2001). In fungi, no similar reports have been published. However, several AS events have been identified experimentally (Davis et al. 2000; Lodato et al. 2003). AS in some fungi (such as baker's yeast) might be less frequent than in mammals. This is probably due to the fact that fewer genes in these fungi contain introns. This may not hold true for other fungal genomes with more frequent intron-containing genes. For instance, 2,670 genes in fission yeast (*S. pombe*) contain at least one intron. Among them, 717 contain 2 introns,

360 contain 3 introns, 162 contain 4 introns, and 151 contain 4 to 15 introns (P. Zhang, unpublished data). It is reasonable to assume that AS could occur more frequently in fungi such as *S. pombe*.

The possible patterns of alternative splicing for a gene can be very complicated and the complexity increases rapidly as number of introns in a gene increases. Experimental approaches to identify AS are inefficient. The availability of EST data enables an alternative approach based on bioinformatics. The EST-based algorithm for AS prediction remains the most popular one, though investigation showed that not all exons predicted by ESTs might be functional (Sorek et al. 2004).

The general strategy for *in silico* alternative splicing prediction is to find large insertions or deletions within a set of ESTs sharing a large portion of aligned sequences. The early algorithms are simple and not fully automated. However, the basic strategies established by that early work are important for later algorithms. Burke et al. (1998). developed an algorithm to partition UniGene clusters. Some of those sub-clusters may represent alternative splicing isoforms. Wolfsberg and Landsman (1997) developed a semi-automatic approach using a gapped alignment program with a gap extension penalty of 0. Brett et al. (2000) identified alternative splicing events by aligning ESTs to their corresponding mRNA sequences. All those early algorithms identify only alternative splicing events instead of isoforms.

Later algorithms are complicated and report isoforms. Take TAP as an example. Kan et al. (2001) developed TAP (Transcript Assembly Program) which reports both predicted gene structures and alternative spliced isoforms. TAP uses genomic sequences as references. The algorithm is as follows. 1) Identifying genomic loci through BLAST searches of mRNA sequences against genomic sequences, 2) extracting sequences for genomic loci and extending the sequences at both ends up to 20 kb, 3) searching the genomic sequences (repeat sequences have been masked) against ESTs from dbEST, 4) aligning high score ESTs to the genomic sequences, 5) extracting splicing pairs (A splicing pair is two boundaries of alignment gap with GT-AG consensus or with more than two ESTs aligned at both ends of the gap), 6) assembling splicing pairs according to their coordinates, 7) determining gene boundaries (splicing pair predictions are generated to this point), 8) generating predicted gene structures by aligning mRNA sequences to genomic templates, 9) comparing splicing pair predictions and gene structure predictions to find alternative spliced isoforms. The algorithm was enhanced by adding statistical analysis in their later work (Kan et al. 2002).

The similar genomic sequence-mRNA-EST alignment strategy was developed for other several AS predicting projects, such as the algorithm developed by Modrek et al. (2001). The algorithm was used to produce a human alternative splicing database ASAP (Alternative Splicing Annotation Project) (Lee et al. 2003).

4.5 Gene Expression Profiling

ESTs are derived from actively expressed genes, so the complexity of ESTs from a non-normalized library represents gene expression profile of the tissue or organ at the time of RNA extraction. Since highly redundant ESTs for some organisms are available,

EST-based gene profiling will be a feasible alternative to traditional approaches, such as microarray and SAGE (serial analysis of gene expression). The principle for the EST-based gene expression profiling is simple - counting the "same" ESTs from an EST dataset. In practice, the process is much more complicated since ESTs derived from the same gene often represent different portions of the gene. The following are some examples of implemented algorithms.

4.5.1 TissueInfo

TissueInfo (http://icb.med.cornell.edu/services/tissueinfo/query, Skrabanek and Campagne 2001) is a pipeline of EST-based gene expression profiling. Originally, TissueInfo was used to establish human gene expression profile. Mouse gene expression profiles are also available in the current version. The enhanced version can be accessed at the above web site. The data source used by TissueInfo is dbEST. A relational database was used to store and manipulate the EST data. There are three steps for the algorithm. 1) Assigning tissue, organ, or organism codes to ESTs. This step was performed by automated extraction of related information and manual curation. 2) Grouping ESTs based on the gene for which they are derived. This step is performed by similarity search (BLAST or MegaBlast) of nucleotide (ESTs or mRNAs) or protein queries against dbEST. ESTs the satisfied predefined similarity parameters were grouped. 3) Calculation of ESTs based on query sequences and tissue code. 4) Hierarchical organization of tissue codes. In their original report, it was claimed that TissueInfo was accurate for 69% of identified tissue specificities and for 80% of expression profiles.

4.5.2 ExProView

ExProView (Expression Profile Viewer) (Larsson et al. 2000) is a Java-based expression profile viewer generated based on EST and SAGE data (http://biobase.biotech.kth.se/thp1a/Startsida_till_PEK2_exproview.htm). The algorithm used to generate expression profiles is outlined as follows. 1) Making a reference sequence BLAST database. The NCBI UniGene and EGAD (Expressed Gene Anatomy Database) datasets were used in the original work. 2) Similarity search of ESTs and other sequence tags (SAGE) against the reference sequence BLAST database. 3) Determining abundances of ESTs or SAGE tags. The primary user interface is virtual chip which shows relative abundances of gene expression.

4.5.3 DigiNorthern

DigiNorthern (http://falcon.roswellpark.org/DN/, Wang and Liang 2003) is a tool for viewing EST-based gene expression profiles using one or two sequences as the query. The query sequences are searched against EST databases using BLAST. The hits which meet a configurable parameter set are grouped according to the library information and ESTs are counted. The results are reported in numerical format and virtual Northern images.

4.5.4 ExQuest

Like DigiNorthern, ExQuest (Brown et al. 2004) is an EST-based expression profile database which can be queried with mRNA or DNA sequences. However, ExQuest uses a different algorithm. ExQuest first parses EST entries from dbEST into a local database (EFEL, EST Field Extraction Library) organized in a manner of Gary's Anatomy (hierarchical tissue 'bins'). Query sequences are searched against dbEST with MegaBlast. Accession numbers and identities are extracted from the output. Identity values are used to filter out low similarity hits. Accession numbers are used to fetch EST entries for the BLAST database. Data fields are extracted and filtered through the EFEL database and expression profiles are generated. ExQuest is configured for Windows XP or Windows 2000.

4.5.5 EST-based expression profiling in fungi

To our knowledge, no systematic tools or databases related to fungal EST-based expression profiling have been reported. However, many fungal EST projects conduct some degree of expression profile analysis on the ESTs produced in-house. For instance, Kamoun at al. (1999) have shown that in *Phytophora infestans* two elicitin-like EST clusters, *inf5* and *inf6* are among the 10 most abundant clusters, including those for elongation factor, ribosomal proteins, and actin. Qutob et al. (2000) have compared expression profiles in infected soybean, mycelium, and zoospore of *Phytophthora sojae*. They have found that the expression profiles under the three conditions are quite different. In infected soybean, the most abundant ESTs include formate dehydrogenase, alcohol β-chain of ATP synthase, dehydrogenase, glutamate dehydrogenase, and RIC1. In mycelium, the most abundant ESTs are RIC1, annexin, guanine nucleotide binding protein, thioredoxin peroxidase, superoxide deismutase, and two hypothetical proteins. In zoospore, elicitin, ABC transporter, GTP cyclohydrolase, ribonucleotide reductase, actin A, membrane glycoprotein, HAM34 protein, mucin, transcription factor protein, and three hypothetical proteins are among the most redundant EST clusters. Zhu et al. (2001) have compared clock-controlling genes using ESTs from morning and evening cDNA libraries of *Neurospora crassa*. Two different expression profiles of the clock-controlling genes have been identified. Several other fungal EST projects also conducted limited expression profile analysis (Karlsson et al. 2003; Nugent et al. 2004).

4.6 Gene Indexing

As we discussed above, ESTs have been used for various applications. However, if unprocessed ESTs are used for these applications, difficulties are encountered due to high-error rates, sequence contaminations, fragmentary representation of transcripts, and poor annotations with ESTs. To overcome the negative impact of these errors and inconsistencies, the EST data need further processing and organization. One of such effort is to group ESTs and other related information into distinct gene classes. These gene classes are usually collected by source organisms of ESTs and the collection is called a gene index (catalog) (Aaronson et al. 1996; Burke et al. 1999; Jongeneel 2000).

Gene indexing is a process to generate a gene index. The starting data for gene indexing are EST and/or annotated gene (mRNA) data. However, algorithms used for different gene indexing projects vary. The process can be divided into 6 major steps:

1) Data collection

Gene indexing requires large EST and annotated gene datasets. The data can be downloaded from public database such as GenBank (see Section 2) or generated in-house through high throughput sequencing.

2) Data preparation

ESTs collected from public databases or in-house sequencing can contain errors, contaminants, repeats, microsatellites, low-complexity sequences. They should be removed, corrected, or masked. Sequence cleaning is a process of removing sequence errors and contaminations which is discussed in Section 5. Stand-alone repeats and low-complexity sequence masking programs are available, such as RepeatMasker (Smit and Green at http//ftp.genome.washington.edu/RM/RepeatMasker.html) and MaskerAid (Bedell et al. 2000).

3) Clustering

Clustering is a process that partitions sequences into groups (clusters) according to their similarities. The similarity cutoff should be great enough to guarantee that all members in a cluster belong to the same gene. Short ESTs should be excluded since they may be partitioned into wrong clusters. Clustering can be supervised or unsupervised. Supervised clustering is performed by using reference sequences (e. g. genome sequences) as guide, while unsupervised clustering is performed without such reference sequences. D2_CLUSTER (Burke et al. 1999) is a popular EST clustering program, which has been integrated into several EST software packages.

4) Assembly

In assembly, alignments of clustered sequences are generated. Consensus sequences can be generated directly from the alignments or after further optimization. Assembly can be conducted in 'strict' or 'loose' modes. In strict mode (such as TIGR_ASSEMBLER, Sutton et al. 1995) accurate consensus sequences with minimal chimerism and other contaminations are produced. However, some valuable information, such as alternate gene forms, is lost. On the other hand, with the 'loose' mode, alternative gene form information is sustained in assemblies but chimerism and contamination can not be efficiently removed. Phrap (http://www.genome.washington.edu/uwgc/analysistools/phrap.htm, P. Green, unpublished) and CAP3 (http://genome.cs.mtu.edu/sas.html, Huang and Madan 1999) are two popular programs for assembling and consensus sequence generation.

5) Annotation

The clusters in gene indices are annotated by searching consensus sequences against functionally annotated public nucleotide and protein databases. They may further be classified by Gene Ontology and mapped to related metabolic or signaling pathways.

6) Storing and visualization

Most gene index data are stored in relational databases and visualized through web interfaces.

4.6.1 Major gene index databases
4.6.1.1 UniGene

NCBI UniGene project started in the mid-1990's in response to rapid growth of EST entries deposited in GenBank (Schuler et al. 1996; Schuler 1997; Wheeler et al. 2000; Wheeler et al. 2001; Wheeler et al. 2002; Wheeler et al. 2003; Wheeler et al. 2004). Unigene database is updated after each GenBank release. The building information for UniGene is available at NCBI website (http://www.ncbi.nlm.nih.gov/UniGene/build.html/). Only the clustering strategy is reviewed here since it is important to the end users.

The source data of UniGene include ESTs and cDNA/CDS sequences in GenBank. The clustering is performed by several steps as follows. 1) Initial cluster generation from similarity comparison using MEGABLAST. 2) Creation of anchored clusters which contain at least a polyadenylation signal or two 3' ESTs and discarding cDNA/CDS sequences partitioned in more than two clusters. 3) Adding clone-associated 5' ESTs to clusters. 4) Re-comparison of singletons or non-anchored clusters with anchored clusters with reduced stringency and merging clusters.

Clusters in UniGene are identified by unique cluster IDs. Note that UniGene clusters may be changed between updates, since clusters can be retired for several reasons: sequences being retracted by submitters, cluster joining, and cluster split. Retired IDs are not reused. UniGene clusters can also be accessed by GenBank accession numbers of sequences in the cluster.

Unlike other gene indices (such as TIGR gene indices), no consensus sequences are available in UniGene. However, the longest sequence in a cluster is selected and separately documented (in file called Hs.seq.uniq.Z). Note that in clusters without full-length cDNA sequence or annotated genomic sequences, the longest sequence may represent only a portion of the gene.

Initially, UniGene contained only human gene clusters. Later, more and more organisms were added. At time of writing, UniGene clusters for 35 organisms, for which there are 7,000 or more EST entries in GenBank, have been created. All those clusters can be searched through NCBI Entrez website. The flat files of the clusters can be downloaded from NCBI FTP server (ftp://ftp.ncbi.nih.gov/repository/UniGene).

4.6.1.2 Merck Gene Index

The Merck Gene Index (MGI) is a human gene index using human ESTs generated through the Merck-sponsored EST project (Eckman et al. 1998). The original MGI used a loose assembly method and the 3' sequences as indices. MGI can be accessed through Merck Gene Index Browser. The database can be searched by three methods: text keywords, EST accession number, and index class ID. The index class is reported by four modules: clone data module, EST data module, protein similarity data module, non-EST nucleic acid similarity data module. MGIv2.0 is an enhanced version of the original MGI (Yuan et al. 2001). MGIv2.0 has been enhanced in several aspects, including throughput, quality control, and clustering. An Oracle relational database was used for data management. The 5' EST data was not used for clustering but stored in the database. The user interface has also been enhanced in MGIv2.0. Java-based user interface allows user to query the database for varying information, including consensus sequences generated by Phrap. For more information about MGIv2.0, contact J. Yuan at Jeffrey_yuan@merk.com.

4.6.1.3 TIGR gene indices

The TIGR gene indices (Liang et al. 2000; Quackenbush et al. 2000; Quackenbush et al. 2001) are multi-organism gene indices which include 31 animals, 29 plants, 9 fungi, and 15 protists as of January, 2005. TIGR gene indices are built with strict criteria. Alternative gene forms are partitioned into different clusters. The consensus sequences (called TC, tentative consensus) are generated for all clusters. The starting data of TIGR gene indices are ESTs (from NCBI dbEST) and ETs (expressed transcripts, from TIGR EGAD database, http://www.tigr.org/tdb/egad/egad.html). Cleaned ESTs, ETs, and TCs and singletons from previous build are compared pair-wise to identify overlaps using MEGABLAST. Clusters are generated by the following criteria: 1) a minimum 40 base pair match, 2) greater than 94% identity in the overlap region, 3) a maximum unmatched overhang of 30 base pairs.

Each cluster is assembled using CAP3 (Huang and Madan 1999) to generate initial TCs. The final TCs are produced by the second round of assembling on the initial TCs. The resulting TC set and singletons are loaded to species-specific databases for annotation.

Annotation is conducted by searching non-redundant protein database using TCs and some singletons as queries and shown in the Tentative Annotation entry. If there are ETs in the assembly of a TC, the annotations of those ETs are concatenated and shown in the Expressed Transcripts entry. The protein hits are shown in Similarity search results entry. The assignment convention used for annotation is shown in Table 1.

Table 1 Assignment convention used for TIGR gene indices.

Criteria	Qualifier
100% > percent identity >= 90%	Homologue to
90% > percent identity >= 70%	Similar to
percent identity < 70%	Weakly similar to
protein coverage > 98%	Complete
protein coverage <= 98%	Partial

TIGR gene indices are available from TIGR website (http://www.tigr.org/tdb/tgi/). The data and related tools can also be downloaded from TIGR FTP server (ftp://ftp.tigr.org/pub/data/tgi/). It should be noted that TC numbers of TIGR gene indices are short-lived; they are only available within a release. No old TC numbers are used in new releases.

4.6.1.4 Other fungus-specific gene index data

Due to the availability of EST clustering/assembling tools, gene indices can be easily generated from EST data, provided that there are enough ESTs. Consequently, many fungal EST projects provide their own gene indices (see Section 3.1). However, due to the lack of standards of the contents of gene indices and of the way to store, search, and retrieve gene index information, it might be not convenient to use those dispersed gene index data for genomics studies, especially for comparative genomics studies.

4.6.2 Using gene indices in genomics

Gene indexing provides a non-redundant view of transcripts of a particular organism. Two sets of data are available in gene indices which are very useful for genomics studies: consensus sequences and EST/RNA clusters. The former is useful for gene discovery, genome annotation, and comparative genomics. The latter is useful for expression profiling, genetic marker discovery and genetic mapping, and alternative splicing prediction. Gene indices can also be useful to other studies not mentioned in this chapter, such as estimation of gene number of an organism (Liang et al. 2000) annotation of array-based expression data (Svensson et al. 2003) mRNA motif discovery and annotation (Bakheet et al. 2003), translation termination site analysis (Ozawa et al. 2002) and gene-specific probe design (Chang and Peck 2003).

5. EST DATA QUALITY MANAGEMENT

ESTs are generated from high-throughput sequencing projects. These single-pass sequences are error-prone. EST data deposited to databases may or may not pass proper quality control measures. EST sequence errors can be roughly grouped into two types: sequencing errors and contaminations. Sequence errors are generated during reverse transcription, sequencing reaction, and base calling. These errors usually result in substitutions, single nucleotide insertions and deletions. ESTs can also contain sequence contaminations. Sequence contaminations include full or partial ESTs sequences from vectors, genomic sequences, or other organisms. The common contaminated sequences are vector/linker sequences, chimeras, cDNAs or genomic sequences from other organisms which are pathogens, endophytes, or parasites of subject organisms, or microorganisms such as *E. coli* which contaminate the target cDNAs in various steps. Measures used to remove or to correct the sequence errors (sequence cleaning) and contaminations are necessary steps for almost all EST data-mining projects.

In practice, strategies used to deal with the problems mentioned above may vary from project to project. General outlines are given bellow.

5.1. Low Quality Sequences

Low quality sequences can be easily identified during base-calling process using base-calling programs such as Phred (Ewing and Green 1998; Ewing et al. 1998). or TraceTuner (Paracel) (see section 2.4) and then they can be removed or trimmed by a quality control program such as LUCY (Chou and Holmes 2001). Trace data can be obtained from public databases (see Section 3.1). However, there are no trace files available for most EST entries in public EST databases. Removal of low quality sequences from EST sequences is not trivial, since different quality standards may have been applied to different EST projects. The most popular strategy is to cluster and assemble redundant EST sequences and to locate low quality region. The strategy has been implemented in several EST cluster/assembly packages discussed in the section 3.

5.2. Single Nucleotide Substitution, Insertion, and Deletion

Single nucleotide substitution, insertion, and deletion can result in false stop codons and virtual frame shifts. These will produce serious consequences when translating ESTs into proteins. One strategy to deal with such errors is to use EST consensus sequences instead raw ESTs, since the errors are corrected in clustering/assembling processes. Statistics-based solutions are also available, such as ESTScan. ESTScan (Iseli et al. 1999) is a HMM (Hidden Markov Model) based program. ESTScan can efficiently extract coding regions from ESTs with accurate correction of frame shift errors.

5.3. Chimerism

Chimerism refers to the artificial fusion of two or more different unrelated sequences. Chimeras can result from concatenated cDNA clones or incorrect lane tracking in the sequencing process. Currently, the most efficient chimerism detection and removing approach is EST clustering with or without reference sequences (Burke et al. 1998). Since the portions of most chimeras are from different chromosomes or from distant regions of the same chromosome, comparison of ESTs with genomic sequences can help identify chimerism. Chimerism can also be identified by EST clustering. If all members of a cluster except one can be partitioned into sub-clusters, the member is likely a chimera. Of course, the approach can only detect chimeras in clusters containing highly redundant members. Depending on the needs of project, the chimeras can be removed automatically or manually with further confirmation.

5.4. Heterogeneous Sequence Contamination

Most of heterogeneous sequence contamination removal programs are based on two strategies: similarity comparison and statistical analysis. Vector and linker contamination can be efficiently removed with the first strategy. This is done by similarity search of EST sequences against vector databases, such as the UniVec database available at NCBI (http://www.ncbi.nlm.nih.gov/VecScreen/UniVec.html), with BLAST, such as RAPID (Miller et al. 1999). An alternative algorithm is used in LUCY (Chou and Holmes 2001; Li and Chou 2004).

Although other heterogeneous sequence contamination can be removed by the similarity comparison strategy (Hsiang and Goodwin 2003), the process is quite inefficient and may be incomplete. Several alternative approaches have been developed. White et al. (1993) have developed an algorithm based on the statistical analysis of composition of oligomers. Unlike BLAST search, the algorithm can identify contamination from unknown sources. Maor et al. (2003) have developed a tool PF-IND to separate plant and fungal sequences. PF-IND is based on the differences in codon usage bias between plants and fungi. Applied to a test dataset of 100 sequences, PF-IND can identify 96 ESTs while BLASTX can only identify 66.

5.5.   Genomic DNA and Pre-mRNA Contamination

Contamination of genomic DNA and pre-mRNA from the organism itself can result in false gene prediction and false alternative splicing prediction. Identification of such contamination is difficult; both similarity search and statistical algorithms do not supply a solution. Sorek and Safer (2003) have developed an algorithm to address the problem based on analysis of EST libraries as a whole. The algorithm identifies contaminated libraries rather than contaminated sequences. However, no automated tool for the removal of contaminated sequences is available, instead, guidelines of how to deal with contaminated libraries have been suggested by the authors.

6. CONCLUSIONS

In this chapter, EST data mining tools and primary applications in genomics research have been reviewed. Since the advent of EST technology in early 1970's, huge amount of data have been accumulated. Publicly accessible EST data can be retrieved from major public databases, such as dbEST and EMBL. Small, dispersed EST datasets specific to fungi can also be accessed from different web sites. Due to the large size of EST datasets, specific tools are required for retrieving, managing, and mining EST data. Most of these tools and packages are freely available to academic users.

ESTs can be used in various applications, such as gene discovery, genetic marker discovery, genetic mapping, genome annotation, gene indexing, alternative splicing prediction, etc. Dedicated bioinformatics tools have been developed for such applications. In comparison to genomic sequence data, ESTs are relatively easy and inexpensive to generate. On the other hand, ESTs are more error-prone. For most applications, these errors need to be removed or corrected. Algorithms and tools have been developed to deal with these errors and other contaminations.

Acknowledgements: We would like to thank Drs. Roscoe Klinck (Centre de génomique fonctionnelle de Sherbrooke, Université de Sherbrooke) and Reginald Storms (Centre for Structural and Functional Genomics, Concordia University) for critical appraisal of the manuscript. We also would like to thank Dr. Adrian Tsang (Centre for Structural and Functional Genomics, Concordia University) for supporting the preparation of the manuscript. X. Min is supported by a fungal enzyme genomics project funded by Genome Canada and Genome Quebec and would like to thank other members of the project for their assistance in the development of sequence analysis and annotation tools.

REFERENCES

Aaronson JS, Eckman B, Blevins RA, Borkowski JA, Myerson J, Imran S and Elliston KO (1996). Toward the development of a gene index to the human genome: an assessment of the nature of high-throughput EST sequence data. Genome Res 6:829-845.

Adams MD, Kelley JM, Gocayne JD, Dubnick M, Polymeropoulos MH, Xiao H, Merril CR, Wu A, Olde B, Moreno RF, Kerlavage AR, McCombie WR and Venter JC (1991). Complementary DNA sequencing: Expressed sequence tags and human genome project. Science 252:1651-1656.

Allen JE, Pertea M and Salzberg SL (2004). Computational gene prediction using multiple sources of evidence. Genome Res 14:142-148.

Areshchenkova T and Ganal MW (2002). Comparative analysis of polymorphism and chromosomal location of tomato microsatellite markers isolated from different sources. Theor Appl Genet 104:229-

235.

Ayoubi P, Jin X, Leite S, Liu X, Martajaja J, Abduraham A, Wan Q, Yan W, Misawa E and Prade RA (2002). PipeOnline 2.0: automated EST processing and functional data sorting. Nucleic Acids Res 30:4761-4769.

Bakheet T, Williams BR and Khabar KS (2003). ARED 2.0: an update of AU-rich element mRNA database. Nucleic Acids Res 31:421-423.

Bedell JA, Korf I and Gish W (2000). MaskerAid: a performance enhancement to RepeatMasker. Bioinformatics 16:1040-1041.

Boguski MS, Lowe TM and Tolstoshev CM (1993). dbEST--database for "expressed sequence tags". Nat Genet 4:332-333.

Brendel V, Xiang L and Zhu W (2004). Gene structure prediction from consensus spliced alignment of multiple ESTs matching the same genomic locus. Bioinformatics 20:1157-1169.

Brett D, Hanke J, Lehmann G, Haase S, Delbruck S, Krueger S, Reich J and Bork P (2000). EST comparison indicates 38% of human mRNAs contain possible alternative splice forms. FEBS Lett 474:83-86.

Brown AC, Kai K, May ME, Brown DC and Roopenian DC (2004). ExQuest, a novel method for displaying quantitative gene expression from ESTs. Genomics 83:528-539.

Brown S, Chang JL, Sadee W and Babbitt PC (2003). A semiautomated approach to gene discovery through expressed sequence tag data mining: discovery of new human transporter genes. AAPS PharmSci 5:E1.

Buetow KH, Edmonson MN and Cassidy AB (1999). Reliable identification of large numbers of candidate SNPs from public EST data. Nat Genet 21:323-325.

Burge C and Karlin S (1997). Prediction of complete gene structures in human genomic DNA. J Mol Biol 268:78-94.

Burke J, Davison D and Hide W (1999). d2_cluster: a validated method for clustering EST and full-length cDNA sequences. Genome Res 9:1135-1142.

Burke J, Wang H, Hide W and Davison DB (1998). Alternative gene form discovery and candidate gene selection from gene indexing projects. Genome Res 8:276-290.

Camon E, Magrane M, Barrell D, Lee V, Dimmer E, Maslen J, Binns D, Harte N, Lopez R and Apweiler R. (2004). The Gene Ontology Annotation (GOA) Database: sharing knowledge in Uniprot with Gene Ontology. Nucleic Acids Res 32:D262-D266.

Chang PC and Peck K (2003). Design and assessment of a fast algorithm for identifying specific probes for human and mouse genes. Bioinformatics 19:1311-1317.

Chou HH and Holmes MH (2001). DNA sequence quality trimming and vector removal. Bioinformatics 17:1093-1104.

Christoffels A, van Gelder A, Greyling G, Miller R, Hide T and Hide W (2001). STACK: Sequence Tag Alignment and Consensus Knowledgebase. Nucleic Acids Res 29:234-238.

Davis CA, Grate L, Spingola M and Ares M Jr (2000). Test of intron predictions reveals novel splice sites, alternatively spliced mRNAs and new introns in meiotically regulated genes of yeast. Nucleic Acids Res 28:1700-1706.

Delcher AL, Harmon D, Kasif S, White O and Salzberg SL (1999). Improved microbial gene identification with GLIMMER. Nucleic Acids Res 27:4636-4641.

Diatchenko L, Lau YC, Campbell AP, Chenchik A, Moqadam F, Huang B, Lukyanov S, Lukyanov K, Gurskaya N, Sverdlov E, and Siebert PD (1996). Suppression subtractive hybridization: A method for generating differentially regulated or tissue-specific cDNA probes and libraries. Proc Natl Acad Sci USA 93:6025-6030.

Diatchenko L, Lukyanov S, Lau YF and Siebert PD (1999). Suppression subtractive hybridization: a versatile method for identifying differentially expressed genes. Methods Enzy 303:349-380.

Drabenstot SD, Kupfer DM, White JD, Dyer DW, Roe BA, Buchanan KL and Murphy JW (2003). FELINES: a utility for extracting and examining EST-defined introns and exons. Nucleic Acids Res 31:e141.

Eckman BA, Aaronson JS, Borkowski JA, Bailey WJ, Elliston KO, Williamson AR and Blevins RA (1998). The Merck Gene Index browser: an extensible data integration system for gene finding, gene characterization and EST data mining. Bioinformatics 14:2-13.

Emanuelsson O and von Heijne G (2001). Prediction of organellar targeting signals. Biochim Biophys Acta 1541:114-119.

Eujayl I, Sledge MK, Wang L, May GD, Chekhovskiy K, Zwonitzer JC and Mian MA (2004). Medicago truncatula EST-SSRs reveal cross-species genetic markers for Medicago spp. Theor Appl Genet 108:414-422.

Ewing B and Green P (1998). Base-calling of automated sequencer traces using Phred. II. Error probabilities. Genome Res 8:186-194.

Ewing B, Hillier L, Wendl M and Green P (1998). Base-calling of automated sequencer traces using Phred. I. Accuracy assessment. Genome Res 8:175-185.

Faria-Campos AC, Cerqueira GC, Anacleto C, Carvalho CMB and Ortega JM (2003). Mining microorganism EST databases in the quest for new proteins. Gent Mol Res 2:169-177.

Felipe MSS, Andrade RV, Petrofeza SS, Maranhão AQ, Torres FAG, Albuquerque P, Arraes FBM, Arruda M, Azevedo MO, Baptista AJ, Bataus LAM, Borges CL, Campos EG, Cruz MR, Daher BS, Dantas A, Ferreira M, Ghil G, Jesuino RSA, Kyaw CM, Leitão L, Martins CR, Moraes LMP, Neves EO, Nicola AM, Alves ES, Parente JA, Pereira M, Poças-Fonseca MJ, Resende R, Ribeiro BM, Saldanha RR, Santos SC, Silva-Pereira I, Silva MAS, Silveira E, Simões IC, Soares RBA, Souza DP, De-Souza MT, Andrade EV, Xavier MAS, Veiga HP, Venancio EJ, Carvalho MJA, Oliveira AG, Inoue MK, Almeida NF, Walter MEMT, Soares CMA and Brígido MM (2003). Transcriptome characterization of the dimorphic and pathogenic fungus Paracoccidioides brasiliensis by EST analysis. Yeast 20:263-271.

Florea L, Hartzell G, Zhang Z, Rubin GM and Miller W (1998). A computer program for aligning a cDNA sequence with a genomic DNA sequence. Genome Res 8:867-974.

Freimoser FM, Screen S, Hu G and St Leger R (2003). EST analysis of genes expressed by the zygomycete pathogen Conidiobolus coronatus during growth on insect cuticle. Microbiol 149:1893-1900.

Gajer P, Schatz M, and Salzberg SL (2004). Automated correction of genome sequence errors. Nucleic Acids Res 32:562-569.

Galagan JE, Calvo SE, Borkovich KA, Selker EU, Read ND, Jaffe D, FitzHugh W, Ma LJ, Smirnov S, Purcell S, Elkins T, Engels R, Wang S, Nielsen CB, Bulter J, Endrizzl M, Qui D, Ianaklev P, Bell-Pedersen D, Nelson MA, Werner-Washburne M, Selitrennikoff CP, Kinsey JA, Braun EL, Zelter A, Schulte U, Kothe GO, Jedd G, Mewes W, Staben C, Marcotte E, Greenberg D, Roy A, Foley K, Naylor J, Stange-Thomann N, Barrett R, Gnerre S, Kamal M, Kamvysselis M, Mauceli E, Bielke C, Rudd S, Frishman D, Krystofova S, Rasmussen C, Metzenberg R, Perkins DD, Kroken S, Cogoni C, Macina G, Catcheside D, Li W, Pratt RJ, Osmani SA, DeSouza CPC, Glass L, Orbach MJ, Berglund JA, Voelker R, Yarden O, Plamann M, Seller S, Dunlap J, Radford A, Aramayo R, Natvig DO, Alex LA, Mannhaupt G, Ebbole DJ, Freitag M, Paulsen I, Sachs MS, Lander ES, Nusbaum C and Birren B (2003). The genome sequence of the filamentous fungus Neurospora crassa. Nature 422:859-868.

Gavin AJ, Scheetz TE, Roberts CA, O'Leary B, Braun TA, Scheffield VC, Soares MB, Robinson JP and Cadavant TL (2002). Pooled library tags for EST-based gene discovery. Bioinformatics 18:1162-1166.

Gemund C, Ramu C, Altenberg-Greulich B and Gibson TJ (2001). Gene2EST: a BLAST2 server for searching expressed sequence tag (EST) databases with eukaryotic gene-sized queries. Nucleic Acids Res 29:1272-1277.

Graveley BR (2001). Alternative splicing: increasing diversity in the proteomic world. Trends Genet 17:100-107.

Haas BJ, Delcher AL, Mount SM, Wortman JR, Smith RK, Hannick LI, Maiti R, Ronning CM, Rusch DB, Town CD, Salzberg SL and White O (2003). Improving the Arabidopsis genome annotation using maximal transcript alignment assemblies. Nucleic Acids Res 31:5654-5666.

Harcus YM, Parkinson J, Fernández C, Daub J, Selkirk ME, Blaxter ML and Maizels RM (2004). Signal sequence analysis of expressed sequence tags from the nematode Nippostrongylus brasiliensis and the evolution of secreted proteins in parasites. Genome Biol 5:R39.

Hatzigeorgiou AG, Fiziev P and Reczko M (2001). DIANA-EST: a statistical analysis. Bioinformatics 17:913-919.

Hennig S, Groth D and Lehrach H (2003). Automated Gene Ontology annotation for anonymous sequence data. Nucleic Acids Res 31:3712-3715.

Hsiang T and Goodwin PH (2003). Distinguishing plant and fungal sequences in ESTs from infected plant tissues. J Microbiol Methods 54:339-351.

Hotz-Wagenblatt A, Hankeln T, Ernst P, Glatting KH, Schmidt ER and Suhai S (2003). ESTAnnotator: A tool for high throughput EST annotation. Nucleic Acids Res 31:3716-3719.

Huang X, Adams MD, Zhou H and Kerlavage AR (1997). A tool for analyzing and annotating genomic sequences. Genomics 46:37-45.

Huang X and Madan A (1999). CAP3: A DNA sequence assembly program. Genome Res 9:868-877.

Huson DH, Reinert K, Kravitz SA, Remington KA, Delcher AL, Dew IM, Flanigan M, Halpern AL, Lai Z, Mobarry CM, Sutton GG and Myers EW (2001). Design of a compartmentalized shotgun assembler for the human genome. Bioinformatics 17 Suppl 1:132-139.

Iseli C, Jongeneel CV and Bucher P (1999). ESTScan: a program for detecting, evaluating, and reconstructing potential coding regions in EST sequences. Proc Int Conf Intell Syst Mol Biol :138-148.

Itoh M, Okuji YK, Goto S, Urushihara H, Maeda M and Tanaka Y (2001). Analysis of Dictyostelium discoideum cDNA obtained from multicellular and unicellular stages. Genome Informatics 12: 400-401.

Johansson T, Le Quere A, Ahren D, Soderstrom B, Erlandsson R, Lundeberg J, Uhlen M and Tunlid A (2004). Transcriptional response of Paxillus involutus and Betula pendula during formation of ectomycorrhizal root tissue. Mol Plant Microbe Interact 17:202-215.

Jongeneel CV (2000). The need for a human gene index. Bioinformatics 16:1059-1061.

Kamoun S, Hraber P, Sobral B, Nuss D and Govers F (1999). Initial assessment of gene diversity for the oomycete pathogen Phytophthora infestans based on expressed sequences. Fungal Genet Biol 28:94-9106.

Kan Z, Rouchka EC, Gish WR and States DJ (2001). Gene structure prediction and alternative splicing analysis using genomically aligned ESTs. Genome Res 11:889-900.

Kan Z, States D and Gish W (2002). Selecting for functional alternative splices in ESTs. Genome Res 12:1837-1845.

Kantety RV, La Rota M, Matthews DE and Sorrells ME (2002). Data mining for simple sequence repeats in expressed sequence tags from barley, maize, rice, sorghum and wheat. Plant Mol Biol 48:501-510.

Karlsson M, Olson A and Stenlid J (2003). Expressed sequences from the basidiomycetous tree pathogen Heterobasidion annosum during early infection of scots pine. Fungal Genet Biol 39:51-59.

Kent WJ (2002). BLAT – The BLAST-like alignment tool. Genome Res 12:656-664.

Kota R, Rudd S, Facius A, Kolesov G, Thiel T, Zhang H, Stein N, Mayer K and Graner A (2003). Snipping polymorphisms from large EST collections in barley (Hordeum vulgare L.). Mol Genet Genomics 270:24-33.

Larsson M, Stahl S, Uhlen M and Wennborg A (2000). Expression profile viewer (ExProView): a software tool for transcriptome analysis. Genomics 63:341-353.

Lee C, Atanelov L, Modrek B and Xing Y (2003). ASAP: the Alternative Splicing Annotation Project. Nucleic Acids Res 31:101-105.

Lee BTK, Tan TW and Ranganathan S (2003). MGAlignIt: a web service for the alignment of mRNA/EST and genomic sequences. Nucleic Acids Res 31:3533-3536.

Li S and Chou HH (2004). LUCY2: an interactive DNA sequence quality trimming and vector removal tool. Bioinformatics: May 6.

Liang L, Ding YQ, Li X, Yang GZ, Xiao J, Lu LC and Zhang JH (2004). Construction of a metastasis-associated gene subtracted cDNA library of human colorectal carcinoma by suppression subtraction hybridization. Worlkd J Gastroentrerol 10:1301-1305.

Liang F, Holt I, Pertea G, Karamycheva S, Salzberg SL and Quackenbush J (2000). An optimized protocol for analysis of EST sequences. Nucleic Acids Res 28:3657-3665.

Lindlof A (2003). Gene identification through large-scale EST sequence processing. Appl Bioinformatics 2:123-129.

Liu S, Stoesz SP and Pickett CB (1998). Identification of a novel human glutathione S-transferase using bioinformatics. Arch Biochem Biophys 352:306-313.

Lodato P, Alcaino J, Barahona S, Retamales P and Cifuentes V (2003). Alternative splicing of transcripts from crtI and crtYB genes of Xanthophyllomyces dendrorhous. Appl Environ Microbiol 69:4676-4682.

Lynn DJ, Lloyd AT and O'Farrelly C (2003). In silico identification of components of the Toll-like receptor (TLR) signaling pathway in clustered chicken expressed sequence tags (ESTs). Vet Immunol Immunopathol 93:177-184.

Mao C, Cushman JC, May GD and Weller JW (2003). ESTAP an automated system for the analysis of EST data. Bioinformatics 19:1720-1722.

Maor R, Kosman E, Golobinski R, Goodwin P and Sharon A (2003). PF-IND: probability algorithm and software for separation of plant and fungal sequences. Curr Genet 43:296-302.

Marth GT, Korf I, Yandell MD, Yeh RT, Gu Z, Zakeri H, Stitziel NO, Hillier L, Kwok PY and Gish WR (1999). A general approach to single-nucleotide polymorphism discovery. Nat Genet 23:452-456.

Matukumalli LK, Grefenstette JJ, Sonstegard TS and Van Tassell CP (2004). EST-PAGE-managing and analyzing EST data. Bioinformatics 20:286-288.

Michalickova K, Bader GD, Dumontier M, Lieu H, Betel D, Isserlin R and Hogue CW (2002). SeqHound: biological sequence and structure database as a platform for bioinformatics research. BMC Bioinformatics 3:32.

Miller C, Gurd J and Brass A (1999). A RAPD algorithm for sequence database comparisons: application to the identification of vector contamination in the EMBL databases. Bioinformatics 15:111-121.

Miller RT, Christoffels AG, Gopalakrishnan C, Burke J, Ptitsyn AA, Broveak TR and Hide WA (1999). A comprehensive approach to clustering of expressed human gene sequence: the sequence tag alignment and consensus knowledge base. Genome Res 9:1143-1155.

Mironov AA and Pevzner PA (1999). SST versus EST in gene recognition. Microb Comp Genomics 4:167-172.

Mita K, Morimyo M, Okano K, Koike Y, Nohata J, Kawasaki H, Kadono-Okuda K, Yamamoto K, Suzuki MG, Shimada T, Goldsmith MR and Maeda S (2003). The construction of an EST database for Bombyx mori and its application. Proc Natl Acad Sci USA 100:14121-14126.

Modrek B, Resch A, Grasso C and Lee C (2001). Genome-wide detection of alternative splicing in expressed sequences of human genes. Nucleic Acids Res 29:2850-2859.

Morales P and Thurston CF (2003). Efficient isolation of genes differentially expressed on cellulose by suppression subtractive hybridization in Agaricus bisporus. Mycol Res. 107:401-407.

Mott R (1977). EST_GENOME: a program to align spliced DNA sequences to unspliced genomic DNA. Comput Appl Biosci 13:477-478.

Mulder NJ, Apweiler R, Attwood TK, Bairoch A, Barrell D, Bateman A, Binns D, Biswas M, Bradley P, Bork P, Bucher P, Copley RR, Courcelle E, Das U, Durbin R, Falquet L, Fleischmann W, Griffiths-Jones S, Haft D, Harte N, Hulo N, Kahn D, Kanapin A, Krestyaninova M, Lopez R, Letunic I, Lonsdale D, Silventoinen V, Orchard SE, Pagni M, Peyruc D, Ponting CP, Selengut JD, Servant F, Sigrist CJ, Vaughan R and Zdobnov EM (2003). The InterPro Database, 2003 brings increased coverage and new features. Nucleic Acids Res 31:315-318.

Nadershahi A, Fahrenkrug SC and Ellis LBM (2004). Comparison of computational methods for identifying translation initiation sites in EST data. BMC Bioinformatics 5:14.

Neumann MJ and Dobinson KF (2003). Sequence tag analysis of gene expression during pathogenic growth and microsclerotia development in the vascular wilt pathogen. Fungal Genet Biol 38: 54-62.

Nielsen H, Engelbrecht J, Brunak S and von Heijne G (1997). Identification of prokaryotic and eukaryotic signal peptides and prediction of their cleavage sites. Protein Eng 10:1-6.

Nishikawa T, Ota T and Isogai T (2000). Prediction whether a human cDNA sequence contains initiation codon by combining statistical information and similarity with protein sequences. Bioinformatics 16:960-967.

Nugent KG, Choffe K and Saville BJ (2004). Gene expression during Ustilago maydis diploid filamentous growth: EST library creation and analyses. Fungal Genet Biol 41:349-360.

Ozawa Y, Hanaoka S, Saito R, Washio T, Nakano S, Shinagawa A, Itoh M, Shibata K, Carninci P, Konno H, Kawai J, Hayashizaki Y and Tomita M (2002). Comprehensive sequence analysis of translation termination sites in various eukaryotes. Gene 300:79-87.

Paquola AC, Nishyiama MY Jr, Reis EM, da Silva AM and Verjovski-Almeida S (2003). ESTWeb: bioinformatics services for EST sequencing projects. Bioinformatics 19:1587-1588.

Parkinson J, Guiliano DB and Blaxter M (2002). Making sense of EST sequences by CLOBBing them. BMC Bioinformatics 3:31.

Pertea G, Huang X, Liang F, Antonescu V, Sultana R, Karamycheva S, Lee Y, White J, Cheung F, Parvizi B, Tsai J and Quackenbush J (2003). TIGR Gene Indices clustering tools (TGICL): a software system for fast clustering of large EST datasets. Bioinformatics. 19:651-652.

Putney SD, Herlihy WC and Schimmel P (1983). A new troponin T and cDNA clones for 13 different muscle proteins, found by shotgun sequencing. Nature 302:718-721.

Quackenbush J, Liang F, Holt I, Pertea G and Upton J (2000). The TIGR gene indices: reconstruction and representation of expressed gene sequences. Nucleic Acids Res 28:141-145.

Quackenbush J, Cho J, Lee D, Liang F, Holt I, Karamycheva S, Parvizi B, Pertea G, Sultana R and White J (2001). The TIGR Gene Indices: analysis of gene transcript sequences in highly sampled eukaryotic species. Nucleic Acids Res 29:159-164.

Qutob D, Hraber PT, Sobral BW and Gijzen M (2000). Comparative analysis of expressed sequences in Phytophthora sojae. Plant Physiol 123:243-254.

Ray A, Macwana S, Ayoubi P, Hall LT, Prade R and Mort AJ (2004). Negative Subtraction Hybridization: An efficient method to isolate large numbers of condition-specific cDNAs. BMC Genomics 5:22.

Rozen S and Skaletsky H (2000). Primer3 on the WWW for general users and for biologist programmers. Methods Mol Biol 132:365-386.

Sacadura NT and Saville BJ (2003). Gene expression and EST analyses of Ustilago maydis germinating teliospores. Fungal Genet Biol 40: 47-64.

Salamov A, Nishikawa T and Swindells MB (1998). Assessing protein coding region integrity in cDNA sequencing projects. Bioinformatics 14:384-390.

Sanger F, Nicklen S and Coulson AR (1977). DNA sequencing with chain-terminating inhibitors. Proc Natl Acd Sci USA 74:5463-5467.

Scheetz TE, Trivedi N, Roberts CA, Kucaba T, Berger B, Robinson NL, Birkett CL, Gavin AJ, O'Leary B, Braun TA, Bonaldo MF, Robinson JP, Sheffield VC, Soares MB and Casavant TL (2003). ESTprep: preprocessing cDNA sequence reads. Bioinformatics 19:1318-1324.

Schuler GD (1997). Pieces of the puzzle: expressed sequence tags and the catalog of human genes. J Mol Med 75:694-698.

Schuler GD, Boguski MS, Stewart EA, Stein LD, Gyapay G, Rice K, White RE, Rodriguez-Tome P, Aggarwal A, Bajorek E, Bentolila S, Birren BB, Butler A, Castle AB, Chiannilkulchai N, Chu A, Clee C, Cowles S, Day PJ, Dibling T, Drouot N, Dunham I, Duprat S, East C, Edwards C, Fan JB, Fang N, Fizames C, Garrett C, Green L, Hadley D, Harris M, Harrison P, Brady S, Hicks A, Holloway E, Hui L, Hussain S, Louis-Dit-Sully C, Ma J, MacGilvery A, Mader C, Maratukulam A, Matise TC, McKusick KB, Morissette J, Mungall A, Muselet D, Nusbaum HC, Page DC, Peck A, Perkins S, Piercy M, Qin F, Quackenbush J, Ranby S, Reif T, Rozen S, Sanders C, She X, Silva J, Slonim DK, Soderlund C, Sun WL, Tabar P, Thangarajah T, Vega-Czarny N, Vollrath D, Voyticky S, Wilmer T, Wu X, Adams MD, Auffray C, Walter NAR, Brandon R, Dehejia A, Goodfellow PN, Houlgatte R, Hudson Jr JR, Ide SE, Iorio KR, Lee WY, Seki N, Nagase T, Ishikawa D, Nomura N, Phillips C, Polymeropoulos MH, Sandusky M, Schmitt K, Berry R, Swanson K, Torres R, Venter JC, Sikela JM, Beckmann JS, Weissenbach J, Myers RM, Cox DR, James MR, Bentley D, Deloukas P, Lander ES and Hudson TJ (1996). A gene map of the human genome. Science 274:540-546.

Schultz J, Doerks T, Ponting CP, Copley RR and Bork P (2000). More than 1,000 putative new human signaling proteins revealed by EST data mining. Nat Genet 25:201-204.

Sims AH, Robson GD, Hoyle DC, Oliver SG, Turner G, Prade RA, Russell HH, Dunn-Coleman NS and Gent ME (2004). Use of expressed sequence tag analysis and cDNA microarrays of the filamentous fungus Aspergillus nidulans. Fungal Genet Biol. 41: 199-212.

Skrabanek L and Campagne F (2001). TissueInfo: high-throughput identification of tissue expression profiles and specificity. Nucleic Acids Res 29:102-102.

Soanes DM, Skinner W, Keon J, Hargreaves J and Talbot NJ (2002). Genomics of phytopathogenic fungi and the development of bioinformatic resources. Mol Plant Microbe Interact 15:421-427.

Soares MB, Bonaldo MDF, Jelene P, L. Su L, Lawton L and Efstratiadis A (1994). Construction and characterization of a normalized cDNA library. Proc Natl Acad Sci USA 91:9228-9232.

Sorek R and Safer HM (2003). A novel algorithm for computational identification of contaminated EST libraries. Nucleic Acids Res 31:1067-1074.

Sorek R, Shamir R and Ast G (2004). How prevalent is functional alternative splicing in the human genome? Trends Genet 20:68-71.

Stein LD, Mungall C, Shu S, Caudy M, Mangone M, Day A, Nickerson E, Stajich JE, Harris TW, Arva A, Lewis S (2002). The generic genome browser: a building block for a model organism system database. Genome Res 12:1599-1610.

Sutton GS, White O, Adams MD and Kerlavage AR (1995). TIGR assembler: a new tool for assembling large shotgun sequencing projects. Genome Sci Technol 1: 9-19.

Svensson BA, Kreeft AJ, van Ommen GJ, den Dunnen JT and Boer JM (2003). GeneHopper: a web-based search engine to link gene-expression platforms through GenBank accession numbers. Genome Biol 4:R35.

Tatusov RL, Fedorova ND, Jackson JD, Jacobs AR, Kiryutin B, Koonin EV, Krylov DM, Mazumder R, Mekhedov SL, Nikolskaya AN, Rao BS, Smirnov S, Sverdlov AV, Vasudevan S, Wolf YI, Yin JJ and Natale DA (2003). The COG database: an updated version includes eukaryotes. BMC Bioinformatics 4:41.

Thiel T, Michalek W, Varshney RK and Graner A (2003). Exploiting EST databases for the development and characterization of gene-derived SSR-markers in barley (Hordeum vulgare L.). Theor Appl Genet 106:411-422.

Torto TA, Li S, Styer A, Huitema E, Testa A, Gow NA, van West P and Kamoun S (2003). EST mining and functional expression assays identify extracellular effector proteins from the plant pathogen Phytophthora. Genome Res 13:1675-1685.

Trail F, Xu JR, Miguel PS, Halgren RG and Kistler HC (2003). Analysis of expressed sequence tags from Gibberella zeae (anamorph Fusarium graminearum). Fungal Genet Biol 38: 187-197.

Usuka J. Zhu W and Brendel V (2000). Optimal spliced alignment of homologous cDNA to a genomic DNA template. Bioinformatics 16:203-211.

van Zijll de Jong E, Guthridge KM, Spangenberg GC and Forster JW (2003). Development and characterization of EST-derived simple sequence repeat (SSR) markers for pasture grass endophytes. Genome 46:277-290.

Voiblet C, Duplessis S, Encelot N and Martin F. (2001). Identification of symbiosis-regulated genes in Eucalyptus globulus-Pisolithus tinctorius ectomycorrhiza by differential hybridization of arrayed cDNAs. Plant J 25:181-191.

Wang J and Liang P (2003). DigiNorthern, digital expression analysis of query genes based on ESTs. Bioinformatics 19:653-654.

Wheelan SJ, Church DM and Ostell JM (2001). Spidey: a tool for mRNA-to-genomic alignments. Genome Res 11:1952-1957.

Wheeler DL, Chappey C, Lash AE, Leipe DD, Madden TL, Schuler GD, Tatusova TA and Rapp BA (2000). Database resources of the National Center for Biotechnology Information. Nucleic Acids Res 28:10-14.

Wheeler DL, Church DM, Lash AE, Leipe DD, Madden TL, Pontius JU, Schuler GD, Schriml LM, Tatusova TA, Wagner L and Rapp BA (2001). Database resources of the National Center for Biotechnology Information. Nucleic Acids Res 29:11-16.

Wheeler DL, Church DM, Lash AE, Leipe DD, Madden TL, Pontius JU, Schuler GD, Schriml LM, Tatusova TA, Wagner L and Rapp BA (2002). Database resources of the National Center for Biotechnology Information: 2002 update. Nucleic Acids Res 30:13-16.

Wheeler DL, Church DM, Federhen S, Lash AE, Madden TL, Pontius JU, Schuler GD, Schriml LM, Sequeira E, Tatusova TA and Wagner L (2003). Database resources of the National Center for Biotechnology. Nucleic Acids Res 31:28-33.

Wheeler DL, Church DM, Edgar R, Federhen S, Helmberg W, Madden TL, Pontius JU, Schuler GD, Schriml LM, Sequeira E, Suzek TO, Tatusova TA, and Wagner L (2004). Database resources of the

National Center for Biotechnology Information: update. Nucleic Acids Res 32 Database issue:D35-40.

White O, Dunning T, Sutton G, Adams M, Venter JC and Fields C (1993). A quality control algorithm for DNA sequencing projects. Nucleic Acids Res 21:3829-3838.

Wittenberger T, Schaller HC and Hellebrand S (2001). An expressed sequence tag (EST) data mining strategy succeeding in the discovery of new G-protein coupled receptors. J Mol Biol 307:799-813.

Wolfsberg TG and Landsman D (1997). A comparison of expressed sequence tags (ESTs) to human genomic sequences. Nucleic Acids Res 25:1626-1632.

Yuan J, Liu Y, Wang Y, Xie G and Blevins R (2001). Genome analysis with gene-indexing databases. Pharmacol Ther 91:115-132.

Zhu H, Nowrousian M, Kupfer D, Colot HV, Berrocal-Tito G, Lai H, Bell-Pedersen D, Roe BA, Loros JJ and Dunlap JC (2001). Analysis of expressed sequence tags from two starvation, time-of-day-specific libraries of Neurospora crassa reveals novel clock-controlled genes. Genetics 157:1057-1065.

Zdobnov E and Apweiler R (2001). InterProScan – an integration platform for the signature-recognition methods in InterPro. Bioinformatics 17:847-848.

**Applied Mycology
and Biotechnology**
An International Series
Volume 5. Genes and Genomics

3

ELSEVIER

Fungal Intervening Sequences

Stefanie Pöggeler
Lehrstuhl für Allgemeine und Molekulare Botanik, Ruhr-Universität Bochum,
Universitätsstr. 150, D-44780 Bochum, Germany (stefanie.poeggeler@rub.de).

Since the late 1970s it had become clear that the coding sequence of many eukaryotic genes
is disrupted by genetic elements, intervening sequences, which must be removed prior to
host gene function. Intervening sequences can be classified into introns and inteins. Introns
are excised from the primary RNA transcript by a process termed splicing, whilst inteins are
transcribed and translated together with their host protein and are removed from the
unprocessed protein. Based on sequence homology, secondary structure and the splicing
mechanism, introns can be classified into spliceosomal mRNA introns, group I introns and
group II introns. Recent annotations of several fungal genomes revealed that introns and
inteins are integral elements of fungal genes. These intervening sequences perform various
important functions. They are carriers of transcription regulatory elements, contain signals
for mRNA stability and export from the nucleus and participate in gene evolution.

1. INTRODUCTION

Sequencing of fungal genomes revealed that most fungal transcripts contain
introns which must be accurately spliced to yield functional transcripts (Goffeau et
al. 1996; Galagan et al. 2003; Borkovich et al. 2004). The main type of introns,
considered as 'spliceosomal', is found in nuclear protein coding genes transcribed by
RNA polymerase II. Introns are also present in nuclear tRNA transcripts. Two other
types of introns, group I and group II introns, are found in nuclear rRNA genes and
in mitochondrial genes. Introns within nuclear pre-mRNAs are removed by two
sequential transesterifications (reviewed in: Sharp 1994). In a series of reactions, the
first step involves cleavage of the 5' splice site and covalent ligation of the first intron
nucleotide (+1G). to the branch point adenosine located within the intron. The lariat-
exon intermediate formed in this reaction is subsequently cleaved behind the last
nucleotide of the intron (-1G). the two exons are ligated together and the intron lariat
is released and degraded (Fig. 1a). Intron-splicing of nuclear mRNA introns occurs
in a spliceosomal complex that contains multiple small nuclear RNAs (snRNAs).
their associated small nuclear ribonucleo-proteins (snRNP) and other trans-acting
protein factors (Nilsen 2003). A comparison between fission yeast, budding yeast
and mammals indicated that many spliceosomal factors present in all three
organisms have been well conserved throughout evolution (Käufer and Potashkin 2000).

The splicing mechanism of introns in nuclear tRNAs is quite different from splicing of spliceosomal introns, because it is catalysed by three enzymes, all with an intrinsic requirement for ATP hydrolysis (reviewed in: Abelson et al. 1998).

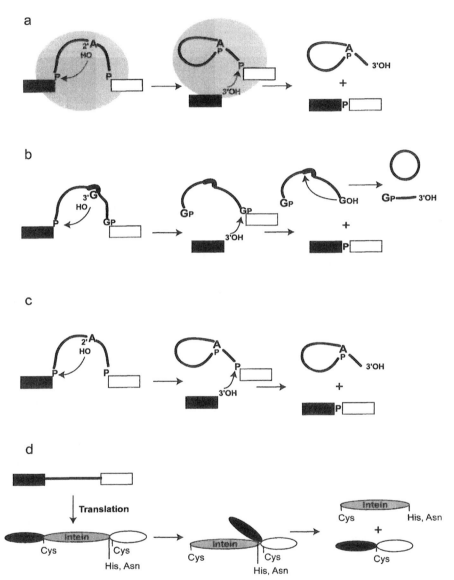

Fig. 1. Comparison of nuclear pre-mRNA splicing (a)., group I-intron splicing (b)., group II-intron splicing (c). and protein splicing (d).. The stippled circle represents the multicomponent-spliceosomal complex that promotes splicing of nuclear spliceosomal introns. Exons are indicated as black and white bars, respectivel, and exteins as black and white ellipsoids, respectively. Introns are indicated as line. A bold line denotes a guanosine-binding site of the group I intron.

Group I introns and group II introns are both capable of self-splicing *in vitro*, and sometimes encode proteins that aid in splicing of the introns *in vivo* (Saldanha et al. 1993). Group I introns exhibit variable nucleotide sequences but have a conserved three-dimensional (3D) core. Several studies identified nine conserved elements of secondary structure designated P1 through P9 (Waring et al. 1985; Michel and Westhof 1990; Golden et al. 1998). Splicing of group I introns occurs by transesterification with an external guanosine as initiating nucleophile. The process is catalysed by RNA structures within the intron, which contain a guanosine-binding site (Cech 1990). The excised linear group I intron is subsequently circularised (Fig. 1b).

Group II introns have a distinctive 3D architecture with six domains, termed domains dI –dVI. Similar to splicing of nuclear pre-mRNA introns, group II intron splicing proceeds by two transesterification steps (Fig. 1c). The 2′hydroxyl of a bulged adenosine in dVI attacks the 5′ splice site, followed by a nucleophilic attack on the 3′splice site by the 3′OH of the upstream exon. The excised intron is a lariat with a 2′-5′ linkage and a 6-7 nt tail (Bonen and Vogel 2001).

In contrast to introns, inteins are internal proteins transcribed and translated together with their host protein. They are present in a wide range of organisms, including fungi (Kane et al. 1990). Only at the protein level the intein is excised precisely, and its flanking sequences, termed N- and C-exteins, joined with a peptide bond to produce the mature spliced protein (Perler et al. 1994). Protein splicing is a posttranslational processing event that releases an internal protein sequence from a protein precursor. The mechanism of protein splicing typically has four steps (Fig. 1d) including two acyl rearrangements of a peptide bond adjacent to cysteine or serine at the two splice site junctions and a cyclization of an asparagine coupled to peptide cleavage (Chong et al. 1996; Paulus 2000; Pietrokovski 2001; Gogarten et al. 2002).

This review focuses on fungal intervening sequences and will include the characteristics, as well as the distribution of fungal spliceosomal introns and mitochondrial group I and group II introns, nuclear group I introns and inteins. It summarizes the recent progress in understanding how introns can influence gene expression and how introns participate in genome evolution.

2. DISTRIBUTION OF INTERVENING SEQUENCES IN FUNGAL GENOMES

Much about fungal genome organization has been revealed since the completion of genome sequence of the yeast *Saccharomyces cerevisiae* (Goffeau et al. 1996). Over the last eight years, several more fungal genomes have been sequenced, amongst them the genomes of the filamentous ascomycete *Neurospora crassa*, the basidiomycetes *Ustilago maydis*, *Coprinus cinereus* and *Phanerochaete chrysosporium* as well as other ascomycetes and basidiomycetes (Galagan et al. 2003; Borkovich et al. 2004; Martinez et al. 2004). Through a combination of bioinformatic and molecular analysis it became clear that many fungal genes are interrupted by intervening sequences.

2.1 Nuclear Spliceosomal Introns

Unlike most eukaryotic genomes, the genome of the yeast *S. cerevisiae* and closely related species of the same genus have only few introns. In *S. cerevisiae*, about 286 of the 6000 genes contain introns (Table 1; http://hgwdev.cse.ucsc.edu/cgi-bin/

yirIntrondb). which are small on average and usually limited to one intron per gene (Spingola et al. 1999). Only four genes contain two introns. Introns in *S. cerevisiae* are primarily located near the 5' end of the gene. The existence of most spliceosomal introns in the genome of *S. cerevisiae* remains unsupported by molecular evidence. Davis et al. (2000) tested, by means of RT-PCR, the predictions for more than one third (87 introns). of all known or suspected introns in the yeast genome. In their study they were able to support only 61 of these predictions, yielding a successful prediction rate of 80 %. Recently it was shown that the comparison of whole genome sequences of several *Saccharomyces* species greatly improves the annotation of genome sequences, since such analyses led to the recognition of ~ 60 new introns (Cliften et al. 2003; Kellis et al. 2003).

Within the complete *S. cerevisiae* genome sequence, 61 nuclear tRNA genes are interrupted by introns. The introns are 14-60 nucleotides in length and interrupt the anticodon loop immediately 3' to the anticodon (Abelson et al. 1998; Lopez and Seraphin 2000).

Similar to *S. cerevisiae*, the genome of the filamentous ascomycete *Ashbya gossypii* is very compact with a size of 9.2 megabases encoding 4718 proteins. The presence of only 221 introns in the entire *A. gossypii* genome, many at identical positions in *S. cerevisiae* homologues, contributes to the compact nature of the *A. gossypii* genome (Table 1, Dietrich et al. 2004). In the genome of the yeast *Schizosaccharomyces pombe* introns are much more frequent with a total of 4730 introns. These are distributed among 43 % of the genes. Introns of *S. pombe* varied from 29 to 819 nucleotides with a mean length of 81 nucleotides (Table 1). Most of the intron-containing *S. pombe* genes have one to six introns, the largest number of introns found within a single *S. pombe* gene is 15. The position of the introns within the genes has a bias for the 5'end (Wood et al. 2002).

Exon | 5' Splice site
S. cerevisiae	G	G U A U G U
N. crassa	G	G U A A G U

Branch point
S. cerevisiae	U A C U A A C A
N. crassa	W R C U R A C M

3' Splice site | Exon
S. cerevisiae	U A Y A G	
N. crassa	U A Y A G	

Fig. 2. Conserved intron signals in *S. cerevisiae* and *N. crassa*. Abbreviations: W (U/T)., R (A/G)., M (A/C)., Y(U/C).

The whole-genome sequences of other ascomycetes revealed that the gene complement of *Neurospora crassa*, *Magnaporthe grisea*, *Fusarium graminearum* and *Aspergillus nidulans* displays a greater structural complexity than that of *S. pombe*.

Genes of these filamentous ascomycetes posses a predicted 17,118 to 25,895 introns (1.7 to 2.69 introns per gene) compared to 4730 introns (0.98 intron per gene) in *S. pombe* (Table 1). However, as with the yeasts *S. cerevisiae* and *S. pombe*, the introns of filamentous ascomycetes do appear to be biased toward the 5' regions of genes (http://www.broad.unit.edu/annotation/fungi).

Conserved sequences are present at the 5'and 3' ends of fungal spliceosomal introns, at their branch point and, to a lesser extent, in flanking exon sequences (Fig. 2). These have been determined for *S. cerevisiae* and *N. crassa* as well as other mycelial fungi (Bruchez et al. 1993; Edelman and Staben 1994; Pöggeler 1997; Jekosch and Kück 1999; Lopez and Seraphin 1999). Intron consensus sequences of *N. crassa* are more variable than those of *S. cerevisiae*. Consensus sequences of *N. crassa* and the closely related ascomycete *Sordaria macrospora* are almost identical (Nowrousian et al. 2004).

Table 1. Statistics of fungal introns

Species	Intron number/ genome	Genome coverage of introns (%).	Length min/max (bp).	Length median/ mean (bp).	Introns per gene (mean).
Saccharomyces cerevisiae[1]	286	0.5	1/1002	135/219	0.05
Schizosaccharomyces pombe[2]	4730	3.0	29/819	48/81	0.98
Ashbya gossypii[3]	221	n.d.	n.d.	n.d.	0.04
Neurospora crassa[1]	17118	6.1	20/2074	84/135	1.70
Magnaporthe grisea[1]	20965	7.7	50/7456	94/143	1.89
Fusarium graminearum[1]	25895	6.7	29/1271	56/93	2.22
Aspergillus nidulans[1]	25443	8.6	13/1331	63/101	2.69
Aspergillus fumigatus[4]	n.d.	n.d.	n.d.	n.d./78	3.0
Ustilago maydis[1]	4900	3.2	19/1122	95/127	0.75
Phanerochaete chrysosporium[5]	n.d.	n.d.	n.d.	54/117	n.d.
Podospora anserina[*6]	n.d	n.d.	n.d.	n.d./73	1.5
Dictyostelium discoideum[*7]	n.d.	n.d.	n.d.	n.d./177	1.28

Data were taken from: [1] http://www.broad.unit.edu/annotation/fungi; [2](Wood et al. 2002).; [3](Dietrich et al. 2004).; [4](Pain et al. 2004).; [5](Martinez et al. 2004). ; [6] (Silar et al. 2003).; [7](Glöckner et al. 2002).; n.d., not determined; *, partial sequence

2.2 Mitochondrial Group I and Group II Introns

Mitochondrial introns are classified in group I and group II introns according to their splicing pathways, their secondary structure, and the presence of short conserved sequences (Lambowitz and Belfort 1993). In fungi, group I and group II introns are inserted in many different mitochondrial genes with a strong preference for protein-coding genes, most frequently within the *coxI* and *cob* genes (Paquin et al. 1997).

An additional benefit of the whole genome sequencing projects of different fungal species was the completion of the sequences for mitochondrial genomes. The size range for fungal mitochondrial DNAs (mtDNAs). is between 19 and 176 kb (Kennell et al. 2004).

The mitochondrial genome of the *S. cerevisiae* strain used for nuclear genome sequencing assembles into a circular map of 85,779 bp. The yeast mitochondrial genome contains genes for cytochrome c oxidase subunits I, II and III (*cox1*, *cox2* and *cox3*)., ATP synthase subunits 6, 8, and 9 (*atp6*, *atp8*, and *atp9*)., apochytochrome b (*cob*)., a ribosomal protein (*var1*)., genes for 21S and 15S (*rnl* and *rns*). ribosomal RNAs (rRNAs)., 24 tRNAs, the 9S RNA component of RNase P and several short ORFs encoding hypothetical proteins of unknown function (Foury et al. 1998). Introns are present in the *cox1* gene (7 introns)., the *cob* gene (5 introns). and the *rnl* gene (1 intron)., which belong to both, group I and group II introns. Introns of both groups can be present within a single gene (Table 2). Most of the introns contain ORFs, some of which are translated independently or in frame, that means cotranslational, with their upstream exons to produce maturases involved in RNA splicing, or endonuclease or reverse transcriptases (RT) involved in intron-mobility (Lambowitz and Belfort 1993; Belfort and Perlman 1995).

The complete mitochondrial sequences from the two budding yeasts *Saccharomyces castellii* and *Saccharomyces servazzii*, consisting of 25,753 and 30,782 bp, respectively, revealed that mtDNAs of these two yeasts are much more compact than the one of *S. cerevisiae*.

Table 2. Localization and characteristics of mitochondrial introns from *S. cerevisiae* and *N. crassa*

Gene / Intron	Intron group	ORF/Function/Mobility
Saccharomyces cerevisiae[1]		
cox1 / ai1	II	Scai1/ maturase and reverse transcriptase /Yes
cox1 / ai2	II	Scai2/ maturase and reverse transcriptase /Yes
cox1 / ai3	I	I-SceIII/ endonuclease/ Yes
cox1 / ai4	I	I-SceII/ endonuclease/ Yes
cox1 / ai5α	I	Scai5α/ endonuclease/ Yes
cox1 / ai5β	I	Scai5β/ unknown/ No
cox1 / ai5γ	II	-
cob / bi1	II	-
cob / bi2	I	Scbi2/ maturase/ Yes
cob / bi3	I	Scbi3/ maturase/ No
cob / bi4	I	Scbi4/ maturase/ No
cob / bi5	I	-
rnl / ω	I	I-SceI/endonuclease/Yes
Neurospora crassa[2]		
atp6	I	atp6 intron protein/ maturase-endonuclease/?
nad1	I	nad1 intron protein/ endo-exonuclease/?
nad3	I	nad3 intron protein/ unknown/?
nad4	I	nad4 intron protein/ endonuclease/?
nad4L	I	nad4L intron protein/ maturase-endonuclease/?
nad5/i1	I	nad5 intron 1 protein/ maturase-endonuclease/?
nad5/i2	I	nad5 intron 2 protein/ maturase-endonuclease/?
cob	I	cob intron 1 protein/ endonuclease/?
cob	I	cob intron 2 protein/ Maturase/?
rnl	I	VAR1/ ribosomal 5S protein/ No

[1] Information concerning classification of the introns and their coding capacity and mobility was taken from (Foury et al. 1998).[2]; information concerning classification of the introns and their coding capacity is taken from http://mips.gsf.de/proj/neurospora/; ?, not determined

Although having almost the same coding potential than that of *S. cerevisiae*, they contain fewer introns and intergenic sequences. A few genes are interrupted by group I introns, but group II introns, some of the introns otherwise found in *S. cerevisiae* mtDNA, are not present (Langkjaer et al. 2003).

In the dimorphic yeast *Yarrowia lipolytica*, the *cox1* and *cob* gene contain at least five introns each. Unlike the *S. cerevisiae* cox3 gene, the *Y. lipolytica* homologue carries an intron. In addition, introns were found in two genes encoding NADH dehydrogenase subunits, *nad1* and *nad5* (Casaregola et al. 2000). The compact mitochondrial genome of the budding yeast *Pichia canadensis* (*Hansenula wingei*). contains only two introns, both of which encode a polypeptide. One intron lies within the *rnl* gene while the other is situated in the *cox1* gene (Sekito et al. 1995). Similarly, in the compact 19,431 bp mtDNA of the yeast *Schizosaccharomyces pombe* group I and group II introns are only present in the *cox1* gene and the *cob* gene (Accession number X54421). The mtDNA of two additional fission yeasts, *Schizosaccharomyces octosporus* and *Schizosaccharomyces japonicus var. japonicus*, have also been sequenced. Whereas the mtDNA of *S. pombe* compact, the mtDNA of *S. octosporus* is 44 227 bp, and that of *S.japonicus var. japonicus* is over 80 kb. The size variation of these mtDNAs is due largely to non-coding regions and not due to the presence of introns (Bullerwell et al. 2003).

The mitochondrial genome of *N. crassa* consists of 64,8840 bp, and contains genes for two rRNAs, at least 27 tRNAs, and 26 ORFs. A total of 10 introns is present in the mitochondrial DNA of *N. crassa*. All of them contain the conserved sequences and potential secondary structures characteristic of group I introns, although group II introns have been found in some natural isolates (Kennell et al. 2004). All of the *N. crassa* mitochondrial introns contain open reading frames (ORFs). either encoding maturases involved in RNA splicing or endonucleases involved in gene conversion. The intron of the *N. crassa rnl* gene codes for ribosomal protein S5, a homologue of the *S. cerevisiae* Var1p protein (Bullerwell et al. 2000; Borkovich et al. 2004). In contrast to *N. crassa*, the complete mtDNA sequence of the closely related ascomycete *Podospora anserina* contains 31 group I introns and two group II introns, many of them encoding proteins with maturase, endonuclease or reverse transcriptase functions (Cummings et al. 1990). In addition to ascomycetes, the whole mitochondrial genome of the chytridiomycete fungus *Allomyces macrogynus* has been determined. In the 57,473 bp mtDNA a total of 28 introns of both groups are present, some of which contain ORFs coding for potential endonucleases or reverse transcriptases (Paquin and Lang 1996). Sequencing of the mitochondrial *rnl* gene of the basidiomycete *Agrocybe aegerita* revealed that this gene is the longest LSU rRNA gene described so far due to a high number of introns. The 13,526 bp *A. aegerita rnl* gene harbours seven group I introns and one group II introns (Gonzalez et al. 1999).

2.2 Nuclear Group I Introns

More than 1000 group I introns have been identified in fungal rDNA. The majority of them (~896) were found in the nuclear rDNA (http://www.rna. icmb. utexas.edu). and many of them are able to self-splice and to produce RNA circles *in vitro* (Haugen et al. 2004).

Several of the rDNA introns are optional among strains of a particular species, and some have been shown to be true mobile elements, since they encode a homing endonuclease (Belfort and Roberts 1997). In contrast to spliceosomal rDNA introns, group I introns are located at a limited number of insertion sites in the highly conserved regions of the small and large subunit rRNA genes, e.g. at position 788 in the small subunit rRNA (Haugen et al. 2004). It was suggested that group I intron fixation may be even more constrained by the exon context than spliceosomal introns, perhaps because group I introns are more dependent on specific exon sequences to facilitate proper folding and excision (Bhattacharya et al. 2003). There are numerous examples of multiple group I introns in a single rRNA gene. As many as eight introns have been identified in the small subunit rRNA gene of the lichen ascomycete *Lecanora dispersa* (Gargas et al. 1995). and in the large subunit rRNA gene of the myxomycete *Fulgio septica*. Phylogenetic analysis indicated that most *Fuligo* introns were only distantly related to each other and were independently gained in ribosomal DNA during evolution (Lundblad et al. 2004).

2.3 Nuclear Intein

Inteins are self-catalytic protein splicing elements that disrupt a host gene and its protein product (Liu 2000). Protein splicing elements were first described in fungi. In 1990 two groups reported an in-frame insertion in the *Saccharomyces cerevisiae VMA* gene encoding a vacuolar ATPase. The insertion was unrelated to the sequence of homologue ATPases. Surprisingly, denaturing gel electrophoresis demonstrated that the mature VMA protein had an electrophoretic mobility that was similar to homologues lacking the protein insertion (Hirata et al. 1990; Kane et al. 1990). Subsequently, it was demonstrated that the insertion was still present in the mRNA and translated together with the VMA, and that the insertion spliced itself out of the protein (Kane et al. 1990). Meanwhile inteins have been found in all three domains of life: Eukaryotes, Bacteria and Archaea.

Inteins are present in proteins with diverse functions including metabolic enzymes, enzymes involved in DNA repair, DNA and RNA polymerases, spliceosomal factors, proteases, ribonucleotide reductases and the vacuolar-type ATPase (Pietrokovski 2001). Beside catalysing their own splicing, many inteins include a site-specific DNA endonuclease. The intein splicing domain is ~ 140 amino acids long and is sufficient for carrying out protein splicing. The endonuclease domains of inteins are not involved in protein splicing and have structures similar to those of intron-encoded endonucleases (Belfort and Roberts 1997). Inteins containing an endonuclease domain are termed large inteins. Some inteins, known as mini-inteins, consist of only the self-splicing domain.

As shown in Table 3, fungal inteins have been detected in *VMA* homologues and in homologues of the *prp8* gene, encoding a conserved spliceosomal protein (Jamieson et al. 1991). A comprehensive database of intein sequences can be found at InBase (Intein Database http://www.neb.com/inteins/intein_intro.html). (Perler 2002).

Two recent analyses of the VMA inteins from different strains of *S. cerevisiae*, saccharomycete yeasts and other yeast species, demonstrated that horizontal transmission of VMA inteins has been a regular occurrence in their evolutionary history (Koufopanou et al. 2002; Okuda et al. 2003). A short intein without an endonuclease domain was shown to be present in the *prp8* gene of the fungal pathogen *Cryptococcus neoformans* (Butler et al. 2001). Sequence analysis of the whole genome sequence of the mycelial ascomycetes *Aspergillus nidulans* and *Aspergillus fumigatus* revealed that both fungi contain a large intein including an endonuclease domain in a putative *prp8* homologue, whilst other mycelial ascomycetes such as *N. crassa, Podospora anserina, Magnaporthe grisea* and *Fusarium graminearum* harbour intein-less *prp8* genes in their genome (Table. 3; Pöggeler, unpublished).

Table 3. Fungal inteins

Species	Intein	Extein	Intein-Size, Endonuclease
Saccharomyces cerevisiae	Sce VMA	VMA, Vacuolar ATPase	454 aa, +
Saccharomyces castellii	Sca VMA	VMA, Vacuolar ATPase	517 aa, +
Saccharomyces cariacanus[1]	Sca VMA	VMA, Vacuolar ATPase	454 aa, +
Saccharomyces dairenensis	Sda VMA	VMA, Vacuolar ATPase	501 aa, +
Saccharomyces exiguus	Sex VMA	VMA, Vacuolar ATPase	502 aa, +
Saccharomyces unisporus	Sun VMA	VMA, Vacuolar ATPase	414 aa, +
Candida tropicalis	Ctr VMA	VMA, Vacuolar ATPase	471 aa, +
Candida glabrata[2]	Cgl VMA	VMA, Vacuolar ATPase	415 aa, +
Kluyveromyces lactis	Kla VMA	VMA, Vacuolar ATPase	410 aa, +
Kluyveromyces polysporus	KpoVMA	VMA, Vacuolar ATPase	433 aa, +
Torulaspora globosa	Tgl VMA	VMA, Vacuolar ATPase	456 aa, +
Torulaspora pretoriensis	Tpr VMA	VMA, Vacuolar ATPase	455 aa, +
Zygosaccharomyces bailii	Zba VMA	VMA, Vacuolar ATPase	456 aa, +
Zygosaccharomyces bisporus	Zbi VMA	VMA, Vacuolar ATPase	450 aa, +
Zygosaccharomyces rouxii	Zro VMA	VMA, Vacuolar ATPase	450 aa, +
Cryptococcus bacillisporus	Cba PRP8	PRP8, pre-mRNA splicing factor	171 aa, -
Cryptococcus neoformans, Serotype A	Cne PRP8	PRP8, pre-mRNA splicing factor	171 aa, -
Cryptococcus neoformans, Serotype AD	Cne PRP8	PRP8, pre-mRNA splicing factor	171 aa, -
Aspergillus nidulans[3]	Ani PRP8	PRP8, pre-mRNA splicing factor	605 aa, +
Aspergillus fumigatus[4]	Afu PRP8	PRP8, pre-mRNA splicing factor	819 aa, +

Data were taken from http://www.neb.com/inteins/intein_intro.html (Perler 2002)., [1]CAC86344 [2]BAC66648.1,[3]http://www.broad.mit.edu/annotation/fungi/aspergillus/, [4]http://www.tigr.org/tdb/e2k1/afu1/

3. MOBILE INTRONS AND INTEINS

A variety of introns and inteins can invade DNA sequences by virtue of proteins encoded within them. The intron-encoded products, in case of group I introns, are endonucleases or maturases which promote splicing of the corresponding intron and multifunctional proteins with reverse transcriptase (RT). maturase, and endonuclease activities in the case of group II introns (Belfort and Perlman 1995; Pöggeler and Kempken 2004). The mobility of inteins is conferred by endonucleases (Chevalier and Stoddard 2001). The lateral transfer of an intervening sequence (either an intron or intein) into a homologous intron-less/intein-less allele is termed homing (Dujon 1989). and has to be distinguished from the transposition process to non-allelic sites.

3.1 Mobility of Group I Introns

Movement of fungal introns was first described for the mitochondrial group I intron of the large rRNA gene of the budding yeast *Saccharomyces cerevisiae*. In the 1970s, the genetic marker ω was found to transfer to strains lacking the marker when crossed to ω+ strains (Dujon et al. 1976). Subsequent analysis indicated that the homing of the LSU rRNA intron is initiated by an intron-encoded endonuclease, which catalyses a double strand break at the intron insertion site. Repair of this double strand-break by a mechanism of gene conversion, which uses the intron-containing gene as matrix, leads to the invasion of the intron into the intronless allele (Szostak et al. 1983; Jacquier and Dujon 1985; Macreadie et al. 1985; Colleaux et al. 1986). According to the current nomenclature, this intron-encoded homing endonuclease is called I-*Sce* I (Dujon 1989).

Homing appears to be widespread in fungi, since many group I introns of mitochondrial genes encode polypeptides having considerable similarities to homing endonucleases. Four families of homing endonucleases have been identified containing highly conserved amino acid sequences, including the LAGLI-DADG, the GIY-YIG, the His-Cys box, and the H-N-H endonucleases (Chevalier and Stoddard 2001).. Most of the fungal group I introns endonucleases belong to the LAGLI-DADG family and are characterized by two copies (P1 and P2) of the LAGLI-DADG motif (Belfort and Roberts 1997).

The reason for the wide distribution of LAGLI-DADG ORFs appears to be due to their ability to invade unrelated types of intervening sequences. Therefore, it has been proposed that group I introns have become mobile by the acquisition of ORFs encoding highly sequence specific endonucleases and that intron-encoded endonucleases could behave as autonomous mobile elements (Loizos et al. 1994). This hypothesis is supported by the finding that proteins harboring the LAGLI-DADG motif are not exclusively encoded in group I introns but are also located as free-standing genes (Kostriken et al. 1983; Watabe et al. 1983; Sharma et al. 1992). or inteins (Gimble 2000). Direct evidence for the hypothesis of the intrinsic mobility of an intronic LAGLI-DADG ORF was first demonstrated in the mitochondrial genome of *Podospora anserina* when Sellem and Belcour (1997). were able to demonstrate the efficient mobility of the mitochondrial nad1-i4-orf1 between two *Podospora* strains. Mobility of fungal GIY-YIG ORFs, a family of intronic endonucleases, with a conserved GIY-(X_{10-11}).-YIG motif, was also demonstrated (Saguez et al. 2000).

3.2 Mobility of Group II Introns

Mobility of self-splicing group II introns is most extensively analysed in the yeast *Saccharomyces cerevisiae*. Introns ai1 and ai2 of the mitochondrial *coxI* gene (Table 2). are mobile elements that insert site specifically in intronless alleles. The homing mechanism used by group II introns is termed retrohoming and differs from the mechanism employed by group I introns (Moran et al. 1995; Eskes et al. 1997). In contrast to group I introns, mobility of group II introns depends on an intron-encoded multifunctional protein with three activities and requires an efficient splicing of the intron. The splicing relies on the maturase activity of the intron-encoded protein. The protein remains subsequently associated with the excised intron RNA lariat to form a ribonucleoprotein (RNP). particle, that has RT and site-specific endonuclease activities. The RNP recognizes the intronless allele via base pairing between the protein-bound RNA and the target DNA. The intron RNA lariat catalyses its own insertion into the sense strand by reverse splicing. Subsequently, the antisense strand is cleaved by the endonuclease domain of the intron-encoded protein. Finally the protein's reverse transcriptase domain synthesizes the DNA using the invading RNA template. The cleaved strand serves as primer for the first strand cDNA synthesis in a reaction known as target DNA-primed reverse transcription (TPRT).. Insertion is completed by host repair enzymes (Kennell et al. 1993; Zimmerly et al. 1995; Yang et al. 1996; Zimmerly et al. 1999).

In addition to this homing pathway, two additional homing pathways have been described for the ai2 intron of *S. cerevisiae*. These include an RT-independent pathway which leads to the insertion of the intron by double-strand break repair recombination initiated by cleavage of the target DNA by the intron-encoded endonuclease and an insertion mechanism that involves synthesis of a full-length cDNA copy of the inserted intron RNA, with completion by a repair process independent of homologous recombination (Moran et al. 1995; Eskes et al. 2000).

Apart from the well analysed group II introns of *S. cerevisiae*, group II introns encoding multifunctional RTs have also been identified in the mitochondrial genome of the yeasts *S. pombe* and *Kluyveromyces lactis*, various filamentous ascomycetes, and the Chytridiomycete *Allomyces macrogynus* (Paquin et al. 1997). All of them are characterized by the conserved domains RT, X, and most of them contain the Zn domain (Toor et al. 2001). The RT domain consists of seven conserved subdomains typically found in reverse transcriptases of retroviruses (Xiong and Eickbush 1990). Domain X is involved in the maturase function, whilst the Zn domain contributes to the endonuclease activity (Mohr et al. 1993; Shub et al. 1994). A recent phylogenetic analysis of 71 ORFs related to group II intron RTs of fungal, algal, plant and bacterial origin revealed a bacterial origin of mobile group II introns (Toor et al. 2001). An investigation of group II intron RNA structures and the intron-encoded RTs demonstrated coevolution of the RNA structure with their intron-encoded RT. Based on this data a new model for the evolution of group II introns was predicted by Toor et al. (2001). The major structural forms of group II introns were developed through coevolution with the intron-encoded protein rather than as independent catalytic RNAs. This implicates that most ORF-less group II introns are derivatives of ORF-containing introns.

3.3 Mobility of Inteins

Similar to mobile group I introns, mobile inteins appear to be derived from invasive endonuclease genes. Fungal intein endonucleases belong to the LAGLI-DADG family and are characterized by copies of this dodecapeptide motif. Due to their endonuclease activity fungal inteins appear to spread by the same mechanism as was described for introns that harbour a homing endonuclease (Gogarten et al. 2002).

The VMA-intein endonuclease of *S. cerevisiae* is very well characterized. It contains an endonuclease called PI-*SceI*, which initiates the mobility of the intein by cleaving at intein-less alleles of the *VMA1* gene (Gimble and Thorner 1992). Subsequently to purification of PI-*SceI*, it was demonstrated by genetic and biochemical studies that the enzyme makes numerous base-specific and phosphate backbone contacts with its 31 bp asymmetrical recognition site (Gimble and Thorner

1993; Gimble and Wang 1996). Site directed mutagenesis experiments and deletion of the endonuclease domain have shown that the endonuclease activity is not required for protein-splicing function (Gimble and Stephens 1995; Chong and Xu 1997). X-ray crystallography confirmed a two domain structure (self-splicing and endonuclease domain). of PI-Sce. However, despite the apparent structural autonomy of the protein splicing and endonuclease domain, the two domains appear to collaborate by interacting with the homing site DNA (Duan et al. 1997). A high resolution structure of PI-*Sce*I confirmed that the splicing domain is involved in DNA binding and recognition of the specific DNA substrate (Werner et al. 2002).

Recently, the evolutionary state of VMA-derived endonucleases from 12 yeast species was addressed by assaying their endonuclease activities. Only two enzymes have been shown to be active, PI-*Zba*I from *Zygosaccharomyces bailii* and PI-*Sca*I from *Saccharomyces cariocanus*. PI-*Zba*I cleaves the *Z. bailii* recognition sequence significantly faster than the *Saccharomyces cerevisiae* site, which differs at six nucleotide positions. A mutational analysis has indicated that PI-*Zba*I cleaves the *S. cerevisiae* substrate poorly due to the absence of a DNA-protein contact that is established by PI-*Sce*I. These findings demonstrated that intein homing endonucleases evolve altered specificities as they adapt to recognize alternative target sites (Posey et al. 2004).

4. FUNCTION AND APPLICATIONS OF FUNGAL INTERVENING SEQUENCES

The biological function of intervening sequences has been the topic of a large huge of review articles. While some regard intervening sequences as selfish DNA (Cavalier-Smith 1985). others cite evidence for a beneficial role of intervening sequences (Federova and Federov 2003; Le Hir et al. 2003).

4.1 Introns in mRNA Metabolism and Encode Small Non-coding RNAs

In recent years, it has become increasingly clear that introns are not just unintelligible sequences that must be removed upon transcription to create mRNAs with intact open reading frames but instead can influence many stages of mRNA metabolism including initial transcription of the gene, polyadenylation, nuclear export, translation and decay of the mRNA (Le Hir et al. 2003). Several of these functions have also been repoted for introns of the yeast *S. cerevisiae* (Lei and Silver 2002; Preker et al. 2002; Rodriguez-Navarro et al. 2002).

Although only 5 % of the yeast genes contain introns, many of these are highly expressed and account for a disproportionate 27 % of the mRNAs. In particular, ~ 70 % of genes encoding for ribosomal proteins contain an intron and account for 90 % of all mRNA transcripts from intron containing genes (Ares et al. 1999; Lopez and Seraphin 1999). Therefore, intron removal is an important activity in *S. cerevisiae*. It has been speculated that intron-dependent regulatory events can help to control expression of ribosomal proteins and that introns may improve expression of genes that contain them (Ares et al. 1999). Moreover, it has been demonstrated in yeast that, once transcribed, the splicing signals in an intron can further stimulate transcription by enhancing polymerase II initiation and processivity (Furger et al. 2002).

There have been numerous reports of inefficient expression of recombinant genes when their introns are removed, in the basidiomycete *Schizophyllum commune*. The cDNA sequence of the *Agaricus bisporus* hydrophobin gene *ABH1* under the regulation sequences of the *S. commune SC3* gene gave no expression in *S. commune*, whilst the genomic coding sequence produced high levels of *ABH1* mRNA in *S. commune* (Lugones et al. 1999). Similarly, the addition of an intron into the transcriptional unit of the gene encoding the green fluorescent protein (GFP). and into the bacterial hygromycin B gene was demonstrated to positively effect heterologous gene expression in *S. commune* (Lugones et al. 1999; Scholtmeijer et al. 2001).

Splicing of an intron is also involved in the adaptive response, which couples protein folding load in the endoplasmic reticulum (ER). with the ER folding capacity. Like all eukaryotic cells, fungi respond to the accumulation of unfolded proteins in the endoplasmic reticulum (ER). by activating a transcriptional induction program termed the unfolded protein response (UPR). The transcription factor Hac1p, responsible for the UPR in *S. cerevisiae*, is tightly regulated by a post-transcriptional mechanism. HAC1 mRNA must be spliced in response to ER stress to produce Hac1p, which then activates transcription via direct binding to the cis-acting UPR element (UPRE). present in the promoter regions of its target genes (Zhang and Kaufman 2004). In the mycelial fungus *Aspergillus niger*, the ER stress results in the splicing of an unconventional 20-nt intron from the *A. niger hacA* mRNA and is associated with truncation of the 5' end of the *hacA* mRNA by 230 nt (Mulder et al. 2004).

In addition to their function in gene expression, introns have the capability to encode non-coding small nucleolar RNAs. snoRNAs are guiding molecules for precise chemical modifications and nucleolar maturation of rRNA. At least seven snoRNAs are intron-encoded in *S. cerevisiae* (Villa et al. 2000). Most of these intron-encoded snoRNAs are present in genes encoding proteins involved in ribosome assembly. They can either be maturated via a splicing-dependent pathway or a secondary splicing-independent one. In the splicing-dependent pathway, exonucleases digest the flanking sequences of the debranched host intron and produce the correct ends of the snoRNA. The splicing-independent pathway involves endonucleolytic cleavage within the host intron followed by exonucleolytic maturation (Maxwell and Fournier 1995; Weinstein and Steitz 1999; Terns and Terns 2002).

4.2 Introns in Gene Evolution

Since the discovery of discontinuous genes in pro- and eukaryotes, there has been much speculation about the evolutionary origin of introns. Thus far, it has remained an open question whether introns are early or late inventions during evolution. Initially, introns have been proposed to be very ancient genetic elements which existed at the beginning of life before the divergence of eukaryotes and prokaryotes (Doolittle 1978). A contrary view on the origin of introns claimes that introns appeared relatively recently in the genomes of eukaryotes and that mobile genetic elements are the ancestors of introns. It is believed that polypeptide-encoding introns were inserted into intron-less genes at a very late stage during evolution. In this view introns play a minor role in early gene evolution (Orgel and Crick 1980;

Cavalier-Smith 1985). One argument for this theory is that polypeptide-encoding introns are still able to transpose. Intron transposition can either occur via a process of reverse splicing by the intron encoded reverse transcriptase or via excision/insertion at the DNA level by an endonuclease (Sharp 1985; Roger and Doolittle 1993). However, most introns have lost their mobility after insertion. Protein-encoding introns of fungal mitochondrial genomes may therefore be the evolutionary link between mobile transposons and introns.

The alternative intron-early theory argues that already the common ancestor of eukaryotes, eubacteria and archaebacteria had discontinuous genes. During evolution, intron-less genes might then have evolved by a subsequent loss of introns. The discovery of autocatalytic group I and group II introns implies that RNA molecules carried introns in an ancient "RNA-world" (Gilbert 1986; Cech 1989). The exon shuffling theory argues that ancient genes were assembled from mini-exons due to the recombination of intron sequences. This exon-shuffling process gave rise to novel mosaic genes, which contain exons of different unrelated genes. The exon-shuffling process was supposed to increase the speed of genome evolution (Gilbert 1978; Long 2001).

The presence of introns in a gene provides that gene an opportunity to generate alternative coding messages through alternative splicing. Alternative splicing results from the use of alternative 5' splice sites, alternative 3' splice sites, optional exons, mutually exclusive exons and retained introns (Federova and Federov 2003) thus resulting in an expression of different mRNAs from a single gene. Therefore, another role of introns may be to generate increased protein diversity by alternative splicing (Gravely 2001). The synthesis of alternatively spliced mRNAs is a well known process in higher eukaryotes. Between a third and half of all human genes appear to undergo alternative splicing (Croft et al. 2000; Modrek and Lee 2002). With regard to yeasts and filamentous fungi, there have been some reports of alternatively splicing. In the budding yeast *S. cerevisiae* two genes (YKL186C/*MTR2* and YML034W). were identified which encode alternative spliced mRNAs; *MTR2* produces at least five different spliced mRNAs (Davis et al. 2000). In the yeast fission *S. pombe*, the *prp10* gene encoding a conserved spliceosome-associated protein was shown to encode two isoforms derived from alternatively splicing of the transcript (Habara et al. 1998). Alternative splicing has also been observed in basidiomycetes and filamentous ascomycetes (Kempken and Kück 1996; Thornewell et al. 1997; Pöggeler and Kück 2000; Lodato et al. 2003; Yadav et al. 2003).

4.3 Application of Fungal Intervening Sequences in Biotechnology

For genome mapping and gene therapy applications there has been considerable interest in the identification of new enzymes with high specificity. In contrast to class II restriction enzymes that recognize sites of usually ≤6 bp, homing-endonucleases, encoded by group I and group II introns as well as inteins, recognize sequences that span 12-40 bp of DNA. This capability offers the opportunity to use homing-endonucleases as rare-cutting enzymes for analysing and manipulating genomes, for mapping, gene cloning and targeting, as well as for studying double-strand-break repair in diverse biological systems (Belfort and Roberts 1997).

Owing to its large recognition sequence (18 nucleotides). and high specificity, I-*Sce*I has been shown to be useful for recombinant DNA technology (Thierry et al.

1991) or for the physical mapping of genomes, rapid sorting of genomic libraries (Thierry and Dujon 1992; Thierry et al. 1995) and genome organization analysis (Jumas-Bilak et al. 1995; Mahillon et al. 1998). Intron-encoded endonucleases have also been shown to be applicable as a powerful tools to investigate DNA repair and genome stability (Jasin 1996; Taghian and Nickoloff 1997; Anglana and Bacchetti 1999).

Group II intron homing sites span about 30-50 bp, from about position –25 to –20 upstream of the target site to position +10 to +19 downstream (Guo et al. 1997; Yang et al. 1998; Singh and Lambowitz 2001). As recognition occurs through base pairing between the DNA target site and the RNA of the RNP, modification of the RNA sequence directly leads to novel substrate binding. This was first demonstrated for the bacterial *Lactococcus lactis* L1.LtrB intron, which was modified to insert at new sites (Mohr et al. 2000). In addition, retargeted group II introns can be used for highly specific chromosomal gene disruption in *E. coli* and other bacteria at frequencies of 0.1-22%. Furthermore, they can be used to introduce targeted chromosomal breaks, which can be repaired by transformation with a homologous DNA fragment, enabling the introduction of point mutations (Karberg et al. 2001). In the near future, group II intron RNP particles might also be applicable to target eukaryotic chromosomal genes. This would greatly facilitate analysis of eukaryotic organisms that lack efficient homologous recombination systems and could have widespread applications in genetic engineering and gene therapy (Doudna and Cech 2002; Sullenger and Gilboa 2002; Sullenger 2003).

The wide variety of homing endonucleases which have been identified in mitochondrial introns of mycelial fungi and fungal inteins offer a strong basis for engineering novel rare cutting restriction enzymes.

Due to their protein splicing function inteins provide an additional source for innovative biotechnology tools. First used for protein purification, inteins are now used to express cytotoxic proteins, to segmentally modify or label proteins, to cyclize proteins or peptides, or to ligate proteins (Perler and Adam 2000; Xu and Evans 2003).

5. CONCLUSION

Genomic sequencing projects for several fungal species have been completed and many more are under way (http://www.fgsc.net/). Correct identification of all intervening sequences is necessary to discern the coding potential of a fungal genome. So far, the existence of many introns predicted in the genomes of fungal species remains unsupported by molecular evidence. Some of this information can be recovered by directly comparing cDNA libraries to genomic sequences or by comparing patterns of conservation between closely related genomes that may allow exons to be discerned from the more divergent introns (Cliften et al. 2003; Kellis et al. 2003; Nowrousian et al. 2004). A compilation of intron sequences can help not only to analyse the chromosomal organization of fungal genomes, but also to define consensus sequences and nucleotide contents, features that are crucial for systematic gene identification. Recently, the first genome-wide view of splicing was presented for the yeast *Saccharomyces cerevisiae*. The development of an intron-specific microarray has permitted the analysis of the splicing of all intron-containing transcripts and has been shown to be useful to analyse which factors affect pre-mRNA splicing (Clark et al. 2002; Barras and Beggs 2003). Because the mechanisms of splicing are highly conserved throughout eukaryotes, knowledge of fungal splicing could even provide insights into splicing defects that are implicated in human diseases.

Acknowledgments: Research of the author is funded by the Ruhr-University Bochum and the Deutsch Forschungsgemeinschaft. I thank Ulrich Kück, Severine Mayrhofer, Nicole Nolting and Minou Nowrousian, for critical reading of the manuscript.

REFERENCES

Abelson J, Trotta CR and Li H (1998). tRNA splicing. J Biol Chem 273:12685-12688

Anglana M and Bacchetti S (1999). Construction of a recombinant adenovirus for efficient delivery of the I-SceI yeast endonuclease to human cells and its application in the in vivo cleavage of chromosomes to expose new potential telomeres. Nucl Acids Res 27:4276-4281

Ares MJ, Grate L and Pauling MH (1999). A handful of intron-containing genes produces the lion's share of yeast mRNA. Nucl Acids Res 5:1138-1139

Barras JD and Beggs J (2003). Splicing goes global. Trends Genet 19:295-298

Belfort M and Perlman PS (1995). Mechanisms of intron mobility. J Biol Chem 270:30237-30240

Belfort M and Roberts RJ (1997). Homing endonucleases: keeping the house in order. Nucl Acids Res 25:3379-3388

Bhattacharya D, Simon D, Huang J, Cannone JJ and R.R. G (2003). The exon context and distribution of Euascomycetes rRNA spliceosomal introns. BMC Evol Biol 3:7

Bonen L and Vogel J (2001). The ins and outs of group II introns. Trends Genet 17:322-331

Borkovich KA, Alex LA, Yarden O, Freitag M, Turner GE, Read ND, Seiler S, Bell-Pedersen D, Paietta J, Plesofsky N, Plamann M, Goodrich-Tanrikulu M, Schulte U, Mannhaupt G, Nargang FE, Radford A, Selitrennikoff C, Galagan JE, Dunlap JC, Loros JJ, Catcheside D, Inoue H, Aramayo R, Polymenis M, Selker EU, Sachs MS, Marzluf GA, Paulsen I, Davis R, Ebbole DJ, Zelter A, Kalkman ER, O'Rourke R, Bowring F, Yeadon J, Ishii C, Suzuki K, Sakai W and Pratt R (2004). Lessons from the Genome Sequence of *Neurospora crassa*: Tracing the path from genomic blueprint to multicellular organism. Microbiol Mol Biol Rev 68:1-108

Bruchez JJP, Eberle J and Russo VEA (1993). Regulatory sequences in the transcription of *Neurospora crassa* genes: CAAT box, TATA box , introns poly(A). tail formation sequences. Fungal Genet Newsl 40:89-96

Bullerwell CE, Burger G and Lang BF (2000). A novel motif for identifying rps3 homologs in fungal mitochondrial genomes. Trends Biochem Sci 25.:363-365

Bullerwell CE, Leigh J, Forget L and Lang BF (2003). A comparison of three fission yeast mitochondrial genomes. Nucl Acids Res 31:759-768

Butler MI, Goodwin TJ and Poulter RT (2001). A nuclear-encoded intein in the fungal pathogen *Cryptococcus neoformans*. Yeast 18:1365-1370

Casaregola S, Neuveglise C, Lepingle A, Bon E, Feynerol C, Artiguenave F, Wincker P and Gaillardin C (2000). Genomic exploration of the hemiascomycetous yeasts: 17. *Yarrowia lipolytica*. FEBS Lett 487:95-100

Cavalier-Smith T (1985). Selfish DNA and the origin of introns. Nature 315:283-284

Cech TR (1989). RNA chemistry. Ribozyme self-replication? Nature 339:507-508

Cech TR (1990). Self-splicing of group I introns. Ann Rev Biochem 59.:543-568

Chevalier BS and Stoddard BL (2001). Homing endonucleases: structural and functional insight into the catalyst of intron/intein mobility. Nucleic Acids Res 29:3757-3774

Chong S, Shao Y, Paulus H, Benner J, Perler FB and Xu MQ (1996). Protein splicing involving the *Saccharomyces cerevisiae* VMA intein. The steps in the splicing pathway, side reactions leading to protein cleavage, and establishment of an in vitro splicing system. J Biol Chem 271:22159-22168.

Chong S and Xu MQ (1997). Protein splicing of the *Saccharomyces cerevisiae* VMA intein without the endonuclease motifs. J Biol Chem 272:15587-15590

Clark TA, Sugnet CW and Ares MJ (2002). Genomewide analysis of mRNA processing in yeast using splicing-specific microarrays. Science 296:907-910

Cliften P, Sudarsanam P, Desikan A, Fulton L, Fulton B, Majors J, Waterston R, Cohen BA and
Johnston M (2003). Finding functional features in Saccharomyces genomes by phylogenetic
footprinting. Science 301:71-76

Colleaux L, d'Auriol L, Betermier M, Cottarel G, Jacquier A, Galibert F and Dujon B (1986). Universal
code equivalent of a yeast mitochondrial intron reading frame is expressed into E. coli as a specific
double strand endonuclease. Cell 44:521-533

Croft L, Schandorff S, Clark F, Burrage K, Arctander P and Mattick JS (2000). ISIS, the intron
information system, reveals the high frequency of alternative splicing in the human genome. Nat
Genet 4:340-341

Cummings DJ, McNally KL, Domenico JM and Matsuura ET (1990). The complete DNA sequence of
the mitochondrial genome of Podospora anserina. Curr Genet 17:375-402

Davis CA, Grate L, Spingola M and Ares MJ (2000). Test of intron predictions reveals novel splice
sites, alternatively spliced mRNAs and new introns in meiotically regulated genes of yeast. Nucl
Acids Res 28:1700-1706

Dietrich FS, Voegeli S, Brachat S, Lerch A, Gates K, Steiner S, Mohr C, Pöhlmann R, Luedi P, Choi S,
Wing RA, Flavier A, Gaffney TD and Phillipsen P (2004). The Ashbya gossypii genome as tool for
mapping ancient the Saccharomyces cerevisiae genome. Science 304:304-307

Doolittle WF (1978). Genes in pieces: were they ever together? Nature 272:581-582

Doudna JA and Cech TA (2002). The chemical repoertoire of natural ribozymes. Nature 418:222-228

Duan X, Gimble FS and Quiocho FA (1997). Crystal structure of PI-SceI, a homing endonuclease with
protein splicing activity. Cell 89:555-564

Dujon B (1989). Group I introns as mobile genetic elements: facts and mechanistic speculations - a
review. Gene 82:91-114

Dujon B, Bolotin-Fukuhara C, Coen D, Deutsch J, Netter P, Slonimski PP and Weill L (1976).
Mitochondrial genetics. XI mutaions at the mitochondrial locus omega affecting the recombination
of mitochondrial genes in Saccharomyces cerevisiae. Mol Gen Genet 143:131-165

Edelman S and Staben C (1994). A statistical analysis of sequence features within genes from
Neursopora crassa. Exp Mycol 18:70-81

Eskes R, Liu L, Ma H, Chao ML, Dickson L, Lambowitz AM and Perlman PS (2000). Multiple Homing
pathways used by yeast mitochondrial group II introns. Mol Cell Biol 20:8432-8466

Eskes R, Yang J, Lambowitz AM and Perlman PS (1997). Mobility of mitochondrial group II introns:
engineering a new site specificity and retrohoming via full reverse splicing. Cell 88:865-874

Federova L and Federov A (2003). Introns in gene evolution. Genetica 118:123-131

Foury F, Roganti T, Lecrenier N and Purnelle B (1998). The complete sequence of the mitochondrial
genome of Saccharomyces cerevisiae. FEBS Lett 440:325-331

Furger A, O'Sullivan JM, Binnie A, Lee BA and Proudfoot NJ (2002). Promoter proximal splice sites
enhance transcription. Genes Dev 16:2792-2799

Galagan JE, Calvo SE, Borkovich KA, Selker EU, Read ND, Jaffe D, FitzHugh W, Ma LJ, Smirnov S,
Purcell S, Rehman B, Elkins T, Engels R, Wang S, Nielsen CB, Butler J, Endrizzi M, Qui D, Ianakiev
P, Bell-Pedersen D, Nelson MA, Werner-Washburne M, Selitrennikoff CP, Kinsey JA, Braun EL,
Zelter A, Schulte U, Kothe GO, Jedd G, Mewes W, Staben C, Marcotte E, Greenberg D, Roy A,
Foley K, Naylor J, Stange-Thomann N, Barrett R, Gnerre S, Kamal. M., Kamvysselis M, Mauceli E,
Bielke C, Rudd S, Frishman D, Krystofova S, Rasmussen C, Metzenberg RL, Perkins DD, Kroken S,
Cogoni C, Macino G, Catcheside D, Li W, Pratt RJ, Osmani SA, DeSouza CP, Glass L, Orbach MJ,
Berglund JA, Voelker R, Yarden O, Plamann M, Seiler S, Dunlap J, Radford A, Aramayo R, Natvig
DO, Alex LA, Mannhaupt G, Ebbole DJ, Freitag M, Paulsen I, Sachs MS, Lander ES, Nusbaum C
and Birren B (2003). The genome sequence of the filamentous fungus Neurospora crassa. Nature
422:859-868

Gargas A, DePriest PT and Taylor JW (1995). Positions of multiple insertions in SSU rDNA of lichen-
forming fungi. Mol Biol Evol 12:208-218

Gilbert W (1978). Why exons are in pieces? Nature 271:501

Gilbert W (1986). The origin of life. The RNA world. Nature 319:618

Gimble FS (2000). Invasion of a multidude of genetic niches by mobile endonuclease genes. FEMS
Microbiol Lett 185:99-107

Gimble FS and Stephens BW (1995). Substitutions in conserved dodecapeptide motifs that uncouple
the DNA binding and DNA cleavage activities of PI-SceI endonuclease. J Biol Chem 270:5849-5856

Gimble FS and Thorner J (1992). Homing of a DNA endonuclease gene by meiotic gene conversion in *Saccharomyces cerevisiae*. Nature 357:301-306

Gimble FS and Thorner J (1993). Purification and characterization of VDE, a site-specific endonuclease from the yeast *Saccharomyces cerevisiae*. J Biol Chem 268:21844-21853

Gimble FS and Wang J (1996). Substrate recognition and induced DNA distortion by the PI-SceI endonuclease, an enzyme generated by protein splicing. J Mol Biol 263:163-180

Glöckner G, Eichinger L, Szafranski K, Pachebat JA, Bankler AT, Dear PH, Lehmann R, Baumgart C, Parra G, Abril JF, Guló R, Kumpf K, Tunggal B, Cox E, Quail MA, Platzer M, Rosental A and Noegel AA (2002). Sequence analysis of chromosome 2 of *Dictyostelium discoideum*. Nature 418:79-85

Goffeau A, Barrell BG, Bussey H, Davis RW, Dujon B, Feldmann H, Galibert F, Hoheisel JD, Jacq C, Johnston M, Louis EJ, Mewes HW, Murakami Y, Philippsen P, Tettelin H and Oliver SG (1996). Life with 6000 Genes. Science 274:546-567

Gogarten JP, Senejani AG, Zhaxybayeva O, Olendzenski L and Hilario E (2002). Inteins: structure, function, and evolution. Annu Rev Microbiol 56:263-287

Golden BL, Gooding AR, Podell ER and Cech TR (1998). A preorganized active site in the cristal structure of the Tetrahymena ribozyme. Science 282:259-264

Gonzalez P, Barroso G and Labarere J (1999). Molecular gene organisation and secondary structure of the mitochondrial large subunit ribosomal RNA from the cultivated Basidiomycota *Agrocybe aegerita*: a 13 kb gene possessing six unusual nucleotide extensions and eight introns. Nucl Acids Res 27:1754-1761

Gravely BR (2001). Alternative splicing: increasing diversity in the proteomic world. Trends Genet 17:100-107

Guo H, Zimmerly S, Perlman PS and Lambowitz AM (1997). Group II intron endonucleases use both RNA and protein subunits for recognition of specific sequences in double-stranded DNA. EMBO J 16:6835-6848

Habara Y, Urushiyama S, Tani T and Ohshima Y (1998). The fission yeast prp10(+). gene involved in pre-mRNA splicing encodes a homologue of highly conserved splicing factor, SAP155. Nucl Acids Res 26:5662-5669

Haugen P, Runge HJ and Bhattacharya D (2004). Long-term evolution of the S788 fungal nuclear small subunit rRNA group I introns. RNA 10:1084-1096

Hirata R, Ohsumk Y, Nakano A, Kawasaki H, Suzuki K and Anraku Y (1990). Molecular structure of a gene, VMA1, encoding the catalytic subunit of H(+).-translocating adenosine triphosphatase from vacuolar membranes of *Saccharomyces cerevisiae*. J Biol Chem 265:6726-6733

Jacquier A and Dujon B (1985). An intron-encoded protein is active in a gene conversion process that spreads an intron into a mitochondrial gene. Cell 41:383-394

Jamieson DJ, Rahe B, Pringle J and Beggs J (1991). A suppressor of a yeast splicing mutation (prp8-1). encodes a putative ATP-dependent RNA helicase. Nature 349:715-717

Jasin M (1996). Genetic manipulation of genomes with rare-cutting endonucleases. Trends Genet 12:224-228

Jekosch K and Kück U (1999). Codon bias in the ß-lactam producer *Acremonium chrysogenum*. Fungal Genet Newslett 46:11-13

Jumas-Bilak E, Maugard C, Michaux-Charachon S, Allardet-Servent A, Perrin A, O'Callaghan D and Ramuz M (1995). Study of the organization of the genomes of *Escherichia coli*, *Brucella melitensis* and *Agrobacterium tumefaciens* by insertion of a unique restriction site. Microbiology 141:2425-2432

Kane PM, Yamashiro CT, Wolczyk DF, Neff N, Goebl M and Stevens TH (1990). Protein splicing converts the yeast TFP1 gene product to the 69-kD subunit of the vacuolar H(+).-adenosine triphosphatase. Science 250:651-657

Karberg M, Guo H, Zhong J, Coon R, Perutka J and AM. L (2001). Group II introns as controllable gene targeting vectors for genetic manipulation of bacteria. Nat Biotechnol 19:1162-1167

Käufer NF and Potashkin J (2000). Analysis of the splicing machinery in fission yeast: a comparison with budding yeast and mammals. Nucl Acids Res 28:3003-3310

Kellis M, Patterson N, Endrizzi M, Birren B and Lander ES (2003). Sequencing and comparison of yeast species to identify genes and regulatory elements. Nature 423:241-254

Kempken F and Kück U (1996). restless, an active Ac-like transposon from the fungus *Tolypocladium inflatum*: structure, expression, and alternative RNA splicing. Mol Cell Biol 11:6563-6572

Kennell JC, Collins RA, Griffiths AJF and Nargang FE (2004). Mitochondrial genetics of *Neurospora*. In: Kück U (ed). The Mycota, 2 edn. Springer-Verlag, Berlin, Heidelberg, pp 95-112

Kennell JC, Moran JV, Perlman PS, Butow RA and Lambowitz AM (1993). Reverse transcriptase activity associated with maturase-encoding group II introns in yeast mitochondria. Cell 73:133-146

Kostriken R, Strathern JN, Klar AJ, Hicks JB, Heffron F (1983). A site-specific endonuclease essential for mating-type switching in *Saccharomyces cerevisiae*. Cell 35:167-174

Koufopanou V, Goddard MR and Burt A (2002). Adaptation for horizontal transfer in a homing endonuclease. Mol Biol Evol 19:239-246

Lambowitz AM and Belfort M (1993). Introns as mobile genetic elements. Annu Rev Biochem 62:587-622

Langkjaer RB, Casaregola S, Ussery DW, Gaillardin C and Piskur J (2003). Sequence analysis of three mitochondrial DNA molecules reveals interesting differences among Saccharomyces yeasts. Nucl Acids Res 31:3081-3091

Le Hir H, Nott A and Moore M (2003). How introns influence and enhance eukaryotic gene expression. Trends Biochem Sci 28:215-220

Lei EP and Silver PA (2002). Intron status and 3'-end formation control cotranscriptional export of mRNA. Genes Dev 16:2761-2766

Liu X (2000). Protein-splicing intein: Genetic mobility, origin, and evolution. Ann Rev Genet 34:61-76

Lodato P, Alcaino J, Barahona S, Retamales P and Cifuentes V (2003). Alternative splicing of transcripts from crtI and crtYB genes of *Xanthophyllomyces dendrorhous*. Appl Environ Microbiol 69:4676-4682

Loizos N, Tillier ER and Belfort M (1994). Evolution of mobile group I introns: recognition of intron sequences by an intron-encoded endonuclease. Proc Natl Acad Sci USA 91:11983-11987

Long M (2001). Evolution of novel genes. Curr Opin Genet Dev 6:673-680

Lopez PJ and Seraphin B (1999). Genomic-scale quantitative analysis of yeast pre-mRNA splicing: implications for splice-site recognition. RNA 5:1135-1137

Lopez PJ and Seraphin B (2000). YIDB: the Yeast Intron DataBase. Nucl Acids Res 28:85-86

Lugones LG, Scholtmeijer K, Klootwijk R and Wessels JG (1999). Introns are necessary for mRNA accumulation in *Schizophyllum commune*. Mol Microbiol 32:681-689

Lundblad EW, Einvik C, Ronning S, Haugli K and Johansen S (2004). Twelve group I Introns in the same pre-rRNA transcript of the myxomycete *Fuligo septica*: RNA processing and evolution. Mol Biol Evol 7:1283-1293

Macreadie IG, Scott RM, Zinn AR and Butow RA (1985). Transposition of an intron in yeast mitochondria requires a protein encoded by that intron. Cell 41:395-402

Mahillon J, Kirkpatrick HA, Kijenski HL, Bloch CA, Rode CK, Mayhew GF, Rose DJ, Plunkett Gr, Burland V and Blattner FR (1998). Subdivision of the *Escherichia coli* K-12 genome for sequencing: manipulation and DNA sequence of transposable elements introducing unique restriction sites. Gene 223:47-54

Martinez D, L.F. L, Putmann N, Maarten D, Gelpke S, Huang K, Chapman J, Helfenbein KG, Ramaiya P, Detter JC, Laimer F, Coutinho PM, Henrissat B, Berka R, Cullen D and Rokhsar D (2004). Genome sequence of the lignocellulose degrading fungus *Phanerochaete chrysosporium* strain RP78. Nature Biotechnology 22:695-700

Maxwell ES and Fournier MJ (1995). The small nucleolar RNAs. Ann Rev Biochem 64:897-934

Michel F and Westhof E (1990). Modelling of the three-dimensional architecture of group I catalytic introns based on comparative sequence analysis. J Mol Biol 216:585-610

Modrek B and Lee C (2002). A genomic view of alternative splicing. Nat Genet 30:13-19

Mohr G, Perlman PS and Lambowitz AM (1993). Evolutionary relationships among group II intron-encoded reverse proteins and identification of a conserved domain that may be related to maturase function. Nucl Acids Res 21:4991-4997

Mohr G, Smith D, Belfort M and AM. L (2000). Rules for DNA target-site recognition by a lactococcal group II intron enable retargeting of the intron to specific DNA sequences. Genes Dev 14:559-573

Moran JV, Zimmerly S, Eskes R, Kennell JC, Lambowitz AM, Butow RA and Perlman PS (1995). Mobile group II introns of yeast mitochondrial DNA are novel site-specific retroelements. Mol Cell Biol 15:2828-2838

Mulder HJ, Saloheimo M, Penttila M and Madrid SM (2004). The transcription factor HACA mediates the unfolded protein response in *Aspergillus niger*, and up-regulates its own transcription. Mol Genet Genomics 271:130-140

Nilsen TW (2003). The spliceosome: the most complex macromolecular machine in the cell. BioEssays 25:1147-1149

Nowrousian M, Würtz C, Pöggeler S and Kück U (2004). Comparative sequence analysis of *Sordaria macrospora* and *Neurospora crassa* as a means to improve genome annotation. Fungal Genet Biol 41:285-292

Okuda Y, Sasaki D, Nogami S, Kaneko Y, Ohya Y and Anraku Y (2003). Occurrence, horizontal transfer and degeneration of VDE intein family in Saccharomycete yeasts. Yeast 20:563-573

Orgel LE and Crick FH (1980). Selfish DNA: the ultimate parasite. Nature 284:604-607

Pain A, Woodward J, Quail MA, Anderson MJ, Clark R, Collins M, Fosker N, Fraser A, Harris D, Larke N, Murphy L, Humphray S, O'Neil S, Pertea M, Price C, Rabbinowitsch E, Rajandream MA, Salzberg S, Saunders D, Seeger K, Sharp S, Warren T, Denning DW, Barrell B and Hall N (2004). Insight into the genome of *Aspergillus fumigatus*: analysis of a 922 kb region encompassing the nitrate assimilation gene cluster. Fungal Genet Biol 41:443-453

Paquin B, Laforest MJ, Forget L, Roewner I, Wang Z, Longcore J and Lang BF (1997). The fungal mitochondrial genome project: evolution of fungal mitochondrial genomes and their gene expression. Curr Genet 31:380-395

Paquin B and Lang BF (1996). The mitochondrial DNA of *Allomyces macrogynus*: the complete genomic sequence from an ancestral fungus. J Mol Biol 225:688-701

Paulus H (2000). Protein splicing and related forms of protein autoprocessing. Annu Rev Biochem 69:447-496

Perler FB (2002). InBase: the Intein Database. Nucl Acids Res 30:383-384

Perler FB and Adam E (2000). Protein splicing and its applications. Curr Opin Biotechnol 11:377-383

Perler FB and Davis EO, Dean GE, Gimble FS, Jack WE, Neff N, Noren CJ, Thorner J, Belfort M (1994). Protein splicing elements: inteins and exteins-a definition of terms and recommended nomenclature. Nucl Acids Res 22:1125-1127

Pietrokovski S (2001). Intein spread extinction in evolution. Trends Genet 17:465-472

Pöggeler S (1997). Sequence characteristics within nuclear genes from *Sordaria macrospora*. Fungal Genet Newsl 44:41-44

Pöggeler S and Kempken F (2004). Mobile genetic elements in mycelial fungi. In: Kück U (ed). Genetics and Biotechnology. Springer, Heidelberg, Berlin, pp 165-198

Pöggeler S and Kück U (2000). Comparative analysis of the mating-type loci from *Neurospora crassa* and *Sordaria macrospora*: identification of novel transcribed ORFs. Mol Gen Genet 263:292-301

Posey KL, Koufopanou V, Burt A and Gimble FS (2004). Evolution of divergent DNA recognition specificities in VDE homing endonucleases from two yeast species. Nucl Acids Res 32:3947-3956

Preker PJ, Kim KS and Guthrie C (2002). Expression of the essential mRNA export factor Yra1p is autoregulated by a splicing-dependent mechanism. RNA 8:969-980

Rodriguez-Navarro S, Strasser K and Hurt E (2002). An intron in the YRA1 gene is required to control Yra1 protein expression and mRNA export in yeast. EMBO Rep 3:438-442

Roger AJ and Doolittle WF (1993). Molecular evolution. Why introns-in-pieces? Nature 364:289-290

Saguez C, Lecellier G and Koll F (2000). Intronic GIY-YIG endonuclease gene in the mitochondrial genome of *Podospora curvicolla*: evidence for mobility. Nucl Acids Res 28:1299-1306

Saldanha R, Mohr G, Belfort M and Lambowitz AM (1993). Group I and group II introns. FASEB J 7:15-24

Scholtmeijer K, Wosten HA, Springer J and Wessels JG (2001). Effect of introns and AT-rich sequences on expression of the bacterial hygromycin B resistance gene in the basidiomycete *Schizophyllum commune*. Appl Environ Microbiol 67:481-483

Sekito T, Okamoto K, Kitano H and Yoshida K (1995). The complete mitochondrial DNA sequence of *Hansenula wingei* reveals new characteristics of yeast mitochondria. Curr Genet 28:39-53

Sellem CH and Belcour L (1997). Intron open reading frames as mobile elements and evolution of a group I intron. Mol Biol Evol 14:518-526

Sharma M, Ellis RL and Hinton DM (1992). Identification of a family of bacteriophage T4 genes encoding proteins similar to those present in group I introns of fungi and phage. Proc Natl Acad Sci USA 89:6658-6662

Sharp PA (1985). On the origin of RNA splicing and introns. Cell 42:397-400

Sharp PA (1994). Split genes and RNA splicing. Cell 77:805-815

Shub DA, Goodrich-Blair H and Eddy SR (1994). Amino acid sequence motif of group I intron endonucleases is conserved in open reading frames of group II introns. Trends Biochem Sci 19:402-404

Silar P, Barreau C, Debuchy R, Kicka S, Turcq B, Sainsard-Chanet A, Sellem C, Billault A, Cattolico L, Duprat S and Weissenbach J (2003). Characterization of the genomic organization of the region bordering the centromere of chromosome V of *Podospora anserina* by direct sequencing. Fungal Genet Biol 39:250-263

Singh NN and Lambowitz AM (2001). Interaction of a group II intron ribonucleoprotein endonuclease with its DNA target site investigated by DNA footprinting and modification interference. J Mol Biol 309:361-386

Spingola M, Grate L, Haussler D and Ares MJ (1999). Genome-wide bioinformatic and molecular analysis of introns in *Saccharomyces cerevisiae*. RNA 5:221-234

Sullenger BA (2003). Targeted genetic repair: an emerging approach to genetic therapy. J Clin Invest 112:310-311

Sullenger BA and Gilboa E (2002). Emerging clinical applications of RNA. Nature 418:252-257

Szostak JW, Orr-Weaver TL, Rothstein RJ, W.F. S (1983). The double-strand-break repair model for recombination. Cell 33:25-35

Taghian DG and Nickoloff JA (1997). Chromosomal double-strand breaks induce gene conversion at high frequency in mammalian cells. Mol Cell Biol 17:6386-6393

Terns MP and Terns RM (2002). Small nucleolar RNAs: versatile trans-acting molecules of ancient evolutionary origin. Gene Expr 10:17-39

Thierry A and Dujon B (1992). Nested chromosomal fragmentation in yeast using the meganuclease I-*Sce* I: a new method for physical mapping of eukaryotic genomes. Nucl Acids Res 20:5625-5631

Thierry A, Gaillon L, Galibert F and Dujon B (1995). Construction of a complete genomic library of *Saccharomyces cerevisiae* and physical mapping of chromosome XI at 3.7 kb resolution. Yeast 11:121-135

Thierry A, Perrin A, Boyer J, Fairhead C, Dujon B, Frey B and Schmitz G (1991). Cleavage of yeast and bacteriophage T7 genomes at a single site using the rare cutter endonuclease I-*Sce* I. Nucl Acids Res 19:189-190

Thornewell SJ, Peery RB and Skatrud PL (1997). Cloning and characterization of CneMDR1: a *Cryptococcus neoformans* gene encoding a protein related to multidrug resistance proteins. Gene 201:21-29

Toor N, Hausner G and Zimmerly S (2001). Coevolution of group II intron RNA structures with their intron-encoded reverse transcriptase. RNA 7:1142-1152

Villa T, Ceradini F and I. B (2000). Identification of a novel element required for processing of intron-encoded box C/D small nucleolar RNAs in *Saccharomyces cerevisiae*. Mol Cell Biol 20:1311-1320

Waring RB, Ray JA, Edwards SW, Scazzocchio C and Davies RW (1985). The Tetrahymena rRNA intron self-splices in *E. coli*: in vivo evidence for the importance of key base-paired regions of RNA for RNA enzyme function. Cell 40:371-380

Watabe H, Iino T, Kaneko T, Shibata T and Ando T (1983). A new class of site-specific endodeoxyribonucleases. Endo.Sce I isolated from a eukaryote, *Saccharomyces cerevisiae*. J Biol Chem 258:4633-4635

Weinstein LB and Steitz JA (1999). Guided tours: from precursor snoRNA to functional snoRNP. Curr Opin Cell Biol 11:378-384

Werner E, Wende W, Pingoud A and Heinemann U (2002). High resolution crystal structure of domain I of the *Saccharomyces cerevisiae* homing endonuclease PI-SceI. Nucl Acids Res 15:3962-3971

Wood V, Gwilliam. R., Rajandream M-A, Lyne M, Lyne R, Stewart A, Sgouros J, Peat N, Hayles J, Baker S, Basham D, Bowman S, Brooks K, Brown D, Brown S, Chillingworth T, Churcher C, Collins M, Connor R, Cronin A, Davis P, Feltwell T, Fraser A, Gentles S, Goble A, Hamlin N, Harris D, Hidalgo J, Hodgson G, Holroyd S, Hornsby T, Howarth S, Huckle EJ, Hunt S, Jagels K, James K, Jones L, Jones M, Leather S, McDonald S, McLean J, Mooney P, Moule S, Mungall K, Murphy L, Niblett D, Odell C, Oliver K, O'Neil S, Pearson D, Quail MA, Rabbinowitsch E, Rutherford K, Rutter S, Saunders D, Seeger K, Sharp S, Skelton J, Simmonds M, Squares R, Squares S, Stevens K, Taylor K, Taylor RG, Tivey A, Walsh. S., Warren T, Whitehead S, Woodward

J, Volckaert G, Aert R, Robben J, Grymonprez B, Weltjens I, Vanstreels E, Rieger M, Schäfer M, Müller-Auer S, Gabel C, Fuchs M, Dusterhoft A, Fritzc C, Holzer E, Moestl D, Hilbert H, Borzym K, Langer I, Beck A, Lehrach H, Reinhardt R, Pohl TM, Eger P, Zimmermann W, Wedler H, Wambutt R, Purnelle B, Goffeau A, Cadieu E, Dreano S, Gloux S, Lelaure V, Mottier S, Galibert F, Aves SJ, Xiang Z, Hunt C, Moore K, Hurst SM, Lucas M, Rochet M, Gaillardin C, Tallada VA, Garzon A, Thodem G., Daga RR, Cruzado L, Jimenez J, Sanchez M, del Rey F, Benito J, Dominguez A, Revuelta JL, Moreno S, Armstrong J, Forsburg SL, Cerutti. L., Lowe T, McCombie WR, Paulsen I, Potashkin J, Shpakovski GV, Ussery D, Barrell BG, Nurse P and Cerrutti L (2002). The genome sequence of *Schizosaccharomyces pombe*. Nature 415:871-880

Xiong Y and Eickbush TH (1990). Origin and evolution of retroelements based upon their reverse transcriptase sequence. EMBO J 9:3353-3362

Xu MQ and Evans TCJ (2003). Purification of recombinant proteins from *E. coli* by engineered inteins. Methods Mol Biol 205:43-68

Yadav JS, Soellner MB, Loper JC and Mishra PK (2003). Tandem cytochrome P450 monooxygenase genes and splice variants in the white rot fungus *Phanerochaete chrysosporium*: cloning, sequence analysis, and regulation of differential expression. Fungal Genet Biol 38:10-21

Yang J, Mohr G, Perlman PS and Lambowitz AM (1998). Group II intron mobility in yeast mitochondria: target DNA-primed reverse transcription activity of al1 and reverse splicing into DNA transpostion sites *in vitro*. J Mol Biol 282:505-523

Yang J, Zimmerly S, Perlman PS and Lambowitz AM (1996). Efficient integration of an intron RNA into double-stranded DNA by reverse splicing. Nature 381:332-335

Zhang K and Kaufman RJ (2004). Signaling the unfolded protein response from the endoplasmic reticulum. J Biol Chem 279:25935-25938

Zimmerly S, Guo H, Perlman PS and Lambowitz AM (1995). Group II intron mobility occurs by target DNA-primed reverse transcription. Cell 82:545-554

Zimmerly S, Moran JV, Perlman PS and Lambowitz AM (1999). Group II intron reverse transcriptase in yeast mitochondria. Stabilization and regulation of reverse transcriptase activity by the intron RNA. J Mol Biol 289:473-490

ELSEVIER

**Applied Mycology
and Biotechnology**
An International Series
Volume 5. Genes and Genomics
© 2005 Elsevier B.V. All rights reserved

4

Gene Silencing as a Tool for the Identification of Gene Function in Fungi

Annette Pickford and Giuseppe Macino
Dipartimento di Biotecnologie Cellulari e Ematologia, Universita' di Roma "La Sapienza",
Viale Regina Elena 324, 00161 Rome, Italy (macino@bce.med.uniroma1.it).

The use of RNA-mediated gene silencing for genetic analysis has revolutionized functional genomics, providing a tool to carry out genome-wide studies of gene function. With respect to traditional methods of genetic analysis whereby loss-of-function alleles are created by mutagenesis at the DNA level, gene silencing mediated by RNA is a post-transcriptional process that reduces the expression of a specific gene by the sequence-specific degradation of homologous mRNAs. This method of gene silencing, as well as being highly efficient and cost effective, is especially relevant for the analysis of essential genes that may escape phenotypic screening following random mutagenesis, as strains lacking an essential function are not viable.

1. INTRODUCTION

More than sixty years ago George Beadle identified the non-pathogenic filamentous fungus *Neurospora crassa* as an ideal candidate to study gene function. Several factors contributed to the choice of this fungus for genetic analysis with respect to higher plants or animals: it is a haploid organism, therefore genetic analysis is not hampered by the complications of dominance and recessivity; it grows rapidly with a short generation time and may be easily manipulated to study the segregation and recombination of mutants. The strategy adopted in the pioneering experiments carried out by Beadle and Tatum in 1941 was to create random mutations in the genome by exposing *Neurospora* to X irradiation or UV light and screen for strains in which mutations had inactivated a genetic function. The functions of genes encoding nutritional requirements were identified by screening for strains that failed to grow on minimal medium, but could grow on a medium supplemented with a specific nutritional compound.

Forward genetic analysis based on the creation random mutations in the genome followed by phenotypic screening for loss-of-function and mapping of the mutated

Corresponding author: Giuseppe Macino

genes has, until recently, been the mainstay of genetic analysis in fungi. As well X-rays and UV light, other DNA-damaging agents including DNA alkylating agents and azides may be used to induce mutations in the genome, and screening for resistance or sensitivity to these agents has enabled many genes involved in DNA repair to be identified. Random insertional mutagenesis may also be used to inactivate gene function and can be carried out by transforming a fungus with a plasmid carrying exogenous DNA sequences or transposons that may become integrated at random locations the fungal genome as a single copy or in random multicopy arrays, disrupting endogenous genes. Screening for partial or gain-of-function as a result of random mutagenesis of a known mutant strain may be used as a further method of forward genetic analysis. Random mutagenesis, however, has several disadvantages: it is not necessarily gene-specific and is not suitable to identify essential genes, as strains carrying null alleles of an essential function are not viable and therefore escape phenotypic screening. It has also been shown that random mutagenesis may consistently miss certain genes and in the case of insertional mutagenesis, certain regions of the genome are not susceptible to insertion. Furthermore insertion may lead to complete, incomplete or no functional reduction depending on the site of integration.

Gene-specific methods of mutagenesis are therefore preferable for accurate genetic analysis, but are only possible for known genetic sequences. The most widely used method of reverse genetic analysis is by homologous recombination, whereby an endogenous gene is replaced with a disrupted or deleted exogenous sequence, thus creating a knockout strain. Homologous recombination may be achieved by transforming a fungus with a plasmid carrying a deleted or non-functional sequence flanked by DNA sequences with homology to endogenous genomic sequences. The use of marker genes such as green fluorescent protein (GPF), β-galactosidase (β-gal) or transposon sequences or epitopes to tag a deleted sequence, facilitates the identification of the inactivated gene under analysis. Homologous recombination is efficient in budding yeast and has been used to create a collection of gene-deletion mutations for almost the entire *Saccharomyces cerevisiae* genome (Ross Macdonald et al. 1999; Giaever et al. 2002) however, it occurs with very low frequency in other fungi and cannot be used to analyze the function of an essential gene, as knockout strains are not viable. Moreover, targeted gene deletion can only be applied to predicted open reading frames (ORFs); overlapping ORFs are more difficult to distinguish and functionally redundant genes do not produce an altered phenotype when deleted. An alternative method of reverse genetic analysis is possible in *N. crassa* which possesses a unique sequence-specific mechanism of mutation that inactivates duplicated DNA sequences during the premeiotic phase of sexual reproduction by creating numerous CT-GA transition mutations in the duplicated sequence (Selker 1990). This process called Repeat-Induced Point mutation (RIP) probably evolved as a genomic defense mechanism to avoid gene duplications being passed on to progeny. RIP has been proposed as a method for genetic analysis also in view of the fact that not all transition mutations determine deleterious amino acid substitutions, and therefore a milder degree of gene inactivation may be useful for the analysis of essential functions (Barbato et al. 1996).

Whereas RIP inactivates homologous genes by creating mutations during the sexual phase, transformation of *Neurospora* with a transgene bearing sequence homology to an endogenous gene during vegetative growth, triggers a mechanism of gene silencing that reduces gene expression at a post-transcriptional level by degrading homologous mRNAs. This post-transcriptional gene silencing (PTGS) mechanism known as quelling (Romano and Macino 1992) has been shown to operate not only in *N. crassa*, but also in other fungi such as *Cryptococcus neoformans* (Liu et al. 2002), the slime mould *Dictyostelium discoideum* (Martens et al. 2002) and *Magnaporthe oryzae* (Kadotani et al. 2003) and may therefore be used as an easy and more versatile method of reverse genetic analysis with respect to traditional mutagenesis-based methods described above. The advantages of quelling for genomic analysis are (i) it is sequence specific and may be used to target any sequence in the genome; (ii) it is rapid and easy to carry out; (iii) transformation with more than one gene sequence may enable multiple phenotypes to be assessed; (iv) reversion of silencing due to loss of transgenes in time enables phenotypical recovery to assessed; (v) it may be used to analyse essential functions by the creation of knock-down strains in which gene expression is reduced rather totally inactivated. Quelling is the form of RNA-mediated gene silencing operating in fungi and has presented an invaluable tool with which to approach the formidable task of deciphering the vast quantities of genetic information now available from genome sequencing projects. The impact of RNA-silencing both in terms of time and cost-effectiveness with respect to traditional mutagenesis-based methods of genetic analysis, not only on genetic analysis, but also on therapeutic applications, is such that in 2002 it was heralded by *Science* as "Breakthrough of the Year" (Couzin 2002). This review describes the RNA-mediated gene silencing mechanism in general and its use to study the genomes of the non-pathogenic fungi used as model organisms including *N. crassa* and *Schizosaccharomyces pombe* and of pathogenic fungi such as *C. neoformans* and *M. oryzae*. The prospect of using RNA-mediated gene silencing to study the genomes of other fungi now being sequenced, represents an important step forward not only to speed up the elucidation of gene function, but also as a means to design of antifungal drugs and to carry out gene therapy.

2. RNA-MEDIATED GENE SILENCING

RNA-mediated gene silencing refers to a post-transcriptional gene silencing (PTGS) phenomenon that determines a reduction in the expression levels of a specific gene in response to the presence in a cell of homologous nucleic acid sequences. Silencing is due to the sequence-specific degradation of homologous mRNAs, mediated by small double-stranded RNAs processed from the silencing trigger molecule. This silencing mechanism probably evolved in not only in fungi, but also in plants and animals, as a means of defending genomic integrity, degrading the translation products of invading DNA that may be represented by transposons, selfish DNA or viruses (Zamore 2002). As well as its role as a genomic immune system (Plasterk 2002) RNA has been shown to play a major role in the cell in developmental regulation, by blocking the translation of target mRNAs in the cytoplasm and in regulating methylation and heterochromatin formation in the nucleus silencing gene expression at a translation level. While these latter aspects of RNA-mediated silencing are discussed in detail in Bartel (2004) and Matkze et al. (2004) here we focus on the PTGS activity that degrades homologous mRNAs and its exploitation to "knock-down" (rather than "knock-out") gene expression creating the equivalent of hypomorphic rather than null alleles of a gene.

In fungi and in plants, gene silencing was discovered as the unexpected result of an attempt to increase pigment biosynthesis by the introduction of a transgenic copy of an endogenous gene. In fungi the silencing effect was called quelling due to the fact that transformation of N. crassa with a copy of the albino-1 (al-1) gene involved in carotene biosynthesis determined an albino rather than orange phenotype (Romano and Macino 1992), while in plants transformation of petunia with an extra copy of the chalcone synthase (chsA) gene involved in anthocyanin biosynthesis resulted in white or variegated flowers (Napoli et al. 1990). The phenomenon in plants was termed co-suppression due to the fact that the expression of both the introduced transgene the homologous endogenous gene was reduced as a result of transformation. Both quelling and co-suppression were demonstrated to be general phenomena as co-suppression was also reported in tomato with the polygalacturonase gene responsible for fruit ripening (Smith et al. 1990) and in petunia with another petal pigmentation gene dihydroflavonol-4-reductase (dfr) (van der Krol et al. 1990). In N. crassa quelling was also achieved with two other carotene biosynthesis genes al-2 and al-3 (Cogoni et al. 1996) and with white collar wc-1 and wc-2 genes involved in light signal transduction (Ballario et al. 1996; Linden and Macino 1997). Silencing was shown to be a reversible post-transcriptional phenomenon affecting the level of mature mRNA of both the transgene and the homologous endogenous al-1 gene in fungi (Cogoni et al. 1996) and of the β-1,3 glucanase gene in tobacco plants (de Carvalho et al. 1995). Reversion was shown to be due to loss of transgenes, correlated with an increase of previously reduced levels of steady-state mRNA of both endogenous and ectopic genes. A diffusible silencing signal produced from the transgene was hypothesized to be responsible for propagating the silencing effect in forced heterokaryons between a silenced and a non-silenced strain in N. crassa (Cogoni et al. 1996, 1997a) and in plants from a silenced transgenic stock to a non-silenced transgenic scion (Palaqui et al. 1997; Voinnet et al. 1998). In the process named silencing acquired systemically (SAS) in plants, the silencing signal was thought to diffuse through plasmodesmata between cells and via the plant's vascular system to other parts of the plant (Voinnet et al. 1998, Fagard and Vaucheret 2000).

PTGS was also observed in animals in the nematode Caenorhabditis elegans. In worms transfected with an antisense construct of the par-1 gene with the aim of controlling gene expression, silencing was unexpectedly found to be equally induced by a sense construct of the same gene (Guo and Kemphues 1995). In order to investigate both the delivery and structure of RNAs causing interference of gene expression, RNA corresponding to a non-essential myofilament protein UNC-22A was directly injected into the worm (Fire et al. 1998). The severe twitching phenotype due to loss-of-function of the *unc-22A* gene was substantially greater as a result of injection of double-stranded rather than either sense or antisense RNA filaments, demonstrating for the first time that dsRNA is the true trigger molecule of the silencing effect which was called dsRNA interference (RNAi). Furthermore, the non-stochiometric ratio between the level of silencing and the quantity of dsRNA injected, indicated that an amplification step may be involved in the process. This experiment was a landmark in the understanding of PTGS and indicated that dsRNA may also be involved in co-suppression and quelling. RNAi was also shown to be diffusible, as it could be induced either by simply soaking worms in dsRNA or by feeding worms *Escherichia coli* bacteria expressing the dsRNA (Timmons and Fire 1998). The identification of a transmembrane protein SID-1 in *C. elegans* may be responsible for the diffusion of the silencing signal (Winston et al. 2000). RNAi was subsequently shown to be active not only *in C. elegans*, but also in other animals such as *Drosophila melanogaster* (Kennerdale and Carthew 1998; Misquitta and Paterson 1999), *Trypanosoma brucei* (Ngo et al. 1998) and *Planaria* (Alvarado and Newmark 1999). RNAi has further been shown to be active in *Arabidopsis thaliana* plants (Waterhouse et al. 1998; Chuang and Meyerowitz, 2000) and in vertebrates, namely in mice (Wianny and Zernicka-Goetz 2000) and in zebrafish (Li et al. 2000; Wargelius et al. 1999). Demonstration that co-suppression and RNAi are mechanistically related was provided by the use of RNA extracts of a co-suppressed tobacco plant to induce RNAi of the same reporter gene in *C. elegans* (Boutla et al. 2002). The analogy of PTGS with a mechanism of resistance to viral infection identified in plants named Virus-Induced Gene silencing (VIGS) in which dsRNA is present as an intermediate in viral replication (English et al. 1996; Ratcliff et al. 1997, 1999) indicates that these phenomena may all have evolved as a defense strategy against invading nucleic acids with dsRNA as the common trigger molecule. In support of this notion dsRNA has recently been demonstrated to have an active role in VIGS (Tenllado and Diaz-Ruiz 2001). Viruses in turn have evolved a counter attack mechanism involving proteins known as viral suppressors of gene silencing (VSGSs) that act at different levels of the silencing process (Brigneti et al. 1998; Anandalakshmi et al. 1998; Voinnet and Baulcombe 1999). Further demonstration of the conservation of gene silencing across the kingdoms is the ability of the VSGS b2 protein encoded by the flock house virus (FHV) that infects *Drosophila* cells, to suppress RNA-induced silencing in plants (Li et al. 2001, 2002).

Elucidation of the mechanism behind these silencing phenomena began with the identification of genes that determined a loss-of-silencing phenotype when mutated. In *N. crassa* random mutagenesis of a strain exhibiting a quelled (albino) phenotype as a result of transformation with an *al-1* transgene, was irradiated with UV light and series of quelling-defective mutants (orange phenotype) were isolated. The quelling-defective mutants where assigned into three complementation groups (Cogoni and Macino 1997b). Subsequently, three *quelling-defective* (*qde*) genes were identified: *qde-1* encoding an RNA-dependent RNA polymerase (RdRP) (Cogoni and Macino 1999a) *qde-2* encoding an Argonuate–like protein (Catalanotto et al. 2000) and *qde-3* encoding a DNA helicase (Cogoni and Macino 1999b). The identification of *qde-1* was the first indication of the involvement an RdRP in gene silencing and suggested a possible explanation to the production of dsRNA in *Neurospora* as a result of transformation with a transgene.

The identification of homologous genes involved in PTGS in plants, fungi and animals (see Table 1) strongly indicated that gene silencing was indeed a universal phenomenon of genome defense across the kingdoms (Cogoni and Macino 2000). However, elucidation of the mechanism involved, has been made possible by the discovery of key elements, subsequently shown to be common to each of these silencing phenomena. Observations in the fruit fly *D. melanogaster* and in the plant *A. thaliana* have been crucial to deciphering gene silencing. Small antisense RNA molecules were isolated both from co-suppressed plants (Hamilton and Baulcombe 1999) and from

Table 1. Genes Involved in RNA-Mediated Gene Silencing

Type of Protein/Domain	Organism	Gene	Reference
RNA-dependent RNA Polymerase (RdRP)	*C. elegans*	*ego-1*	Smardon et al. 2000
	C. elegans	*rrf-1,rrf-2, rrf-3*	Sijen et al. 2001
	N. crassa	*qde-1*	Cogoni and Macino 1999a
	A. thaliana	*sgs-2/sde-1*	Mourrain et al. 2000 Dalmay et al. 2000
	D. discoideum	*RrpA*	Martens et al. 2002
Dicer Ribonuclease ATP-dependent RNA helicase PAZ domain + two RNAse III domains + dsRNA-binding domain	*D. melonogaster*	Dicer-1	Bernstein et al. 2001
	D. melonogaster	Dicer-2	Lee et al. 2003
	C. elegans	*dcr-1*	Ketting et al. 2001 Grishok et al. 2001
	N. crassa	*dcl-1,dcl-2*	Catalanotto et al. 2004
	Homo sapiens	Dicer	Provost et al. 2002
	D. discoideum	Dicer	Novotny et al. 2001
Dicer dsRNA Binding Protein	*D. melanogaster*	R2D2	Liu et al. 2003
	C. elegans	Rde-4	Tabara et al. 2000
Argonaute PPD Protein PAZ domain + PIWI domain	*C. elegans*	*rde-1*	Tabara et al. 1999
	N. crassa	*qde-2*	Catalanotto et al. 2000
	A. thaliana	*ago-1*	Bohmert et al. 1998 and Fagard et al. 2000
	D. melanogaster	*ago-2*	Hammond et al. 2001
RISC dsRNA Binding Protein	*D. melanogaster*	VIG	Caudy et al. 2002
	D. melanogaster	FMR	Caudy et al. 2002
RISC Endonuclease	*D. melanogaster*	TudorSN	Caudy et al. 2003
RNaseD	*C. elegans*	*mut-7*	Ketting et al. 1999
RNA helicase	*D. melanogaster*	*Dmp68*	Ischizuka et al. 2003
	A. thaliana	*sde-3*	Dalmay et al. 2001
	C. elegans	*drh-1/drh-2*	Tabara et al. 2002
	C. elegans	*smg-3*	Domeier et al. 2000
	C. elegans	*mut-14*	Tijsterman et al. 2002
	C. reinhardtii	*mut-6*	Wu-Scharf et al. 2000
DNA helicase	*N. crassa*	*qde-3*	Cogoni and Macino 1999b

Drosophila in which RNAi had been induced by the introduction of dsRNA (Hammond et al. 2000). In *Drosophila* the small RNAs with sequence homology to the silenced target mRNA, were found associated to a multi-protein nuclease complex which was named the RNA-Induced Silencing Complex (RISC) (Hammond et al. 2001). RISC was shown to be an Argonaute-like protein (AGO-2) providing for the first time a biochemical link to QDE-2, RDE-1 and AGO-1 shown to be essential for quelling in *N. crassa* (Cogoni et al. 1997a, Catalanotto et al. 2000). RNAi in *C.elegans* (Tabara et al.1999) and co-suppression in *A. thaliana* (Fagard et al. 2000) respectively. This link was confirmed by the isolation of siRNAs homologous to transgenic sequences, associated with QDE-2 in *N. crassa* (Catalanotto et al. 2002). The identification of an RNase III-like endonuclease named Dicer (Bernstein et al. 2001) capable of digesting dsRNA into small 21-23 nucleotide RNA duplexes with characteristic 2-nucleotide 3' overhangs (Elbashir et al. 2001a, 2001b) was another landmark discovery, providing an explanation for the presence of 21-23 nt small-interfering RNAs (siRNAs) associated with RNAi in *Drosophila*, confirming the prediction of the involvement of RNase III in siRNA processing (Bass 2000). The observation that target mRNA is also cleaved into discreet 21-23 nt segments (Zamore et al. 2000) suggested that the siRNAs as part of the RISC, were guiding such degradation with the antisense strand of the siRNA pairing with homologous target mRNA (Parrish et al. 2000).

The model of post-transcriptional gene silencing that has emerged from these observations is a two-step process mediated by siRNAs that are generated from Dicer cleavage of dsRNAs in the cytoplasm. This first step of the silencing process is bypassed in the case of RNAi induced directly by siRNAs. In the second step the siRNA effector molecules are incorporated into the RISC to direct the sequence-specific degradation of homologous mRNAs (Fig. 1).

Other RNA-mediated processes within the cell have recently been shown to have elements in common with the gene silencing pathway. Dicer ribonuclease appears to be a central player in these overlapping mechanisms processing not only siRNAs from dsRNA molecules, but also hairpin RNAs formed from 70 nt non-coding microRNA precursor molecules into small temporal RNAs (stRNAs) that suppress translation of target developmental genes (Grishok et al. 2001, Hutvagner et al. 2001, Ketting et al. 2001). The imperfect sequence complementarity of stRNAs to their target sequence is the key to their acting as translational suppressors rather than mediators of mRNA degradation. Regulation of gene expression at a transcriptional level by methylation and heterochromatin formation may also be mediated by siRNAs in plants (Fagard and Vaucheret 2000), *Drosophila* (Pal-Bhadra 2002, 2004) and in fission yeast *S. pombe* (Volpe et al. 2003, Hall et al. 2002, 2003). The most recently discovered RNA-mediated process linked to heterochromatin formation and to the metabolism of repetitive elements is in *Tetrahymena thermophila* where 28 nt RNAs correlate with redundant DNA sequences eliminated during development of the somatic macronucleus (Yao et al. 2003). Many accessory proteins possibly involved in common steps of these different pathways have been identified and are described in detail in recent reviews (Pickford and Cogoni 2002; Agrawal et al. 2003). For the purpose of this review, the main elements of the two-step

RNA-mediated gene silencing mechanism are described in the following paragraphs.

2.1 The Silencing Trigger Molecule

RNAi may be induced directly by the introduction of dsRNA molecules, or indirectly by the introduction of vectors either expressing RNAs of opposite polarity that hybridize as dsRNA or single-stranded RNAs that fold into hairpin or panhandle dsRNA. For effective RNAi the sense and antisense trigger RNA strands must form a duplex and include a region of identity with target mRNAs. The effectiveness of hairpin RNAs as trigger molecules resides in the fact that they structurally mimic the 70 nt

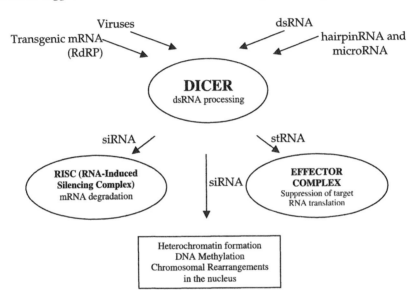

Fig. 1. RNA-mediated gene silencing

precursor microRNAs that are targeted by Dicer for cleavage into stRNAs.   With respect to the sense RNA strand, sequence modifications of the antisense are not well tolerated, indicating a distinctive role for each trigger molecule (Parrish et al. 2000). In mammals efficient RNAi has been induced directly by the introduction of synthetic siRNAs (Elbashir et al. 2001c; Caplen et al. 2001) in order to avoid stimulation of the interferon response that inhibits all gene expression by long dsRNA molecules (Stark et al. 1998).

In transgene-induced PTGS, the production of a dsRNA has still to be clarified. As many transgenes are integrated into the genome as multiple copies in tandem or as inverted repeats, intramolecular pairing may create hairpin or cruciform structures that are in effect double-stranded RNA molecules. Although a correlation between tandemly arranged transgenic loci and PTGS has been observed in plants (Que et al. 1998) and in fungi (Romano and Macino 1992). tandem repeats *per se* do not necessarily trigger silencing in fungi (Cogoni et al. 1996). Since single transgenes are capable of inducing silencing (Bernstein 2001a). transgene expression above a threshold level may be necessary to activate silencing (Vaucheret and Fagard 2001). A plausible model is that the aberrant nature of the exogenous mRNA transcript (for example premature termination) is likely to be the signal for the unprimed synthesis of dsRNA carried out by an RNA-dependent RNA polymerase (RdRP) to create the dsRNA trigger molecule in fungi, plants and some animals. In fact, an RdRP encoded by the quelling-defective gene *qde-1* was shown to be essential to quelling in *N. crassa* (Cogoni and Macino 1999a). RdRP had already been isolated in tomato (Schiebel et al. 1998) however, the identification of *qde-1* was the first demonstration of a link between an RdRP and PTGS. Homologues of *qde-1* involved in PTGS have been identified as *sde-1/sgs 2* in *A. thaliana* plants (Dalmay *et al.* 2000; Mourrain et al. 2000). and *ego-1* in *C. elegans* (Smardon et al. 2000). In support of the role of RdRP in transforming transgenic RNA transcripts into dsRNA substrates for Dicer, is the fact that siRNAs do not accumulate in *qde-1* mutants in *N. crassa* (Catalanotto et al. 2002) and in plants *sgs-2* is not required for PTGS induced with inverted repeats (Beclin et al. 2002). Purified recombinant QDE-1 has been demonstrated to carry out polymerization on an ssRNA template either by back priming or *de novo* initiation mode, producing full-length copies of template RNA. A surprising observation was that RdRP can also initiate *de novo* synthesis internally to produce 9-21 nt copies of input ssRNA. The length of these short oligonucleotides may depend on the size of regions of RNA free from stable secondary structure Makeyev and Bamford (2002).

The fact that only catalytic quantities of dsRNA had triggered efficient RNAi in the experiments of Fire et al. 1998, led to the hypothesis of an amplification step carried out by an RNA-dependent RNA polymerase (RdRP) (Sijen et al. 2001). In this amplification step RdRP may be responsible for the synthesis of secondary dsRNAs primed by siRNA on the mRNA template that may then be processed into secondary siRNAs. A similar amplification step has been identified in *Drosophila* by Lipardi et al. (2001). however, no RdRP gene has yet been identified in *Drosophila* or in humans. A specific role in siRNA amplification has been attributed to *rrf-1* one of the three RdRP homologues in *C. elegans* (*rrf-1, rrf-2* and *rrf-3*), maybe to generate secondary siRNAs not only to sustain the degradation of homologous mRNAs, but also as mediators of systemic silencing (Sijen et al. 2001). Likewise, *RrpA*, one of the three RdRP homologues (*RrpA, RrpA* and *DosA*) in *Dictyostelium discoideum* has also been implicated in the amplification of 23 nt siRNAs (Martens et al. 2002).

2.2 Dicer dsRNA Ribonuclease

Dicer is a dsRNA ribonuclease similar to bacterial Class III RNaseIII (Nicholson 1999) that comprises an N-terminal ATP-dependent RNA helicase domain, a PAZ (piwi-argonaute-zwille) motif, and two C-terminal endonuclease domains followed by a single dsRNA binding domain. The PAZ domain of 130 amino acids is conserved in the piwi protein in *Drosophila*, in the argonaute and zwille proteins in *Arabidopsis* (Cerutti et al. 2000). Dicer ribonuclease, first identified in *Drosophila* by Bernstein et al. (2000) processes dsRNA trigger molecules into short 21-23 dsRNA fragments known as small interfering RNAs (siRNAs) that mediate the sequence-specific degradation of homologous mRNAs (Elbashir et al. 2001a, 2001b). Crystallographic studies of RNAseIII (Blaszczyk et al. 2001) indicate that Dicer may act as a homodimer in antiparallel conformation generating two ATP-dependent staggered breaks, 22 nucleotides (nt) apart on each dsRNA. The siRNAs produced by such cleavage are characterized by 2 nt 3′ overhanging ends and 5′ phosphate and 3′ hydroxyl termini (Zamore 2001; Zhang, 2004). The PAZ domain of Dicer is implicated in the transfer of the siRNAs to the silencing effector complex called the RNA-Induced Silencing Complex (RISC) that carries out the sequence-specific degradation of target mRNAs (Ma et al. 2004). Transfer of the siRNAs may be facilitated by R2D2 (Liu et al. 2003) an accessory protein with two dsRNA binding sites that co-purifies with DCR2, the Dicer homologue in *Drosophila* that specifically processes siRNAs (Lee et al. 2003). The R2D2 homologue RDE-4 in *C. elegans* (Tabara et al. 1999, 2002) has also been shown not only to interact with Dicer in *C. elegans* (Ketting et al. 2001) but also with RDE-1, its AGO-2 homologue, supporting a similar role in bridging the initiator and effector steps of RNAi in *C. elegans*. Dicer enzymes have also been identified in humans (Provost et al. 2002), in fungi in *D. discoideum* (Novotny et al. 2001) and *N. crassa* (Catalanotto et al. 2004). While RD2D has been shown to preferentially interact with the DCR-2 isoform in *Drosophila*, in *N. crassa* the two Dicer proteins DCL-1 and DCL-2 have been demonstrated to be redundant. Lack of siRNAs in Dicer double mutants has demonstrated the involvement of Dicer in transgene-induced PTGS to process dsRNA generated by the action of RdRP on transgenic transcripts (Catalanotto et al. 2004).

2.3 The RNA-Induced SilencingComplex (RISC)

RISC is a multi-protein complex including an Argonaute-like endonuclease (Hammond et al. 2000, 2001) characterized by a 300 amino acid PIWI domain and a PAZ domain also present in Dicer proteins (Cerutti et al. 2000; Carmell et al. 2002). The PAZ domain has been proposed to be involved in the transfer of siRNAs from Dicer to RISC (Baulcombe 2001) and the structural basis for this interaction has recently been deduced from the 2.6 Å crystal structure of the PAZ domain from human Argonaute eIF2I (Ma et al. 2004). The PAZ domain of Dicer may act as an siRNA-end-binding module after dsRNA cleavage,while the Argonaute-PAZ constitutes an anchoring site for the 3′ end of the siRNA within the effector complex. Binding of the siRNA duplex to the two PAZ domains is independent, with specific recognition of siRNAs by PAZ provided by a binding pocket for the characteristic 2 nt 3′ overhangs of siRNAs (Lingel et al. 2004). In addition to the AGO-2 endonuclease, other components of the RISC in *Drosophila* include two RNA binding proteins named VIG and dFMR (homologous to the human Fragile X protein), Tudor SN nuclease (Caudy et al. 2002, 2003) an RNA helicase Dmp68 and ribosomal proteins L5 and L11 as well 5S rRNA (Ischizuka et al.

2002). RISC is activated by the ATP-dependent unwinding of the siRNA (Nykanen et al. 2001) with the antisense "guide" strand of the siRNA responsible for targeting mRNA cleavage (Martinez et al. 2002; Schwarz et al. 2003). The molecular mechanism of target mRNA cleavage has been recently reported by Martinez and Tuschl (2004).

3. RNA-MEDIATED GENE SILENCING AS A TOOL FOR FUNCTIONAL GENOMICS

The enormous potential of using RNA-mediated gene silencing as a tool for functional genomics is represented by the fact that virtually any sequence may be targeted for sequencing as a long as a dsRNA with sequence homology to the desired sequence is introduced into the organism. Additional advantages of silencing by RNA with respect to conventional methods are that a reduction of gene expression may be monitored with time and in the case of an essential gene at various time points until death. Furthermore, RNA-mediated gene silencing may be used to create double knock-downs to analyse epistasis enhancement or suppression, or multiple knock-downs in the case of gene families. Functionally redundant multicopy genes can be knocked down with one transgene construct by targeting a conserved domain. Combinatorial RNAi with two dsRNAs has successfully been used to analyse the function of two fertility determinants in *C. elegans* (Kuznicki et al. 2000) and in *Drosophila* to evaluate how a family of genes regulates axon guidance and synaptogenesis (Schmid et al. 2002). RNAi also represents a powerful tool to assess the function of proteins synthesized from alternatively spliced genes and has been used in *Drosophila* to analyse alternatively spliced mRNA isoforms of the *Dscam* gene (Celotto and Graveley 2002). Although some disadvantages such as the possible effects of transitive RNAi (Nishikura 2001; Vaistij et al. 2002) and non-uniformity of effectiveness do exist, as some cells may boost expression levels when low amounts of a protein are sensed, this is far outweighed by the extreme efficiency and low cost of this method with respect to traditional methods. RNAi is now routinely used for functional genomics and genome-wide screening with systemic RNAi has been used to analyse gene function in *C. elegans*, (Fraser et al. 2000; Gonczy et al. 2000; Maeda et al. 2001; Kamath et al. 2003) in *Drosophila* (Kiger et al. 2003) and mammals (Paddison et al. 2004; Berns et al. 2004).

The following paragraphs outline steps of the procedure to use RNA-mediated gene silencing as a tool for functional genomics:

3.1 Analysis of Sequence Data

Raw data from gene sequencing projects must be analysed to predict putative genes. Many programmes are available for this analysis at the website of the National Center for Biotechnology Information (http://www.ncbi.nih.gov).

3.2 Generation and Delivery of Trigger Molecules

An appropriate trigger molecule must be introduced into the organism under study. This may be achieved by transformation or transfection either with an expression vector that generates complementary RNAs from tandem promoters that hybridize *in vivo* or one long RNA strand that folds back to form hairpin or panhandle RNA (Waterhouse et al. 1998; Wang et al. 2000) or by the direct introduction of either long dsRNA as in the case of the pioneering experiments in C. *elegans* (Fire et al. 1998). In each case the dsRNA may be processed by the endogenous Dicer enzyme to generate siRNAs. Alternatively, siRNAs that can induce silencing directly may be generated by vector systems (Donze and Picard 2002; Brummelkamp et al. 2002) or produced *in vitro* as a result of RNAse III digestion of dsRNA (Kawasaki et al. 2003). Although much of the available information on the design and delivery of RNAi triggers is referred to mammalian cells, recent reviews by Dykxhoorn et al. (2003). Scherer and Rossi (2003) and Dorsett and Tuschl (2004) describe various methods to generate RNAi trigger molecules and give a comparison of plasmid-based versus synthetic siRNA-based silencing. Guidelines for the design of short hairpin RNAs (shRNA) and expression vectors to generate hairpin constructs is reported by McManus et al. (2002). Wesley et al. (2004) and Ui-Tei et al. (2004). A summary of *in vitro* and *in vivo* methods for introducing dsRNA into an organism, are shown in Figure 2.

It is known that the efficacy of siRNAs that target different regions of a gene can vary considerably (Holen et al. 2002). Factors influencing effectiveness include base composition of the siRNA, secondary structure of the target mRNA and binding of RNA-binding proteins. Bioinformatics methods are now available to determine the region of the target gene to induce the most efficient silencing response and several organizations are now involved in trigger molecule design. The Whitehead siRNA (short interfering RNA) Selection Web Server (http://jura.wi.mit.edu/bioc/siRNA) automates the design of short oligonucleotides to act as siRNAs as there are many combinations of possible 21mers for each gene and there is experimental evidence that

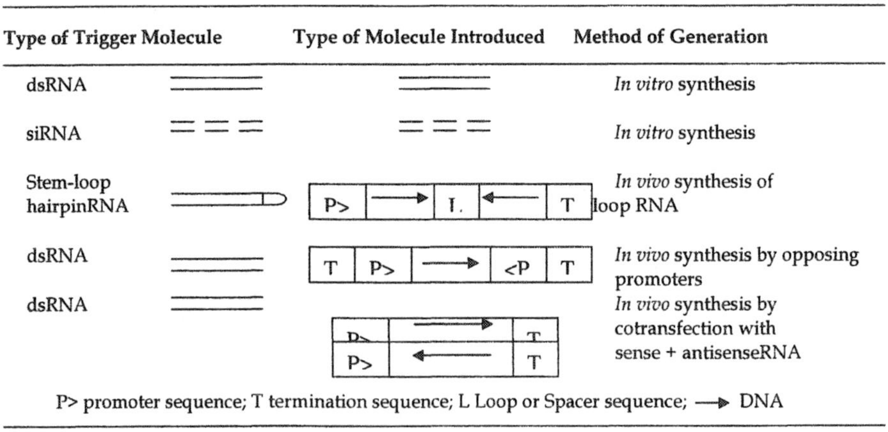

Type of Trigger Molecule	Type of Molecule Introduced	Method of Generation
dsRNA		*In vitro* synthesis
siRNA		*In vitro* synthesis
Stem-loop hairpinRNA	P> → I. ← T	*In vivo* synthesis of loop RNA
dsRNA	T P> → <P T	*In vivo* synthesis by opposing promoters
dsRNA	P> → T / P> ← T	*In vivo* synthesis by cotransfection with sense + antisenseRNA

P> promoter sequence; T termination sequence; L Loop or Spacer sequence; → DNA

Fig. 2. Generation of dsRNA trigger molecules.

not all 21mers in a gene have the same effectiveness in knocking down gene function. Information that may be found at this site includes published design rules and information about uniqueness of the 21mers within the genome, thermodynamic stability of the double stranded RNA duplex, GC content, presence of SNPs and other features that may contribute to the effectiveness of a siRNA (Yuan *et al.* 2004). Other support for trigger molecule design is UniGene (http://www.ncbi.nlm.nih.gov/entrez/query.fcgi?db=unigene) for the selection of a representative mRNA sequence for each target and RepeatMasker (http://ftp.genome.washington.edu/cgi-bin/RepeatMasker) for masking repetitive sequences.

Libraries of gene perturbing reagents can be created for genome-wide analysis. A restriction enzyme-generated siRNA (REGS) system has been recently reported to create a siRNA library from any gene or pool of double-stranded cDNAs. The process uses restriction enzymes to digest double-stranded cDNAs into multiple fragments of no more than 21 nt that are then made into palindromic structures using hairpin loops and isothermal rolling circle amplification (RCA) and cloned into expression vectors (Sen et al. 2004). Libraries of shRNA have recently been developed for assessment of mammalian genomes (Paddison et al. 2004; Berns et al. 2004).

3.3 Phenotypical Assay

Many biological processes may be identified by viewing colony size or intensity, death of a cell versus proliferation, nutrient metabolism, and resistance versus sensitivity to chemicals/radiation. A phenotypical assay by visual screening may also be carried out using microscopy. Motorized microscopes and automated imaging systems can facilitate screening of more subtle phenotypes. Automated imaging can provide data not only on a large number of samples but also on multiple phenotypes. The progressive effects of RNA-mediated gene silencing can be monitored by screening at different time points. This may also be carried out using time-lapse microscopy. A molecular "bar code", as developed for large-scale genetic screens in yeast (Shoemaker et al. 1996). can be incorporated in the trigger molecules. Cell microarrays and multiwell plates developed for high-throughput screening can also be used to monitor relative levels of molecular bar codes to assess pools of silencing constructs.

3.4 Analysis of Perturbed Phenotypes

Analysis of phenotypes that have been perturbed as a result of RNA-mediated gene silencing may be carried out on the basis of quantitative or qualitative evaluation parameters. Comparative analysis with genomic analysis data from other organisms can be informative. For example the RNAi Database (http://www.rnai.org) is a central archive of data from analyses in *C. elegans* available to the public and constitutes a model database for comparative large-scale phenotypic analysis in different organisms. A novel tool called Phenoblast can search the RNAi database using combination queries ranking genes on the basis of the overall phenotypical similarity. (Gunsalus *et al.* 2004). Confirmation of results obtained from the screening of silenced strains may be carried out by engineering true knock-outs or by engineering an over-expressing strain to create a reverse phenotype.

4. FUNCTIONAL GENOMICS IN FUNGI

Pathogenic fungi are responsible for many diseases both in plants and animals and the identification of genes that are critical for growth or those that may be targeted by anti-fungal drugs, is of extreme social and economical importance. Genome sequencing projects have been undertaken for several pathogenic and non-pathogenic fungi used as model organisms. The Fungal Genome Stock Center (http://www.fgsc.net) provides information on all fungi and Table 2 lists the web sites for specific fungi.

Not all fungi have yet been shown to be amenable to RNA-mediated gene silencing; however, comparative analysis of genetic data from fungi used as model organisms such as *N. crassa* and *S. pombe* in which quelling and RNAi can be used to down-regulate gene expression, will be fundamental to understanding the function of homologous genes in other fungi. Identification of homologous genes for components of the RNA-mediated gene silencing machinery may be indicative of an active gene silencing system, although not all fungi possess such machinery, for example *S. cerevisiae*. Comparative genetic analysis of *N. crassa* with that of another filamentous fungus *P. anserina* used as a model organism to study fundamental aspects of cell biology such as ageing, differentiation and cell death, may also be facilitated by the use of quelling. To date, antisense RNA has been used to silence gene function to carry out functional genomics in *Candida albicans* (De Backer et al. 2001) with the aim of identifying genes critical for growth that may represent antifungal drug targets. It is possible that that the efficacy of this antisense RNA method depends on an active RNA-mediated gene silencing system in *C. albicans*. Likewise anti-sense inhibition of gene function has been carried out in *Aspergillus oryzae* (Kitamoto et al. 1999) indicating that RNAi may also be effective in this and in other members of the *Aspergillus* family. Genetic analysis of the non-pathogenic fungus *A. nidulans* will be fundamental for comparative studies with its pathogenic counterparts *A. niger*, *A. flavus* and *A. fumigatus* that causes more infectious disease worldwide than any other mould, to elucidate the genetic basis of pathogenicity and identify targets for antifungal drugs. Also genetic analysis of phytopathogenic fungi that damage crops such as *Fusarium graminearum* that causes head blight of wheat and barley, *Ustilago maydis* that induces tumor of maize and teosinte and *M. oryzae* that causes rice blast disease, will hopefully enable the identification of genes that may be targeted to eradicate these diseases.

4.1 RNA-Mediated Gene Silencing in Non-Pathogenic Fungi
4.1.1 *Neurospora crassa*

Much of the early data on PTGS came from observations in *N. crassa* and the term "quelling" was based on the observation that carotene pigment levels were reduced or "quelled" as a result of transformation with a transgene bearing with sequence homology to an endogenous carotene pigment gene. Quelling was observed in approximately 30% of transformants and was shown to be relatively unstable with reversion to wild type due to excision of transgenic sequences during vegetative growth (Romano and Macino 1992). The identification of factors that overcome these limitations, by increasing both the efficiency of silencing and the stability of the silenced phenotype, make quelling the ideal tool carry out genetic analysis.

Table 2. Fungal genome projects and their corresponding URLs.

Aspergillus nidulans
http://www.fgsc.net/aspergenome.htm
Coprinus cinereus
http://www.broad.mir.edu/annotation/fungi/corinus_cinereus/
Coccidioides immiti
http://www.broad.mir.edu/annotation/fungi/coccidioides_immiti/
Cryptococcus neoformans
http://www.broad.mir.edu/annotation/fungi/cryptococcus_neoformans/index.html
Fusarium graminearum
http://www.broad.mir.edu/annotation/fungi/fusarium/index.html
Magnaporthe grisea
http://www.broad.mir.edu/annotation/fungi/mangaporthe/
Neurospora crassa
http://www.fgsc.net/Neurospora/neuros.htm
Podospora anserina
http://www.podospora.igmors.u-psud.fr/index.html
Phanerochaete chrysosporium
http://genome.jgi-psf.org/whiterot1.home.html
Ustilago maydis
http://www.broad.mir.edu/annotation/fungi/ustilago_mayd/

RdRP encoded by the *qde-1* gene in *Neurospora* has been shown to be essential for transgene-induced silencing (Cogoni et al. 1997a, 1999a) but dispensable upon direct expression of dsRNA (Catalanotto et al. 2004) indicating that the principle role of QDE-1 is in the conversion of transgenic RNA to dsRNA. A threshold level of dsRNA has been postulated to be necessary to trigger silencing, as strains harboring only a few copies of a transgene do not exhibit quelling (Cogoni et al. 1996). Over-expression of *qde-1* has been shown to compensate for low copy number by augmenting RdRP activity to reach a threshold level of dsRNA, even in strains that would not normally be silenced due to the fact that they harbor only one or two copies of a transgene. The frequency of silencing, scored by the appearance of an albino phenotype following transformation with either an *al-1* or *al-2* transgene, increased from 22% to 92% when *qde-1* over-expression was induced. Likewise, over-expression of *qde-1* increases the stability of quelling, as loss of transgene copies that normally correlates with reversion to a wild type phenotype, is compensated by increased RdRP activity. Over-expression of *qde-1* therefore influences both the activation and maintenance of quelling and in both cases correlates with an increase in the quantity of siRNAs in silenced strains (Forrest et al. 2004).

Both the efficiency and stability of silencing may also be increased with the use of hairpin RNA trigger molecules. Transformation of *Neurospora* with a vector that may be induced to expresses intron containing self-complementary RNA has been shown to trigger highly efficient silencing. The frequency of silencing in strains transformed with a plasmid carrying an inverted-repeat (IR) construct of the *al-1* gene was 87%. This increase in efficiency with respect to that normally observed in transgene-induced silencing, may be explained by the fact that silencing does not depend either on RdRP activity or on nuclear events possibly involving a RecQ DNA helicase encoded by the *qde-3* gene to enable transcription of the transgenic repeats. In fact a high frequency of silencing that correlated with an increase in siRNAs, was maintained in both *qde-1* and *qde-3* quelling-defective mutants, indicating that the IR construct constitutes a dsRNA substrate for Dicer, bypassing up-stream events of the silencing pathway. Furthermore, strains transformed with the IR construct displayed higher phenotypic stability (scored by frequency of reversion) with respect the tandem arrangements of transgenes that are prone to recombination and excision events. The inclusion of a spliceable intron in the IR construct gave a higher frequency of quelling than a similar construct carrying an unspliceable intron and an IR construct with a hairpin stem length of 600 nt was able to trigger efficient silencing whereas a stem of only 200 nt was not, indicating further parameters for the construction of efficient trigger molecules for quelling (Goldoni et al. 2004). The genomic sequence of *N. crassa* has now been completed (Galagan et al. 2003) and the improvement of both the efficiency and stability of silencing by the use of RdRP over-expression or dsRNA constructs, together with the possibility to modulate transgene expression with inducible promoters, represents a means of carrying out functional genomics in this fungus with unprecedented ease and versatility.

4.1.2 *Schizosaccharomyces pombe*

Genome sequencing of *S. pombe* has also been completed (Wood et al. 2002). In *S. pombe* the identification from sequence data of genes encoding Argonaute (*ago-1*) Dicer (*dcr-1*) and RdRP (*rdp-1*) prompted investigations into whether dsRNA could trigger silencing also in this fungus. The presence of siRNAs in a strain expressing a panhandle RNA construct of the *aes2* a gene, previously shown to enhance antisense-RNA silencing, demonstrated for the first time that RNAi operates in fission yeast and that antisense-RNA and dsRNA silencing pathways can be modulated by over-expression of the same enhancer protein (Raponi and Arndt 2003). Deletion of elements of the RNA-mediated gene silencing machinery in *S. pombe* revealed that it is also involved in epigenetic silencing at centromeres (Volpe et al. 2003) and in heterochromatin formation at the silent mating type region (Hall et al. 2002). Furthermore RNAi has been shown to be required for correct segregation of chromosomes during mitosis and meiosis due to loss of cohesin at centromeres (Hall et al. 2003). In a separate study hairpin RNA was shown to induce production of siRNAs and the assembly of silent chromatin at the target locus (Schramke and Allshire 2004). These observations are of extreme importance to understanding gene regulation at the transcriptional level.

4.2 RNA-Mediated Gene Silencing in Pathogenic Fungi

4.2.1 *Cryptococcus neoformans*

C. neoformans is an encapsulated fungal pathogen responsible for fatal meningitis in immunocompromised individuals. The genome of *C. neoformans* is currently being sequenced. It is the first pathogenic fungus in which RNA-mediated gene silencing has been demonstrated to operate. Expression of double-stranded RNA corresponding to portions of the cryptococcal *CAP59* and *ADE2* genes results in reduced mRNA levels for those genes, with phenotypic consequences similar to that of gene disruption. As gene replacement is hampered by both the tendency of *Cryptococcus* to modify exogenous DNA and its high frequency of nonhomologous recombination, the use of RNAi is important for functional analysis of genes of interest to enable efficient exploration of genes discovered by genome sequencing (Lui et al. 2002).

4.2.2 *Magnaporthe oryzae*

The phytopathogenic fungus *M. oryzae (grisea)* is the cause of rice blast disease that destroys a significant portion of world rice crops each year. *M. oryzae* is closely related to *N. crassa* and has been shown to be amenable to RNA-mediated gene silencing demonstrated by silencing of a GFP reporter gene by hairpin RNA. Silencing correlated with the accumulation of siRNAs in silenced transformants. Interestingly siRNAs or three different sizes ranging from 19-23 nt were identified for the first time in this fungus (Kadotani et al. 2003) possibly indicating a separate role for each class of siRNA as has been proposed in *Arabidopsis* where short siRNAs (21-22 nt) are necessary for mRNA degradation and long siRNAs (24-26 nt) correlate with systemic silencing and methylation of homologous mRNAs (Hamilton et al. 2002).

5. CONCLUSIONS

The advent of RNA-mediated gene silencing has changed the face of genetic analysis and investigation into the silencing mechanism has revealed several dsRNA-mediated cellular processes that operate both in the cytoplasm and in the nucleus. Double-stranded RNA can selectively destroy homologous mRNAs, suppress gene expression, inhibit gene transcription by heterochromatin formation and methylation or cause chromosomal rearrangements. Each of the process may be targeted for specific purposes that range from genetic analysis to gene therapy. The use of RNA-mediated gene silencing for functional genomics may be used to better understand the gene silencing processes themselves and the identification of all the components of each pathway will enable better exploitation of these endogenous mechanisms to fight disease.

REFERENCES

Alvarado AS and Newmark PA (1999). Double stranded RNA specifically disrupts gene expression during planarian regeneration. Proc Natl Acad Sci USA 96:5049-5054.

Anandalakshmi R, Pruss GJ, Ge X, Marathe R, Mallory AC, Smith TH and VanceVB (1998). A viral suppressor of gene silencing in plants. Proc Natl Acad Sci USA 95:13079-13084.

Ballario P, Vittorioso P, Magrelli A, Talora C, Cabibbo A and Macino G (1996). White colar 1, a central regulator of blue-light responses in *Neurospora crassa*. EMBO J 15:1650-1657.

Bartel DP (2004). MicroRNAs: genomics, biogenesis, mechanism, and function. Cell 116:281-297.

Barbato C, Calissano M, Pickford A, Romano N, Sandmann G, and Macino G (1996). Mild RIP-an alternative method for in vivo mutagenesis of the *albino-3* gene in *Neurospora crassa*. Mol Gen Genet 252:353-361.

Bass BL (2000). Double-stranded RNA as a template for gene silencing. Cell 101:235-238.

Baulcombe D (2001). Diced defence. Nature 409:295-296.

Beclin C, Boutet S, Waterhouse P and Vaucheret H (2002). A branched pathway for transgene-induced RNA silencing in plants. Curr Biol 12:684-688.

Berns K, Hijmans EM, Mullenders J, Brummelkamp TR, Velds A, Heimerikx M, Kerkhoven RM, Madiredjo M, Nijkamp W, Weigelt B, Agami R, Ge W, Cavet G, Linsley PS, Beijersbergen RL and Bernards R (2004). A large-scale RNAi screen in human cells identifies new components of the p53 pathway. Nature 428:431-437.

Bernstein E., Caudy AA, Hammond SM and Hannon GJ (2001a). Role for a bidentate ribonuclease in the initiation step of RNA interference. Nature 409:363-366.

Bernstein E, Denli AM and Hannon GJ (2001b). The rest is silence RNA 7:1509-1521.

Blaszczyk J, Trppea JE, Bubunenko M, Routzahn KM, Waugh DS, Court DL and Ji X (2001). Crystallographic and modeling studies of RNase III suggest a mechanism for double-stranded RNA cleavage. Structure 9:1225-1236.

Bohmert K, Camus I, Bellini C, Bouchez D, Caboche M and Benning C (1998). *AGO1* defines a novel locus of *Arabidopsis* controlling leaf development. EMBO J 17:170-180.

Boutet S,Vazquez F, Liu J, Beclin C,Fagard M, Gratias A, Morel JB, Crete P, Chen X and Vaucheret H (2003). *Arabidopsis* HEN1. A genetic link between endogenous miRNA controlling development and siRNA controlling transgene silencing and virus resistance. Curr Biol 13:843-848.

Boutla A., Kalantidis K., Tavernarakis N., Tsagris M. and Tabler M (2002). Induction of RNA interference in *Caenorhabditis elegans* by RNAs derived from plants exhibiting post-transcritional gene silencing. Nucl Acid Res 30:1688-1694.

Brummelkamp TR, Bernards R and Agami R (2002). A system for stable expression of short interfering RNAs in mammalian cells. Science 296:550-553.

Brigneti G, Voinnet O, Li WX, Ji LH, Ding SW and Baulcombe DC (1998). Viral pathogenicity determinants are suppressors of transgene silencing in *Nicotiana benthamiana*. EMBO J 17:6739-6746.

Caplen N J, Parrish S, Imani F, Fire A and Morgan RA (2001). Specific inhibition of gene expression by small double-stranded RNA in invertebrate and vertebrate systems. Proc Natl Acad Sci USA 98:9742-9747.

Carmell MA, Xuan Z, Zhang MQ and Hannon GJ (2002). The Argonaute family: tentacles that reach into RNAi, developmental control, stem cell maintenance, and tumorigenesis. Genes Dev 16:2733-2742.

Catalanotto C, Azzalin G, Macino G and Cogoni C (2000). Gene silencing in worms and fungi. Nature 404:245.

Catalanotto C, Azzalin G, Macino G and Cogoni C (2002). Involvement of the *qde* genes in the gene silencing pathway in *Neurospora*. Genes Dev 16:790-795.

Catalanotto C, Pallotta M, ReFalo P, Sachs MS, Vayssie L, Macino G and Cogoni C (2004). Redundancy of the two *dicer* genes in transgene-induced posttranscriptional silencing in *Neurospora crassa*. Mol Cell Biol 24:2536-2545.

Caudy AA, Myers M, Hannon G J and Hammond SM (2002). FragileX-related protein and VIG associate with RNA interference machinery. Genes Dev 16:2491-2496.

Caudy AA, Ketting R, Hammond SM, Denli AM, Bathoorn AMP, Tops BBJ, Silva JM, Myers MM, Hannon G and Plasterk RHA (2003). A micrococcal nuclease homologue in RNAi effector complexes. Nature 425:411-414.

Celotto AM and Graveley BR (2002). Exon-specific RNAi: a tool for dissecting the functional relevance of alternative splicing. RNA 8:718-724.

Cerruti L, Mian N and Bateman A (2000). Domains in gene silencing in cell differentiation proteins: the novel PAZ domain and redefinition of the Piwi domain. Trends Biochem Sci 25:481-482.

Cerutti H (2003). RNA interference: traveling in the cell and gaining functions? Trends Genet 19:39–46.

Cogoni C, Irelan JT, Schumacher M, Schmidhauser TJ, Selker EU and Macino G (1996). Transgene silencing of al-1 gene in vegetative cells of *Neurospora* is mediated by a cytoplasmic effector and does not depend on DNA-DNA interaction or DNA methylation. EMBO J 15:3153–3163.

Cogoni C and Macino G (1997a). Isolation of quelling-defective (*qde*) mutants impaired in post-transcriptional transgene-induced gene silencing in *Neurospora crassa*. Proc Natl Acad Sci USA 94:10233–10238.

Cogoni C and Macino G (1999a). Gene silencing in *Neurospora crassa* requires a protein homologous to RNA-dependent RNA polymerase. Nature 399:166–169.

Cogoni C and Macino G (1997b). Conservation of transgene-induced posttranscriptional gene silencing in plants and fungi. Trends Plant Sci 2:438–443.

Cogoni C and Macino G (1999b). Posttranscriptional gene silencing in *Neurospora* by a RecQ DNA helicase. Science 286:2342–2344.

Cogoni C and Macino G (2000). Post-transcriptional gene silencing across kingdoms. Curr Opin Genet Dev 10:638–643.

Couzin J (2002). Breakthrough of the year. Small RNAs make a big splash. Science 298:2296-2297.

Dalmay T, Hamilton A, Rudd S, Angell S and Baulcombe DC (2000). An RNA-dependent RNA polymerase gene in *Arabidopsis* is required for posttranscriptional gene silencing mediated by a transgene but not by a virus. Cell 101:543–553.

Dalmay T, Horsefield R, Braunstein TH and Baulcombe DC (2001). *SDE3* encodes an RNA helicase required for post-transcriptional gene silencing in *Arabidopsis*. EMBO J 20:2069–2077.

De Backer MD, Nelissen B, Logghe M, Viaene J, Loonen I, Vandoninck S, de Hoogt R, Dewaeie S, Simons FA, Verhasselt P, Venhoof G, Contreras R and Luyte WHM (2001). An antisense-based functional genomics approach for identification of genes critical for growth in *Candida albicans*. Nature 19:235-241.

De Backer MD, Raponi M, Arndt GM (2002). RNA-mediated gene silencing in non-pathogenic and pathogenic fungi. Curr Opin Microbiol 5:323-329.

de Carvalho F, Niebel F, Frendo P, Van Montagu M and Cornelissen M (1995). Post-transcriptional cosuppression of β-1,3-glucanase genes does not affect accumulation of transgene nuclear mRNA. Plant Cell 7:347-358.

Donze O and Picard D (2002). RNA interference in mammalian cells with siRNA synthesized with T7 RNA polymerase. Nucleic Acid Res 30:E46

Dormeier ME, Morse DP Knight SW, Portereiko M, Bass BL and Mango SE (2000). A link between RNA Interference and nonsense-mediated decay in *Caenorhabditis elegans*. Science 289:1928-1931.

Dorsett Y and Tuschl T (2004). siRNAs: applications in functional genomics and potential as therapeutics. Nat Rev Drug Discov 3:318-329.

Dykxhoorn DM, Novina CD and Sharp PA (2003). Killing the messenger: short RNAs that silence gene expression. Nat Rev Mol Cell Biol 4:457–467.

Elbashir SM, Lendeckel W and Tuschl T (2001a). RNA interference is mediated by 21- and 22-nucleotide RNAs. Genes Dev 15:188–200.

Elbashir SM, Martinez J, Patkaniowska A, Lendeckel W and Tuschl T (2001b). Functional anatomy of siRNAs for mediating efficient RNAi in *Drosophila melanogaster* embryo lysate. EMBO J. 20:6877–6888.

Elbashir SM, Harborth J, Lendecknel W, Yalcin A, Weber K and Tuschl T (2001c). Duplexes of 21-nucleotide RNAs mediate RNA interference in mammalian cell culture. Nature 411:494–498.

Elbashir SM, Harborth J, Weber K and Tuschl T (2002). Analysis of gene functions in somatic mammalian cells with small interfering RNAs. Methods 26:199–213.

English JJ, Mueller E and Baulcombe DC (1996). Suppression of virus accumulation in transgenic plants exhibiting silencing of nuclear genes. Plant Cell 8:179–188.

Fagard M, and Vaucheret H (2000). (Trans) gene silencing in plants: how many mechanisms? Ann. Rev Plant Physiol Plant Mol Biol 51:167–194.

Fagard M and Vaucheret H (2000a). Systemic silencing signal(s). Plant Mol Biol 43:285–293.

Fagard M, Boutet S, Morel J-B, Bellin C and Vaucheret H (2000b). AGO1, QDE-2, and RDE-1 are related proteins required for post-transcriptionalgene silencing in plants, quelling in fungi, and RNA interference in animals. Proc Natl Acad Sci USA 97:11650–11654.

Fire A, Xu S, Montgomery MK, Kostas SA, Driver SE and Mello CC (1998). Potent and specific genetic interference by double-stranded RNA in C. elegans. Nature 391:806–811.

Forrest EC, Cogoni C and Macino G (2004). The RNA-dependent RNA polymerase, QDE-1, is a rate-limiting factor in post-transcriptional gene silencing in Neurospora crassa. Nucl Acids Res 32:2123–2128.

Fraser AJG, Kamath RS, Zipperten P, Campos MM, Sohrmann M and Ahringer J (2000). Functional genomic analysis of C. elegans chromosome I by systemic RNA interference. Nature 408:325–330.

Galagan et al. (2003). Neurospora crassa genome sequence. Nature 422:859–868.

Giaever G, Chu AM, Ni L, Connelly C, Riles L, Veronneau S, Dow S, Lucau-Danila A, Anderson K, Andre B et al. (2002). Functional profiling of the Saccharomyces cerevisiae genome. Nature 418:387–391.

Goldoni M, Azzalin G, Macino G and Cogoni C (2004). Efficient gene silencing by expression of double stranded RNA in Neurospora crassa. Fungal Genet Biol Fungal Genet Biol 41:1016-1024

Gonczy P, Echerverri C, Oegema K, Coulron A, Jones SJM, Copley RR, Duperon J, Oegema J, Brehm M, Carrin E, Hannak E, Kirkham M, Pichler S, Flohrs K, Gossen A, Liedel S, Alleaume AM, Martin C, Ozlu N, Bork P, and Hyman AA (2000). Functional genomic analysis of cell division in C. elegans with RNAi of genes on chromosome III. Nature 408:331–335.

Grishok A, Pasquinelli AE, Conte D, Li N, Parrish S, Ha I, Baillie DL, Fire A, Ruvkun G and Mello CC (2001). Genes and mechanisms related to RNA interference regulate expression of the small temporal RNAs that control C. elegans developmental timing. Cell 106:23–34.

Grishok A, Tabara H and Mello CC (2000). Genetic requirements for inheritance of RNAi in C. elegans. Science 287:2494–2497.

Gunsalus KC, Yueh WC, MacMenamin P, Piano F (2004). Nucl Acid Res 32:406-410

Guo S, Kemphues KJ (1995) par-1, a gene required for establishing polarity in C. elegans embryos, encodes a putative Ser/Thr kinase that is asymmetrically distributed. Cell 81:611-620.

Hall IM, Shankaranarayana GD, Noma K-I, Ayoub N, Cohen A and Grewal SL (2002). Establishment and maintenance of a heterochromatin domain. Science 297:2232–2237.

Hall IM, Noma K and Grewal SI (2003). RNA interference machinery regulates chromosome dynamics during mitosis and meiosis in fission yeast. Proc Natl Acad Sci USA 100:193–198.

Hamilton AJ, and DC Baulcombe (1999). A species of small antisense RNA in posttranscriptional gene silencing in plants. Science 286:950–952.

Hamilton A, Voinnet O, Chappell L and Baulcombe D (2002). Two classes of short interfering RNA in RNA silencing. EMBO J 21:4671–4679.

Hammond SM, Caudy AA and Hannon GJ (2001). Post-transcriptional gene silencing by double-stranded RNA. Nat Rev Genet 2:110–119.

Hammond SM, Berstein E, Beach D and Hannon GJ (2000). An RNA-directed nuclease mediates post-transcriptional gene silencing in Drosophila cells. Nature 404:293–296.

Hammond SM, Boettcher S, Caudy AA, Kobayashi R and Hannon GJ (2001). Argonaute2, a link between genetic and biochemical analyses of RNAi. Science 293:1146–1150.

Hannon G J (2002). RNA interference. Nature 418:244–251.

Holen T, Amarzguioui M, Wiiger MT, Babaie E and Prydz H (2002). Positional effects of short interfering RNAs targeting the human coagulation trigger tissue factor. Nucl Acids Res 30:1757–1766.

Hutvagner G, McLachlan J, Pasquinelli AE, Balint E, Tuschl T and Zamore PD (2001). A cellular function for the RNA-interference enzyme Dicer in the maturation of the let-7 small temporal RNA. Science 293:834–838.

Ischizuka A, Siomi MC and Siomi H (2002). A Drosophila fragile X protein interacts with components of RNAi and ribosomal proteins. Gene Dev 16:2497–2508.

Kadotani N, Nakayashiki H, Tosa Y and Mayama S (2003). RNA silencing in the phytopathogenic fungus Magnaporthe oryzae. Mol Plant Microbe Interact 16:769-776.

Kamath RS, Fraser AG, Dong Y, Poulin G, Durbin R,Gotta M, Kanapin A, Bot NI, Moreno S, Sohrmann M, Welchman DP, Zipperlen P and Ahringer J (2003). Systematic functional analysis of the Caenorhabditis elegans genome with RNAi. Nature 421:231–237.

Kawasaki H, Suyama E, Iyo M and Taira K (2003). siRNAs generated by recombinant human Dicer induce specific and significant but target site-independent gene silencing in human cells. Nucl Acids Res 31:981-987.

Kennerdell JR and Carthew RW (1998). Use of dsRNA-mediated genetic interference to demonstrate that frizzled and frizzled 2 act in the wingless pathway. Cell 95:1017-1026.

Ketting, RF, Fischer SE, Bernstein E, Sijen T, Hannon GJ and Plasterk RHA (2001). Dicer functions in RNA interference and in synthesis of small RNA involved in developmental timing in *C. elegans*. Genes Dev 15:2654-2659.

Ketting, RF, Haverckamp THA, Van Luenen HGAM, and Plasterk RHA (1999). *mut-7* of *C. elegans*, required for transposon silencing andRNA interference, is a homolog of Werner syndrome helicase and RNase D. Cell 99:133-141.

Kiger AA, Baum B, Jones S, Jones MR, Coulson A, Escheverri C, Perrimon N (2003). A functional genomic analysis of cell morphology using RNA-interference. J Biol 2:27.

Kitamoto N, Yoshino S, Ohmiya K and Tsukagoshi N (1999). Sequence analysis, overexpression and antisense inhibition of a beta-xylosidase gene ,*xylA*, from *Aspergillus oryzae* KBN616. Appl Environ Microbiol 65:20-24.

Kuznicki et al.(2000) Combinatorial RNA interference indiates GLH-4 can compensate for GLH-1: these two P granule components are critical for fertility in *C. elegans*. Development 127:2907-2916.

Lee YS, Nakahara K, Pham JW, Kim K, He Z, Sontheimer EJand Carthew RW (2004). Distinct roles for *Drosophila* Dicer-1 and Dicer-2 in the siRNA/miRNA silencing pathways. Cell 117:69-81.

Li WX and Ding SW (2001). Viral suppressors of RNA silencing. Curr Opin Biotechnol 12:150-154.

Li H, Li WX and Ding SW (2002) Induction and suppression of RNA silencing by an animal virus. Science 296:1319-1321

Li YX, Farrell MJ, Liu R, Mohanty N, Kirby ML (2000). Double-stranded RNA injuction produces null phenotypesin zebrafish. Dev Biol 217:394-405.

Linden H and Macino G (1997). White collar 2, a partner in blue-light signal transduction, controlling expression of light-regulated genes in *Neurospora crassa*. EMBO J 16:98-109.

Lingel A, Simon B, Izaurralde E, Sattler M (2004). Nucleic acid 3'-end recognition by the Argonaute2 PAZ domain. Nat Struct Mol Biol 11:576-577.

Lipardi C, Wei Q and Paterrson BM (2001). RNAi as random degradation PCR: siRNA primers convert mRNA into dsRNA that are degraded to generate new siRNAs. Cell 101:297-307.

Liu H, Cottrell TR, Pierini LM, Goldman WE and Doering TL (2002). RNA interference in the pathogenic fungus *Cryptococcus neoformans*. Genetics 160:463-470.

Liu Q, Rand TA, Kalidas S, Du F, Kim HE, Smith DP, Wang X (2003). R2D2, a bridge between the initiation and effector steps of the Drosophila RNAi pathway. Science 301:1921-1925.

Ma J-B, Ye K and Patel DL (2004). Structural basis for overhang-specific small interfering RNA recognition by the PAZ domain. Nature 429:318-322.

Maeda I, Kohara Y, Yamamoto M, Sugimoto A (2001). Large-scale analysis of gene function in *Caenorhabditis elegans* by high through-put RNAi. Curr Biol 11:171-176.

Makeyev, E and Bamford DH (2002). Cellular RNA-dependent RNA polymerase involved in posttranscriptional gene silencing has two distinct activity modes. Mol Cell 10:1417-1427.

Martens H, Novotny J, Oberstrass J, Steck TL, Postlethwait P and Nellen W (2002). RNAi in *Dictyostelium*: the role of RNA-directed RNA polymerases and double-stranded RNase. Mol Biol Cell 13:445-453.

Martinez J, Patkaniowska A, Urlaub H, Luhrmann R and Tuschl T (2002). Single-stranded antisense siRNAs guide target RNA cleavage in RNAi. Cell 110:563-574.

Martinez J and Tuschl T (2004). RISC is a 5' phosphomonoester-producing RNA endonuclease. Genes Dev 18:975-980.

Matzke M, Aufsatz W, Kanno T, Daxinger L, Papp I, Mette MF, Matzke AJ (2004). Genetic analysis of RNA-mediated transcriptional gene silencing. Biochim Biophys Acta 1677:129-141.

McManus MT, Petersen CP, Haines BB, Chen J, Sharp PA (2002). Gene silencing using micro-RNA designed hairpins. RNA 8:842-850.

Mette MF, Aufsatz W, van der Winden J, Matzke MA and Matzk AJ (2000). Transcriptional silencing and promoter methylation triggered by double-stranded RNA. EMBO J 19:5194-5201.

Misquitta L and Paterson BM (1999). Targeted disruption of gene function in *Drosophila* by RNA interference (RNA-i): a role for nautilus in embryonic somatic muscle formation. Proc Natl Acad Sci USA 96:1451-1456.

Mourrain P, Beclin C, Elmayan T, Feuerbach F, Godon C, Morel J-B, Jouette D, Lacombe AM, Nikic S, Picault N, Remoue K, Sanial M, Vo T-A and Vaucheret H (2000). *Arabidopsis SGS2* and *SGS3* genes are required for posttranscriptional gene silencing and natural virus resistance. Cell 101:533-542.

Napoli C, Lemieux C and Jorgensen R (1990). Introduction of chimeric chalcone synthase gene into *Petunia* results in reversible cosuppression of homologous genes in trans. Plant Cell 2:279-289.

Ngo H, Tschudi C, Gull K and Ullu E (1998). Double-stranded RNA induces mRNA degradation in *Trypanosoma brucei*. Proc Natl Acad Sci USA 95:14687-14692.

Nicholson AW (1999). Function, mechanism and regulation of bacterial ribonucleases. FEMS Microbiol Rev 23:371-390.

Nishikura K (2001). A short primer on RNAi: RNA-directed RNA polymerase acts as a key catalyst. Cell 107:415-418.

Novotny J, Diegel S, Schirmacher H, Mohlre A, Hildebrandt M, Oberstrass J and Nellen W (2001). *Dictyostelium* dsRNase. Methods Enzymol 324:193-212

Nykanen A, Haley B and ZamorePD (2001). ATP requirement and small interfering RNA structure in the RNA interference pathway. Cell 107:309-321.

Paddison PJ, Silva JM, Conklin DS, Schlabach M, Li M, Aruleba S, Balija V, O'Shaughnessy A, Gnoj L, Scobie K, Chang K, Westbrook T, Cleary M, Sachidanandam R, McCombie WR, Elledge SJ and Hannon GJ (2004). A resource for large-scale RNA-interference-based screens in mammals. Nature 428:427-431.

Palauqui JC, Elmayan T, Pollien JM and Vaucheret H (1997). Systemic acquired silencing: transgene specific posttranscriptional gene silencing is transmitted by grafting from silenced stocks to nonsilenced scions. EMBO J 16:4738-4745.

Pal-Bhadra M, Bhadra U and Birchler JA (2002). RNAi-related mechanisms afftect both transcriptional and posttranscriptional transgene silencing in *Drosophila*. Mol Cell 9:315-327.

Pal-Bhadra M, Leibovittch BA, Gandhi SG, Rao M, Bhadra U, Birchler JA and Elgin SC (2004) Heterochromatic silencing and HP1 localization in *Drosophila* are dependent on the RNAi machinery. Science 303:669-672.

Parrish S, Fleenor J, Xu S, Mello C and Fire A (2000). Functional anatomy of a dsRNA trigger. Differential requirement for the two trigger strands in RNA interference. Mol Cell 6:1077-1087.

Plasterk RHA (2002). RNA silencing: The genome's immune system. Science 296:1263-1265.

Pickford A and Cogoni C (2002). RNA-mediated gene silencing. Cell Mol Life Sci 60:871-882.

Provost P, Dishart D, Doucet J, Frendewey D, Samuelsson B and Radmark O (2002). Ribonuclease activity and RNA binding of recombinant human Dicer. EMBO J 21:5864-5874.

Que Q and Jorgensen RA (1998). Homology-based control of gene expression patterns in transgenic petunia flowers. Dev Genet 22:100-109.

Raponi, M and Arndt GM (2003). Double-stranded RNA-mediated gene silencing in fission yeast. Nucl Acids Res 31:4481-4489.

Ratcliff F, Harrison BD and Baulcombe DC (1997). A similarity between virus defense and gene silencing in plants. Science 276:1558-1560.

Ratcliff FG, MacFarlane SA and Baulcombe DC (1999).Gene silencing without DNA. RNA-mediated cross-protection between viruses. Plant Cell 11:1207-1216.

Romano N and Macino G (1992). Quelling: transient inactivation of gene expression in *Neurospora crassa* by transformation with homologous sequences. Mol Microbiol 6:3343-3353.

Ross-Macdonald P, Coelho PS, Roemer T, Agarwal S, Kumar A, Jansen R, Cheung KH, Sheehan A, Symoniatis D, Umansky L, Heidtman M, Nelson FK, Iwasaki H, Hager K, Gerstein M, Miller P, Roeder GS and Snyder M (1999). Large-scale analysis of the yeast genome by transposon tagging and gene disruption. Nature 402:413-418.

Scherer L and Rossi JJ (2003). Approaches for the sequence-specific knockdown of mRNA. Nat Biotechnol 21:1457-1465.

Schiebel W, Pelissier T, Riedel L, Thalmier S, Schiebel R, Kempe D, Lottspeich F, Sanger HL and Wassenegger M (1998). Isolation of an RNAdirectedRNA polymerase-specific cDNA clone from tomato. Plant Cell 10:2087-2101.

Schmid A, Schindelholz B and Zinn K (2002). Combinatorial RNAi: a method for evaluating the functions of gene families in *Drosophila*. Trends Neurosci 25:71-74.

Shoemaker DD, Lashkari DA, Morris D, Mittmann M and Davis RW (1996). Quantitative phenotypic analysis of yeast deletion mutants using a highly parallel molecular bar-coding strategy. Nature Genet 14:450-456.

Schramke V and Allshire R (2004). Those interfering little RNAs! Silencing and eliminating chromatin. Curr Opin Genet Dev 14:174-180.

Schwarz DS, Hutvagner G, Haley B and Zamore PD (2002). Evidence that siRNAs function as guides, not primers, in the *Drosophila* and human pathways. Mol Cell 10:537-548.

Schwarz DS, Hutvagner G, Tingting D, Zuoshang X, Aronin N and Zamore PD (2003). Asymmetry in the assembly of the RNAi enzyme complex. Cell 115:199-208.

Selker EU (1990). Premeiotic instability of repeated sequences in *Neurospora crassa*. Ann Rev Genet 24:579-613.

Sen G, Wehrman TS, Myers JW and Blau HM (2004). Restriction enzyme-generated siRNA (REGS) vectors and libraries. Nat Genet 36:183-189.

Sijen T, Fleenor J, Simmer F, Thijssen KL, Parrish S, Timmons L, Plasterk RH and Fire A (2001). On the role of RNA amplification in dsRNA-triggered gene silencing. Cell 107:465-476.

Smardon A, Spoerke JM, Stacey SC, Klein ME, Mackin N and Maine EM (2000). EGO-1 is related to RNA-directed RNA polymerase and functions in germ-line development and RNA interference in *C. elegans*. Curr Biol 10:169-178.

Smith CJS, Watson CF, Bird CR, Ray J, Schoch W and Grierson D (1990). Expression of a truncated tomato polygalacturonase gene inhibits expression of endogenous gene in transgenic plants. Mol Gen Genet 224: 477-481.

Smith NA, Singh SP, Wang MB, Stoutjesdijk PA Green AG and Waterhouse PM (2000). Total silencing by intron spliced hairpin RNA. Nature 407:319-320.

Stark GR, Kerr I, Williams BR, Silverman RH and Schreiber RD (1998). How cells respond to interferons. Ann Rev Biochem 67:227-264.

Tabara H, Sarkissian M, Kelly WG, Fleenor J, Grishok A, Timmons, Fire A and Mello CC (1999). The *rde-1* gene, RNA interference and transposon silencing in *C. elegans*. Cell 99:123-132.

Tabara H, Yigit E, Siomi H and Mello CC (2002). The dsRNA binding protein RDE-4 interacts with RDE-1, DCR-1, and a DExH-box helicase to direct RNAi in *C. elegans*. Cell 109:861-871.

Tenllado F and Diaz-Ruiz JR (2001). Double-stranded RNA-mediated interference with plant virus infection. J Virol 75:12288-12297.

Tijsterman M, Ketting RF, Okihara KL, Sijen T and Plasterk RHA (2002). RNA helicase MUT-14-dependent gene silencing triggered in *C. elegans* by short antisense RNAs. Science 295:694-697.

Timmons L and Fire A (1998). Specific interference by ingested dsRNA. Nature 395:854.

Tuschl T, Zamore PD, Lehmann R, Bartel DP and Sharp PA (1999). Targeted mRNA degradation by double-stranded RNA *in vitro*. Genes Dev 13:3191-3197.

Ui-Tei K, Naito Y, Takahashi F, Haraguchi T, Ohki-Hamazaki H, Juni A, Ueda R and Saigo K (2004). Guidelines for the selection of highly effective siRNA sequences for mammalian and chick RNA interference. Nucl Acids Res 32:936-948.

Vaistij FE, Jones L and Baulcombe DC (2002). Spreading of RNA targeting and DNA methylation in RNA silencing requires transcription of the target gene and a putative RNA-dependent RNA polymerase. Plant Cell 14:857-867.

Van Blokland R, vander Geest N, Mol JNM and Kooter JM (1994). Transgene-mediated suppression of chalcone synthase expression in *Petunia hybrida* results from an increase in RNA turnover. Plant J 6:861-877.

Vander Krol, AR, Mur LA, Beld M, Mol JNM and Stuitje AR (1990). Flavanoid genes in petunia: addition of limited number of gene copies may lead to a suppression of gene expression. Plant Cell 2:291–299.

Vaucheret H and Fagard M (2001). Transcriptional gene silencing in plants: targets, inducers and regulators. Trends Genet 17:29–35.

Voinnet O, Vain P, Angell S and Baulcombe DC (1998). Systemic spread of sequence-specific transgene RNA degradation in plants is initiated by localized introduction of ectopic promoterless DNA. Cell 95:177–187.

Voinnet O, Pinto YM and Baulcombe DC (1999). Suppressor of gene silencing: a general strategy used by diverse DNA and RNA viruses of plants. Proc Natl Acad Sci USA 96:14147–14152.

Volpe, TA, Kidner C, Hall IM, Teng G, Grewal SLS and Martienssen RA (2003). Regulation of heterochromatic silencing and histone H3 Lysine-9 methylation by RNAi. Science 297:1833–1837.

Wang Z, Morris JC, Drew ME and England PT (2000). Inhibition of *Trypanosoma brucei* gene expression by RNA interference using an integratable vector with opposing T7 promotoers. J Biol Chem 275:40174–40179.

Wargelius A, Ellingsen S and Fjose A (1999). Double-stranded RNA induces specific developmental defects in zebrafish embryos. Biochem Biophys Res Commun 263:156–161.

Waterhouse, PM, Graham MW, and Wang MB (1998). Virus resistance in gene silencing in plants can be induced by simultaneous expression of sense and antisense RNA. Proc Natl Acad Sci USA 95:13959–13964.

Wesley SV, Helliwell C, Wang MB and Waterhouse P (2004). RNA interference, editing and modification methods and protocols. In: Methods in Molecular Biology Vol. 265 pp 117–130.

Wianny F and Zernicka-Goetz M (2000). Specific interference with gene function by double-stranded RNA in early mouse development. Nat Cell Biol 2:70–75.

Winston WM, Molodowitch C and Hunter CP (2002). Systemic RNAi in *C. elegans* requires the putative transmembrane protein SID-1. Science 295:2456–2459.

Wood *et al.* (2002). The genome sequence of *Schizosaccharomyces pombe*. Nature 415:871–880.

Wu-Scharf D, Jeong B, Zhang C and Cerutti H (2000). Transgene and transposon silencing in *Chlamydomonas reinhardtii* by a DEAH-box RNA helicase. Science 290:1159–1162.

Yang D, Buchholz F, Huang Z, Goga A, Chen C-Y, Brodsky FM and Bishop JM (2002). Short RNA duplexes produced by hydrolysis with RNase III mediate effective RNA interference in mammalian cells. Proc Natl Acad Sci USA 99:9942–9947.

Yuan B, Latek R, Hossbach M, Tuschl and Lewitter F (2004). SiRNA selection server: an automated siRNA oligonucleotide predition server. Nucl Acids Res 32:130-134.

Zhang H, Kolb FA, Jaskiewicz L, Westhof E, Filipowicz W (2004). Single processing center models for human dicer and bacterial RNase III. Cell 118:57-68.

Yao MC, Fuller P and Xi X (2003). Programmed DNA deletion as an RNA-guided system of genome defense. Science 300:1581–1584.

Zamore PD, Tuschl T, Sharp PA and Bartel DP (2000). RNAi: double-stranded RNA directs the ATP-dependent cleavage of mRNA at 21- to 23-nucleotide intervals. Cell 101:25–33.

Zamore PD (2001). Thirty-three years later, a glimpse at the ribonuclease III active site. Mol Cell 8:1158-1160.

Zamore PD (2002). Ancient pathways programmed by small RNAs. Science. 296:1265-1269.

Applied Mycology
and Biotechnology
An International Series
Volume 5. Genes and Genomics
© 2005 Elsevier B.V. All rights reserved

5

ELSEVIER

The DNA Damage Response of Filamentous Fungi: Novel Features Associated with a Multicellular Lifestyle

Camile P. Semighini[1], Gustavo H. Goldman[2] and Steven D. Harris[1]
[1]Plant Science Initiative and Department of Plant Pathology, University of Nebraska, N234 Beadle Center, Lincoln, NE 68588-0660, USA; [2]Faculdade de Ciências Farmacêuticas de Ribeirão Preto, Universidade de São Paulo, São Paulo, Brazil (sharri1@unlnotes.unl.edu).

Living organisms have to protect the integrity of their genomes in order to survive. A mechanism called the DNA damage response was developed during evolution to ensure the maintenance of genome integrity. Damage to DNA induces several cellular responses, including the repair of DNA lesions, the activation of cell cycle checkpoints, and cell death if the damage is not repaired. In this review, we first summarize known features of the DNA damage response, emphasizing those elements that function as sensors of DNA damage, transducers of the damage signal, or effectors that mediate the response. Subsequently, we focus on aspects of the fungal DNA damage response that distinguish it from the well-characterized yeast models. These include the characterization of conserved repair proteins that first appear in the multicellular fungi, the potential role of hyphal cell death in the fungal DNA damage response, the challenge of regulating the response in a multinucleate cell, and the novel interaction between cell cycle checkpoints and hyphal morphogenesis. Although, in some cases, our analyses may be somewhat speculative, we propose that these aspects of the fungal DNA damage response may serve as valuable models that yield important insight into analogous processes in other multicellular eukaryotes.

1. INTRODUCTION

Maintaining the correct genetic information is essential to the survival of a cell and also to the preservation of a species. Although DNA is a stable molecule, it is constantly under the threat of damage from both endogenous and exogenous sources. Therefore, from the early stages of evolution, living organisms had developed a protective system

---

Corresponding author: Steven D. Harris

to guarantee the integrity of their genomes, called DNA damage response (DDR). Cells spend a considerable amount of metabolic energy on the activation of a complex network in response to DNA damage. This network consists of interacting pathways that promote DNA repair and cell cycle checkpoint activation, and usually share common components with other functions, including replication, transcription, recombination and gene silencing. If the DNA lesion is not repaired, cells can undergoes death, thereby ensuring that damaged genomes are not transmitted to future generations. The clonal expansion of cells in unrepaired damage resulted in mutations is underlies the pathology of cancer.

In the past few years, the understanding of DDR pathways has increased due to several studies performed in budding and fission yeasts, nematodes, *Drosophila* and mammalian systems. Although considerable conservation was found in those pathways, some changes occurred during evolution. This review will summarize the similarities and, more importantly, discuss the differences that characterize fungal DDR.

2- DNA DAMAGE RESPONSE PROTEINS

Proteins involved in DDR are classified as (i) sensors, which detect DNA damage and generate a signal that initiates the response; (ii) transducers, which transmit the damage signal from the sensors and amplify it; and (iii) effectors, which usually are activated by the transducers and start executing repair or other responses, such as cell cycle arrest. Figure 1 provides an example of conserved sensors, transducers and effectors found in *Saccharomyces cerevisiae*, *Aspergillus nidulans* and *Homo sapiens*.

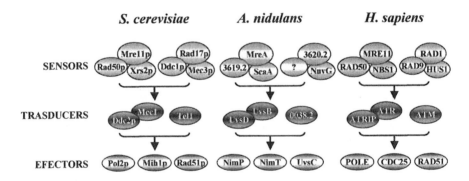

Fig. 1. Conserved DNA Damage Response in *Saccharomyces cerevisiae*, *Aspergillus nidulans* and *Homo sapiens*; proteins in this signaling cascade are classified as sensors that detect DNA damage, transducers that transmit the damage signal or effectors that promote the different responses to the damage.

Sensing DNA damage is the earliest and least understood step in the DDR. Sensors were initially believed to recognize specific types of DNA lesions, perhaps by directly binding to damaged bases or broken chromosomes, thereby generating a signal that activates a common transduction pathway that relays the damage signal to the appropriate effectors. However, recent findings suggest that different types of DNA damage are processed to a common intermediate that activates the DDR. The proposed candidate for this intermediate is single-stranded DNA (ssDNA), which is very fragile, existing *in vivo* only in association with replication protein A (RPA) (Zou and Elledge, 2003).

Genetic studies in yeast and animals have demonstrated that two protein complexes, RAD17-RFC2-5 and RAD9-RAD1-HUS1, function as sensors of DNA damage (for a review see Nyberg et al., 2002). RAD17 interacts with replication factor C (RFC) subunits 2, 3, 4 and 5 to form a pentameric structure, whereas the proliferating cell nuclear antigen (PCNA) homolog RAD9, hydroxyurea-sensitive 1 (HUS1) and RAD1 form a heterotrimeric ring around the DNA (Stergiou and Hengartner, 2004). Zou et al. (2002) showed that human RAD9 is recruited to chromatin in a RAD17-dependent manner in response to DNA damage. The same group showed that RPA stimulates the binding of the RAD17-RFC2-5 complex to ssDNA and that RPA-coated ssDNA is an intermediary recognized by this complex to initiate checkpoint signaling (Zou and Elledge, 2003).

DNA double-strand breaks (DSBs) in eukaryotic cells can be repaired by non-homologous end-joining (NHEJ) or homologous recombination (HR). Immediately after DSB formation, the Mre11 complex associated with the ends of the broken chromosome, and is believed to keep the sister chomatids or DNA ends in register until repair of the DSB is complete (for a review see Petrini and Stracker, 2003). The same authors proposed that in addition to its role in DSB repair, the Mre11 complex is a DSB sensor. Interestingly, an essential step before DSB repair is conversion of the double-stranded gap to 3' ssDNA tails. In yeast, null alleles of any of the genes encoding the proteins that form the Mre11 complex (*MRE11*, *RAD50* and *XRS2*) reduced the level of DSB processing (Ivanov et al. 1994, Tsubouchi and Ogawa, 1998). indicating the Mre11 complex as responsible for the initial DSB resection step. However, the Mre11 protein only exhibits 3' to 5' exonuclease activity, suggesting that other protein partners associate with this complex to facilitate the processing of DSBs (de Jager, 2001). Another persistent question is whether RPA caps the ssDNA tails formed during DSB repair. Carr (2003) suggested that RPA-ssDNA generated after DSBs is responsible for activation of the ATM kinase, reinforcing the hypothesis that ssDNA is a common intermediate recognized by all DNA damage sensors. Recently, Costanzo et al. (2004) showed using cell-free *Xenopus* eggs extracts that the Mre11 complex binds to fragmented DNA, forming a DNA-MRE11 complex that directly promotes ATM activation.

Proteins containing the BRCT domain, including BRCA1, are also considered DNA damage sensors. Although they do not directly bind DNA lesions, these proteins perform an important role in coordinating the DDR by physically linking sensors to transducers and effectors. BRCA1 itself associates with multiple DNA-repair proteins, forming the BRCA1-associated genome-surveillance complex, called BASC (Wang et al., 2000). Among the proteins associated with BRCA1 in BASC are the MRE11 complex, the mismatch repair heterodimeric complex MSH2-MSH6, the DNA helicase BLM (Bloom's-syndrome protein), the ATM kinase, RFC and PCNA. These proteins and complexes are thought to associate with each other in a highly dynamic manner, which may provide the versatility needed to rapidly sense damaged DNA and initiate the appropriate response (for a review see Tutt and Ashworth, 2002).

Much is known about the transducers, which are composed of two classes of proteins possessing conserved motifs. The first class consists of phosphoinositide 3-kinase related kinases (PIKKs), which include ATM (Ataxia Telangiectasia Mutated) and ATR (ATM-related). The second class consists of two families of serine/threonine protein kinases, known as the Chk1 and Chk2 kinases. ATM and ATR function at the nexus of the DDR, and appear to be responsible for the phosphorylation of multiple downstream effectors, including Chk1 and Chk2, which in turn are thought to potentiate the phosphorylation of other downstream targets (Nyberg et al. 2002). Studies in yeast and mammals have also identified ATRIP (Rad26/Ddc2 in yeast), which may function as a regulatory subunit of ATM or ATR that recruits them to damaged DNA and facilitates their association with appropriate effectors (Cortez, 2001). The substrates of ATM, ATR and the CHK kinases are generally responsible for DNA repair, transcriptional regulation, cell-cycle control and induction of cell death, and include, amongst others, RAD51, E2F1, Cdc25 and p53. It is important to emphasize that although the role of many of these effectors is reasonably well established, particularly in mammals and yeast, the mechanisms by which the DNA damage signal modifies their activity are not well known.

3. DNA DAMAGE RESPONSE IN FUNGI

Many of the proteins involved in DDR were first identified and characterized in *Saccharomyces cerevisiae*. The contribution of yeast to the characterization of the DDR was further enhanced by the completion of the yeast genome sequence project and the release of multiple genomics-based resources (i.e., whole-genome microarrays, deletion mutants) (Jazayeri and Jackson, 2002). These resources enabled the comprehensive identification of genes required for the response to numerous types of DNA damaging agents (Bennett et al. 2001, Birrell et al., 2001, Chang et al. 2002) and also provided a global overview of alterations in gene expression triggered by these agents (Gasch et al. 2001). Notably, these studies revealed that gene expression does not necessarily correlate with the requirement for gene function within the DDR (Birrell et al. 2002).

On the other hand, analysis of the DDR in filamentous fungi has largely been confined to the characterization of *Aspergillus nidulans* and *Neurospora crassa* mutants sensitive to genotoxic agents (for a review see Goldman et al. 2002). Only recently, the genome sequences of both fungi became available (Birren et al. 2003, Galagan et al. 2003) which permitted the identification of conserved fungal orthologues of key DDR components from yeast and mammals (see Table 1, also note that Goldman and Kafer (2004) and Borkovich et al. (2004) reviewed the conserved DDR orthologous found in *A. nidulans* and *N. crassa* genomes, respectively). Ninomaya et al. (2004) recently provided a good example of how a genome sequence can be exploited to further understand the fungal DDR. These authors disrupted the annotated genes referred to as NCU08290.1 and NCU00077.1 in *N. crassa*. As indicated in Table 1, these genes encode presumptive homologues of human Ku70 and Ku80, respectively, which function in the NHEJ pathway of DSB repair. As expected, the resulting mutants were sensitive to agents that trigger DSB formation, but they did not exhibit any additional growth or developmental defects. Accordingly, the authors were able to use these mutants as transformation recipients, where they achieved 100% homologous integration of exogenous DNA (compared to 30% in wild type controls) due to loss of the NHEJ pathway. Therefore, the Ku disruption strains have immense practical value as transformation recipients that enable high-throughput gene targeting in *N. crassa*.

Unlike yeast, filamentous fungi are multicellular. They form multicellular hyphae that interact to form a dense, highly branched mycelium. They also produce elaborate multicellular developmental structures that facilitate spore dispersal. This feature, combined with the renowned genetic tractability of filamentous fungi, means that they can be used to explore one of the remaining voids in our understanding of the DDR; in particular, how does the DDR interact with core developmental programs to ensure the maintenance of genome integrity? Is the DDR modified during development to emphasize cell death as opposed to potentially costly efforts at DNA repair? Are cell cycle checkpoint functions compromised by developmental signals to maintain rapid rates of cell division? To highlight the promise that filamentous fungi hold for addressing these issues, we focus on three conserved protein families that possess homologues in filamentous fungi, but not yeast, or have undergone expansion in filamentous fungi comparable to that seen in mammals. Each of these protein families has a significant role in the DDR, and we find it intriguing that their conservation within the fungal kingdom appears to be restricted to those fungi with a multicellular lifestyle. Therefore, using filamentous fungi as model organisms to characterize these protein families may offer the unique chance to better understand their function in mammals.

3.1 RECQ Helicase Homologues

Many biological processes, such as replication, recombination, and transcription, require the transient separation of the complementary DNA strands. Duplex DNA is unwound through the activity of enzymes known as DNA helicases (reviewed by Tuteja and Tuteja, 2004). These enzymes are classified into families according to their conserved hallmark helicase motifs and ATP-binding domains, such as the Walker A and B box motifs. One example is the RECQ helicase family, required for the maintenance of genomic stability (for reviews see Bjergbaek et al. 2002, Hickson 2003, Khakhar et al. 2003). Genomes of prokaryotes and unicellular eukaryotes encode just one RECQ helicase, whereas multiple homologues often exist in vertebrates. Five members of the RECQ family have been identified in human and mutations in three of these genes are responsible for autosomal recessive diseases characterized by genomic instability, cancer susceptibility and premature ageing. For instance, mutations in BLM, WRN and RECQ4 helicases are associated with the respective human diseases: Bloom syndrome (BS), Werner syndrome (WS) and Rothmund-Thomson syndrome (RTS) (reviewed by Mohaghegh and Hickson, 2001).

Table 1. Non-exhaustive list of orthologous DDR proteins in *Aspergillus nidulans, Neurospora crassa, Saccharomyces cerevisia, Schizosaccharomyces pombe* and metazoans

Protein Function	A. nidulans	N. crassa	S. cerevisiae	S. pombe	metazoans
SENSORS					
RFC1-like	AN8876.2	NCU00517.1	Rad24p	Rad17	RAD17
PCNA-like	None	NCU00470.1	Ddc1p	Rad9	RAD9
	AN3620.2	NCU00942.1	Rad17p	Rad1	RAD1
	AN6616.2 (NuvG)	NCU03820.1	Mec3p	Hus-1	HUS-1
DSB repair	AN0556.2 (MreA)	NCU08730.1 (Mus-23)	Mre11p	Rad32	MRE11
	AN3619.2	NCU00901.1 (Uvs-6)	Rad50p	Rad50	RAD50
	AN3372.2 (ScaA)	NCU04329.1	Xrs2p	None	NBS1
BRCT-containing	AN8255.2	NCU08879.1 (NcRhp9)	Rad9p	Rhp9	BRCA1
	AN6896.2	NCU09503.1	Dpb11p	Cut5	TopBP1
TRANSDUCERS					
PI3-kinases (PIKK)	AN0038.2	NCU00274.1 (Mus-21)	Tel1p	Tel1	ATM
	AN6975.2 (UvsB)	NCU01625.1 (Mus-9)	Mec1p	Rad3	ATR
PIKK regulation	AN5165.2 (UvsD)	NCU09644.1 (Uvs-3)	Ddc2p	Rad26	ATRIP
Checkpoint kinases	AN5494.2	NCU08346.1 (Un-1)	Chk1p	Chk1	CHK1
	AN4279.2	NCU02814.1	Dun1p	Cds1	CHK2
EFECTORS					
Checkpoint control	AN5822.2 (AnkA)	NCU04326.1	Swe1p	Wee1	WEE1
	AN3941.2 (NimT)	NCU02496.1	Mih1p	Cdc25	CDC25
	AN7499.2	NCU05250.1	Pds5p	Pds5	AS3
	AN4571.2	NCU04321.1	Mrc1p	Mrc1	Claspin
	AN7007.2 (MdrA)	NCU09552.1	Tof1p	Siw1	None
DSB repair	AN4407.2 (RadC)	NCU04275.1 (Mus-11)	Rad52p	Rad22	RAD52
	AN1237.2 (UvsC)	NCU02741.1 (Mei-3)	Rad51p	Rhp51	RAD51
	AN2087.2 (MusN)	NCU08598.1 (Qde-3)	Sgs1p	Rqh1	BLM/WRN
	AN4432.2 (OrqA)	NCU03337.1 (RecQ-2)	None	None	RECQL5
	AN5796.2	NCU07957.1	None	Rad60	None
	AN3118.2	NCU07457.1	Mus81p	Mus81	MUS81
	AN6878.2	B14D6.80	Mms4p	Eme1	MMS4
	AN1051.2	NCU01771.1	Rad57p	Rph57	XRCC3
	AN5527.2/AN0855.2	NCU02348.1 (Mus-25)	Rad54p	Rhp54	RAD54L/B
	AN6728.2	NCU08806.1	Rad55p	Rph55	XRCC2
	AN7753.2	NCU08290.1	Ku70p	Ku70	Ku70
	AN4552.2	NCU00077.1	Ku80p	Ku80	Ku80
	AN0097.2	NCU06264.1	Dnl4p	Dnl4	LIGASE4

Abasic sites repair	AN3947.2	NCU10044.1	Apn1p	Apn1	CeAPN-1
	AN4736.2	NCU01961.1	Apn2p	Apn2	APE1
	AN8713.2	NCU07440.1 (Mus-38)	Rad1p	Rad16	ERCC4
	AN4331.2	NCU07066.1 (Mus-44)	Rad10p	Swi10	ERCC1
	AN2764.2	NCU02288.1	Rad27p	Rad2	FEN1
	AN6682.2	NCU03040.1	Ogg1	None	OGG1
	AN7653.2	NCU06654.1	Ntg1p	Nth	NTH1
	AN0097.2	NCU06264.1	Cdc9	Cdc17	LIGASE 3
	AN5751.2	NCU08151.1	Tpp1p	ScPnkp	PNKP
	AN3129.2 (PrpA)	NCU08852.1	None	None	PARP
	AN0604.2	NCU08850.1 (Mus-18)	None	Uve1	None
Replication repair	AN3067.2 (NimP)	NCU04548.1	Pol2p	Cdc20	POLE
	AN2739.2	NCU07441.1	None	None	POLQ
	AN4555.2	NCU00081.1	Top3p	Top3	Top3
	AN6303.2 (UvsF)	NCU06767.1	Rfc1p	Rfc1	RFC1
	AN2969.2	NCU02687.1	Rfc2p	Rfc2	RFC2
	AN8064.2	NCU08427.1	Rfc3p	Rfc3	RFC3
	AN6517.2	NCU01295.1	Rfc4p	Rfc4	RFC4
	AN6300.2	NCU06769.1	Rfc5p	Rfc5	RFC5
	AN7309.2 (UvsH/NuvA)	NCU05210.1 (Uvs-2)	Rad18p	Rhc18p	RAD18
	AN4789.2 (UvsI)	NCU01951.1 (Upr-1)	Rev3p	None	REV3L
	AN2163.2	NCU06577.1	Mad2p	Mad2	REV7
	AN5344.2 (UvsJ)	NCU09731.1 (Mus-8)	Rad6p	Rhp6	UBE2A
	AN3797.2	NCU04733.1	Srs2p	Srs2	None

The first RECQ helicase characterized was the *E. coli* RecQ, which is involved in double-strand break repair and acts as a suppressor of illegitimate recombination (Tuteja and Tuteja, 2004). Budding and fission yeasts contain a sole RECQ helicase homologue: Sgs1 in *S. cerevisiae* and Rqh1 *Sc. pombe*. Cells lacking Sgs1 presented a similar phenotype to cells from BS and WS patients: hyper-recombination and enhanced cellular senescence. Interestingly, the expression of human BLM or WRN gene in an *SGS1* deficient yeast cells suppressed some of the associated phenotypes (Bjergbaek et al. 2002). On the other hand, *rqh1* null cells are sensitive to hydroxyurea (HU) and unable to recover from HU arrest due to a high level of recombination (Stewart et al., 1997 cited in Kitao et al. 1998).

As indicated on Table 1, filamentous fungi present two RECQ helicase homologues: a larger one and a smaller one. The smaller homologues primarily consist of the RECQ helicase domain and possess significant homology with human RECQ5 (Appleyard et al. 2000), whereas the larger homologues resemble Sgs1p, Rqh1, and BLM (Hofmann and Harris, 2001). In *Neurospora crassa*, the larger RECQ helicase (Qde-3) was shown to be specifically required for a post-transcriptional gene silencing mechanism termed quelling (Cogoni and Macino, 1999). The same group also characterized the smaller

RECQ helicase in *Neurospora crassa* (RecQ-2), showing that it is not involved in quelling but may collaborate with Qde-3 to repair DNA damage caused by mutagenic agents (Pickford et al. 2003). Therefore, Qde-3 might have a dual role: either acting alone as part of the quelling mechanism, or associated with the RecQ-2 to repair DNA lesions during replication. On the other hand, both *Aspergillus nidulans* RECQ homologues were shown to be involved in the DDR (Hofmann and Harris, 2001). The larger one (MusN) was proposed to be involved in a broader range of functions involving the DDR than the smaller one (OrqA). Specifically, MusN might have a role on protecting stalled replication forks from undergoing promiscuous recombination events that could trigger genome instability (Hofmann and Harris, 2001).

A phylogenetic tree was constructed using the RECQ helicase domains of known human, fungal and *E.coli* homologues, with the aim of determining the evolutionary relationship amongst those proteins. As shown in Figure 2, *E.coli* RecQ and human RECQ4 helicase are distinct from the other homologues. As the prokaryotic progenitor of this helicase family, it is not surprising that *E.coli* RecQ has diverged extensively from the others homologues. This protein should be able to perform all functions that were presumably split amongst specialized homologues that subsequently evolved. Regarding RECQ4, this protein might possess a yet-to-be-characterized function that distinguishes it from the other homologues. In support of this idea, the expression of RECQ4 was shown to be tissue-specific (Kitao et al. 1998), and is the 6th most highly over-expressed gene recovered from the analysis of the transcriptome associated with human tumors (Velculescu et al. 1999 cited in Bjergbaek et al. 2002). All the other RECQ homologues descended from a common ancestor, being classified in three clusters. Human RECQ5 formed a separate cluster with the filamentous fungal OrqA, Qde-3 and RecQ-2 homologues. Appleyard et al. (2000) characterized the *A. nidulans* RECQ5 homologue that was later re-named OrqA, and found that its amino acid sequence is more similar to the human RECQ5, supporting our phylogenetic data. RECQ1, BLM, Sgs1, MusN and Rqh1 formed the second cluster whereas WRN evolved separately as the third cluster. This result is consistent with the fact that BLM function resembles that of Sgs1 more closely than that of WRN, since only the former suppressed all the phenotypes of the *SGS1* null cells (Heo et al. 1994 cited in Bjergbaek et al. 2002). Similarly, Stewart et al. (1997). suggested that the *rqh1* gene product resembles BLM in function (Kitao et al. 1998).

Our phylogenetic analysis demonstrates that filamentous fungi possess an additional RECQ homologue when compared to yeast. The significance of the observation that it clusters with RECQ4 and RECQ5 is unclear, since the role of these proteins in the human DDR, if any, remains uncertain. Moreover, it is not known if the fungal homologues evolved by descent from RECQ4 and RECQ5, or if this is simply a case of convergent evolution. It is possible that the representative members of each fungal cluster possess overlapping functions. For example, in *A. nidulans*, both *musN* and *orqA* are expressed in hyphae, and extra copies of *orqA* can partially suppress the methyl-methanesulfonate (MMS) sensitivity caused by a *musN* mutation (Hofmann and Harris, 2001). On the other hand, these two RECQ helicases could display distinct spatial and/or temporal expression patterns during asexual or sexual development. Further studies are necessary to better understand the possible tissue and functional specificity of the fungal RECQ homologues.

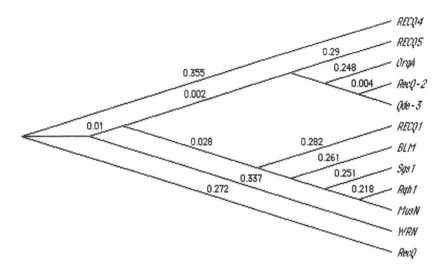

Fig.2. Phylogenetic tree representing the similarity among the RECQ helicases. The evolutionary relationship between RECQ helicase family genes was analyzed using the ClustalW algorithm to originate a Pairwise alignment of the sequences. The conserved helicase domains were then used to perform a Phylogenetic reconstruction using the Neighbor Joining method and the Best Tree mode of the MacVector software (Oxford Molecular Ltd). The numbers represent the relative genetic distance among the following amino acid sequences obtained from the GenBank database: *E. coli* ReqQ (M30198), *S. cereviseae* Sgs1 (U22341), *Sc. pombe* Rqh1 (Y09426), *A. nidulans* MusN (AAF72650) and OrqA (AAK53404), *Neurospora crassa* Qde-3 (AAF31695) and RecQ-2 (NCU03337.1) and human RECQ1 (M30198), BLM (U39817), WRN (L76937), RECQ4 (AB006532) and RECQ5 (AB006533).

3.2- DNA Polymerase Q (POLQ) homologues

The genomes of eukaryotic cells predict the existence of multiple DNA polymerases, which are proposed to have specialized roles in DNA replication and repair (Seki et al. 2003). The POLQ family is found in multicelular eukaryotes, but not in yeast (see Table 1), and presents a unique structure among known proteins: its C-terminal portion encodes a DNA polymerase related to the A family of DNA polymerases while its N-terminal portion encodes the seven motifs found in the large superfamily II of DNA and RNA helicases (Harris et al. 1996).

The first member of the POLQ family (Mus308) was identified in a large-scale screen for mutagen-sensitive mutations on the third chromosome of *Drosophila melanogaster*

(Boyd et al. 1981, cited in Sekelsky et al. 1998). The *mus308* gene was the only one amongst the 11 identified complementation groups that, when mutated, caused sensitivity to the crosslinking agent nitrogen mustard, but not to noncrosslinking mutagens such as MMS and UV light. Therefore, Mus308 is believed to be involved in the repair of interstrand DNA cross-linking damage. The hypermutability of *mus308* mutants when exposed to point mutagen N-ethyl-N-nitrosourea suggests a general role for Mus308 in postreplicational repair, which is responsible for the removal of non-excised adducts (Aguirrezabalaga et al. 1995). Thus, interstrand crosslink repair may represent only one of the roles of the Mus308 protein in DNA repair *in vivo*. Biochemical characterization of Mus308 revealed that it possesses both DNA polymerase and ATPase activities, and localizes to nuclei during both oogenesis and embryogenesis (Pang et al. 2001). Mus308 shows highest expression during late embryogenesis (Pang et al. 2001).

In humans, a full-length POLQ cDNA was recently isolated (Seki et al. 2003). Also, two mammalian *POLQ* paralogs have been reported: *HEL308* encodes a helicase showing homology to the N-terminal portion of POLQ (Marini and Wood, 2002) and *POLN* encodes an A family DNA polymerase similar to the POLQ C-terminus (Marini et al. 2003). Like Mus308, the human POLQ, possessed both DNA polymerase and single-stranded DNA-dependent ATPase activities (Seki et al. 2003). The same authors showed that POLQ expression is highest in testis, which was later confirmed by Kawamura et al. (2004). This result might suggest a specialized POLQ function in testis, such as a role in spermatogenesis. However, taken together with the fact that Mus308 was detected in ovary (Pang et al. 2001). the high POLQ/Mus308 expression in gonads might reflect the high DNA repair and recombination activity in those organs. No direct evidence has been presented to show that human POLQ functions like Mus308 to confer resistance to DNA-damaging agents similar to Mus308. Shima et al. (2003) identified a mutant mouse strain called *chaos1* (chromosome aberration occurring spontaneously 1), which exhibited elevated spontaneous and chemically induced levels of micronuclei in reticulocytes. Genetic mapping of mouse chromosome 16 suggested that *POLQ* is a candidate gene for the *chaos1* mutation. Kawamura et al. (2004) examined *POLQ* expression in malignant tissues and found tumor-associated over-expression in a wide range of human cancers, suggesting that *POLQ* aberrant expression of *POLQ* might contribute to tumor progression. Taken together, these findings suggest a possible role for mammalian POLQ in maintaining the genetic stability. Although a POLQ/Mus308 homologue is present in the genome sequences of both *A. nidulans* and *N. crassa* (see Table 1), the predicted fungal *POLQ* gene products show homology only to the helicase domains (see Figure 3). Therefore these genes are more similar to *HEL308* which is considered a *POLQ* paralog. HEL308 is a single-stranded DNA-dependent ATPase and DNA helicase that translocates on DNA with 3' to 5' polarity and efficiently displaces 20- to 40-mer duplex oligonucleotides (Marini and Wood, 2002). This translocation direction along the DNA (3' to 5') fits a model whereby coupling of helicase and polymerase activity coordinates duplex unwinding and polymerization. Marini et al. (2003). suggested that HEL308 might work together with POLN to fulfill both helicase and polymerase activity of POLQ. Consistent with this possibility, the expression pattern of *POLN* and *HEL308* partially overlaps: both are expressed in heart, muscle and testis, but only *HEL308* is expressed in ovary (Marini et al., 2003). We were unable to identify obvious *POLN* homologues in the genome sequences of several filamentous fungi. It still remains to be determined if the fungi POLQ parologues possess helicase activity, and if these proteins work together with a

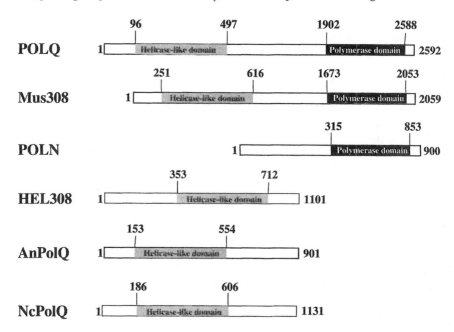

Fig. 3. Domain structures of representative POLQ homologues. The DNA helicase-superfamily II domains are showed in gray , whereas the DNA polymerase A domains are represented in black. The amino acid sequences obtained from the GenBank database were submitted to a NCBI Conserved Domain Search, and the results are depicted above. POLQ: Human DNA Polymerase Q (O75417); Mus308: *Drosophila melanogaster* Mus308 (AAB67306); POLN: human DNA Polymerase N (AAN52116.1); HEL308: human DNA Helicase (NP_598375); AnPOLQ: *Aspergillus nidulans* PolQ (AN2739.2) and NcPolQ: *Neurospora crassa* POLQ (NCU07441.1).

DNA polymerase to mimic POLQ activities. Moreover, because POLQ homologues are not found in yeast, it is interesting to speculate on why they are conserved in filamentous fungi. For example, they may be involved in a specific mode of cross-link repair that is associated with asexual or sexual development.

3.3. Poly (ADP-ribose) Polymerase (PARP) Homologues

Poly(ADP-ribose) polymerases (PARP) constitute a protein family that catalyses the transfer of ADP-ribose from NAD^+ to specific acceptor proteins. After binding the first ADP-ribose to specific amino acid residues, such as glutamate, aspartate or lysine,

PARP subsequently catalyzes the formation of an elongated, branched poly-ADP-ribose chain. Poly-ADP-ribosylation is a post-translational modification that has been reported to play an important role in diverse cellular processes such as the DDR, replication, transcriptional regulation, telomere maintenance and protein degradation (for reviews see D'Amours et al. 1999, Chiarugi, 2002, Bouchard et al. 2003).

Recently, the list of PARP enzymes in humans has expanded to include eighteen presumptive homologues, most of which are poorly characterized (Ame et al., 2004). The presence of so many homologues in both metazoans and plants has complicated the use of genetic approaches to fully comprehend PARP function. However, the presence of a single PARP homologue in the genomes of filamentous fungi, including *A. nidulans* and *N. crassa* (Thrane et al. 2004, Goldman and Kafer, 2004). affords a unique opportunity to characterize the ancestral function of this highly conserved protein. In particular, because there are no PARP homologues in yeast, filamentous fungi represent the lowest genetically tractable eukaryotes with a single PARP enzyme.

Amongst the PARP family members, PARP-1 is the best characterized. It contains three domains: an amino-terminal DNA binding domain (DBD), an auto-modification domain and a carboxy-terminal catalytic domain (see Figure 4). The latter domain encompasses a 50 amino acid block that is found in all PARP homologues (Smith, 2001, Virag and Szabo, 2002). The auto-modification domain contains the BRCT motif that is characteristic of many proteins involved in DNA metabolism. Notably, the fungal PARP homologues do not possess the PARP zinc finger-based DNA binding domain (Figure 4). Presumably, an interacting partner protein that is capable of binding DNA provides this function in fungi.

PARP-1 knockout mice and cells treated with chemical inhibitors of PARP function are both hypersensitive to genotoxic agents (reviewed by Shall and de Murcia, 2000 and Tong et al. 2001). Although those results implicate PARP-1 in the DDR, its precise role remains unclear. Ziegler and Oei (2001) proposed a model that correlates poly-ADP-ribosylation of chromatin with DNA repair. In the presence of DNA damage, specifically single- and double-strand breaks, PARP-1 binds to DNA lesions through its +

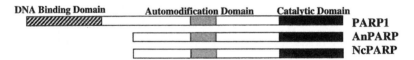

Fig. 4. Schematic representation of PARP1, AnPARP and NcPARP proteins. PARP homologues can present three types of domains: DNA binding domain (diagonal stripes), automodification domain (gray) and the PARP catalytic domain.

DBD. This triggers activation of the catalytic domain, resulting in the poly-ADP-ribosylation of acceptor proteins, including PARP-1 itself. The negative charge of the ribosyl chain added to histones would cause chromatin dissociation, thereby granting repair proteins access to the lesion. Also, the auto-modification of PARP-1 would result in its dissociation from DNA, exposing the damage to be repaired. Once the lesion is repaired, the removal of ADP-ribose polymers from histones and other target proteins would allow re-formation of chromatin. In mammals, this last step is catalysed by poly (ADP)-ribose glycohydrolase (PARG). Because there are no obvious homologues of PARG in the genomes of filamentous fungi, it remains unclear how they would down-regulate PARP function according to this model. Finally, it should be noted that this model could also explain the involvement of PARP-1 in transcriptional control.

In addition to its signaling role within the DDR, PARP1 has also been directly implicated in the base excision repair (BER) pathway, which repairs products of base oxidation and alkylation. *In vitro* experiments demonstrated that PARP1 interacts with BER components such as POLβ and XRCC1 (Caldecott et al. 1996, Masson et al. 1998, Dantzer et al. 2000). PARP-1 also bound with high affinity to an oligonucleotide containing a synthetic abasic site, which represents a BER intermediate (Lavrik et al. 2001). In addition, the key BER enzyme apurinic/apyrimidinic endonuclease (APE) interacts with the same repair intermediate as PARP-1 (Cistulli et al. 2004). Either APE and PARP1 compete for binding this intermediate, or, as suggested by Allison et al. (2003), PARP-1 could displace APE from the repair site to enable the next step in processing the lesion.

There is no reason to presume that the role of PARP in DDR signaling and BER is not conserved in filamentous fungi. However, this likely does not explain why filamentous fungi possess a PARP homologue, whereas yeast do not; particularly since all other DDR signaling components and BER enzymes are apparently conserved. Instead, it is possible that PARP may have an additional function that is unique to the DDR of multicellular organisms. A potential candidate for this function is DNA damage-induced cell death, which is a crucial option available to these organisms, but not necessarily to unicellular yeasts. The role of PARP in the regulation of cell death, and its implications for the fungal DDR are described in the next section.

4. PROGRAMMED CELL DEATH IN YEAST AND FILAMENTOUS FUNGI

When cells are unable to properly repair DNA damage, the DDR induces cell death. In metazoans, two major forms of cell death have been described: apoptosis and necrosis. Apoptosis, also called programmed cell death, is an energy-dependent process by which a cell actively destroys itself. On the other hand, necrosis has been considered a more severe but passive process by which the cell dies in response to excessive exposure to different insults (see Kanduc et al. 2002 for review on apoptosis and necrosis and their respective characteristics). DNA damage can trigger both apoptotic and necrotic cell death. PARP-1 has emerged as an important determinant of cell survival in response to DNA damage and also as modulator of the type of cell death (reviewed by Chiarugi, 2002, Virag and Szabo, 2002, Bouchard et al. 2003).

Chiarugi (2002) proposed that PARP1 controls cell fate upon exposure to DNA damage. Its damage-induced activation triggers the DDR, and proper repair of the damage permits survival. However, if the damage is not repaired, poly-ADP-ribosylation would then promote cell death. Based on findings that ATP and NAD+ levels are important determinants of whether a cell dies by necrosis or apoptosis, PARP-1 was hypothesized to be a molecular switch between the two types of cell death (Virag and Szabo, 2002, Bouchard et al. 2003). By consuming NAD+ generated as a result of cellular metabolism, PARP1 activity can trigger ATP depletion and induce cell death by necrosis. By contrast, during apoptosis, PARP1 is inactivated to maintain the ATP levels necessary to accomplish this energy dependent process (Virag and Szabo, 2002, Bouchard et al. 2003). Recently, activation of PARP-1 in response to massive DNA damage has been shown to regulate apoptosis-inducing factor (AIF), an early effector of caspase-independent apoptosis (reviewed by Hong et al. 2004). PARP-1 activation quickly induces the translocation of AIF from mitochondria to the nucleus, where it mediates chromatin condensation and high-molecular-weight DNA fragmentation. The mechanism responsible for PARP-1-dependent release of AIF from mitochondria remains to be identified, but it might involve poly-ADP-ribosylation signaling (Hong et al. 2004).

In the past few years, it has been demonstrated that yeast cells undergo apoptosis in response to numerous stimuli, including DNA damage (Del Carratore et al. 2002, Qi et al. 2003). as well as oxidative stress (Madeo et al. 1999). low doses of acetic acid (Ludovico et al. 2001), mating pheromone (Severin and Hyman, 2002). and salt stress (Huh et al. 2002). In addition, mutations affecting CDC48 and ASF1 also trigger apoptosis (Madeo et al. 1997, Yamaki et al. 2001). At least in the case of oxidative stress, yeast apoptosis is mediated by the caspase-like protease, or metacaspase, Yca1p (Madeo et al. 2002). Nevertheless, the role of programmed cell death in the yeast life cycle remains controversial, as it is not clear how apoptosis could contribute to the evolutionary fitness of a unicellular organism (Burhans et al. 2003). Herker et al. (2004). provided a possible solution to this problem by suggesting that apoptosis of aged yeast cells caused the release of substances into the media that promote the survival of younger, presumably fitter, cells, thereby ensuring the viability of a clonal population under stressful conditions. More recently, the role of apoptosis in the yeast DDR was questioned, and it was proposed that the initial observations were an artefactual reflection of cell lysis (Wysocki and Kron, 2004).

As in yeast, it has been established that filamentous fungi are capable of undergoing programmed cell death. For example, sphingoid bases and farnesol trigger apoptosis in *A. nidulans* (Cheng et al. 2003, C. Semighini and S. Harris, unpublished results). and, in both *A. fumigatus* and *Candida albicans*, apoptosis occurs in response to oxidative stress or exposure to the antifungal drug amphotericin B (Phillips et al. 2004, Mousavi and Robson, 2004). In addition, filamentous fungi appear to undergo programmed cell death in response to heterokaryon incompatibility reactions (reviewed by Glass et al. 2000). At least two distinct metacaspases have been characterized in *A. nidulans*, *A. fumigatus*, and *N. crassa* (Cheng et al. 2003, Mousavi and Robson, 2003, Thrane et al. 2004). where they are apparently required for apoptosis. Moreover, caspase-independent, but AIF-dependent, apoptosis has also been reported in *A. nidulans* (Cheng et al. 2003). Collectively, these observations demonstrate that filamentous fungi possess the machinery necessary for undergoing apoptosis. Nevertheless, the role of apoptosis in the fungal DDR remains to be firmly established (Goldman et al. 2002).

By analogy to multicellular organisms, we suggest that PARP has an integral role in the fungal DDR. For example, PARP signaling may function as a "rheostat" that calibrates the DDR to the level of damage. In the presence of tolerable levels of damage, PARP signals could facilitate DNA repair and promote viability. By contrast, when excessive levels of damage have been inflicted, PARP signals may trigger cell death. In a fungus such as *A. nidulans*, the programmed death of an active hyphal tip cell would have minimal consequences for colony growth, because a new tip would simply be generated from a previously quiescent subapical cell (Harris, 1997). Based on the paradigm established in animal cells, PARP-induced hyphal cell death would likely be caspase-independent, but AIF-dependent. The genetic tractability of *A. nidulans* and *N. crassa* will make it possible to test this model, as well as to identify other novel downstream effectors of PARP signaling.

5. NUCLEAR AUTONOMOUS RESPONSES TO DNA DAMAGE

One of the most intriguing aspects of fungal cell biology is the multinucleate nature of hyphal cells. Nuclear numbers range anywhere from close to one hundred in some ascomycetes, to as few as two in basidiomycete dikaryons. This poses an interesting challenge to hyphal cells: do individual nuclei within a multinucleate cell display an autonomous response to DNA damage? In fungi that display asynchronous mitoses, such as *N. crassa* and *Ashbya gossypii*, the answer to this question is likely to be yes. Because nuclei in these fungi already display autonomous behavior, they are likely to be protected from a damaged neighbor. On the other hand, a damaged nucleus presents a considerable problem in fungi that undergo parasynchronous mitoses, such as *A. nidulans*. Is the damaged nucleus insulated from the mitotic wave, or can it prevent further propagation of the wave? Moreover, regardless of mitotic lifestyle, is it possible that a single damaged nucleus can activate the DDR in adjacent undamaged nuclei? It may be possible to derive some insight into these questions by considering previous studies in yeast and animal cells.

To determine whether the DNA damage checkpoint signal is diffusible from one nucleus to another, Demeter et al. (2000). constructed binucleate *S. cereviseae* heterokaryons and induced an irreparable DSB in one of the nuclei. Whereas the undamaged nucleus entered mitosis, the damaged nucleus underwent cell cycle arrest. These experiments demonstrate that DNA damage in one nucleus induces a checkpoint-mediated arrest of that nucleus only, without preventing the division of a second undamaged nucleus in the same cytoplasm. Furthermore, in sea urchin zygotes, a checkpoint-arrested nucleus cannot be driven into mitosis by an adjacent mitotic nucleus (Sluder et al. 1995). Collectively, these experiments suggest that in hyphal cells, the DNA damage signal is likely to be nuclear autonomous, and that a damaged nucleus may indeed be insulated from an approaching mitotic wave. By analogy to the synchronous mitoses that occur in the syncytial embryos of *Drosophila* (Sullivan et al. 1993) a damaged nucleus that fails to respond to a mitotic wave may be discarded from the population of dividing nuclei. This would effectively prevent a single damaged nucleus from disrupting the normal pattern of mitosis in growing hyphae. What remains to be determined is the number of nuclei that must be damaged to arrest a multinucleate hyphal cell.

6. THE DNA DAMAGE RESPONSE AND HYPAL MORPHOGENESIS

Filamentous fungi form colonies consisting of tubular cells called hyphae, which can be uninucleate or multinucleate, depending on species and developmental stage. *Aspergillus nidulans* hyphae undergo a well-defined series of morphogenetic events that has been used to study cell cycle regulation (for a reviews see Doonan, 1992, Osmani and Mirabito, 2004). *A. nidulans* colonies are established from an asexual conidiospore, which contains a single nucleus arrested in the G1 phase of the cell cycle (Bergen & Morris, 1983). Germination starts with isotropic expansion of the spore and re-entry of the nucleus into mitosis. Upon completion of the first nuclear division, a polarized axis of growth is established and a germ tube is emitted (Harris, 1999). After the germling achieves a specific size and complete two more mitotic divisions, cytokinesis takes place through the formation of a septum (Harris et al. 1994, Wolkow et al. 1996, Harris, 1997). Subsequently, each additional round of nuclear division results in the formation of an additional septum (Harris et al. 1994). Successive rounds of nuclear division and septum formation produce an elongated multicellular hyphae, where two distinct compartments can be recognized: (i) an apical compartment that continuously extends as it progresses through an active mitotic cycle and (ii) sub-apical compartments that enter a period of mitotic quiescence and growth arrest (Harris, 1997, Kaminskyj and Hamer, 1998). Branching of the sub-apical compartments generates new hyphal tip that repeats the entire morphogenetic cycle. This iterative pattern of growth allows filamentous fungi such as *A. nidulans* to rapidly colonize new substrates.

The *A. nidulans* duplication cycle requires the coordination of nuclear division with hyphal morphogenesis and growth (Harris, 1997). In principle, by blocking nuclear division, DNA damage could indirectly prevent hyphal morphogensis. Indeed, this is likely true during spore germination, where spores possessing irreparably damaged nuclei do not undergo mitosis of form germ tubes (Harris, 1999). Similarly, branch formation can also be halted if adjacent nuclei are driven into a checkpoint that prevents mitotic entry (Dynesen and Nielsen, 2003). However, studies in *A. nidulans* have revealed a striking link between DNA damage and septum formation that appears to act independently of mitosis. In particular, low levels of DNA damage caused by mutagenic agents or mutations that perturb DNA metabolism (*sepB, sepI, sepJ, bimA*) result in the activation of a checkpoint that blocks septation but not nuclear division (Harris and Kraus, 1998, Wolkow et al. 2000, Kraus and Harris, 2001). Cell cycle checkpoint controls in *A. nidulans* are similar to those of yeast. For example, NimX[Cdc2] is a cyclin-dependent kinase (CDK) whose activation triggers mitotic entry. The tyrosine kinase AnkA[Wee1] promotes an inhibitory phosphorylation on Tyr-15 of NimX[Cdc2], which is removed by the tyrosine phosphatase NimT[Cdc25] (Ye et al. 1996). In the presence of DNA damage, AnkA[Wee1] phosphorylates NimX[Cdc2] and its activity is inhibited, preventing entry into mitosis (Ye et al. 1996). These same components also control the septation checkpoint (Kraus and Harris, 2001). In particular, mutations that abolish inhibitory NimX^{Cdc2} Tyr-15 phosphorylation allow septum formation to proceed despite the presence of DNA damage. These observations suggest that DNA damage-induced inhibition of NimX activation may affect the activity of a CDK substrate required for septation. Additional studies have shown that both the UvsB^{ATM}-UvsD^{ATRIP} and MreA^{MRE11}-ScaA^{NBS1}-Rad50 complexes are required to inhibit septum formation in the presence of DNA damage (Kraus and Harris, 2001, Semighini et al. 2003). These

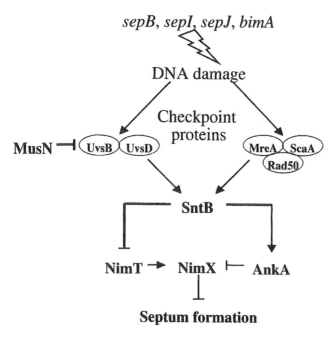

Fig. 5. A model for the regulation of septum formation in *Aspergillus nidulans*. DNA damage activates checkpoint proteins that influence SntB to reduce NimX action through the induction of AnkA and inhibition of NimT, resulting in the delay of the septum formation.

complexes presumably modify CDK function by somehow affecting the regulators AnkA^{wee1} and/or NimT^{cdc25} (Figure 5). SntB may be a key intermediate between the checkpoint complexes and the CDK module, because *sntB* and *ankA* mutants share many phenotypes, including failure to block septation, and display a genetic interaction

(Kraus and Harris, 2001). Finally, genetic analysis of the *musN201* mutation suggested that the DNA damage-induced inhibition of septum formation is normally a transient effect, and that the ability to septate is ultimately restored upon recovery from the DDR (Hoffman and Harris, 2000).

7. CONCLUSIONS

The availability of complete genome sequences of several filamentous fungi, including *A. nidulans* and *N. crassa*, marks a pivotal transition in our understanding of the fungal DDR. We now know the full complement of genes that are homologous to known DDR components as determined in yeast and animal. Not surprisingly, the core DDR components are conserved; however, a small number of animal DDR proteins (i.e., PARP) are also present in filamentous fungi, but not yeast. We propose that these proteins have a specific role within the DDR that is associated with the multicellular lifestyle shared by fungi and animals. Moreover, fungi such as *A. nidulans* and *N. crassa* provide a unique opportunity to characterize the function of these proteins and determine how it is integrated with the remainder of the DDR. Perhaps more importantly, certain features of the fungal DDR, such as the links to morphogenesis and the possible role of cell death, offer the potential to serve as novel targets for the therapeutic manipulation of fungal growth. For example, filamentous fungi presumably use a novel PARG-independent mechanism to down-regulate PARP. Future characterization of this mechanism may reveal that growth can be inhibited by hyper-activation of PARP. Additional studies that address the regulatory circuitry that links the DDR to mitosis, morphogenesis, and colony growth should also identify features that can be manipulated to our benefit. Indeed, now that we have most of the pieces of the fungal DDR puzzle, the future challenge is to determine how they fit together.

Acknowledgements: DNA damage research in GHG's lab is supported by Fundacao de Amparo a Pesquisa do Estado de Sao Paulo (FAPESP) and Conselho Nacional de Desenvolvimento Cientifico e Tecnologico (CNPq) and in SDH's lab is sponsored by the American Cancer Society and the Nebraska Research Foundation.

REFERENCES
Aguirrezabalaga I, Sierra LM and Comendador MA (1995). The hypermutability conferred by the *mus308* mutation of *Drosophila* is not specific for cross-linking agents. Mutat Res. 336(3): 243-250
Ame JC, Spenlehauer C and de Murcia G (2004). The PARP superfamily. Bioessays. 26(8): 882-893
Appleyard, MV, McPheat WL and Stark MJ (2000). A *recQ* family DNA helicase gene from *Aspergillus nidulans*. DNA Seq. 11(3-4): 315-319
Allinson SL, Dianova II and Dianov GL (2003). Poly(ADP-ribose) polymerase in base excision repair: always engaged, but not essential for DNA damage processing. Acta Biochim Pol. 50(1): 169-179
Bennett CB, Lewis LK, Karthikeyan G, Lobachev KS, Jin YH, Sterling JF, Snipe JR and Resnick MA (2001). Genes required for ionizing radiation resistance in yeast. Nat Genet. 29(4): 426-434
Bergen LG and Morris NR (1983). Kinetics of the nuclear division cycle of *Aspergillus nidulans*. J Bacteriol. 156(1): 155-160

Bernstein C, Bernstein H, Payne CM and Garewal H (2002). DNA repair/pro-apoptotic dual-role proteins in five major DNA repair pathways: fail-safe protection against carcinogenesis. Mutat. Res. 511 (2): 145-178

Bijergbaek L, Cobb JA and Grasser SM (2002). RecQ helicases and genomic stability: lessons from model organisms and human disease. Swiss Med. Wkly. 132: 433-442

Birrell GW, Giaever G, Chu AM, Davis RW and Brown JM (2001). A genome-wide screen in Saccharomyces cerevisiae for genes affecting UV radiation sensitivity. Proc Natl Acad Sci U S A. 98(22): 12608-12613

Birrell GW, Brown JA, Wu HI, Giaever G, Chu AM, Davis RW and Brown JM (2002). Transcriptional response of Saccharomyces cerevisiae to DNA-damaging agents does not identify the genes that protect against these agents.Proc Natl Acad Sci U S A. 99(13): 8778-8783

Birren B, Nusbaum C, Abebe A, Abouelleil A, Adekoya E, Ait-zahra M, Allen N, Allen T, An P, Anderson M, Anderson S, Arachchi H, Armbruster J, Bachantsang P, Baldwin J, Barry A, Bayul T, Blitshsteyn B, Bloom T, Blye J, Boguslavskiy L, Borowsky M , Boukhgalter B, Brunache A, Butler J , Calixte N, Calvo S, Camarata J et al. (2003). Genome sequence of Aspergillus nidulans. Whitehead Institute/MIT Center for Genome Research, Cambridge, MA (http://www.broad.mit.edu)

Borkovich KA, Alex LA, Yarden O, Freitag M, Turner GE, Read ND, Seiler S, Bell-Pedersen D, Paietta J, Plesofsky N, Plamann M, Goodrich-Tanrikulu M, Schulte U, Mannhaupt G, Nargang FE, Radford A, Selitrennikoff C, Galagan JE, Dunlap JC, Loros JJ, Catcheside D, Inoue H, Aramayo R, Polymenis M, Selker EU, Sachs MS, Marzluf GA, Paulsen I, Davis R, Ebbole DJ, Zelter A, Kalkman ER, O'Rourke R, Bowring F, Yeadon J, Ishii C, Suzuki K, Sakai W and Pratt R (2004). Lessons from the genome sequence of Neurospora crassa: tracing the path from genomic blueprint to multicellular organism. Microbiol Mol Biol Rev. 68(1):1-108

Bouchard VJ, Rouleau M and Poirier GG (2003). PARP-1, a determinant of cell survival in response to DNA damage. Exp Hematol. 31(6): 446-454

Burhans WC, Weinberger M, Marchetti MA, Ramachandran L, D'Urso G and Huberman JA (2003). Apoptosis-like yeast cell death in response to DNA damage and replication defects. Mutat Res. 532(1-2): 227-243

Carr, AM (2003) Beginning at the end. Science 300(5625): 1512-1513

Caldecott KW, Aoufouchi S, Johnson Pand Shall S (1996). XRCC1 polypeptide interacts with DNA polymerase beta and possibly poly (ADP-ribose) polymerase, and DNA ligase III is a novel molecular 'nick-sensor' in vitro.Nucleic Acids Res. 24(22): 4387-4394

Chang M, Bellaoui M, Boone C and Brown GW (2002). A genome-wide screen for methyl methanesulfonate-sensitive mutants reveals genes required for S phase progression in the presence of DNA damage. Proc Natl Acad Sci U S A. 99(26): 16934-16939

Cheng J, Park T-S, Chio L-C, Fischl AS and Ye XS (2003). Induction of apoptosis by sphingoid long-chain bases in Aspergillus nidulans. Mol. Cell. Biol. 23(1): 163-177

Chiarugi A (2002) Poly(ADP-ribose) polymerase: killer or conspirator? The 'suicide hypothesis' revisited. Trends Pharmacol. Sci. 23(3): 122-129

Cistulli C, Lavrik OI, Prasad R, Hou E and Wilson SH (2004). AP endonuclease and poly(ADP-ribose) polymerase-1 interact with the same base excision repair intermediate. DNA Repair 3(6): 581-591

Cogoni C and Macino G (1999). Posttranscriptional gene silencing in Neurospora by a RecQ DNA helicase. Science, 286 (5448): 2342-2344

Cortez D, Guntuku S, Qin J and Elledge SJ. ATR and ATRIP: partners in checkpoint signaling. Science 294(5547): 1713-1716

Costanzo V, Paull T, Gottesman M and Gautier J (2004). Mre11 assembles linear DNA fragments into DNA damage signaling complexes. PLoS Biol. 2(5): 0600-0609

D'Amours D, Desnoyers S, D'Silva I and Poirier GG (1999). Poly(ADP-ribosyl)ation reactions in the regulations of nuclear functions. Biochem. J. 342(Pt 2): 249-268

Dantzer F, de La Rubia G, Menissier-De Murcia J, Hostomsky Z, de Murcia G and Schreiber V (2000). Base excision repair is impaired in mammalian cells lacking Poly(ADP-ribose) polymerase-1. Biochemistry. 39(25): 7559-7569

136

de Jager M, van Noort J, van Gent DC, Dekker C, Kanaar R and Wyman C (2001). Human Rad50/Mre11 is a flexible complex that can tether DNA ends. Mol Cell. 8(5): 1129-1135

Del Carratore R, Della Croce C, Simili M, Taccini E, Scavuzzo M and Sbrana S (2002). Cell cycle and morphological alterations as indicative of apoptosis promoted by UV irradiation in *S. cerevisiae*. Mutat Res. 513(1-2):183-191

Demeter J, Lee SE, Haber JE and Stearns T (2000). The DNA damage checkpoint signal in budding yeast is nuclear limited. Mol Cell. 6(2): 487-492

Doonan, JH (1992). Cell division in *Aspergillus*. J. Cell Sci. 103 (Pt 3): 599-611

Dynesen J and Nielsen J (2003). Branching is coordinated with mitosis in growing hyphae of *Aspergillus nidulans*. Fungal Genet Biol. 40(1): 15-24

Galagan JE, Calvo SE, Borkovich KA, Selker EU, Read ND, Jaffe D, FitzHugh W, Ma LJ, Smirnov S, Purcell S, Rehman B, Elkins T, Engels R, Wang S, Nielsen CB, Butler J, Endrizzi M, Qui D, Ianakiev P, Bell-Pedersen D, Nelson MA et al. (2003). The genome sequence of the filamentous fungus *Neurospora crassa*. Nature 422(6934): 859-868

Gasch AP, Huang M, Metzner S, Botstein D, Elledge SJ and Brown PO (2001). Genomic expression responses to DNA-damaging agents and the regulatory role of the yeast ATR homolog Mec1p. Mol Biol Cell 12(10): 2987-3003

Glass NL, Jacobson DJ and Shiu PKT (2000). The genetics of hyphal fusion and vegetative imcompatibility in filamentous asco mycete fungi. Annu Rev Genet 34: 165-186

Goldman GH, McGuire SL, Harris SD (2002). The DDR in filamentous fungi. Fungal Genet Biol. 35(3): 183-195

Goldman GH and Kaffer E (2004). *Aspergillus nidulans* as a model system to characterize the DDR in eukaryotes. Fungal Genet Biol. 41(4): 428-442

Harris SD, Morrell JL and Hamer JE (1994). Identification and characterization of *Aspergillus nidulans* mutants defective in cytokinesis. Genetics. 136(2): 517-532

Harris PV, Mazina OM, Leonhardt EA, Case RB, Boyd JB and Burtis KC (1996). Molecular cloning of *Drosophila mus308*, a gene involved in DNA cross-link repair with homology to prokaryotic DNA polymerase I genes. Mol Cell Biol. 16(10): 5764-5771

Harris SD (1997).The duplication cycle in *Aspergillus nidulans*. Fungal Genet Biol. 22(1): 1-12

Harris SD and Kraus PR (1998). Regulation of septum formation in *Aspergillus nidulans* by a DNA damage checkpoint pathway. Genetics. 148(3): 1055-1067

Harris SD (1999). Morphogenesis is coordinated with nuclear division in germinating *Aspergillus nidulans* conidiospores. Microbiology 145(Pt 10): 2747-2756

Herker E, Jungwirth H, Lehmann KA, Maldener C, Frohlich KU, Wissing S, Buttner S, Fehr M, Sigrist S and Madeo F (2004). Chronological aging leads to apoptosis in yeast. J Cell Biol. 164(4): 501-507

Hickson ID (2003). RecQ helicases: caretakers of the genome. Nat Rev Cancer 3(3): 169-178

Hofmann AF and Harris (2000). The *Aspergillus nidulans uvsB* gene encodes an ATM-related kinase required for multiple facets of the DNA damage response. Genetics 154(4): 1577-1586

Hofmann and Harris (2001). The *Aspergillus nidulans musN* gene encodes a RecQ helicase that interacts with the PI-3K-related kinase UVSB. Genetics 159(4): 1595-1604

Hong SJ, Dawson TM and Dawson VL (2004). Nuclear and mitochondrial conversations in cell death: PARP-1 and AIF signaling. Trends Pharmacol Sci. 25(5): 259-264

Huh GH, Damsz B, Matsumoto TK, Reddy MP, Rus AM, Ibeas JI, Narasimhan ML, Bressan RA and Hasegawa PM (2002). Salt causes ion disequilibrium-induced programmed cell death in yeast and plants. Plant J. 29(5): 649-659

Ivanov EL, Sugawara N, White CI, Fabre F and Haber JE (1994). Mutations in XRSL, and RAD50 delay but do not prevent mating-type switching in *Saccharomyces cmeuisiae*. Mol. Cell. Biol. 14(5): 3414-3425

Jazayeri A and Jackson SP (2002) Screening the yeast genome for new DNA-repair genes. Genome Biol. 3(4): 1009.1-1009.5

Kanduc D, Mittelman A, Serpico R, Sinigaglia E, Sinha AA, Natale C, Santacroce R, Di Corcia MG, Lucchese A, Dini L, Pani P, Santacroce S, Simone S, Bucci R and Farber E (2002). Cell death: apoptosis versus necrosis. Int J Oncol. 21(1): 165-170

Kaminskyj SG and Hamer JE (1998). *hyp* loci control cell pattern formation in the vegetative mycelium of *Aspergillus nidulans*. 148(2): 669-680

Kawamura K, Bahar R, Seimiya M, Chiyo M, Wada A, Okada S, Hatano M, Tokuhisa T, Kimura H, Watanabe S, Honda I, Sakiyama S, Tagawa M and O-Wang J (2004). DNA polymerase θ is preferentially expressed in lymphoid tissues and upregulated in human cancers. Int J Cancer 109(1): 9-16

Khakhar RR, Cobb JA, Bjergbaek L, Hickson ID and Gasser SM (2003). RecQ helicases: multiple roles in genome maintenance. Trends Cell Biol. 13(9): 493-501

Kitao S, Ohsugi I, Ichikawa K, Goto M, Furuichi Y and Shimamoto A (1998). Cloning of two new human helicase genes of the RecQ family: biological significance of multiple species in higher eukaryotes. Genomics 54(3): 443-452

Kraus PR and Harris SD (2001). The *Aspergillus nidulans snt* genes are required for the regulation of septum formation and cell cycle checkpoints. Genetics 159(2): 557-569

Kraus PR, Hofmann AF and Harris SD (2002). Characterization of the *Aspergillus nidulans* 14-3-3 homologue, ArtA. FEMS Microbiol Lett. 210(1): 61-66

Laun P, Pichova A, Madeo F, Fuchs J, Ellinger A, Kohlwein S, Dawes I, Frohlich KU and Breitenbach M (2001). Aged mother cells of *Saccharomyces cerevisiae* show markers of oxidative stress and apoptosis. Mol Microbiol.39(5): 1166-1173

Lavrik OI, Prasad R, Sobol RW, Horton JK, Ackerman EJ and Wilson SH (2001). Photoaffinity labeling of mouse fibroblast enzymes by a base excision repair intermediate. Evidence for the role of poly(ADP-ribose) polymerase-1 in DNA repair. J. Biol. Chem. 276(27): 25541-25548

Ludovico P, Sousa MJ, Silva MT, Leao C, Corte-Real M (2001). *Saccharomyces cerevisiae* commits to a programmed cell death process in response to acetic acid. Microbiology 147(Pt 9): 2409-2415

Ludovico P, Sansonetty F, Silva MT and Corte-Real M (2003). Acetic acid induces a programmed cell death process in the food spoilage yeast *Zygosaccharomyces bailii*. FEMS Yeast Res. 3(1): 91-96

Madeo F, Frohlich E and Frohlich KU (1997). A yeast mutant showing diagnostic markers of early and late apoptosis. J Cell Biol. 139(3): 729-734

Madeo F, Frohlich E, Ligr M, Grey M, Sigrist SJ, Wolf DH and Frohlich KU (1999). Oxygen stress: a regulator of apoptosis in yeast. J Cell Biol. 145(4): 757-767

Madeo F, Herker E, Maldeher C, Wissing S, Lachelt S, Herlan M, Fehr M, Lauber K, Sigrist SJ, Wesselborg S and Frohlich KU (2002). A caspase-related protease regulates apoptosis in yeast. Mol Cell. 9(4):911-917

Marini F and Wood RD (2002). A human DNA helicase homologous to the DNA cross-link sensitivity protein Mus308. J Biol Chem. 277(10): 8716-8723

Marini F, Kim N, Schuffert A, and Wood RD (2003). POLN, a nuclear PolA family DNA polymerase homologous to the DNA cross-link sensitivity protein Mus308. J Biol Chem. 278(34): 32014-32019

Masson M, Niedergang C, Schreiber V, Muller S, Menissier-de Murcia J and de Murcia G (1998). XRCC1 is specifically associated with poly(ADP-ribose) polymerase and negatively regulates its activity following DNA damage.Mol Cell Biol. 18(6):3563-3571

Mohaghegh P and Hickson ID (2001). DNA helicase deficiencies associated with cancer predisposition and premature aging disorders. Hum. Mol. Genet. 10(7): 741-746

Mousavi SA and Robson GD (2003). Entry into the stationary phase is associated with a rapid loss of viability and an apoptotic-like phenotype in the opportunistic pathogen *Aspergillus fumigatus*. Fungal Genet Biol. 39(3): 221-229

Mousavi SA and Robson GD (2004). Oxidative and amphotericin B-mediated cell death in the opportunistic pathogen *Aspergillus fumigatus* is associated with an apoptotic-like phenotype. Microbiology 150(Pt 6):1937-1945

Ninomiya Y, Suzuki K, Ishii C and Inoue H (2004). Highly efficient gene replacements in *Neurospora* strains deficient for nonhomologous end-joining. Proc Natl Acad Sci U S A. Aug 6 [Epub ahead of print]

Nyberg KA, Michelson RJ, Putnam CW and Weinert TA (2002). Toward maintaining the genome: DNA damage and replication checkpoints. Annu. Rev. Genet. 36:617-656

Osmani SA and Mirabito PM (2004). The early impact of genetics on our understanding of cell cycle regulation in *Aspergillus nidulans*. Fungal Genet Biol. 41(4):401-410

Pang M, McConnell M and Fisher PA (2001). The *mus308* gene codes for a nuclear DNA polymerase/atpase implicated in interstrand cross-link repair. 42nd Annual Drosophila Research Conference, Washington, http://genetics.faseb.org/genetics/dros01/html/f301.htm

Petrini JHJ and Stracker TH (2003). The cellular response to DNA double-strand breaks: defining the sensors and mediators. Trends Cell Biol. 13(9): 458-462

Phillips AJ, Sudbery I and Ramsdale M (2003). Apoptosis induced by environmental stresses and amphotericin B in *Candida albicans*. Proc Natl Acad Sci USA 100(24): 14327-14332

Pickford A, Braccini L, Macino G and Cogoni C (2003). The QDE-3 homologue RecQ-2 co-operates with QDE-3 in DNA repair in *Neurospora crassa*. Curr Genet. 42(4): 220-227

Qi H, Li TK, Kuo D, Nur-E-Kamal A and Liu LF (2003). Inactivation of Cdc13p triggers MEC1-dependent apoptotic signals in yeast. J Biol Chem 278(17): 15136-15141

Seki M, Marini F and Wood RD (2003). POLQ (Pol theta), a DNA polymerase and DNA-dependent ATPase in human cells. Nucleic Acids Res. 31: 6117-6126

Sekelsky JJ, Burtis KC and Hawley RS (1998). Damage control: the pleiotropy of DNA repair genes in *Drosophila* melanogaster. Genetics 148(4): 1587-1598

Semighini CP, von Zeska Kress Fagundes MR, Ferreira JC, Pascon RC, de Souza Goldman MH and Goldman GH (2003). Different roles of the Mre11 complex in the DNA damage response in *Aspergillus nidulans*. Mol Microbiol. 48(6):1693-1709

Severin FF and Hyman AA (2002). Pheromone induces programmed cell death in *S. cerevisiae*. Curr. Biol. 12(13): R233-R235

Shall S and de Murcia G (2000). Poly(ADP-ribose) polymerase-1: what have we learned from the deficient mouse model? Mutant Res. 460(1): 1-15

Shiloh Y (2001). ATM and ATR: networking cellular responses to DNA damage. Curr Opin Genet Dev. 11(1): 71-77

Shima N, Hartford SA, Duffy T, Wilson LA, Schimenti KJ and Schimenti JC (2003) Phenotype-based identification of mouse chromosome instability mutants. Genetics, 163 (3): 1031-1040

Sluder G, Thompson EA, Rieder CL and Miller FJ (1995). Nuclear envelope breakdown is under nuclear not cytoplasmic control in sea urchin zygotes. J Cell Biol. 129(6): 1447-1458

Smith S. (2001). The world of according to PARP. Trends Biochem. Sci., 26(3): 174-179

Stergiou L and Hengartner MO (2004). Death and more: DDR pathways in the nematode*C. elegans*. Cell Death Differ. 11:21-28

Sullivan W, Daily DR, Fogarty P, Yook KJ and Pimpinelli S (1993). Delays in anaphase initiation occur in individual nuclei of the syncytial *Drosophila* embryo. Mol Biol Cell. 4(9): 885-896

Thrane C, Kaufmann U, Stummann BM and Olsson S (2004). Activation of caspase-like activity and poly(ADP-ribose)polymerase degradation during sporulation in *Aspergillus nidulans*. Fungal Genet Biol. 41(3): 361-368

Tong WM, Cortes U andWang ZQ (2001). Poly(ADP-ribose) polymerase: a guardian angel protecting the genome and suppressing tumorigenesis. Biochim Biophys Acta. 1552(1): 27-37

Tuteja N and Tuteja R (2004). Prokaryotic and eukaryotic DNA helicases. Essential molecular motor proteins for cellular machinery. Eur J Biochem. 271(10): 1835-1848

Tutt A and Ashworth A (2002). The relationship between the roles of *BRCA* genes in DNA repair and cancer predisposition. Trends Mol. Med. 8:571-576

Tsubouchi H and Ogawa H (1998). A novel *mre11* mutation impairs processing of double-strand breaks of DNA during both mitosis and meiosis. Mol Cell Biol. 18(1): 260-268

Virag L and Szabo C (2002). The therapeutic potential of poly(ADP-ribose)polymerase inhibitors. Pharmacolo Rev 54(3): 375-429

Wang Y, Cortez D, Yazdi P, Neff N, Elledge SJ and Qin J. (2000). BASC, a super-complex of BRCA1-associated proteins involved in the recognition and repair of aberrant DNA structures. Genes Dev., 14:927-939

Wolkow TD, Harris SD and Hamer JE (1996). Cytokinesis in *Aspergillus nidulans* is controlled by cell size, nuclear positioning and mitosis. J Cell Sci. 109 (Pt 8): 2179-2188

Wolkow TD, Mirabito PM, Venkatram S and Hamer JE (2000). Hypomorphic *bimA*(APC3) alleles cause errors in chromosome metabolism that activate the DNA damage checkpoint blocking cytokinesis in *Aspergillus nidulans*. Genetics 154(1): 167-179

Wysocki R and Kron SJ (2004). Yeast cell death during DNA damage arrest is independent of caspase or reactive oxygen species. J Cell Biol. 166(3): 311-316

Yamaki M, Umehara T, Chimura T and Horikoshi M (2001). Cell death with predominant apoptotic features in *Saccharomyces cerevisiae* mediated by deletion of the histone chaperone ASF1/CIA1. Genes Cells. 6(12):1043-1054

Ye XS, Fincher RR, Tang A, O'Donnell K and Osmani SA (1996). Two S-phase checkpoint systems, one involving the function of both BIME and Tyr15 phosporylation of p34^{cdc2}, inhibit NIMA and prevent premature mitosis. EMBO J., 15(14): 3599-3610

Ziegler M and Oei SL (2001). A cellular survival switch: poly(ADP-ribosyl)ation stimulates DNA repair and silences transcription. Bioessays 23(6): 543-548

Zou L, Cortez D and Elledge SJ (2002). Regulation of ATR substrate selection by Rad17-dependent loading of Rad9 complexes onto chromatin Genes Dev. 16(2): 198-208

Zou L and Elledge SJ (2003). Sensing DNA damage through ATRIP recognition of RPA-ssDNA complexes. Science 300(5625): 1542-1548

Zou L, Liu D and Elledge SJ (2003). Replication protein A-mediated recruitment and activation of Rad17 complexes PNAS 100(24): 13827-13832

**Applied Mycology
and Biotechnology**
An International Series
Volume 5. Genes and Genomics
© 2005 Elsevier B.V. All rights reserved

ELSEVIER

6

The *cfp* Genes (*cfp-1* and *cfp-2*) of *Neurospora crassa*: A Tale of a Bunch of Filaments, One Enzyme, and Two Genes

Esteban D. Temporini[1], Hernan D. Folco[2] and Alberto L. Rosa*
Instituto de Investigación Médica "Mercedes y Martín Ferreyra", INIMEC-CONICET, Córdoba, Argentina; [1]Division of Plant Pathology and Microbiology, Department of Plant Sciences, University of Arizona, Tucson, AZ, USA; [2]Wellcome Trust Centre for Cell Biology, University of Edinburgh, Scotland, United Kingdom.

For over three decades, scientists have been puzzled by the observation of cryptic filamentous structures in the cytoplasm of filamentous fungi. This chapter describes major developments in the characterization of 8 to 10-nm in diameter cytoplasmic filaments from *Neurospora crassa*. We review the studies that demonstrated that these structures are made of the tetrameric protein pyruvate decarboxylase (PDC), a key enzyme catalyzing the branch-point step between the respiration and fermentation pathways in fungi. Similar arrangements of PDC have been found in *N. tetrasperma*, *Podospora anserina* and *Sordaria macrospora*, showing that PDC filaments are ubiquitous structures in fungi. A complex array of signals, and possibly alternative splicing, modulate the expression of the genes *cfp-1* and *cfp-2* encoding PDC in *N. crassa*. Challenging future studies should address the relationship between the supramolecular organization and function of PDC in filamentous fungi.

1. INTRODUCTION

Filamentous fungi, unlike yeasts and slime molds, grow by apical elongation of tubular cells, which extend by the incorporation of new cell wall material at the cell apex (Bartnicki-Garcia and Lippman 1969; Gooday 1971; Katz and Rosenberger 1971). This highly polarized growth is facilitated by a complex array of elements of the cytoskeleton that allows a coordinated interaction of processes like secretion, organelle movement and cell wall deposition (reviewed by Weber and Pitt 2001). For decades, scientists have devoted considerable efforts to characterize the cytoskeleton of filamentous fungi and the physiological processes to which it is associated. This

*Corresponding author: Alberto L. Rosa

resulted in a wealth of information on the proteins constituting the fungal cytoskeleton as well as the molecular mechanisms involved in the regulation of their expression. The main components of the fungal cytoskeleton as we know it today are microtubules, actin, and the motor proteins dynein, kinesin and myosin (reviewed by Xiang and Plamann 2003). In addition, septins have also been described more recently in filamentous fungi (Westfall and Momani 2002).

But for years, a number of cytoplasmic filamentous structures not related to tubulin or actin have fascinated those who study the cytoskeleton of these microorganisms. And although these enigmatic filaments were reported in a number of publications, their biochemical nature, as well as their function in the fungal cell, remained obscure (Anderson and Zachariah 1974; Caesar et al. 1987; Gull 1975; Heath 1987; Hoch and Howard 1980; Howard 1981; Raudaskoski and Koltin 1973). In addition, there was a great deal of discrepancy between observations made by different laboratories concerning the presence of these filaments in the cytoplasm versus the nuclei, their relative abundance in the cell and, in some cases, their occurrence in wild type versus mutant strains (Allen et al. 1974; Beck et al. 1970; Rosa et al. 1990 a, b, c; Wood and Luck 1971).

Some of the "mysteries" around the cryptic cytoplasmic filaments in fungi began to be resolved between 1992 and 1993, when the gene *cfp-1* encoding P59Nc, the 59-kDa polypeptide constituent of cytoplasmic filaments of 8-10 nm in diameter in *Neurospora crassa* (Rosa et al. 1990 a, b, c), was cloned and characterized (Haedo et al. 1992; Alvarez et al. 1993). These studies demonstrated that P59Nc is pyruvate decarboxylase (PDC), which is a key postglycolytic enzyme that catalyses the branch-point step between respiration and fermentation (Konig 1998). In addition to being the first report showing that this enzyme has such a complex supramolecular arrangement *in vivo*, these results demonstrated that the presence and abundance of PDC filaments depended on the metabolic condition of the cells (Alvarez et al. 1993). These findings could explain many of the discrepancies reported on the nature and abundance of cytoplasmic filaments in fungi.

In this chapter, we review the history of the developments that led to our current understanding of the identity and function of the PDC filaments in *N. crassa*. Our knowledge of the *cfp-1* and *cfp-2* genes, the PDC enzyme and the 8-10 nm cytoplasmic filaments has evolved at the pace of technological developments, from basic biochemistry to classical and molecular genetics, to modern genomics. From a bunch of filaments to one enzyme and two genes.

2. THE PRE-GENOMIC ERA: BIOCHEMICAL AND ULTRASTRUCTURAL STUDIES OF P59Nc AND THE 8-10 nm CYTOPLASMIC FILAMENTS OF *N. CRASSA*

2.1. The Cytoplasmic Filaments of Filamentous Fungi

By the mid 1980s, a general picture of the constituents of the fungal cytoskeleton was beginning to emerge, with studies that described and characterized cytoskeletal proteins such as tubulin and actin (Adams and Pringle 1984; Davidse and Flach 1977; Fidel et al. 1988; Gambino et al. 1984; Greer and Schekman 1982; Huffaker et al. 1987;

Kilmartin and Adams 1984; Morris 1986). There was also an increasing awareness of the existence of a number of non-tubulin, non-actin filamentous structures in the cytoplasm of fungal cells that attracted the attention of scientists for years, although their biochemical nature remained uncertain (Anderson and Zachariah 1974; Caesar et al. 1987; Gull 1975; Heath 1987; Hoch and Howard 1980; Howard 1981; Raudaskoski and Koltin 1973).

In *N. crassa*, several authors reported the presence of these cryptic filaments, and although their ultrastructural features were very similar (Figure 1), they were called a number of different names, like "striated inclusions" (Beck et al. 1970). "paracrystals" (Wood and Luck 1971), "actin-like microfilaments" (Allen and Sussmann 1978), or simply "cytoplasmic filaments" (Rosa et al. 1990 a and b). This was perhaps a consequence of the great variability in the relative abundance of these structures, likely based on different metabolic conditions, and the dispute over their detection in wild type versus mutant strains. Although most of these controversial observations can now be explained in light of our current knowledge of the biochemical and genetic nature of the cytoplasmic filaments in *N. crassa* (Alvarez et al. 1993), they constituted an enigma in the early studies of the fungal cytoskeleton. Beck and colleagues (1970), for example, observed that "striated inclusions" accumulated in the cytoplasm of the wild type strain growing in sucrose or glucose. However, the inclusions were undetectable if the mycelia were grown in a non-fermentable carbon source or in media without a carbon source. They hypothesized that the absence of striated inclusions in cells growing under these specific metabolic conditions may reflect the inability of these cells to synthesize

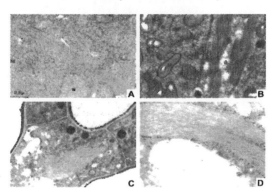

Fig. 1. Electron microscopic pictures of cryptic cytoplasmic filaments from *N. crassa*. A: "striated inclusions" (adopted from: Beck et al. 1970, with permission); B: "paracrystals" (adopted from: Wood and Luck 1971, with permission); C: "actin-like microfilaments" (adopted from: Allen et al. 1974, with permission); D: "cytoplasmic filaments".

and/or assemble the components of the inclusions. In another study, Wood and Luck (1971) demonstrated that the "paracrystals" that accumulate in some mitochondrial mutants of *N. crassa* with defects in respiration could be induced in the wild type strain by treatment with ethidium or euflavine, which are drugs that selectively inhibit mitochondrial function. These authors found that the paracrystals were composed of a nonmitochondrial protein, which is a normal product of cytoplasmic, rather than mitochondrial, protein synthesis. The paracrystals could be dissociated when exposed to Tris-based buffers with pH above 7, and they readily re-associated after dialysis against non-Tris buffers with lower pH. Moreover, the authors hypothesized that the protein of the paracrystal is a single polypeptide that undergoes self-aggregation into tetramers, which arrange into filaments (Wood and Luck 1971).

In 1990, Rosa and colleagues reported the biochemical and ultrastructural characterization of bundles of cytoplasmic filaments in *N. crassa* (Rosa et al. 1990 a). They described the association of a 59-kDa polypeptide with complex arrays of filaments observed by electron microscopy of a highly purified preparation of this polypeptide, obtained after subcellular fractionation and glass permeation chromatography (Figure 2). A similar procedure had been previously used by these authors to purify and characterize clathrin-coated vesicles from *N. crassa* (Rosa et al. 1987). Analyses of the cellular distribution of the 59-kDa polypeptide (termed P59Nc) by immunoelectron microscopy and indirect immunofluorescence utilizing antibodies

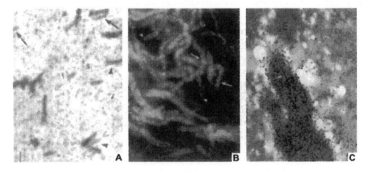

Fig. 2. The 8-10-nm diameter filaments of P59Nc. A: Electron microscopic picture of P59Nc filaments in a highly purified fraction after glass permeation chromatography (adopted from: Rosa et al. 1990a, with permission). B: Immunofluorescence staining of P59Nc filaments (adopted from: Rosa et al. 1990b, with permission). C: Electron microscopy of P59Nc filaments decorated with colloidal gold-conjugated protein A after treatment with anti-P59Nc antibodies (adopted from: Rosa et al. 1990b, with permission).

raised against it, showed that, *in vivo*, it is organized in bundles of a variable number of filaments, with each filament being 8-10 nm in diameter (Figure 2; Rosa et al. 1990 a and b). The filaments were randomly distributed in the cytoplasm and they were not associated in any kind of network. Interestingly, similar to the paracrystals described by Wood and Luck (1971), the P59Nc filaments could be disassembled by exposure to Tris-based buffers with pH 7.5 and re-assembled by shifting the pH to 6.5. The possibility to dissociate and re-associate the filaments by changes in pH suggested a method for fast and efficient purification of P59Nc (Rosa et al. 1990b). Further characterization demonstrated that P59Nc and the 8-10 nm filaments were not immunologically related to tubulin or actin. However, a weak cross-reaction between P59Nc and the anti-IFA monoclonal antibody, which recognizes an antigenic determinant common to many members of the intermediate filament polypeptide family of higher eukaryotes, was observed (Rosa et al. 1990a). Thus, although P59Nc appeared to be unrelated to well established fungal cytoskeletal components like tubulin or actin, its supramolecular organization in the cytoplasm, plus the weak reaction with the anti-IFA antibody suggested the possibility that P59Nc and the 8-10 nm filaments could be distantly related to intermediate filament proteins of higher eukaryotes and play a structural role associated with the cytoskeleton of *N. crassa*.

2.2. *Snowflake* Mutants of *N. crassa* have Abnormal Bundles of P59Nc

The intriguing possibility that the filaments of P59Nc could represent a novel component of the cytoskeleton in filamentous fungi had additional support in studies performed on *snowflake*, a morphological mutant of *N. crassa*. *Snowflake* mutants are characterized by their colonial growth, with altered patterns of hyphal branching, tip elongation, and spore morphology (Perkins et al. 2001; Temporini and Rosa 1993).

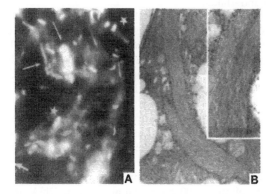

Fig. 3. The *snowflake* mutant of *N. crassa* accumulates abnormally shaped and sized filaments of P59Nc. **A:** immunofluorescence staining. **B:** electron microscopy. (adopted from: Rosa et al. 1990c, with permission)

Interestingly, in 1974, Allen and co-workers reported a massive accumulation of bundles of "actin-like microfilaments" in these mutants (Figure 1C). An apparent correlation was then established between the accumulation of these microfilaments and the morphological phenotype of *snowflake* mutants, which suggested that the filaments may play a structural role implicated in some process related to cell growth and/or morphogenesis (Allen et al. 1974).

The microfilaments of the *snowflake* mutants shown in the electron microscopy pictures published by Allen et al. in 1974 had a striking resemblance to the P59Nc filaments found by Rosa and colleagues in the wild type strain (Rosa et al. 1990 a and b).

Not surprisingly, when the antibodies against P59Nc were used in immunofluorescence and immunoelectron microscopy studies of *snowflake* mutants, the results showed that the microfilaments reported by Allen and colleagues were made of P59Nc (Rosa et al. 1990 c). These studies also revealed that *snowflake* mutants have abnormally shaped and sized bundles of P59Nc filaments (Figure 3), although the polypeptide purified from these mutants appeared to be normal in its subcellular distribution and its relative abundance in total mycelia (Rosa et al. 1990c). It was not clear, however, whether the morphological phenotypes observed in *snowflake* strains were a consequence of the abnormally assembled P59Nc filaments, or vice versa.

3. THE GENETICS BEHIND THE CYTOPLASMIC FILAMENTS OF *N. CRASSA*
3.1. The *cfp-1* Gene and its Genetic Relationship to the *snowflake* Locus

The results obtained analyzing the *snowflake* mutants suggested that *sn*, the genetic locus affected in these mutants, could be the locus for the gene encoding P59Nc. Alternatively, the *sn* gene could encode a modifier of the *in vivo* supramolecular assembly of P59Nc (Rosa et al. 1990c). Answers to these questions started to emerge when the molecular cloning of the gene encoding P59Nc, *cfp-1* (cellular filament polypeptide), was undertaken.

The cloning of *cfp-1* started with a ~250-bp cDNA, isolated after immunoscreening of a mycelial cDNA expression library of *N. crassa*, utilizing anti-P59Nc antibodies (Haedo et al. 1992). This small cDNA was used as a probe to obtain several larger cDNA clones, including a full-length clone of ~2.0 kb (pET2; Haedo et al. 1992). This cDNA clone was confirmed to correspond to P59Nc by comparing its partial nucleotide sequence with a 23-amino-acid partial sequence obtained from the N-terminus of the purified native P59Nc.

With the P59Nc-cDNA available, an important question could now be addressed: are *sn* and *cfp-1* encoded by the same locus? Using pET2 as a probe, the pattern of segregation of RFLPs detected by the P59Nc-cDNA among selected progeny from a cross between two *N. crassa* strains with significantly different genetic backgrounds (*Mauriceville*-1-c and *Multicent*-2; Metzenberg et al. 1984), was compared to the segregation patterns of about 100 genes and anonymous DNA fragments obtained by Metzenberg and Grotelueschen (1989) in the same set of strains. The results obtained revealed that *cfp-1* is located on the right arm of linkage group VII (Haedo et al. 1992). This result ruled out the possibility that *cfp-1* and *sn* were encoded by the same genetic locus, since *sn* is located in the centromeric region of linkage group I (Perkins et al. 2001; Temporini and Rosa 1993). Nonetheless, there was a formal possibility that the *sn* mutations were the result of a translocation of altered forms and/or abnormally expressing versions of the *cfp-1* gene to the centromere region of chromosome I. However, Southern blot analyses of the wild type and *snowflake* strains, as well as RFLP segregation patterns among the progeny of a cross between them, showed that, in the *snowflake* mutants, there is a single *cfp-1* locus with no apparent rearrangements located on the right arm of linkage group VII (Haedo et al. 1992). These findings indicated that the aberrant bundles of P59Nc filaments observed in *snowflake* mutants are not a consequence of mutations in the *cfp-1* gene. Therefore, it was hypothesized that the product of the *sn* locus could act as a post-translational modifier of the P59Nc polypeptide and/or of the *in vivo* capability of the P59Nc filaments to form bundles (Haedo et al. 1992). This hypothesis remains to be addressed experimentally.

3.2. The *cfp-1* Gene Encodes a Pyruvate Decarboxylase

When the combined nucleotide sequences from various *cfp-1* cDNA clones (including the full-length cDNA clone pET2) were analyzed, a putative open reading frame (ORF) of 1,713 bp was identified (Alvarez et al. 1993). This ORF encoded a predicted polypeptide of 62.3 kDa, which is in close agreement to the 59 kDa of P59Nc as estimated by SDS-PAGE (Rosa et al. 1990a). In addition, a comparison between the cDNA and genomic sequences indicated that no introns were present in the *cfp-1* gene. The big surprise, however, came when the predicted amino acid sequence of P59Nc was compared to protein sequences present in the SWISSPROT database. This analysis revealed that P59Nc has no homology to any cytoskeleton-associated protein. The best scores were the enzymes pyruvate decarboxylase from *Zymomonas mobilis* (51.5% identity), indolepyruvate decarboxylase from *Enterobacter cloacae* (33.2% identity) and pyruvate decarboxylase isoenzymes 1 and 2 from *Saccharomyces cerevisiae* (30.4% and 29.1% identity, respectively). These enzymes belong to a family of proteins that require

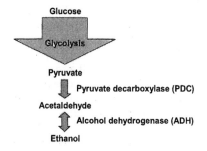

Fig. 4. The pathway of ethanolic fermentation. Sugars, represented by glucose, are converted to pyruvate at the end of glycolysis. Subsequently, PDC catalyses the conversion of pyruvate to acetaldehyde, which is then converted to ethanol by ADH.

the co-factor thiamin diphosphate for catalytic activity (Hohmann and Meacock, 1998; Schorken and Sprenger, 1998). In particular, pyruvate decarboxylase (PDC) catalyses the non-oxidative decarboxylation of pyruvate to acetaldehyde, which is then converted to ethanol by alcohol dehydrogenase (ADH) in the process known as ethanolic fermentation (Figure 4). This pathway has been extensively studied in the yeast *S. cerevisiae* (reviewed by Barnett 2003), as well as in bacteria and plants (Tadege et al. 1999; Ward, 2002).

The high similarity of P59Nc to PDCs suggested the possibility that P59Nc may be a pyruvate decarboxylase. This possibility was confirmed when a pure preparation of P59Nc was demonstrated to have PDC activity (Alvarez et al. 1993). Moreover, it was also determined that the expression of the *cfp-1* gene was regulated in response to the different carbon sources present in the culture medium. Thus, the levels of *cfp-1*-mRNA in mycelia grown in media containing glucose were approximately 10-fold higher than in media containing sucrose, while no cfp-1-mRNA could be detected when the mycelia was grown in ethanol, which is a non-fermentable carbon source. This variation in the levels of cfp-1-mRNA correlated with variations in PDC activity measured in cells grown in the same carbon sources (e.g., the PDC activity detected in mycelia grown in glucose was approximately three times higher than in sucrose, while no PDC activity could be detected in mycelia grown in ethanol). Similar patterns of carbon source-dependent regulation of the expression and enzymatic activity of PDC have been reported in *S. cerevisiae* (Butler and MacConnell 1988; Kellermann and Hollemberg 1988), *Z. mobilis* (Conway et al. 1987), and *Kluyveromyces lactis* (Bianchi et al. 1996; Destruelle et al. 1999). But perhaps the most interesting observation was that the abundance of the 8-10 nm filaments of P59Nc also correlated with the levels of cfp-1-mRNA. In mycelia grown in glucose, the 8-10 nm filaments are more abundant than in mycelia grown in sucrose, but they are completely absent in mycelia grown in ethanol (Alvarez et al. 1993). These results represented the first demonstration that pyruvate decarboxylase has a complex supramolecular arrangement *in vivo*, and it was hypothesized that the filamentous matrix composed of the PDC filaments would provide support to a complex multienzymatic network that may include other glycolytic and post-glycolytic enzymes. This enzymatic network could help to increase the efficiency of the glycolytic flux by facilitating the process of substrate channeling (Miles et al. 1999). Alternatively, the PDC filaments could play a dual role as a catalytic enzyme and as a structural component of the cytoskeleton. The observation that mycelia grown in ethanol are depleted of filaments but show normal patterns of hyphal morphology and branching (Alvarez et al. 1993), however, strongly argues against a possible function of the 8-10 nm filaments in cell morphogenesis. It also implies that the morphological phenotype of the *snowflake* mutants is not a consequence of the abnormally assembled PDC filaments. Instead, it seems more likely that the PDC filaments are abnormally assembled as a consequence of the altered growth of these strains. It is also possible that the abnormal bundles of PDC have an additional "toxic" effect on cell elongation and/or branching, contributing to the morphological phenotype of these mutants.

The findings that the 8-10 nm filaments in *N. crassa* are made of PDC and that their abundance in the cytoplasm depends on the metabolic condition of the cells, provided a long awaited explanation to the puzzling observations of cryptic cytoplasmic filaments in *N. crassa* and other fungi. For example, the "striated inclusions" of *N. crassa* described by Beck and colleagues (1970), which accumulated in mycelia grown in glucose and were absent in mycelia grown in a non-fermentable carbon source, most likely correspond to the 8-10 nm filaments of PDC. Likewise, several lines of evidence make it reasonable to conclude that the "paracrystals" described by Wood and Luck in 1971 corresponded to PDC filaments. First, the paracrystals accumulate in mitochondrial mutants of *N. crassa* with defects in respiration or in the wild type strain after treatment with drugs that inhibit mitochondrial function. In such a respiratory stress, energy production would heavily rely on the fermentative pathway, which depends on the activity of PDC. Therefore, accumulation of PDC filaments is expected to occur in those conditions. Second, the "paracrystals" and the PDC filaments have similar biochemical properties: they can be dissociated when exposed to Tris-based buffers with pH above 7, and re-assembled after dialysis against non-Tris buffers with lower pH. Finally, Wood and Luck hypothesized that the "paracrystals" were made of a single polypeptide that aggregates into tetramers. Interestingly, the 8-10 nm filaments are composed of a single polypeptide (Rosa et al. 1990 a and b), which we now know is PDC, a homotetrameric enzyme (Konig 1998).

Similar filaments made of PDC have also been found in other *Neurospora* species as well as in *Podospora anserina* and *Sordaria macrospora* (see below; Thompson-Coffe et al. 1999). Thus, after years of research a fascinating mystery is being resolved: a large fraction of the cryptic cytoplasmic filaments observed in fungi are made of pyruvate decarboxylase.

4. THE PDC FILAMENTS AND THE SEXUAL CYCLE: THE CYTOSKELETON REVISITED

4.1. PDC Filaments are Associated with the Cytoskeleton of Asci and Spores During the Sexual Cycle of Filamentous Ascomycetes

Since their discovery in 1990 (Rosa et al. 1990a), the P59Nc/PDC filaments have been suspected to play some structural role either as components of the cytoskeleton or in association with some element of the cytoskeleton. However, their biochemical nature - they are made of an active post-glycolytic enzyme-, plus the observations that vegetative *N. crassa* hyphae have normal patterns of growth and branching when depleted of PDC filaments (e.g., growing on ethanol as the sole carbon source; Alvarez et al. 1993), strongly implied that the filaments do not have a structural function required for normal cell development during vegetative growth. However, no information was available about the expression and structural organization of PDC during the sexual cycle.

In 1999, Thompson-Coffe and co-workers used immunofluorescence to show that PDC also forms filaments during the sexual cycle of *N. crassa*. In addition, they showed that PDC has the same filamentous supramolecular arrangement in *N. tetrasperma*, *Podospora anserina* and *Sordaria macrospora* (Thompson-Coffe et al. 1999). However, unlike the filaments of vegetative mycelia, which appear to be randomly distributed in the cytoplasm (Rosa et al. 1990 a and b), in asci and ascospores of *N. crassa* and the other fungi tested the PDC filaments colocalize with the cortical microtubule array (Figure 5, A and B). Moreover, it was shown that in all four species analyzed, treatment with nocodazole resulted in the loss of both cortical microtubule and PDC filament organization (Figure 5, C and D). These results strongly suggested an association between the PDC filaments and the cortical array of microtubules.

Additional support for this hypothesis was provided by the analysis of the *spo644* mutant strain of *S. macrospora*. In this mutant the cortical microtubules are aberrantly arranged, which results in abnormally shaped asci compared to the wild type strain. Interestingly, the PDC filaments of *spo644* are also disorganized, with a random pattern

150

of distribution (Figure 5, E and F). Thus, while in the vegetative cycle the PDC filaments do not show any particular pattern of organization in the cell, during the sexual cycle they are clearly associated with the cortical microtubule array.

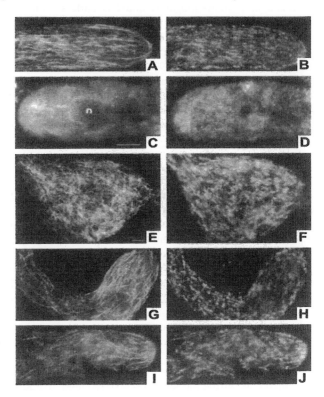

Fig. 5. PDC filaments during the sexual phase of filamentous fungi. Immunofluorescence of microtubules (A, C, E, G, I) and PDC filaments (B, D, F, H, J) during sexual development of *N. crassa* (A-D and G-J) and the *spo644* strain of *S. macrospora* (E, F). The PDC filaments (B) run parallel to the cortical microtubule array (A) of prophase asci. Treatment with nocodazole results in disruption of microtubule (C) and PDC filaments (D) organization. E: Anti-tubulin stain of an ascus of *spo644*. F: Anti-PDC stain of the ascus shown in E. In asci of *snowflake* mutants, the PDC filaments (H) run parallel to the cortical microtubules (G) and are indistinguishable from those of the WT strain (compare G and H to A and B). In asci of crosses performed on ethanol, PDC filaments are readily detected (J) and are associated with the cortical microtubule array (I). (adopted from: Thompson-Coffe et al. 1999, with permission)

The association of glycolytic enzymes to elements of the cytoskeleton has been well described in mammals (Arnold and Pette 1968; reviewed by Beitner 1993, and Pagliaro 1993). This interaction has been proposed to increase the efficiency of glycolysis by placing the different enzymes in close proximity to each other, improving in that way the glycolytic flux. Interestingly, Götz and colleagues (1999) showed that an intact cytoskeleton is required for metabolic reorganization from a gluconeogenic to a glycolytic metabolism in *S. cerevisiae*, suggesting the involvement of the cytoskeleton in a functional organization of glycolytic enzymes in yeast. However, it is difficult to apply this hypothesis to *N. crassa* because during vegetative growth there is no obvious association of the PDC filaments with any element of the cytoskeleton (Rosa et al. 1990a and b). It is possible that there is a specific need for the activity of PDC to be localized to a particular cell compartment during sexual development, which would explain its association with the cytoskeleton only during the sexual cycle. Alternatively, while the catalytic activity of PDC may not be specifically required during the sexual cycle, it is possible that the PDC filaments are necessary for normal asci development. Thus, the PDC filaments would have a structural role during the development of asci and ascospores, in contrast to their dispensability for normal development of vegetative hyphae. Thompson-Coffe and colleagues (1999) hypothesized that the PDC filaments could stabilize the cortical microtubule array, which is known to be required for morphogenesis during sexual development (Thompson-Coffe and Zickler 1992). Since microtubules are not essential for hyphal extension during vegetative growth (Raudaskoski et al. 1994), stabilization of the cortical microtubules, and therefore the PDC filaments, would be dispensable for morphogenesis in vegetative cells. The generation of mutant strains unable to make PDC filaments or expressing altered versions of the protein should help to understand the function of the filaments during the sexual cycle of *N. crassa* and other filamentous fungi.

4.2. The Organization of PDC Filaments in Sexual Tissues is not Affected by the *sn* Mutation and is Independent of the Metabolic Condition of the Cell

Because of the aberrant shape and size of the PDC filaments observed in *snowflake* (Rosa et al. 1990c), these mutants seemed well suited to further analyze the possible role of the PDC filaments in cell morphogenesis during sexual development. Immunofluorescence studies using anti-PDC antibodies showed that the PDC filaments of paraphysal cells (e.g., sterile hyphae associated with asci in the perithecium) of crosses involving *snowflake* strains presented the same abnormal pattern of assembly seen in *snowflake* vegetative mycelia. Likewise, the PDC filaments of wild type paraphysal cells were similar to those seen in wild type vegetative mycelium (Thompson-Coffe et al. 1999). Therefore, the supramolecular organization of PDC in paraphysal cells of both wild type and *snowflake* mutants was indistinguishable from that of their respective vegetative hyphae. Interestingly, and somewhat surprisingly, the organization, shape, and association with the cortical microtubule array of the PDC filaments in asci of *snowflake* were indistinguishable from those of wild type asci (Figure 5, G and H; Thompson-Coffe et al. 1999). Thus, the genetic defect of the *snowflake* mutants appears to selectively affect the assembly of PDC filaments in vegetative mycelium and paraphysal cells but not during the development of asci. These observations suggested that the assembly of the PDC filaments could be controlled by different mechanisms in vegetative and sexual cells. Moreover, the expression of the *cfp-*

1 gene appeared to be differentially regulated during the vegetative and sexual cycles: while the expression of *cfp-1* is repressed in vegetative mycelia grown in ethanol, resulting in a complete absence of PDC filaments and catalytic activity (Alvarez et al. 1993), asci of crosses performed on ethanol showed normal patterns of PDC filaments (Figure 5, I and J; Thompson-Coffe et al. 1999). The recent discovery of *cfp-2*, a *cfp-1* homolog identified in the genome sequence of *N. crassa* (see below), suggested the possibility that during the sexual cycle, *cfp-2*, and not *cfp-1*, is the gene expressing PDC. However, an analysis of the *Neurospora* EST database shows only *cfp-1* sequences present in perithecial libraries (see Table 2). This result support the hypothesis that *cfp-1* is the gene expressing PDC during the sexual cycle, and that its expression is differentially regulated during the vegetative and sexual cycles.

The results mentioned above provided additional support to the hypothesis that, in *N. crassa*, vegetative and sexual cells may utilize PDC filaments in different ways. Therefore, to ensure the presence of the filaments during sexual development, even under conditions in which the catalytic activity of PDC is not required (e.g., when cells are growing in ethanol-containing media), *N. crassa* may have evolved specific mechanisms for the differential control of the expression of the *cfp-1* gene and the assembly of the PDC filaments during the sexual cycle.

5. REGULATION OF THE EXPRESSION OF THE *cfp-1* GENE

It is well known that in bacteria, yeasts, and plants the expression of the genes encoding PDC is heavily regulated at the transcriptional level in response to a diverse array of signals, some of which remain elusive to our understanding (Chambers et al. 1995; Geigenberger 2003; Gunasekaran and Raj 1999; Pronk et al. 1996). Our own results demonstrated that the expression of the *cfp-1* gene in *N. crassa* is induced or repressed in response to the carbon source present in the culture medium (Alvarez et al. 1993; Temporini et al. 2004). In addition, as discussed above, we have also shown that the expression of the *cfp-1* gene is differentially regulated during the sexual cycle by a mechanism that apparently does not respond to the same signals that trigger induction or repression of this gene during vegetative growth (Thompson-Coffe et al. 1999). Thus, a detailed characterization of the transcriptional unit of the *cfp-1* gene was undertaken, with the purpose to unveil some of the regulatory domains and mechanisms that mediate the control of the expression of this gene. In the next few paragraphs we review the results of our analysis of the promoter region of the *cfp-1* gene. We also review a surprising discovery about the role of histone H1 in the regulation of the expression of the *cfp-1* gene. Finally we discuss how the *Neurospora* genome project has contributed to unmask an unforeseen element with a potential impact on the expression of *cfp-1* as well as the recent discovery of *cfp-2* encoding a putative pyruvate decarboxylase.

5.1. The Promoter of the *cfp-1* Gene Contains Several Sequence Motifs Conserved in Carbon-regulated Fungal Genes

Figure 6 shows a diagrammatic representation of the entire transcriptional unit of the *cfp-1* gene. This 3,741 bp fragment (GenBank accession number U65927) contains all sequences necessary for proper expression of the *cfp-1* gene (Temporini et al. 2004). A closer look into the 5' region upstream of the transcriptional start site identified a number of sequence motifs with high homology to well characterized *cis*-acting regulatory elements present in the promoters of many carbon-regulated fungal genes (Figure 6 and Table 1). In particular, this analysis identified sequences with homology to the binding sites for the yeast transcription factors GCR1 (Baker 1991) and RAP1 (Huet and Sentenac 1987), which are important for the transcriptional activation of the yeast *PDC1* gene and other glycolysis-related genes in *S. cerevisiae* (Butler et al. 1990; Chambers et al. 1995; Kellermann and Hollenberg 1988; Lopez and Baker 2000; McNeil et al. 1990; Uemura et al. 1997; Turkel et al. 2003). It has been shown that depending on the sequence context of its binding site, RAP1 is capable of many cellular functions such as activation and repression of transcription, control of telomere length, and organization of repressive chromatin at the telomeres and the silent mating type loci (Morse 2000). Interestingly, most promoters of glycolytic genes in yeast contain binding sites for RAP1 and GCR1 in close proximity and these binding sites display a strong synergism for each other (Buchman et al. 1988 a and b; Chambers et al. 1989 and 1995; Stanway et al. 1989; Bitter et al. 1991; Huie et al. 1992; Willett et al. 1993). Moreover, in the promoter of the *PDC1* gene, as well as several other glycolytic genes, overlapping RAP1/GCR1 motifs have been described (Baker 1991; Butler and McConnell 1988; Chambers et al. 1995; Uemura et al. 1997; Willet et al. 1993). It is believed that RAP1 facilitates the binding of GCR1 to its target, which results in transcriptional activation (López et al. 1998; Tornow et al. 1993; Uemura et al. 1997). Similar to the *PDC1* gene in yeast, in the promoter of the *cfp-1* gene some of the motifs homologous to the RAP1 binding site are in close proximity to, or even overlapping, the motifs with homology to the GCR1 binding sites (Temporini et al. 2004; Figure 6 and Table 1).

Another motif identified in the *cfp-1* gene promoter is the *ERA* element (Figure 6 and Table 1), originally described in the promoter region of the *PDC1* gene in *S. cerevisiae* (Liesen et al. 1996). This element has been proposed to mediate the repression of the expression of the *PDC1* gene in response to ethanol. However, as we discuss later on, this motif does not appear to be involved in ethanol-mediated repression of *cfp-1* expression (Temporini et al. 2004). Finally, a sequence with similarity to the "*pgk* box" of *A. nidulans* glycolytic genes (Punt et al. 1990), is also present in the *cfp-1* promoter (Figure 6 and Table 1).

Fig. 6. Transcriptional unit of the *cfp-1* gene. The most abundant full-length mRNA (*grey rectangle*) is approximately 2.0 kb. An intron, processed by alternative splicing, is represented by a *white rectangle*. The positions of the translational stop codons are indicated for the most abundant *cfp-1* mRNA (TAA, *grey arrow*) and its alternatively spliced transcript (TGA, *white arrow*). Sequences with similarity to the binding sites for GCR1 (*white circles*), RAP1 (*black ovals*), the *ERA* motif (*grey square*), and the "*pgk* box" (*white triangle*) have been identified in the promoter of *cfp-1*. A *white bar* spans the fragment of the *cfp-1* gene promoter (Pcfp) utilized to drive the expression of reporter genes in *N. crassa*. S: *SacII*, E: *EcoRI*. Modified from Temporini et al. 2004.

Table 1. Putative cis-acting regulatory sequences present in the promoter of the *cfp-1* gene

Motif	Consensus	Position at the *cfp-1* promoter
RAP1	A/GA/CACCCANNCAC/TC/T	-313 GCACACCTTCCTT -327 GCAGCCATGCATC -694 GCAACCCCACCAG -956 GCACCCACCACCC
GCR1	CTTCC	-307 CTTCC -1052 CTTCC -1241 CTTCC
ERA	AAATGCATA	-1102 AAATACACA
pgk box	TGA/TGGTG/AT/C	-751 TGAGGCAT

It is interesting that several *cis*-acting sequences that are important for the regulation of the expression of genes in response to carbon sources are well conserved among taxonomically distant fungi. Of particular interest are the cases of the *PDC1* gene from *S. cerevisiae* and *cfp-1* from *N. crassa*. Although the overall similarity in the coding regions of these two genes is not particularly high (see below), the nucleotide sequence of the regulatory *cis*-acting elements and their relative position in their promoters is very similar. These observations suggest that the transcriptional machinery involved in the regulation of the expression of these genes may be conserved between these fungi. Interestingly, a hypothetical protein with similarity to GCR1 and other transcription factors is present in the *N. crassa* genome. This protein, AL389891 (locus NCU02968.1), shares >30% homology with a 300-amino-acid domain of GCR1 (Temporini et al. 2004). Whether this protein plays any role in the transcriptional regulation of the expression of *cfp-1* and other glycolytic genes in *N. crassa* needs to be determined experimentally. Interestingly, the RAP1 and GCR1 binding sites, the *ERA* motif and the *pgk* box are present in the 5′ regions of several glycolysis-related genes in *N. crassa* (Temporini et al. 2004). Thus, it is reasonable to speculate that the mechanisms that regulate the coordinated expression of genes involved in carbon metabolism have a widespread conservation among the fungi. Preliminary experimental evidence in support of this hypothesis has been obtained (Temporini et al. 2004), and those results are discussed below.

5.2. Revelations from the *Neurospora* Genome Project: Alternative Splicing may Contribute to the Regulation of the Expression of the *cfp-1* gene

In the original characterization of the transcriptional unit of the *cfp-1* gene (Alvarez et al. 1993), the nucleotide sequence of the entire transcribed genomic region was found to be co-linear with several *cfp-1*-cDNA sequences isolated from different cDNA libraries. These observations led to the conclusion that the *cfp-1* gene does not contain introns (Alvarez et al. 1993). However, a recent analysis of the *Neurospora* EST database has identified two ESTs corresponding to the *cfp-1* gene (clones NCSC5G5T7 and NCSC3E7T7; accession numbers gi | 4240693 and gi | 4220248, respectively), which are missing a fragment between nucleotides +1,634 to +1,734 (Figure 6; Temporini et al. 2004). The putative alternative transcript, which is only 101 nucleotides shorter than the full-length *cfp-1* mRNA, would encode a protein that lacks 24 amino acids at the C-terminus domain of the protein and would utilize a different stop codon signal (TGA at +1,746 instead of TAA at +1,711; see Figure 6). Interestingly, the clones containing the shorter transcript, which represent 20% (2 out of 10) of the *cfp-1*-ESTs present in the database, come from a cDNA library specific for germinating conidia (Nelson et al. 1997). These observations suggest the intriguing possibility that the *cfp-1* mRNA may be processed by alternative splicing in response to specific developmental signals to produce the hypothetical shorter version of PDC, which may be required for a specific function during spore germination.

5.3. The *cfp-1* Promoter can Drive the Expression of Reporter Genes in Filamentous Fungi in a Carbon Source-dependent Fashion

Filamentous fungi are commonly employed in industrial processes for the production of a wide variety of enzymes and metabolites that are critical for the food and pharmaceutical industries. Over the last three decades, many technological advances have resulted in the development of efficient methodologies for the stable transformation of fungi with chimeric genes (reviewed by Gold et al. 2001), which represent a powerful tool for large-scale production of enzymes and secondary metabolites of industrial importance, as well as for the study of basic physiological processes. There are some instances, however, in which the chimeric genes encode potential deleterious products for the host cell or the particular study being conducted requires a temporary expression of the transformed gene. In these cases, a precise control of the expression of the chimeric genes is highly desirable. Such a controlled expression is often achieved by the utilization of regulated fungal promoters to control the transcription of the gene of interest.

Our observations that the expression of the *cfp-1* gene is regulated in response to the carbon sources present in the culture medium, plus the presence in the *cfp-1* promoter of regulatory elements conserved among fungi, indicated a potential for the utilization of the *cfp-1* gene promoter to control the expression of chimeric fungal genes in a carbon source-dependent fashion. In fact, the promoter of the *cfp-1* gene has been successfully utilized to drive the expression of reporter genes to study various aspects of DNA methylation in *N. crassa* (Barra et al. 1996; Mautino et al. 1996). More recently, we have shown that a drastic modification in the abundance of the *cfp-1* mRNA can be achieved as fast as four hours after switching carbon sources in the culture medium (Temporini et al. 2004). The same pattern of regulated expression was obtained when a fragment of the *cfp-1* promoter (P*cfp*; white bar in Figure 6) was used to direct the transcription of two different reporter genes (Temporini et al. 2004). In the first case, *N. crassa* transformants containing the *hph* gene from *E. coli* (encoding hygromycin B phosphotransferase) under the control of P*cfp*, were very resistant to hygromycin B when growing in glucose-containing media, while poor resistance was observed when the carbon source was ethanol. Intermediate levels of resistance were observed if the carbon source was a mixture of glucose plus ethanol. A second, more quantitative analysis confirmed these results. In these experiments, P*cfp* was fused to the *eth-1* gene of *N. crassa*, encoding S-adenosylmethionine synthetase (AdoMet-S). In transformants containing this fusion, AdoMet-S activity increased 1.6 fold when mycelia were transferred from sucrose- to glucose-containing medium. Mycelia transferred to media containing either ethanol or glucose plus ethanol resulted in 2.8 and 1.3-fold decrease, respectively, of AdoMet-S activity. These results indicate that P*cfp* contains all the regulatory elements required for the proper regulation of the expression of the *cfp-1* gene in *N. crassa*. They also indicate that the *ERA* motif is not required for ethanol-mediated repression of *cfp-1* expression, since P*cfp*, which does not carry this motif, is still able to repress the expression of the reporter genes in response to the presence of ethanol in the culture media. A similar situation has been observed for the *PDC* gene from *Kluyveromyces lactis*, in which the *ERA* motif has been shown to be dispensable for ethanol-mediated repression (Destruelle et al. 1999). Interestingly, P*cfp* also lacks two of the three GCR1 binding sites (Figure 8), indicating that these two sites are not required for proper regulation of *cfp-1* expression. In any case, a more detailed characterization, including site-directed mutagenesis and deletion analyses, should shed some light on the functional relevance of the different motifs found in the *cfp-1* promoter. These studies should also provide valuable information on the molecular mechanisms involved in the control of the expression of the *cfp-1* gene. Nonetheless, these results indicate that the *cfp-1* promoter can be utilized to precisely control the expression of reporter genes in *N. crassa*, in a carbon-source-dependent way. Moreover, a chimeric gene involving the entire *cfp-1* promoter fused to the *E. coli uidA* gene (encoding β-glucuronidase), showed the same carbon regulated expression in *Aspergillus nidulans* (Temporini et al. 2004). Thus, the *cfp-1* gene promoter can also be used in a heterologous host to drive the expression of a reporter gene, in a carbon-source-dependent manner. In addition, these observations complement previous findings on the conservation of regulatory motifs in the promoter of carbon-regulated fungal genes, lending additional support to the hypothesis that the transcriptional machinery of fungi is highly conserved.

The studies on the expression of the *cfp-1* gene suggest that the *cfp-1* promoter could be of great value for biotechnological and research applications requiring the controlled expression of transgenes in fungi (Temporini et al. 2004). The *cfp-1* gene promoter, compared to other regulated promoters, has some advantages to drive the expression of genes in filamentous fungi. Most of the inducible systems available utilize stressful conditions like heat shock or antibiotics to achieve gene activation. With the *cfp-1* promoter instead, a short treatment with a non-toxic and inexpensive metabolite such as glucose is enough to induce gene expression. Reciprocally, the expression of the gene of interest can be quickly repressed by transferring the culture to a growth medium containing ethanol as the sole carbon source. Thus, this promoter offers the possibility to analyze in the same fungal transformant line the effects caused by over- and underexpression of a desired gene. In addition, activation by glucose and repression by ethanol occur very quickly (~4 h after switching carbon sources), which results in cell populations almost synchronized for transgene expression (Temporini et al. 2004).

5.4. Histone H1 is Required for Proper Expression of the *cfp-1* Gene

The history of the *cfp-1* gene and the 8-10 nm filaments of PDC revealed a multitude of surprising results. One of them surfaced during the characterization of the *hH1* gene of *N. crassa*, encoding the linker histone H1 (Folco et al. 2003). In these studies, *hH1* mutants were obtained by RIP and characterized in order to identify the phenotypes associated with the lack of function of histone H1. An analysis of global chromatin structure revealed a subtle alteration in the pattern of nucleosomal DNA digestion of chromatin from the *hH1*[RIP] mutants. However, the mutants showed normal morphology in liquid or solid minimal media at all temperatures tested. In addition, the size, shape and distribution of nuclei in mycelia as well as the number and morphology of conidia were completely normal in these strains. The mutants also showed normal sensitivity to several mutagens and had normal patterns of DNA methylation. The only noticeable morphological difference from the wild type strain was that *hH1*[RIP] mutants showed slower rates of linear growth on all carbon sources tested. Interestingly, although still reduced with respect to the wild type strain, the *hH1*[RIP]/wild type growth ratio was higher in media containing ethanol than in media containing sucrose or glucose (0.91 on ethanol vs. 0.85 on sucrose or glucose; Folco et al. 2003). This selective response to different carbon sources led to the hypothesis that the histone H1 is required for the proper expression of one or more genes involved in carbon metabolism. Thus, the *cfp-1* gene appeared as a logical candidate to explore this possibility. Northern blot analyses showed that *cfp-1* mRNA levels of wild type and *hH1*[RIP] mutants were similar when cells were grown on media containing either sucrose or glucose (Figure 7). As expected, no *cfp-1* mRNA was detected when the wild-type strain was grown under repressing conditions in media containing ethanol. However, the *cfp-1* transcript was readily detected in the *hH1*[RIP] mutants grown under this repressing condition (Figure 7). These observations suggested that histone H1 might play a role in modulating the expression of the *cfp-1* gene. Additional evidence in support of this possibility is provided by the analysis of the chromatin structure in the promoter region of the *cfp-1* gene. MNase-treated chromatin was digested with *Eco*RI (cleavage site at +997; Figure 8A), and analyzed by Southern blot with a *cfp-1* probe (+460 to +997; Figure 8A). As Figure 8B shows, similar patterns of chromatin digestion are observed in wild type and *hH1*[RIP] mutant strains growing under inducing conditions (i.e., glucose-containing media). However, under repressing conditions (i.e., ethanol-containing media), chromatin of the wild-type strain shows one distinctive MNase band. This "repressive" chromatin configuration is not adopted by *hH1*[RIP] mutants growing in ethanol-containing media (Figure 8B and C). Extended chromatin analyses using *Bgl*II (Figure 8A) instead of *Eco*RI confirmed these results (data not shown). Our observations indicate that *hH1*[RIP] mutants are affected in their ability to repress the expression of the *cfp-1* gene in response to the presence of ethanol in the culture media. Thus, histone H1 has an active role in the

158

modulation of the expression of the *cfp-1* gene, most likely as a negative regulator. Moreover, the modulatory effect of H1 does not appear to be related to the chromosomal position of the *cfp-1* gene. Instead, it seems to be specific for this gene,

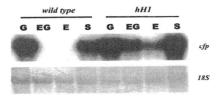

Fig. 7. The *cfp-1*-mRNA is present in *hH1^RIP* mutants grown under repressing conditions. Total RNA was extracted from cells grown 16 hs in media containing sucrose as the carbon source, and then transferred to media containing the indicated carbon sources for an additional 4 hs incubation period. *E*, ethanol; *EG*, ethanol + glucose; *G*, glucose; *S*, sucrose. The 18S rRNA is shown as a loading control. (adopted from: Folco et al. 2003, with permission)

since an ectopic integration of the *hph* reporter gene fused to the *cfp-1* promoter showed the same misregulation under repressing conditions as observed for the endogenous *cfp-1* gene (Folco et al. 2003). Further analyses of the expression of other carbon-regulated genes in an *hH1^RIP* genetic background should help to determine whether the role of *Neurospora* H1 in the modulation of gene expression is common to genes involved in carbon metabolism.

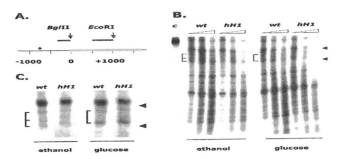

Fig. 8. Absence of histone H1 alters the normal "repressive" configuration of chromatin at the *cfp-1* promoter. (A) Diagram of the *cfp-1* gene promoter showing the position of *Bgl*II and *Eco*RI sites (arrows). Probes utilized in MNase-digestion experiments are indicated by black bars. The region exhibiting the abnormal MNase digestion pattern in chromatin of *hH1^RIP* mutants is indicated with a diamond. (B) MNase digestion pattern of the *cfp-1* promoter chromatin in the wild type and the *hH1^RIP* mutant strain. Chromatin was obtained from mycelia and treated with MNase (2.5, 5.0 and 10 U MNase/g of mycelium). Wild type chromatin, treated under similar conditions without MNase, is shown (c). Distinctive patterns of normal MNase digestion are indicated by a single (inducing conditions; *glucose*) or double (repressing conditions; *ethanol*) bracket. Positions upstream of the *Bgl*II site are indicated by arrows (-1,000 bp: upper arrow; and -600 bp: lower arrow). (C) A higher magnification view of selected lanes from (B) is shown.

5.5. Multiple PDC Genes in *Neurospora* and Other Eukaryotic Organisms

As we mentioned above, the release of the *N. crassa* genome sequence added new pieces to the PDC puzzle. A recent analysis of this data revealed the presence of the gene *cfp-2* (NCU02397.1; accession number gi|32421459) encoding a putative pyruvate decarboxylase (Borkovich et al. 2004). The existence of a second gene encoding PDC in *Neurospora* was unexpected as standard Southern blot analyses using *cfp-1* sequences as probes do not detect any cross hybridizing DNA (Haedo et al. 1992). The failure of *cfp-1* to detect *cfp-2* can be explained by the low overall sequence similarity between the coding regions of these two genes. Although the function of *cfp-2* is still unclear, the existence of a *cfp-2* EST (clone NCM4C12T7) indicates that this gene is expressed. Interestingly, the promoter region of *cfp-2* contains putative *cis*-acting elements, similar to those found in the promoter of *cfp-1* (i.e., GCR1, RAP1, *ERA*, and "*pgk* box"; see Table 2). The presence of similar regulatory sequences would suggest that the expression of these two genes is controlled by similar mechanisms. However, there is a noticeable difference in the number of ESTs found for each gene, with ten ESTs found for *cfp-1* compared to only one for the novel gene *cfp-2*. Whereas, eight out of ten *cfp-1* ESTs are present in libraries from germinating conidia (RNA isolated from germlings 4.5 h after inoculation; Nelson et al. 1997) and the other two in perithecial libraries, the only *cfp-2* EST found is present in a mycelial library (RNA isolated from mycelia grown 24 h; Nelson et al. 1997).

Table 2. Comparative analysis of *cfp-1* and *cfp-2* from *N. crassa*

Gene	Number of EST	Introns[1]	Number of motifs				Protein size (aa)[2]
			RAP1	GCR1	ERA	*pgk* box	
cfp-1	8 conidial 2 perithecial	1 (2)	4	3	1	1	570 (548)
cfp-2	1 mycelial	2 (?)	1	4	1	2	576

[1]Values between brackets correspond to the putative mRNA species generated by alternative splicing of the corresponding gene; [2]The size of a putative protein encoded by the mRNA generated after alternative splicing of *cfp-1* is indicated between brackets.

These findings strongly suggest different roles and temporal regulation for the expression of the genes encoding PDC: *cfp-1* is expressed in the sexual cycle, germinating conidia and vegetative mycelia, whereas *cfp-2* is apparently expressed only in mycelia. As mentioned above, these findings also support the contention that *cfp-1* is the gene expressed in the sexual cycle and whose protein product is associated with the cytoskeleton of asci and spores (Thompson-Coffe et al. 1999).

The occurrence of multiple genes encoding PDC with different patterns of expression appears to be a common feature of eukaryotic organisms. In yeast, three structural genes for PDC have been identified: *PDC1*, *PDC5*, and *PDC6* (Hohmann and Cederberg

1990; Hohmann 1991). *PDC1* is the gene normally expressed during fermentative growth, while *PDC5* is expressed only when *PDC1* is deleted (Hohmannn and Cederberg 1990). *PDC6* is strongly induced under high sugar stress (Erasmus et al. 2003); however, this gene is not required for growth in glucose but is important during growth in ethanol and in galactose (Hohmann 1991).

In plants, ethanolic fermentation, and therefore PDC expression, normally occur in conditions of low oxygen or anoxia (Tadege et al. 1999). However, in tobacco for example, it has been shown that a pollen-specific PDC isoform is highly expressed during pollen development. This occurs concomitantly with high levels of respiration, which indicates that the expression of the pollen-specific PDC isoform is not affected by oxygen availability (Tadege et al. 1997), and suggests that there is an alternative mechanism for the control of the expression of PDC during pollen development. In rice, there are at least four PDC structural genes. Three of those (*PDC1, PDC2* and *PDC4*) have been shown to be induced under anaerobic conditions (Hossain et al. 1994; Huq et al. 1995; Rivoal et al. 1997). The fourth gene, *PDC3*, which was originally postulated to be a pseudogene, has been recently shown to be expressed only during pollen maturation, even under aerobic conditions (Li et al. 2004).

Thus, it seems that in yeast and plants the requirement for PDC in different stages of development and different metabolic conditions has resulted in the evolution of a PDC gene family, most likely by gene duplication and divergence. Different members of this gene family may have different functions in the cell and their expression appears to be controlled by different signals. This evolutionary strategy would ensure that whatever the cellular requirements for PDC function are in any specific developmental or metabolic condition, there would be a PDC gene expressed to meet those cellular requirements.

In *N. crassa*, the existence of RIP (Selker, 2002) may have prevented the evolution of a PDC gene family by gene duplication. Moreover, the possibility that the *cfp-1* and *cfp-2* genes are the result of an ancestral gene duplication event followed by divergence due to RIP is unlikely. An analysis of the rate of dinucleotides in these genes indicates that neither *cfp-1* nor *cfp-2* exhibit the characteristic skewing of dinucleotides produced by RIP (Margolin et al. 1998). Given the fact that only 59 out of 9,200 genes in *N. crassa* show evidence of RIP (Galagan et al. 2003) it appears that gene duplication followed by RIP occurs at a very low frequency in this fungus. Therefore, it appears that *cfp-1* and *cfp-2* have independent evolutionary origins. Moreover, a comparative analysis of multiple PDCs from a variety of organisms, reveals that the PDCs encoded by *cfp-1* and *cfp-2* belong to two different phylogenetic groups (Figure 9). It is interesting that the product encoded by *cfp-1* and hypothetical PDCs from *Schizosaccharomyces pombe* and *Fusarium graminearum* are phylogenetically nested in a clade that groups a few bacterial PDCs and all the plant PDCs (Clade I; Figure 9). On the other hand, the hypothetical protein encoded by *cfp-2* is nested in a clade that groups the PDCs from most bacterial species, other filamentous fungi, and yeasts. Along with the hypothetical product of *cfp-2*, this clade also contains a second hypothetical PDC from *F. graminearum* and PDC1 from *S. pombe* (Figure 9). Interestingly, in *Aspergillus nidulans* there is also a second putative gene encoding pyruvate decarboxylase (accession number XM_412533), in addition to the already characterized *pdcA*, although both protein products are located in clade II.

The observations presented above suggest that eukaryotic organisms have developed multiple PDC-encoding genes, perhaps as a way to adapt to the variety of metabolic conditions they encounter in their respective habitats. In some cases, like plants and yeast, these multiple genes appear to have originated by duplication of an ancestral gene. In others, like *N. crassa*, *F. graminearum* and *S. pombe*, the PDC-encoding genes have evolved independently. The biological significance of these observations has yet to be determined.

6. PYRUVATE DECARBOXYLASE: AN ENZYME WITH "MOONLIGHTING" FUNCTIONS?

In the last few years, a growing body of evidence has emerged showing that certain enzymes do more than just catalyze a specific reaction. "Moonlighting" enzymes are those proteins that have additional functions, which could be enzymatic, structural or regulatory (Copley 2003; Jeffrey 1999). A developing concept with ample consensus is that the catalytic site of an enzyme is just a small part of the protein, and therefore, it is possible that the remaining portions of it can be utilized for additional, moonlighting functions (Copley 2003). Moonlighting functions may be related to the catalytic activity of a protein or may utilize special structural features of the protein that are not involved in its catalytic activity.

Two of the best characterized examples of moonlighting proteins correspond to glycolytic enzymes. In the first case, phosphoglucose isomerase, which catalyzes the interconversion of glucose 6-phosphate and fructose 6-phosphate in the second step of glycolysis, has been shown to play four additional roles. It is the same protein as neuroleukin (Chaput et al. 1988; Faik et al. 1988), which acts as a cytokine involved in the maturation of B cells to secrete antibodies (Gurney et al. 1986a), and is also a nerve growth factor (Gurney et al. 1986b). The third known function for this protein is as autocrine motility factor (AMF), which stimulates cell migration (Watanabe et al. 1996). Finally, phosphoglucose isomerase is a differentiation and maturation mediator that can cause differentiation of human myeloid leukemia cells (Xu et al. 1996).

The second example of a glycolytic enzyme with moonlighting functions is provided by the 37-kDa subunit of human glyceraldehyde-3-phosphate dehydrogenase (GPD). As a tetramer, GPD converts glyceraldehyde-3-phosphate to 1,3-diphosphoglycerate. As a monomer, however, it has been shown to function as a nuclear uracil-DNA glycosylase (Meyer Siegler et al. 1991).

In *N. crassa*, the mitochondrial tyrosyl-tRNA synthetase is an enzyme with moonlighting functions. This protein, in addition to charging tRNA[Tyr], was shown to promote the folding of Group I introns into the active structure required for self-splicing (Mohr et al. 2001). It has been shown that the catalytic core of the intron is structurally very similar to mitochondrial tRNA[Tyr], which could explain the involvement of this enzyme in RNA splicing.

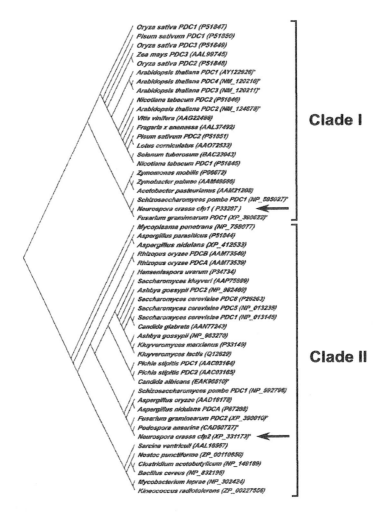

Fig. 9. Phylogenetic tree showing the evolutionary relationships among pyruvate decarboxylases from several organisms. Protein sequences utilized for this analysis were obtained from the publicly available databases and the corresponding accession numbers are given in parentheses. The arrows indicate the positions of the products encoded by *cfp-1* and *cfp-2* from *N. crassa*. Hypothetical proteins are indicated with asterisks. Sequence alignments and phylogenetic analyses were performed with MacVector 7.1.1 (Oxford Molecular).

Gephyrin is a mammalian protein required for anchoring a variety of proteins involved in neurotransmission to the cytoskeleton (Kneussel and Betz 2000). Structurally, gephyrin consists of two domains homologous to the *MogA* and *MoeA* enzymes of *E. coli*, which are involved in the synthesis of the molybdenum cofactor (MoCo), required by most molybdenum-containing enzymes. Interestingly, gephyrin does function in MoCo synthesis and can complement bacterial, plant, and mammalian cells deficient in MoCo (Stallmeyer et al. 1999). The role of gephyrin at post-synaptic membranes is not related to its catalytic activity in MoCo synthesis. In fact, it has been proposed that the structural organization of this protein is ideal to form a scaffold on which other proteins may assemble. Interestingly, the existence of tissue-specific splice variants of gephyrin suggests that alternative splicing may play a role in determining which proteins may interact with it in a particular tissue (Ramming et al. 2000). Our observations that PDC may play a dual role in *Neurospora* (a catalytic enzyme in ethanolic fermentation and a cytoskeleton-associated protein during sexual development), plus similar indications of more than one function for this protein in other organisms, suggest that this enzyme may be a moonlighting protein. A more detailed characterization of the role that PDC plays during the sexual cycle of *Neurospora* is required in order to unambiguously establish if this protein has a structural function in addition to its catalytic activity.

7. CONCLUSIONS

The biochemical and genetic characterization of the 8-10 nm filaments of P59Nc/PDC was initiated more than fifteen years ago, with the objective to unveil the identity and function of these cytoplasmic structures in *N. crassa* and, possibly, in other filamentous fungi. Our journey has taken us through many surprising revelations, which provide a long awaited explanation to the fascinating mystery generated by controversial observations of the presence and abundance of these structures in fungi. We have learned that these filaments are composed of pyruvate decarboxylase and that its expression is regulated by a complex array of signals that allow this protein to be synthesized in different developmental stages, in response to a variety of metabolic conditions, and with potential different functions in each cell type.

Although many questions have been answered, many more have been generated. Of particular interest will be determining the relationship between protein structure and function of the PDC filaments. Are the filaments required for a structural role? Do they provide structural support for the interaction of other glycolytic enzymes? Is the supramolecular structure of PDC required for its catalytic activity? Or is it an evolutionary strategy to utilize the same protein for different functions? Do the *N. crassa* 8-10 nm filaments contain the putative protein product of the *cfp-2* gene? Does the putative product of the shorter version of the *cfp-1* mRNA assemble into filaments? What is the relationship between the *snowflake* mutation and PDC? These and other questions constitute the driving force for the future research in this system. Mutational analyses of PDC should reveal whether there are functional domains in this protein required for functions other than decarboxylation of pyruvate. Also, the possibility that the *cfp-1* gene undergoes regulated alternative splicing opens an entire new avenue of research. Finally, from an evolutionary standpoint, the observation that *N. crassa*, *S. pombe* and *F. graminearum* have two genes encoding PDC that can be placed in two

164

different phylogenetic groups is very intriguing, and the biological relevance of this observation remains to be addressed.

Acknowledgements: We thank our colleagues from the *Neurospora* group at Cordoba, Argentina, as well as our collaborators from other labs for their contributions to the study of PDC and the 8-10 nm filaments. This research has been funded over the years by grants from CONICOR, CONICET, SECYT, CEPROCOR, Fundacion Antorchas (Argentina), Centre for Genetic Engineering and Biotechnology (ICGEB; Triste, Italy), CONICOR-British Council (UK), Thirld World Academy of Sciences (TWAS, Italy), and Howard Hughes Medical Institute (HHMI; USA).

REFERENCES

Adams A and Pringle J (1984). Relationship of actin and tubulin distribution to bud growth in wild type and morphogenetic mutant *Saccharomyces cerevisiae*. J Cell Biol 98:934-945.

Allen E, Lowry R and Sussman A (1974). Accumulation of microfilaments in a colonial mutant of *Neurospora crassa*. J Ultrastruc Res 48:455-464.

Allen E and Sussman A (1978). Presence of an actin-like protein in mycelium of *Neurospora crassa*. J Bacteriol 135:713-716.

Alvarez M, Rosa A, Temporini E, Wolstenholme A, Panzetta G, Patrito L and Maccioni H (1993). The 59kDa polypeptide constituent of 8-10-nm filaments in *Neurospora crassa* is a pyruvate decarboxylase. Gene 130: 253-258.

Anderson R and Zachariah K (1974). On the structure, function, and distribution of filament bundles in apotheciat cells of the fungus *Ascobolus stercorarius*. J Ultrastruc Res 46:375-392.

Arnold H and Pette D (1968). Binding of glycolytic enzymes to structure proteins of the muscle. Eur J Biochem 6:163-171.

Baker H (1991). GCR1 of *Saccharomyces cerevisiae* encodes a DNA binding protein whose binding is abolished by mutations in the CTTCC sequence motif. Proc Natl Acad Sci USA 88:9443-9447.

Barnett J (2003). A history of research on yeasts 5: the fermentation pathway. Yeast 20:509-543.

Barra J, Mautino M and Rosa A (1996). A dominant negative effect of *eth-1'*, a mutant allele of the *Neurospora crassa* S-adenosylmethionine synthetase-encoding gene conferring resistance to the methionine toxic analogue ethionine. Genetics 144:1455-1462.

Bartnicki-Garcia S and Lippman E (1969). Fungal morphogenesis – Cell wall construction in *Mucor rouxii*. Science 165:302-304.

Beck D, Decker G and Greenawalt J (1970). Ultrastructure of striated inclusions in *Neurospora*. J Ultrastruc Res 33: 245-251.

Beitner R (1993). Control of glycolytic enzymes through binding to cell structures and by glucose-1,6-biphosphate under different conditions. The role of Ca^{2+} and calmodulin. Int J Biochem 25:297-305.

Bianchi M, Tizzani L, Destruelle M, Frontali L and Wesolowski-Louvel M (1996). The 'petite-negative' yeast *Kluyveromyces lactis* has a single gene expressing pyruvate decarboxylase activity. Mol Microbiol 19:27-36.

Bitter G, Chang K and Egan K (1991). A multi-component upstream activation sequence of *Saccharomyces cerevisiae* glyceraldehyde-3-phosphate dehydrogenase gene promoter. Mol Gen Genet 231:22-32.

Borkovich K et al. (2004). Lessons from the genome sequence of *Neurospora crassa*: Tracing the path from genomic blueprint to multicellular organism. Microbiol Mol Biol Rev 68:1-108.

Buchman A, Kimmerley W, Rine J and Kornberg R (1988a). Two DNA-binding factors recognize specific sequence at silencers, upstream activating sequences, autonomously replicating sequences, and telomeres in *Saccharomyces cerevisiae*. Mol Cell Biol 8:210-225.

Buchman A, Lue N and Kornberg R (1988b). Connections between transcriptional activators, silencers, and telomeres as revealed by functional analysis of a yeast DNA-binding protein. Mol Cell Biol 8:5086-5099.

Butler G, Dawes I and McConnell D (1990). TUF factor binds to the upstream region of the pyruvate decarboxylase structural gene (*PDC1*) of *Saccharomyces cerevisiae*. Mol Gen Genet 223:449-456.

165

Butler G and McConnell D (1988). Identification of an upstream activation site in the pyruvate decarboxylase structural gene (*PDC1*) of *Saccharomyces cerevisiae*. Curr Genet 14:405-412.

Caesar C, Hoang-Van K, Turian G and Hoch C (1987). Isolation and characterization of coated vesicles from filamentous fungi. Eur J Cell Biol 43:189-194.

Chambers A, Packham E and Graham I (1995). Control of glycolytic gene expression in the budding yeast (*Saccharomyces cerevisiae*). Curr Genet 29:1-9.

Chambers A, Tsang J, Stanway C, Kingsman A and Kingsman S (1989). Transcriptional control of the *Saccharomyces cerevisiae PGK* gene by RAP1. Mol Cell Biol 9:5516-5524.

Chaput M, Claes V, Portetelle D, Cludts I, Cravador A, Burny A, Gras H and Tartar A (1988). The neurotrophic factor neuroleukin is 90 percent homologous with phosphohexose isomerase. Nature 332:454-455.

Conway T, Osman Y, Konnan I, Hoffmann E and Ingram L (1987). Promoter and nucleotide sequences of the *Zymomonas mobilis* pyruvate decarboxylase. J Bacteriol 169:949-954.

Copley S (2003). Enzymes with extra talents: moonlighting functions and catalytic promiscuity. Curr Op Chem Biol 7:265-272.

Davidse L and Flach W (1977). Differential binding of methyl benzimidazol-2-yl carbamate to fungal tubulin as a mechanism of resistance to this antimitotic agent in mutant strains of *Aspergillus nidulans*. J Cell Biol 72:174-193.

Destruelle M, Menghini R, Frontali L and Bianchi M (1999). Regulation of the expression of the *Kluyveromyces lactis PDC1* gene: carbon source-responsive elements and autoregulation. Yeast 15:361-370.

Erasmus DJ, van der Merwe GK and van Vuuren HJ (2003). Genome-wide expression analyses: Metabolic adaptation of *Saccharomyces cerevisiae* to high sugar stress. FEMS Yeast Res 3:375-99.

Faik P, Walker J, Redmill A and Morgan M (1988). Mouse glucose-6-phosphate isomerase and neuroleukin have identical 3' sequences. Nature 332:455-456.

Fidel S, Doonan J and Morris N (1988). *Aspergillus nidulans* contains a single actin gene which has unique intron locations and encodes a gamma actin. Gene 70:283-293.

Folco H, Freitag M, Ramon A, Temporini E, Alvarez M, Garcia I, Scazzocchio C, Selker E and Rosa A (2003). Histone H1 Is required for proper regulation of pyruvate decarboxylase gene expression in *Neurospora crassa*. Eukaryot Cell 2:341-350.

Galagan JE et al. (2003). The genome sequence of the filamentous fungus *Neurospora crassa*. Nature 422:859-68.

Gambino J, Bergen L and Morris N (1984). Effects of mitotic and tubulin mutations on microtubule architecture in actively growing protoplasts of *Aspergillus nidulans*. J Cell Biol 99:830-838.

Geigenberger P (2003). Response of plant metabolism to too little oxygen. Curr Op Plant Biol 6:247-256.

Greer C and Scheckman R (1982). Actin from *Saccharomyces cerevisiae*. Mol Cell Biol 2:1270-1278.

Gold S, Duick J, Redman R and Rodriguez R (2001). Molecular transfromation, gene cloning, and gene expression systems for filamentous fungi. In: G. Khachatourians and D. Arora, eds. Applied Mycology and Biotechnology, Vol. 1. Agriculture and food production. Elsevier Science B. V., pp 13-54.

Gooday G (1971). Autoradiographic study of hyphal growth of some fungi. J Gen Microbiol 67:125-133.

Götz R, Schlüter E, Shoham G and Zimmermann F (1999). A potential role for the cytoskeleton of *Saccharomyces cerevisiae* in a functional organization of glycolytic enzymes. Yeast 15:1619-1629.

Gull K (1975). Cytoplasmic microfilament organization in two basidiomycete fungi. J Ultrastruc Res 50:226-232.

Gunasekaran P and Raj K (1999). Ethanol fermentation technology - *Zymomonas mobilis*. Curr Sci 77:56-68.

Gurney M, Apatoff B, Spear G, Baumel M, Antel J, Bania M and Reder A (1986a). Neuroleukin - a lymphokine product of lectin-stimulated T-cells. Science 234:574-581.

Gurney M, Heinrich S, Lee M and Yin H (1986 b). Molecular cloning and expression of neuroleukin, a neurotrophic factor for spinal and sensory neurons. Science 234:566-574.

Haedo S, Temporini E, Alvarez M, Maccioni H and Rosa A (1992). Molecular cloning of a gene (*cfp*) encoding the cytoplsmic filament protein P59Nc and its genetic relationship to the *snowflake* locus of *Neurospora crassa*. Genetics 131:575-580.

Heath I (1987). Preservation of a labile cortical array of actin filaments in growing hyphal tips of the fungus *Saprolognia ferax*. Eur J Cell Biol 44:10-16.

Hoch H and Howard R (1980). Ultrastructure of freeze-substituted hyphae of the basidiomycete *Laetisaria arvalis*. Protoplasma 103:281-297.

Hohmann S (1991). Characterization of *PDC6*, a third structural gene for pyruvate decarboxylase in *Saccharomyces cerevisiae*. J Bacteriol 173:7963-7969.

Hohmann S and Cederberg H (1990). Autoregulation may control the expression of yeast pyruvate decarboxylase structural genes *PDC1* and *PDC5*. Eur J Biochem 188:615-621.

Hohmann S and Meacock P (1998). Thiamin metabolism and thiamin diphosphate-dependent enzymes in the yeast *Saccharomyces cerevisiae*: genetic regulation. Biochem Biophys Acta 1385:201-219.

Hossain M, Huq E and Hodges T (1994). Sequence of a cDNA from *Oryza sativa* (L) encoding the *pyruvate decarboxylase 1* gene. Plant Physiology 106:799-800.

Howard R (1981). Ultrastructural analysis of hyphal tip cell growth in fungi: Spitzenkörper, cytoskeleton and endomembranes after freeze-substitution. J Cell Sci 48:89-103.

Huet J and Sentenac A (1987). TUF, the yeast DNA-binding factor specific for UAS$_{RPG}$ upstream activating sequence: identification of the protein and its DNA binding domain. Proc Natl Acad Sci USA 84:3648-3652.

Huffaker T, Hoyt M and Botstein D (1987). Genetic analysis of the yeast cytoskeleton. Ann Rev Genetics 21:259-284.

Huie M, Scott E, Drazinic C, Lopez M, Hornstra I, Yang T and Baker H (1992). Characterization of the DNA-binding activity of GCR1 – *in vivo* evidence for two GCR1-binding sites in the upstream activating sequence of TPI of *Saccharomyces cerevisiae*. Mol Cell Biol 12:2690-2700.

Huq E, Hossain M and Hodges T (1995). Cloning and sequencing of a cDNA encoding the *pyruvate decarboxylase 2* gene (accession no. U27350) from rice. Plant Physiology 109:722.

Jeffrey C (1999). Moonlighting proteins. Trends Biochem Sci 24:8-11.

Katz D and Rosenberger R (1971). Hyphal wall synthesis in *Aspergillus nidulans* – Effect of protein synthesis inhibition and osmotic shock on chitin insertion and morphogenesis. J Bacteriol 108:184-190.

Kellermann E and Hollenberg C (1988). The glucose-and ethanol-dependent regulation of *PDC1* from *Saccharomyces cerevisiae* are controlled by two distinct promoter regions. Curr Genet 14:337-344.

Kilmartin J and Adams A (1984). Structural rearrangements of tubulin and actin during the cell cycle of the yeast *Saccharomyces*. J Cell Biol 9:922-933.

Kneussel M and Betz H (2000). Receptors, gephyrin, and gephyrin-associated proteins: novel insights into the assembly of inhibitory postsynaptic membrane specializations. J Physiol 525.1:1-9.

Konig S (1998). Subunit structure, function and organisation of pyruvate decarboxylases from various organisms. Biochem Biophys Acta Protein Structure and Molecular Enzymology 1385:271-286.

Li Y, Ohtsu K, Nemoto K, Tsutsumi N, Hirai A and Nakazono M (2004). The rice pyruvate decarboxylase 3 gene, which lacks introns, is transcribed in mature pollen. J Exp Botany 55:145-146.

Liesen T, Hollenberg C and Heinisch J (1996). *ERA*, a novel *cis*-acting element required for autoregulation and ethanol repression of *PDC1* transcription in *Saccharomyces cerevisiae*. Mol Microbiol 21:621-632.

Lockington RA, Borlace GN and Kelly JM (1997). Pyruvate decarboxylase and anaerobic survival in *Aspergillus nidulans*. Gene 191:61-7.

Lopez M and Baker H (2000). Understanding the growth phenotype of the yeast gcr1 mutant in terms of global genomic expression patterns. J Bacteriol 182:4970-4978.

López M, Smerage J and Baker H (1998). Multiple domains of repressor activator protein 1 contribute to facilitated binding of glycolysis regulatory protein 1. Proc Natl Acad Sci USA 95:14112-14117.

Margolin BS, Garrett-Engele PW, Stevens JN, Fritz DY, Garrett-Engele C, Metzenberg RL and Selker EU (1998). A methylated *Neurospora* 5S rRNA pseudogene contains a transposable element inactivated by repeat-induced point mutation. Genetics 149:1787-97.

Mautino M, Barra J and Rosa A (1996). *eth-1*, the *Neurospora crassa* locus encoding S-adenosylmethionine synthetase: molecular cloning, sequence analysis and *in vivo* overexpression. Genetics 142:789-800.

McNeil J, Dykshoorn P, Huy J and Small S (1990). The DNA-binding protein RAP1 is required for efficient transcriptional activation of the yeast *PYK* glycolytic gene. Curr Genet 18:405-412.

Metzenberg R and Grotelueschen J (1989). Restriction polymorphism maps of *Neurospora crassa*: update. Fungal Genet Newsl 36:51-57.

Metzenberg R, Stevens J, Selker E and Morzycka-Wroblewska E (1984). A method for finding the genetic map position of cloned DNA fragments. Neurospora Newsl 31:35-39.

Meyer-Siegler K, Mauro D, Seal G, Wurser J, Deriel J and Sirover M (1991). A human nuclear uracil DNA glycosylase is the 37-kDa subunit of glyceraldehyde-3-phosphate dehydrogenase. Proc Natl Acad Sci USA 88: 8460-8464.

Miles E, Rhee S and Davies D (1999). The molecular basis of substrate channeling. J Biol Chem 274:12193-12196.

Mohr G, Rennard R, Cherniack A, Stryker J and Lambowitz A (2001). Function of the *Neurospora crassa* mitochondrial tyrosyl-tRNA synthetase in RNA splicing. Role of the isiosyncratic N-terminal extension and different modes of interaction with different Group I introns. J Mol Biol 307:75-92.

Morris N (1986). The molecular genetics of microtubule proteins in fungi. Exp Mycol 10:77-82.

Morse R (2000). RAP, RAP, open up! New wrinkles for RAP1 in yeast. Trends Genet 16:51-53.

Nelson M et al. (1997). Expressed sequences from conidial, mycelial, and sexual stages of *Neurospora crassa*. Fungal Genet Biol 21:348-363.

Pagliaro L (1993). Glycolysis revisited - A funny thing happened on the way to the Krebs cycle. News Physiol Sci 8:219-223.

Perkins D, Radford A and Sachs M (2001). The *Neurospora* Compendium - Chromosomal Loci. Elsevier Science 325 pps.

Pronk J, Steensma H and vanDijken J (1996). Pyruvate metabolism in *Saccharomyces cerevisiae*. Yeast 12:1607-1633.

Punt P, Dingemanse M, Kuyvenhoven A, Soede R, Pouwels P and Van den Hondel C (1990). Functional elements in the promoter region of the *Aspergillus nidulans gpdA* gene encoding glyceraldehyde-3-phosphate dehydrogenase. Gene 93:101-109.

Ramming M, Kins S, Werner N, Hermann A, Betz H and Kirsch J (2000). Diversity and phylogeny of gephyrin: Tissue-specific splice variants, gene structure, and sequence similarities to molybdenum cofactor-synthesizing and cytoskeleton-associated proteins. Proc Natl Acad Sci USA 97:10266-10271.

Raudaskoski M and Koltin Y (1973). Ultrastructural aspects of a mutant of *Schizophyllum commune* with continuous nuclear migration. J Bacteriol 116:981-988.

Raudaskoski M, Mao W and Ylimattila T (1994). Microtubule cytoskeleton in hyphal growth – Response to nocodazole in a sensitive and a tolerant strain of the homobasidiomycete *Schizophyllum commune*. Eur J Cell Biol 64:131-141.

Rivoal J, Thind S, Pradet A and Ricard B (1997). Differential induction of pyruvate decarboxylase subunits and transcripts in anoxic rice seedlings. Plant Physiol 114:1021-1029.

Rosa AL and Maccioni HJF (1987). Clathrin coated vesicles in *Neurospora crassa*. Mol Cell Biochem 77:63-70.

Rosa A, Alvarez M, Lawson D and Maccioni H (1990a). A polypeptide of 59kDa is associated with bundles of cytoplsmic filaments in *Neurospora crassa*. Biochem J 268:649-655.

Rosa A, Peralta Soler A and Maccioni H (1990b). Purification of P59Nc and immunocytochemical studies of the 8- to 10-nm cytoplasmic filaments from *Neurospora crassa*. Exp Mycol 14:360-371.

Rosa A, Alvarez M and Maldonado C (1990c). Abnormal cytoplasmic bundles of filaments in the *Neurospora crassa snowflake* colonial mutant contain P59Nc. Exp Mycol 14:372-380.

Schorken U and Sprenger G (1998). Thiamin-dependent enzymes as catalysts in chemoenzymatic syntheses. Biochem Biophys Acta Protein Structure and Molecular Enzymology 1385:229-243.

Selker E (2002). Repeat-induced gene silencing in fungi. Adv Genet 46:439-450.

Stallmeyer B, Schwarz G, Schulz, J, Nerlich A, Reiss J, Kirsch J and Mendel R (1999). The neurotransmitter receptor-anchoring protein gephyrin reconstitutes molybdenum cofactor biosynthesis in bacteria, plants, and mammalian cells. Proc Natl Acad Sci USA 96:1333-1338.

Stanway C, Chambers A, Kingsman A and Kingsman S (1989). Characterization of the transcriptional potency of sub-elements of the UAS of the yeast *PGK* gene in a PGK mini-promoter. Nucleic Acids Res 17:9205-9218.

Tadege M, Brändle R and Kuhlemeier C (1997). Aerobic fermentation during tobacco pollen development. Plant Mol Biol 35:343–354

Tadege M, Dupuis I and Kuhlemeier C (1999). Ethanolic fermentation: new functions for an old pathway. Trends Plant Sci 4:320-325.

Temporini E, Alvarez M, Mautino M, Folco H and Rosa A (2004). The *Neurospora crassa cfp* promoter drives a carbon source-dependent expression of transgenes in filamentous fungi. J App Microbiol 96:1256-1264.

Temporini E and Rosa A (1993). Pleiotropic and differential phenotypic expression of two *sn* (*snowflake*) mutant allelles of *Neurospora crassa*- Analysis in homokaryotic and heterokaryotic cells. Curr Genet 23:129-133.

Thompson-Coffe C, Borioli G, Zickler D and Rosa AL (1999). Pyruvate decarboxylase filaments are associated with the cortical cytoskeleton of asci and spores over the sexual cycle of filamentous ascomycetes. Fungal Genet Biol 26:71-80.

Thompson-Coffe C and Zickler D (1992). Three microtubule-organizing centers are required for ascus growth and sporulation in the fungus *Sordaria macrospora*. Cell Motil Cytoskel 22:257-273.

Tornow J, Zeng X, Gao W and Santangelo G (1993). GCR1, a transcriptional activator in *Saccharomyces cerevisiae*, complexes with RAP1 and can function without its DNA binding domain. EMBO J 12:2431-2437.

Turkel S, Turgut T, Lopez MC, Uemura H and Baker HV (2003). Mutations in *GCR1* affect *SUC2* gene expression in *Saccharomyces cerevisiae*. Mol Genet Genom 268:825-831.

Uemura H, Koshio M, Inoue Y, Lopez M and Baker H (1997). The role of Gcr1p in the transcriptional activation of glycolytic genes in yeast *Saccharomyces cerevisiae*. Genetics 147:521-532.

Watanabe H, Takehana K, Date M, Shinozaki T and Raz A (1996). Tumor cell autocrine motility factor is the neuroleukin/phosphohexose isomerase polypeptide. Cancer Res 56:2960-2963.

Ward O and Singh A (2002). Bioethanol technology: Developments and perspectives. Adv Appl Microbiol 51:53-80.

Weber R and Pitt D (2001). Filamentous fungi – growth and physiology. In: G. Khachatourians and D. Arora, eds. Applied Mycology and Biotechnology, Vol. 1. Agriculture and Food Production. Elsevier Science B. V., pp 13-54.

Westfall P and Momany M (2002). *Aspergillus nidulans* septin AspB plays pre- and postmitotic role in septum, branch, and conidiophore development. Mol Biol Cell 13:110-118.

Willett C, Gelfman C and Holland M (1993). A complex regulatory element from the yeast gene *ENO2* modulates GCR1-dependent transcriptional activation. Mol Cell Biol 13:2623-2633.

Wood D and Luck D (1971). A paracrystalline inclusion in *Neurospora crassa*. J Cell Biol 51:249-264.

Xiang X and Plamann M (2003). Cytoskeleton and motor proteins in filamentous fungi. Curr Op Microbiol 6:628-633.

Xu W, Seiter K, Feldman E, Ahmed T and Chiao J (1996). The differentiation and maturation mediator for human myeloid leukemia cells shares homology with neuroleukin or phosphoglucose isomerase. Blood 87:4502-4506.

**Applied Mycology
and Biotechnology**
An International Series
Volume 5. Genes and Genomics

7

ELSEVIER

A Search for Developmental Gene Sequences in the Genomes of Filamentous Fungi

David Moore, Conor Walsh and Geoffrey D. Robson
School of Biological Sciences, The University of Manchester, 1.800 Stopford Building,
Manchester M13 9PT, United Kingdom (david.moore@man.ac.uk).

There is now a sufficient number of filamentous fungal genomes in the public databases to
warrant at least initial comparisons with animal and plant genomes. Our interest lies in the
control of multicellular morphogenesis, which is a feature of filamentous ascomycetes and
basidiomycetes. Search of a representative collection of filamentous fungal genomes with
gene sequences generally considered to be essential and highly conserved components of
normal development in animals failed to reveal any homologies. We conclude that fungal
and animal lineages diverged from their common opisthokont line well before the
emergence of any multicellular arrangement, and that the unique cell biology of filamentous
fungi has caused control of multicellular development in fungi to evolve in a radically
different fashion from that in animals and plants.

1. INTRODUCTION

Fungi form a large, arguably the largest, eukaryotic Kingdom that includes
yeasts, moulds and mushrooms and probably comprises more than 1.5 million
species (Hawksworth 1991, 2001). Fungi have spread through all habitats. Many live
parasitically or symbiotically with animals or plants, but they are characteristically
saprotrophs that are able to produce externalized enzymes capable of digesting an
enormous variety of organic and mineral materials in their immediate environment
and absorbing the nutrients so released.

The small genome size and extensive classic genetic analysis of yeasts made these
fungi especially suitable for pioneering genome analysis in the mid-1990s. A
breakthrough in genome research was the determination of the first complete
eukaryotic genome sequence of the well-studied budding yeast *Saccharomyces
cerevisiae*, which was finished in 1996 (Goffeau et al. 1996). However, the yeast
growth form is highly specialized and the small genome of yeasts is not adequately
representative   of   the   genetic,   biochemical   and   behavioural   diversity

Corresponding author: David Moore

of the bulk of the fungal kingdom, which are filamentous hyphal organisms (Bennett 1997). More recently, the collection of fungal genomes available has become more representative with the sequencing of the genomes of several ascomycetes, including the classic genetic model organism *Neurospora crassa* (Galagan et al., 2003; Schulte et al., 2002). Then, in mid-2003, the Whitehead Institute at MIT announced the release of the first mushroom genome, that of the small ink-cap mushroom *Coprinus* (= *Coprinopsis*) *cinereus* (Anon. 2003a). Since then, full or partial genomes of several other filamentous fungi have been sequenced and the genomic data deposited in fungal sequence databases. There is now a sufficiently representative collection of higher fungal genomes in the public databases to warrant at least initial enquiries that address questions of comparison with animal and plant genomes. Our interest lies in the control of multicellular morphogenesis which is a feature of filamentous ascomycetes and basidiomycetes. Unless a mycologist is speaking, fungi are never included in discussions of developmental biology. An example in point was published in the millennium combined issue of TCB, TIBS and TIG (Meyerowitz 1999), an otherwise excellent paper which discussed evolution of the originally common genetic toolkit of major eukaryote clades without using the word fungus! Nevertheless, we heartily subscribe to the author's view that 'comparison of the genes used to serve similar functions shows how organisms can use different genes for similar ends and thereby reveals the principles of development' (Meyerowitz 1999). In this chapter we report some initial searches for fungal sequences that are similar to those known to be both highly conserved and crucially important to the multicellular development of animals.

There is no obvious reason to exclude the third great Kingdom of eukaryotes from such comparisons, because there is plenty of evidence that fungi undertake developmental processes every bit as sophisticated as those seen in animals or plants. This is true for the majority of the Ascomycota that produce small fruiting structures, but reaches a pinnacle of expression with the mushrooms and toadstools of the Basidiomycota. That sliced mushroom on your dinner plate clearly shows regular tissue formation with specialisation of cell and tissue function, and the briefest after-dinner amble outdoors will reveal such a diversity of fungal fruit bodies that it should be no surprise to find that form and structure are just as important in fungal taxonomy as they are in animal and plant systematics. Yet there is every reason to be cautious about assuming comparability too readily, because of the potential effect of the considerable differences in cell biology that exists between the three major Kingdoms of eukaryotes.

The fundamental aspect of cell biology that distinguishes fungi from other major kingdoms is the apical growth of hyphae. Extension growth of the hypha is limited to the apex, but there is more to development than growth at hyphal tips can achieve alone. The vegetative fungal mycelium is an exploratory, invasive organism. Its component hyphae are regulated to grow outwards into new territory and consequently possess controls that ensure that hyphae normally grow away from one another to form the typical 'colony' with an outwardly-migrating growing front. Tissue development requires that different hyphae cooperate in an organised way. For tissue to be formed the invasive outward growth pattern of the vegetative mycelium must be modified so that independent hyphal apices grow towards each other, allowing their hyphae to branch and differentiate in a cooperative fashion. Kinetic analysis has shown clearly that fungal filamentous growth can be interpreted on the basis of a regular cell cycle (Prosser 1995), and Griffin et al. (1974), pointed out that in mycelial fungi, branch formation (by increasing the number of growing points) is the equivalent of cell division in animals, plants and protests. Common to normal morphogenesis in animals and plants alike is the concept of cellular polarity and the developmental consequences of precise positioning of the plane of cleavage (in animals) or wall formation at the cell plate or phragmoplast in plants (e.g. Samuels et al. 1995). The classic examples of embryology in both groups of organisms include instances of asymmetric divisions partitioning 'stem' cells in ways that result in the daughter cells expressing some sort of differentiation relative to one another, though the differentiation is not necessarily expressed immediately.

However, cross-walls in fungal hyphae are almost always formed at right angles to the long axis of the hypha. Except in cases of injury or in hyphal tips already differentiated to form sporing structures, hyphal tip cells are *not* subdivided by oblique cross-walls, nor by longitudinally oriented ones. Even in fission yeast cells forced to produce irregular septation by experimental manipulation, the plane of the septum was always perpendicular to the plane including the longest axis of the cell (Miyata et al. 1986). In general, then, the characteristic fungal response to the need to convert the one-dimensional hypha into a two-dimensional plate or three-dimensional mass cannot depend on a different geometrical arrangement of the septum. The only solution open to the fungal hypha is the formation of branches and even casual observation shows that branching patterns alter greatly during fungal differentiation and tissue development. Consequently, placement of the hyphal branch (that is, its position of emergence and subsequent direction of growth) is the fungal equivalent of the determination of morphogenetic growth by orientation of the plane of division and the new cross-wall as is seen in plants, and the morphogenetic cell migrations that contribute to development of body form and structure in animals (Moore 1998).

Viewed in this light, therefore, Kingdom Fungi is seen as employing processes during morphogenesis that have affinities with both of the other major eukaryotic kingdoms. The presumption that the three kingdoms have evolved their multicellular organisation independently, but using much the same set of genes inherited from their common unicellular ancestor, is the main reason for interest in making sequence comparisons between the genomes.

Currently, it is generally thought that eukaryotes emerged from prokaryotic ancestry a little less than 2 billion years ago (Gupta and Golding 1996; Knoll 1992), and in the present day about 60 lineages of eukaryotes can be distinguished on the basis of their cellular organization (Patterson 1999). Most of these are traditionally classified as protists, but one lineage comprises green algae and the land plants, and another animals and fungi. Understanding of eukaryote phylogenetic relationships is not yet complete (Cavalier-Smith 1993; Katz 1998, 1999; Knoll 1992; Kuma et al. 1995; Kumar and Rzhetsky 1996; Roger 1999; Sogin et al. 1996; Sogin and Silberman 1998). The plant comparison is compelling because recent discussion has stressed the importance of the symbiotic partnership between phototrophs and fungi in early colonization of the land, protein sequence comparisons indicating that major fungal and algal lineages were present one billion years ago (Heckman et al. 2001). Animals and fungi are more directly related, though. It is generally agreed that the Metazoa and choanoflagellates (collar-flagellates) are sister groups, and that these, together with the fungi and chytrids form a single lineage called the opisthokonts. This name opisthokont (Copeland 1956) refers to the posterior (opistho) location of the flagellum (kont) in swimming cells. The term was applied to the (animals + fungi) clade (Cavalier-Smith and Chao 1995) because comparative molecular analysis has indicated that fungi and animals are each other's closest relatives (Baldauf and Palmer 1993; Baldauf 1999; Patterson and Sogin 2000; Sogin and Silberman 1998; Wainright et al. 1993).

So, accepting the opinion that fungi are more closely related to animals than to plants, we decided to search the genomic databases of several tissue-making filamentous fungi for sequences homologous to conserved genes involved in animal development biology. Tissue-making filamentous fungi, that is ascomycetes and basidiomycetes known to form macroscopic fruit bodies, were chosen as it was felt that the cellular mechanics of their development pathways might be similar those of animals.

Several gene networks controlling animal development processes were identified as potential search targets. These were predominantly conserved sequences involved in controlling multicellular structure such as: transmembrane molecules used for cell adhesion and cell signalling; gene regulatory DNA-binding proteins that coordinate expression of other genes; and sequences involved in the processes of movement and/or targeting of moving cells.

We searched for homologues of *Caenorhabditis elegans* sequences involved in apoptosis and in the signalling mechanisms Notch, TGF-β, Wnt (including the MOM and POP genes) and also a Hedgehog sequence from *Drosophila melanogaster* (as *C. elegans* lacks a Hedgehog homologue). As an additional comparison, the genome of the plant *Arabidopsis thaliana* (Bevan et al. 2001) at The Institute for Genome Research (TIGR) was searched as a control for each sequence search. These species were chosen as they represent good model organisms of animals and plants respectively and their complete DNA sequences are known and well annotated. The Wnt, Hedgehog, Notch and TGF-β signalling mechanisms are all absent in *Arabidopsis* but this plant has its own highly developed signalling pathways (Herve et al. 1999; Katz et al. 2004; Reyes and Grossniklaus 2003; Thummler et al. 1995; Walker 1994; Yu and Tang 2004). Filamentous fungal genomes were also explored for sequences from one of these, the ethylene hormone signalling pathway (Chang 1996; Hua and Meyerowitz 1998), to investigate whether fungal signalling shows any similarities to that plant system. *Arabidopsis* also lacks direct homologues of the caspases, enzymes normally involved in apoptosis in animals, but plants do possess distantly related sequences termed metacaspases, which perform a similar function (Watanebe and Lam 2004). As caspase-like activity has recently been associated with entry into the stationary phase by cultures of *Aspergillus fumigatus* (Mousavi and Robson 2003), and because of other observations discussed below that suggest programmed cell death occurs in several fungi, we included caspase sequences in our searches.

The filamentous fungal species selected for genome analysis were: the basidiomycetes *Coprinus cinereus* (syn. *Coprinopsis* (Redhead et al. 2001)) and *Ustilago maydis* (Anon. 2003a, b) at the Whitehead Institute's Centre for Genome Research; *Cryptococcus neoformans* (last updated 29/4/03) at TIGR; *Phanerochaete chrysosporium* at the Department of Energy's (DOE) Joint Genome Institute; and the ascomycetes *Aspergillus nidulans* (Anon. 2003c) and *Neurospora crassa* (Galagan et al. 2003), also at the Whitehead Institute's Centre for Genome Research; and *Aspergillus fumigatus* at TIGR (URLs are detailed in Table 1). These form a representative and accessible collection of tissue-making ascomycete and basidiomycete filamentous fungal genomic databases that are finished or nearing completion.

2. METHODS & MATERIALS
Default values were used throughout unless otherwise stated.

2.1 Locating Query Sequences
GenBank (Benson et al. 2003), maintained by the National Centre for Biotechnology Information (NCBI) and the Nucleotide Sequence Database (Kulikova et al. 2004), maintained by the European Molecular Biology Laboratory (EMBL), were used to locate DNA sequences. Swiss-Prot, release 42.12 of 15-Mar-2004, TrEMBL Release 25.12 of 15-Mar-2004 (Bairoch and Apweiler 2000), and GenBank (Benson et al. 2003) text searches were used to retrieve protein sequences.

2.2 Similarity Search
An e-value of 1e = -1 was used for each BLAST search. BLASTN (version 2.2.8, last updated Jan-05-2004) and BLASTP (version 2.2.8, last updated Jan-05-2004) at the NCBI (Altschul et al. 1997) were used to search both nucleotide and protein sequences for similar sequences. The organism default parameter was altered so that only fugal sequences were searched.

WU BLASTN nucleotide-nucleotide and WU BLASTP protein-protein searches (both version 2.0, last updated Mar-3-2004; (Gish 1996-2004) were performed with the gene and protein sequences respectively at the TIGR *Arabidopsis thaliana* (Bevan et al. 2001) and *Cryptococcus neoformans* sites (see Table 1).

In the case of the TIGR *Aspergillus fumigatus* site (see Table 1), WU BLASTN nucleotide-nucleotide and WU tBLASTn protein-translated nucleotide searches were performed as the WU BLASTp function was not available.

BLASTN nucleotide-nucleotide and BLASTP protein-protein searches (both version 2.2.1, last updated Aug-1-2001; (Altschul et al. 1997) were performed with the gene and protein sequences respectively at the Whitehead Institute's *Neurospora crassa* (update of 24/4/03) and *Aspergillus nidulans* (update of 31/10/03) sites (see Table 1).

BLASTN nucleotide-nucleotide and tBLASTn protein-translated nucleotide searches were performed at the Whitehead Institute's *Ustilago maydis* (28/5/03 update) and *Coprinus* (= *Coprinopsis*) *cinereus* (25/6/03 update) sites and also at the Department of Energy's (DOE) Joint Genome Institute *Phanerochaete chrysosporium* site (v. 1.0, updated 16/2/02), as there was no BLASTp function available (see Table 1).

Table 1. Table displaying all query organisms used (one plant and seven filamentous fungi); where their genomic databases were accessed and their URLs

Organism	Source	URL
Arabidopsis thaliana	TIGR	http://www.tigr.org/tdb/e2k1/ath1/
Aspergillus fumigatus	TIGR	http://www.tigr.org/tdb/e2k1/afu1/
Cryptococcus neoformans	TIGR	http://www.tigr.org/tdb/e2k1/cna1/
Neurospora crassa	Whitehead Institute	http://www.broad.mit.edu/annotation/fungi/neurospora/
Ustilago maydis	Whitehead Institute	http://www.broad.mit.edu/annotation/fungi/ustilago_maydis/
Coprinus cinereus	Whitehead Institute	http://www.broad.mit.edu/annotation/fungi/coprinus_cinereus/index.html
Aspergillus nidulans	Whitehead Institute	http://www.broad.mit.edu/annotation/fungi/aspergillus/
Phanerochaete chrysosporium	DOE Joint Genome Institute	http://genome.jgi-psf.org/whiterot1/whiterot1.home.html

2.3 Further Analysis

The genome locating services provided by the websites in Table 1 were used to retrieve the protein and/or DNA sequence of a potential hit, and Internet sites listed in Table 2 were used for further analysis of protein and DNA sequences.

BLOCKS version 14.0, last updated October 2003 (Henikoff and Henikoff 1994), and PFAM version 12.0, last updated January 2004 (Bateman et al. 2004) were used to query DNA and protein sequences and to ascertain their domain families.

Interpro [13] release 7.0, last updated 27/11/2003 (Mulder et al. 2003) and PFAM keyword searches were used to look up information on protein domains and families.

CLUSTAL-W version 1.82, (Higgins et al. 1994) at the EMBL-EBI was used to construct two automatic multiple sequence alignments of metacaspases from the DNA and protein sequences respectively of a variety of organisms. The CLUSTAL-W phylogram showing tree distances were used to visualize phylogenetic differences between the sequences. A phylogram is a branching diagram (tree) assumed to be an estimate of phylogeny. In these trees, the branch lengths are proportional to the amount of inferred evolutionary change. The CLUSTAL-W output format default was then altered to give PHYLIP output and these results were analyzed using PAUP 4.0*, Beta 10 (Swafford 1998). Phylogenetic analysis was performed using the distance method, and the trees produced were unrooted. In distance-based analysis, neighbour-joining searches (Saitou and Nei 1987) were used. Analysis was performed using the uncorrected ("p") distance measure.

3. RESULTS

3.1 Apoptosis

The seven filamentous fungal genomic databases, along with that of the plant *A. thaliana*, were first searched for sequences displaying similarity to the sequences of three genes involved in apoptosis in the *Caenorhabditis elegans*: ced-3, ced-4 and egl-1, which are homologues of a caspase, Apaf-1 and Bad respectively. Both the DNA and protein sequences of the above genes were used to search each of the query databases. Many low-quality results were returned but none of them were good enough to warrant further investigation.

Table 2. Names and URLs of sites used for location and further analysis of protein and DNA sequences.

Site and/or tool	URL
GenBank	http://www.ncbi.nlm.nih.gov/Genbank/index.html
EMBL Nucleotide Sequence Database	http://www.ebi.ac.uk/embl/
Swiss-Prot & TrEMBL	http://us.expasy.org/sprot/
BLOCKS	http://www.blocks.fhcrc.org/
PFAM	http://www.sanger.ac.uk/Software/Pfam/
Interpro	http://www.ebi.ac.uk/interpro/
CLUSTAL-W	http://www.ebi.ac.uk/clustalw
PAUP*	http://paup.csit.fsu.edu/

The DNA and protein sequences of the yeast metacaspase YOR197w (also known as YCA1) were then used to search each of the query databases. There were no hits returned from the *C. cinereus*, *P. chrysosporium* and *U. maydis* databases. However, two hits with good e-values were found in each of the *A. nidulans*, *N. crassa*, *C. neoformans* and *A. fumigatus* databases. Also hits to several latex-abundant family protein/caspase family protein/metacaspases (AMC) were returned from the *A. thaliana* database.

3.1.1 Metacaspase Multiple Sequence Alignment

Using CLUSTAL-W, two metacaspase multiple sequence alignments, were performed using protein and DNA sequences respectively. Each multiple sequence alignment consisted of: the putative filamentous fungal metacaspases, two yeast metacaspases (one *S. cerevisiae* and one *S. pombe*) and three plant metacaspases (AMC1, 2 and 8 from *A. thaliana*).

The phylogram trees with distances displayed from CLUSTAL-W were used to visualize the phylogenetic distances between each sequence (Figs 1 & 2). The aligned sequences were then analyzed using PAUP 4.0* Beta-10 and unrooted neighbour-joining trees were returned and analyzed (data not shown).

The DNA metacaspase multiple sequence alignment differs from that of the protein in that it contains two extra *A. fumigatus* sequences. The TIGR *A. fumigatus* site was incomplete at the time of writing, and attaining gene and protein sequences proved very tough. Attempts to translate the *A. fumigatus* sequences were unsuccessful as the sequences acquired from the TIGR database were not complete genes and therefore contained many stop codons.

3.2 Signalling Mechanisms

The gene sequences investigated in this category were all from *C. elegans*, except for the Hedgehog gene from the *Drosophila melanogaster* (as the *C. elegans* lacks a Hedgehog homologue).

3.2.1 MOM & POP

The MOM and POP gene sequences investigated here were the MOM-1, 2 and 5 and the POP-1 genes. The normal function of POP-1 is to prevent endoderm formation, thus the Wnt signal is required to prevent POP-1 function (Rocheleau et al. 1997). BLAST searches of the genomic databases of the seven filamentous fungal species and the plant *A. thaliana* were performed for each MOM-1, 2 and 5 and pop-1. Many matches to small fragments were located but no significant matches were returned.

3.2.2 Wnt Signalling Pathway

Here we searched the seven filamentous fungal and the *A. thaliana* genomic databases for homologues of the Wnt-1, Wnt-2, Egl-20 and Lin-44 sequences. Their nucleotide and protein sequences were located and used to search the query databases, but again no quality hits were returned.

3.2.3 Hedgehog Signalling Pathway

The sonic hedgehog gene and protein sequences from the only *Drosophila melanogaster* hedgehog sequence were used to search the query databases. No quality hits were returned.

3.2.4 Notch Signalling Pathway

The notch sequences Glp-1 and Lin-12 were used to search the query databases. There were some low-quality hits. However, upon further investigation it was found that theses hits were all against an ankyrin repeat located in the protein sequences. There were no other quality hits returned.

3.2.5 TGF-β Signalling Pathway

Here the *C. elegans* TGF-β sequences of Daf-4, Daf-7, Dbl-1, Unc-129 and Cet-1 were used to search the query databases. No quality hits were returned.

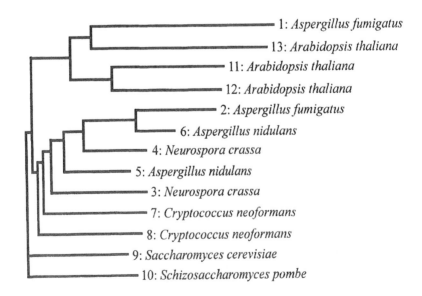

1: *Aspergillus fumigatus*
13: *Arabidopsis thaliana*
11: *Arabidopsis thaliana*
12: *Arabidopsis thaliana*
2: *Aspergillus fumigatus*
6: *Aspergillus nidulans*
4: *Neurospora crassa*
5: *Aspergillus nidulans*
3: *Neurospora crassa*
7: *Cryptococcus neoformans*
8: *Cryptococcus neoformans*
9: *Saccharomyces cerevisiae*
10: *Schizosaccharomyces pombe*

Seq. 1: *Aspergillus fumigatus*	chr_0 | TIGR.5170 | 54
Seq. 2: *Aspergillus fumigatus*	chr_0 | TIGR.5237 | 59
Seq. 3: *Neurospora crassa*	AABX01000434
Seq. 4: *Neurospora crassa*	AABX01000362
Seq. 5: *Aspergillus nidulans*	AF528964
Seq. 6: *Aspergillus nidulans*	AACD01000042
Seq. 7: *Cryptococcus neoformans*	185.m02420
Seq. 8: *Cryptococcus neoformans*	186.m04030
Seq. 9: *Saccharomyces cerevisiae*	NC_001147
Seq. 10: *Schizosaccharomyces pombe*	AF316601
Seq. 11: *Arabidopsis thaliana* (AMC1)	NM_100097
Seq. 12: *Arabidopsis thaliana* (AMC2)	NM_118643
Seq. 13: *Arabidopsis thaliana* (AMC8)	NM_101508

Fig. 1. CLUSTAL-W metacaspase nucleotide sequence phylogram showing tree distances. The top clade contains the three plant (*A. thaliana*) sequences and one of the *A. fumigatus* sequences (sequence number 1). The middle clade contains the rest of the filamentous fungal sequences. Sequences 9 and 10, which are the two yeast metacaspase sequences, essentially form an outgroup as the bottom clade which is quite distinct from the others. Branch lengths are proportional to the amount of inferred evolutionary change.

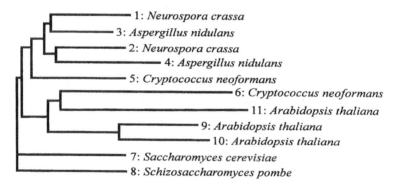

Seq. 1: *Neurospora crassa*	EAA29413
Seq. 2. *Neurospora crassa*	EAA30484
Seq. 3: *Aspergillus nidulans*	AA013381
Seq. 4: *Aspergillus nidulans*	EAA63988
Seq. 5: *Cryptococcus neoformans*	185.m02420
Seq. 6: *Cryptococcus neoformans*	186.m04030
Seq. 7: *Saccharomyces cerevisiae*	YOR197
Seq. 8: *Schizosaccharomyces pombe*	CAA20127
Seq. 9: *Arabidopsis thaliana* (AMC1)	AAP84706
Seq. 10: *Arabidopsis thaliana* (AMC2)	AAP84707
Seq. 11: *Arabidopsis thaliana* (AMC8)	AAP84711

Fig. 2. CLUSTAL-W metacaspase protein sequence phylogram showing tree distances. Again the sequences divide into three main clades. One comprises the three plant proteins (sequences 9, 10 & 11), together with a *C. neoformans* (sequence 6) protein sequence. The two plant metacaspases AMC 1 and 2 (sequences 9 &10) seem to be quite distinct from the other plant metacaspase AMC 8 (sequence 11), which is the one that shows similarity to the *Cryptococcus* sequence. The rest of the filamentous fungal sequences (sequences 1, 2, 3, 4 & 5) fall into a distinct clade, leaving the two yeast sequences (sequences 7 & 8) isolated again, virtually as an outgroup. Branch lengths are proportional to the amount of inferred evolutionary change.

3.2.6 Plant Ethylene Signalling Pathway

The *A. thaliana* sequences of ESR1 and ETR1, which are receptors involved in the plant ethylene hormone signalling pathway, were used to search for homologues in the query fungal databases; but again no quality hits were returned.

4. DISCUSSION

The obvious motive for this sort of survey is to estimate how the different eukaryotic kingdoms have made use of their common genetic heritage since the divergence of the major eukaryotic clades, animals, fungi and plants, about 1×10^9 years ago. There is an essential underlying logic to morphogenesis and it is this that justifies sequence comparisons between organisms that are so very different today. Since their last common ancestor was unicellular, complex multicellular development must have arisen independently in the three kingdoms, so comparison of the way similar functions are controlled can reveal whether and how different cellular mechanisms have been used to solve common developmental demands (Meyerowitz 1999). However, bearing in mind the considerable differences that exist in the basic cell biology of animals, fungi and plants, similarities between the mechanisms used in their developmental biologies may be hard to find.

In this initial survey we have attempted to establish whether fungal multicellular development shows any closer relationship to that of animals than to that of plants by searching filamentous fungal genomic databases for sequences demonstrating similarity to developmental gene sequences. The phylogenetic logic of this approach is that the plant lineage was the first to diverge, leaving animals and fungi in a common clade (the opisthokonts) for potentially several hundred million years before diverging into the distinct kingdoms we know today. It is not unreasonable to argue that the opisthokonts evolved basic strategies for dealing with cellular interactions prior to their divergence and that evidence of this might be found in present day genomes in the form of similarities between sequences devoted to tasks that can be defined broadly as 'developmental'. On the other hand, as 41% of the predicted proteins of the *Neurospora crassa* genome lack significant matches to known proteins from public databases (Galagan et al. 2003) it is not at all impossible that fungi have their own unique development processes using genes found exclusively in fungi.

4.1 Programmed Cell Death

Programmed cell death (apoptosis) is one of the central cellular processes in multicellular eukaryote development (Hentgartner 2000), and there are many observations (discussed below) that imply that programmed sacrifice of hyphal segments is important in fungal development. However, no positive hits were recorded in the query databases to the sequences of the three genes involved in apoptosis in *Caenorhabditis elegans*. This result may be expected as no caspase homologue has yet been found in fungi (Thrane et al. 2004), though sequences characteristic of the distantly related metacaspases have been found in fungi, plants and protozoa (Wu et al. 2003). With these findings in mind, we chose the yeast metacaspase YOR197w (also known as YCA1), which is involved in regulation of apoptosis in *Saccharomyces cerevisiae* (Madeo et al. 2002), to search the filamentous fungal query databases.

C. cinereus, *P. chrysosporium* and *U. maydis* did not contain a metacaspase homologue but *A. nidulans*, *N. crassa*, *C. neoformans* and *A. fumigatus* all contained two. The two *N. crassa* hits corresponded to the two hypothetical metacaspases previously deduced (GenBank Accession No. XP-331176 and XP-330804); and one of the *A. nidulans* hits corresponded to the recently cloned and described casA metacaspase (GenBank Accession No. AAO13381) (Cheng et al. 2003). The many hits encountered in *Arabidopsis thaliana* were to be expected as it is known that plants possess metacaspases.

4.1.1 Metacaspase Multiple Sequence Alignment

In the DNA (Fig. 1) and protein (Fig. 2) multiple sequence alignments of the metacaspase sequences two main regions showing high similarity were observed, which are assumed to correspond with metacaspase apoptotic domains. The trees returned by CLUSTAL-W and PAUP were very similar, comprising clades that contained essentially the same groups of sequences, which were at similar distances from each other.

The DNA sequence phylogram tree returned by CLUSTAL-W (Fig. 1) showed three main clades: one containing the two yeast sequences; one containing the three plant sequences along with one of the *Aspergillus fumigatus* sequences; and the other containing the rest of the filamentous fungal sequences. The two yeast metacaspases from *S. cerevisiae* and *S. pombe* formed an outgroup. In the earlier BLAST searches the *A. fumigatus* sequence that associated with the *Arabidopsis thaliana* metacaspases returned the inferior e-value and is obviously more similar to the plant metacaspases than to those of the other filamentous fungi. However, the plant metacaspase to which it seems most closely related (AMC 8) is somewhat distant from the other two (AMC 1 and AMC 2).

The protein multiple sequence alignment tree displayed similar results. Fig. 2 shows three main clades: with the two yeast metacaspases (sequences 7 and 8) forming an outgroup and the plant metacaspases (sequences 9, 10 and 11), being associated with a filamentous fungal sequence (sequence 6). In this case the fungal sequence found within the plant clade is one of the *C. neoformans* sequences. It is worth noting that the *C. neoformans* sequence that clusters with *Arabidopsis* in Fig. 2 is the most distantly related to the other filamentous fungal sequences in the DNA tree shown in Fig. 1. The main clade in Fig. 2 is comprised of the rest of the filamentous fungal sequences (sequences 1, 2, 3, 4 and 5). Evidently, several fungal genomes contain metacaspase homologues, there being one in ascomycetous yeasts, and two in several filamentous fungi.

In some cases it is likely that such sequences will prove to be responsible for the caspase-like activities that have been associated with sporulation (Thrane et al. 2004) and the onset of stationary phase lysis (Mousavi and Robson 2003) in *Aspergillus* species. There are many other reports that suggest dependence on a programmed cell death (PCD) in fungi. Interestingly, some of these observations relate to *Coprinus* (= *Coprinopsis*) *cinereus* despite the fact that we have been unable to find any caspase or metacaspase homologues in this genome. Recently, for example, (Lu et al. 2003) have presented elegant cytological evidence for 'apoptotic' DNA degradation in basidia of meiotic mutants of the mushroom *C. cinereus*; a compelling comparison with animal apoptosis where specific DNA fragmentation is an essential component. Hasty comparison with vertebrates in which avoidance of antigen release as the cell dies is a primary 'design requirement' to protect the animal against autoimmunity may be a mistake. Protection against autoimmunity is not a consideration in fungi; rather such a process in a mushroom may well be a matter of resource conservation (the *Coprinopsis* mutants cannot complete meiosis so there is presumably some value in recycling the contents of the defective basidia (Money 2003)).

There may well, in fact, be several different mechanisms, reflecting the several different roles, for PCD in fungi. Cell death is a common occurrence in various structures starting to differentiate, for example the formation of gill cavities in the cultivated mushroom *Agaricus bisporus* (Umar and van Griensven 1997, 1998). These authors point out that specific timing and positioning imply that cell death is part of the differentiation process and that fungal PCD could play a role at many stages in development of many species. Individual hyphal compartments can be sacrificed to trim hyphae to create particular tissue shapes. PCD is used, therefore, to sculpture the shape of the fruit body from the raw material provided by the hyphal mass of the fruit body initial and primordium (Moore et al. 1998). Several examples detailed by Umar & van Griensven (1998) feature a PCD programme that involves the sacrificed cells over-producing mucilaginous materials that are released by cell lysis. Probably the most obvious example of fungal PCD is the autolysis that occurs in the later stages of development of fruit bodies of ink-cap mushrooms (many species of *Coprinus*, *Coprinopsis* and *Coprinellus*) that involves specifically-timed production and organised release of a range of lytic enzymes (Iten 1970; Iten and Matile 1970). This autolysis in coprinoid fungi has been interpreted as a mechanism to remove gill tissue from the bottom of the cap to avoid interference with discharge of spores from regions above (Buller 1924, 1931). This interpretation of autolysis being part of the developmental programme of a fungal fruit body predates by 40 years the first use of the phrase programmed cell death (Lockshin 1963; Lockshin et al. 1998).

4.2 Signalling Mechanisms

Cell-cell signalling is essential for many biological processes ranging from developmental patterning to the regulation of cell proliferation and cell death. In addition, almost all cancer cells have mutations in genes involved in signalling. The animal signalling mechanisms investigated here were the Wnt, Hedgehog, Notch and TGF-β. It is known that they are all absent in *Arabidopsis*, but plants compensate with their own highly developed signalling pathways (Meyerowitz 1999).

4.2.1 Wnt signalling pathway

The Wnt signalling pathway is highly conserved and functions during the development of many animals by regulating processes such as cell proliferation, cell polarity, mitotic alignment and the specification of cell fate. Inappropriate expression of the Wnt pathway is implicated in tumorigenesis (van Es et al. 2003). The Wnt genes encode secreted glycoproteins that regulate a huge variety of developmental processes in vertebrates and invertebrates by inducing transcriptional or morphological changes in responding cells. The Wnt family comprises 18 members in humans, four members in *Drosophila* (the best-studied being Wingless), five in *Caenorhabditis elegans* and at least one in *Hydra* (Huelsken and Birchmeier 2001; Kalderon 2002; Seto and Bellen 2004; Thorpe et al. 2000).

The MOM and POP genes investigated here all play a role in the Wnt signal transduction pathway in *C. elegans*. The MOM-1, 2 and 5 and the POP-1 genes are all essential in providing polarity signals that distinguish endoderm from mesoderm in the early embryo (Thorpe et al. 1997). POP-1 distinguishes the fates of anterior daughter cells from their posterior sisters throughout development as part of a MAP kinase-like signalling mechanism (Shin et al. 1999). The MOM-2 gene encodes a member of the Wnt family of proteins, while MOM-5 encodes a protein that resembles the mammalian Wnt receptor (the Frizzled protein). The MOM-1 gene appears to encode a protein needed for the secretion of the Wnt protein. The normal function of POP-1 is to prevent endoderm formation, thus the Wnt signal is required to prevent POP-1 function (Rocheleau et al. 1997). The lack of highly positive hits suggests that filamentous fungi lack the Wnt signal transduction pathway. This notion was further explored by searching for Wnt proteins, but again we had no success. Thus, the lack of hits to several components of the pathway adds weight to the conclusion that the Wnt signalling pathway is absent from fungi.

4.2.2 Hedgehog signalling pathway

Hedgehog proteins belong to a smaller family of secreted signal molecules that act as local transcriptional mediators serving as key organizers of tissue patterning in many developmental processes in animals. Although they are notably absent from some invertebrates, the prime example of this deficiency being *Caenorhabditis elegans*, signalling proteins of the Hedgehog family are generally considered to be essential for patterning and morphogenesis in most multicellular animals (Nybakken and Perrimon 2002; Stark 2002; Tabin and McMahon 1997). There are similarities between the Wnt and Hedgehog pathways, and, like Wnt, misregulation of Hedgehog leads to several disease states in humans (Kalderon 2002; Mullor et al. 2002; Wetmore 2003). No quality hits were returned from our search of the filamentous fungal genomes with Hedgehog gene and protein sequences from *Drosophila melanogaster*, suggesting that fungi also lack the Hedgehog signalling pathway.

4.2.3 Notch signalling pathway

Notch is a large cell-surface receptor that is activated by contact with membrane-bound ligands on neighbouring cells. Signalling through notch receptor proteins is probably the most widely used signalling pathway in animal development. It is involved in several developmental pathways with a widespread role in the determination of cell fates, including defining the two different cell fates that result from asymmetric cell divisions, and in signalling events that establish tissue boundaries (Fleming et al. 1997; Martinez Arias et al. 2002; Weinmaster 2000). Notch signalling also defines the future dorsal-ventral axis in nematode embryos, and is one of the systems implicated in the establishment of a segmental pattern within the vertebrate body plan (Bessho and Kageyama 2003; Dubrulle and Pourquié 2002; Pourquié 1999).

A search for homologues of the *Caenorhabditis elegans* Notch sequence in genomes of filamentous fungi yielded hits only to the ankyrin repeat located within that sequence. According to InterPro, the ankyrin repeat is one of the most common protein-protein interaction motifs in nature and occurs in a large number of functionally diverse proteins from eukaryotes (Mulder et al. 2003). Therefore, we conclude again that the lack of quality returns suggests that the Notch signalling pathway is also absent from filamentous fungi.

4.2.4 TGF-β signalling pathway

The transforming growth factor-β (TGF-β) superfamily consists of a large number of molecules that act either as cytokine hormones, or more commonly as local mediators. During development they regulate pattern formation and influence various cell behaviours, including proliferation, differentiation, extracellular matrix production and PCD (Massague et al. 1997; Souchelnytskyi et al. 2002). Lack of quality hits to the *Caenorhabditis elegans* TGF-β sequence in a search of the genomes of filamentous fungi suggests that this signalling pathway is also absent from these fungi.

4.2.5 Plant ethylene signalling pathway

From the results outlined above, we can conclude that all four of the Wnt, Hedgehog, Notch and TGF-β signalling pathways are absent from the seven query fungi; just as they are absent in plants. In compensation, plants have their own highly developed signalling pathways. In the ethylene signalling pathway the simple two-carbon gas ethylene functions as a plant hormone. It is important at many stages of a plant development including: germination, flower development, fruit ripening and responses to many environmental stimuli (Muller-Dieckmann et al. 1999; Stearns and Glick 2003; Zhao et al. 2002). Ethylene perception and signal transduction into the cell are carried out by a family of membrane-bound receptors of which ETR1 and ESR1 are members. Again no substantive similarities were returned by a search of fungal genomes using the *Arabidopsis thaliana* ETR1 and ESR1 sequences, suggesting that fungi also lack the plant hormone ethylene signalling pathway.

4.3 General Discussion

The sort of survey we present here is predicated on the assumption that there is a fundamental logic to development that is inescapably shared by the different eukaryotic kingdoms. As a unicellular system develops into a multicellular organism, a need for intercellular communication is assumed. As soon as two cells collaborate there will be a need for differential gene expression, for polarity, and for positional awareness. There will be a need to assign cell fates, and to control sequential and synchronous events. Numerous roles can be imagined for both intra- and intercellular trafficking, with a consequential need for mechanisms capable of signal creation, transmission, reception, and eventual action and these are represented among the animal sequences we have used here.

It is further expected that if two extant kingdoms shared a common ancestor that itself evolved at least some way towards multicellularity, the mechanisms then developed would survive through phylogenetic change as sequence homologies between the lineages. This expectation is fulfilled in all those aspects that characterise eukaryotic unicells. Interestingly, this does include some features that are important in morphogenesis. One example is the Armadillo sequence repeats that form a conserved three-dimensional protein structure functioning in intracellular signalling, membrane docking and cytoskeletal regulation (Coates 2003; Wang et al. 2001). Another example is the MADS-box domain, which is found in a diverse range of eukaryotes and reveals conservation of a transcriptional regulator (Alvarez-Buylla et al. 2000; Krüger et al. 1997).

Nevertheless, we have demonstrated here that several major components of animal cell interaction do not have homologies in fungal genomes. There seem to be three inferences that are not necessarily mutually exclusive. First, this circumstance may indicate that fungal and animal lineages diverged from their common opisthokont line well before the emergence of any multicellular arrangement and have consequently evolved all aspects of multicellular management independently.

Second, the cell biology of fungal cells may be so different from that of animals (as outlined in section 1) that responses to the basic 'logical' demands of multicellular development are equally different. In this respect it is significant that a new vector-based mathematical model of hyphal growth (the Neighbour-Sensing model) shows that fruit bodies can be simulated by applying the same regulatory functions to all of the growth points active in a structure at any specific time (Meškauskas et al. 2004). Shape of the fruit body emerges from the concerted response of the entire population of hyphal tips, in the same way, to the same signals. This at least suggests the possibility that control of multicellular development in fungi is radically different from that in animals and plants, and may be indicating that fungal tissues can get by without much cell-to-cell communication. The implication could be that a multicellular system that depends exclusively on apical growth must comply with a far simpler set of rules, and the simplicity of the rule set is reflected in cell signalling mechanisms that tend to employ very basic metabolites rather than hormones or cytokines (Moore 1998; Novak Frazer 1996). Third, it is possible that this initial survey has failed to identify homologies which exist because, perhaps, the animal gene sequences selected, or the fungal genomes examined, or both, were in some way unrepresentative. Only further more comprehensive analysis will decide between these possibilities, but we believe that the first and second inferences combined make the most likely conclusion.

5. CONCLUSION

Searching what we believe to be a representative collection of filamentous fungal genomes with gene sequences generally considered to be essential and highly conserved components of normal development in animals failed to reveal any homologies. We conclude this indicates that fungal and animal lineages diverged from their common opisthokont line well before the emergence of any multicellular arrangement and that the unique cell biology of filamentous fungi has caused control of multicellular development in fungi to evolve in a radically different fashion from that in animals and plants. A more comprehensive analysis should confirm this.

REFERENCES

Altschul SF, Madden TL, Schäffer AA, Zhang J, Zhang Z, Miller W and Lipman DJ (1997). Gapped BLAST and PSI-BLAST: a new generation of protein database search programs. Nucl Acids Res 25:3389-3402.

Alvarez-Buylla ER, Pelaz S, Liljegren SJ, Gold SE, Burgeff C, Ditta GF, Ribas de Pouplana L, Martinez-Castilla L and Yanofsky MF (2000). An ancestral MADS-box gene duplication occurred before the divergence of plants and animals. Proc Natl Acad Sci USA 97:5328-5333.

Anon. (2003a). *Coprinus cinereus* sequencing project (http://www.broad.mit.edu): Centre for Genome Research.

Anon. (2003b). *Ustilago maydis* sequencing project (http://www.broad.mit.edu: Centre for Genome Research.

Anon. (2003c). *Aspergillus* sequencing project (http://www.broad.mit.edu): Centre for Genome Research.

Bairoch A and Apweiler R (2000). The SWISS-PROT protein sequence database and its supplement TrEMBL in 2000. Nucl Acids Res 28:45-48.

Baldauf SL and Palmer JD (1993). Animals and fungi are each other's closest relatives: Congruent evidence from multiple proteins. Proc Natl Acad Sci USA 90:11558-11562.

Baldauf SL (1999). A search for the origins of animals and fungi: Comparing and combining molecular data. Amer Nat 154 (suppl.):S178-S188.

Bateman A, Coin L, Durbin R, Finn RD, Hollich V, Griffiths-Jones S, Khanna A, Marshall M, Moxon S, Sonnhammer ELL, D.J S, Yeats C and S.R E (2004). The Pfam protein families database. Nucl Acids Res 32:D138-D141.

Bennett JW (1997). White Paper: Genomics for filamentous fungi. Fungal Genet Biol 21:3-7.

Benson DA, Karsch-Mizrachi I, Lipman DJ, Ostell J and Wheeler DL (2003). GenBank. Nucl Acids Res 31:23-27.

Bessho Y and Kageyama R (2003). Oscillations, clocks and segmentation. Curr Opin Genet Devel 13:379-384.

Bevan M, Mayer K, White O, Eisen JA, Preuss D, Bureau T, Salzberg SL and Mewes HW (2001). Sequence and analysis of the *Arabidopsis* genome. Curr Opin Plant Biol 4:105-110.

Buller AHR (1924). Researches on Fungi, Vol. 3. London: Longmans Green.

Buller AHR (1931). Researches on Fungi. Vol. 4. London: Longmans Green.

Cavalier-Smith T (1993). Kingdom Protozoa and its 18 phyla. Microbiol Rev 57:953-994.

Cavalier-Smith T and Chao EE (1995). The opalozoan *Apusomonas* is related to the common ancestor of animals, fungi and choanoflagellates. Proc Roy Soc Lond Ser B 261:1-6.

Chang C (1996). The ethylene signal transduction pathway in *Arabidopsis*: an emerging paradigm? Trends Biochem Sci 21:129-133.

Cheng J, Park TS, Chio LC, Fischl AS and Ye XS (2003). Induction of apoptosis by sphingoid long-chain bases in *Aspergillus nidulans*. Mol Cell Biol 23:163-177.

Coates JC (2003). Armadillo repeat proteins: beyond the animal kingdom. Trends Cell Biol 13:463-471.

Copeland HF (1956). The Classification of Lower Organisms. Palo Alto, California: Pacific Books.

Dubrulle J and Pourquié O (2002). From head to tail: links between the segmentation clock and antero-posterior patterning of the embryo. Curr Opin Genet Devel 12:519-523.

Fleming RJ, Purcell K and Artavanis-Tsakonas S (1997). The NOTCH receptor and its ligands. Trends Cell Biol 7:437-441.

Galagan JE, Calvo SE, Borkovich KA, Selker EU, Read ND, Jaffe D, FitzHugh W, Ma LJ, Smirnov S, Purcell S, Rehman B, Elkins T, Engels R, Wang S, Nielsen CB, Butler J, Endrizzi M, Qui D, Ianakiev P, Bell-Pedersen D, Nelson MA, Werner-Washburne M, Selitrennikoff CP, Kinsey JA, Braun EL, Zelter A, Schulte U, Kothe GO, Jedd G, Mewes W, Staben C, Marcotte E, Greenberg D, Roy A, Foley K, Naylor J, Stange-Thomann N, Barrett R, Gnerre S, Kamal M, Kamvysselis M, Mauceli E, Bielke C, Rudd S, Frishman D, Krystofova S, Rasmussen C, Metzenberg RL, Perkins DD, Kroken S, Cogoni C, Macino G, Catcheside D, Li W, Pratt RJ, Osmani SA, DeSouza CP, Glass L, Orbach MJ, Berglund JA, Voelker R, Yarden O, Plamann M, Seiler S, Dunlap J, Radford A, Aramayo R, Natvig DO, Alex LA, Mannhaupt G, Ebbole DJ, Freitag M, Paulsen I, Sachs MS, Lander ES, Nusbaum C and Birren B (2003). The genome sequence of the filamentous fungus *Neurospora crassa*. Nature 422:859-868.

Gish W (1996-2004).http://blast.wustl.edu.

Goffeau A, Barrell BG, Bussey H, Davis RW, Dujon B, Feldmann H, Galibert F, Hoheisel JD, Jacq C, Johnston M, Louis EJ, Mewes HW, Murakami Y, Philippsen P, Tettelin H and Oliver SG (1996). Life with 6000 genes. Science 274:546-567.

Griffin DH, Timberlake WE and Cheney JC (1974). Regulation of macromolecular synthesis, colony development and specific growth rate of *Achlya bisexualis* during balanced growth. J Gen Microbiol 80:381-388.

Gupta RS and Golding GB (1996). The origin of the eukaryotic cell. Trends Biochem Sci 21:166-171.

Hawksworth DL (1991). The fungal dimension of biodiversity - magnitude, significance and conservation. Mycol Res 95:641-655.

Hawksworth DL (2001). The magnitude of fungal diversity: the 1.5 million species estimate revisited. Mycol Res 105:1422-1432.

Heckman DS, Geiser DM, Eidell BR, Stauffer RL, Kardos NL and Hedges SB (2001). Molecular evidence for the early colonization of land by fungi and plants. Science 293:1129-1133.

Henikoff S and Henikoff JG (1994). Protein family classification based on searching a database of blocks. Genomics 19:97-107.

Hentgartner MO (2000). The biochemistry of apoptosis. Nature 407:770-776.

Herve C, Serres J, Dabos P, Canut H, Barre A, Rouge P and Lescure B (1999). Characterization of the Arabidopsis lecRK-a genes: members of a superfamily encoding putative receptors with an extracellular domain homologous to legume lectins. Plant Mol Biol 39:671-682.

Higgins D, Thompson J, Gibson T, Thompson JD, Higgins DG and Gibson TJ (1994). CLUSTAL W: improving the sensitivity of progressive multiple sequence alignment through sequence weighting position-specific gap penalties and weight matrix choice. Nucl Acids Res 22:4673-4680.

Hua J and Meyerowitz EM (1998). Ethylene responses are negatively regulated by a receptor gene family in Arabidopsis thaliana. Cell 94:261-271.

Huelsken J and Birchmeier W (2001). New aspects of Wnt signaling pathways in higher vertebrates. Curr Opin Genet Dev 11:547-553.

Iten W (1970). Zur funktion hydrolytischer enzyme bei der autolysate von *Coprinus*. Ber Schweiz Bot Ges 79:175-198.

Iten W and Matile P (1970). Role of chitinase and other lysosomal enzymes of *Coprinus lagopus* in the autolysis of fruiting bodies. J Gen Microbiol 61:301-309.

Kalderon D (2002). Similaritiies between the Hedgehog and Wnt signaling pathways. Trends Cell Biol 12:523-531.

Katz A, Oliva M, Mosquna A, Hakim O and Ohad N (2004). FIE and CURLY LEAF polycomb proteins interact in the regulation of homeobox gene expression during sporophyte development. Plant J 37:707-719.

Katz LA (1998). Changing perspectives on the origin of eukaryotes. Trends Ecol Evol 13:493-497.

Katz LA (1999). The tangled web: gene genealogies and the origin of eukaryotes. Amer Nat 154 (suppl.):S137-S145.

Knoll AH (1992). The early evolution of eukaryotes: a geological perspective. Science 256:622-627.

Krüger J, Aichinger C, Kahmann R and Bölker M (1997). A MADS-box homologue in Ustilago maydis regulates the expression of pheromone-inducible genes but is nonessential. Genetics 147:1643-1652.

186

Kulikova T, Aldebert P, Althorpe N, Baker W, Bates K, Browne P, van den Broek A, Cochrane G, Duggan K, Eberhardt R, Faruque N, Garcia-Pastor M, Harte N, Kanz C, Leinonen R, Lin Q, Lombard V, Lopez R, Mancuso R, McHale M, Nardone F, Silventoinen V, Stoehr P, Stoesser G, Tuli MA, Tzouvara K, Vaughan R, Wu D, Zhu W and Apweiler R (2004). The EMBL nucleotide sequence database. Nucl Acids Res 32:D27-D30.

Kuma K, Nikoh N, Iwabe N and Miyata T (1995). Phylogenetic position of *Dictyostelium* inferred from multiple protein data sets. J Mol Evol 41:238-246.

Kumar S and Rzhetsky A (1996). Evolutionary relationships of eukaryotic kingdoms. J Mol Evol 42:183-193.

Lockshin RA (1963).Programmed cell death in an insect. Cambridge, MA: Harvard University, PhD thesis.

Lockshin RA, Zakeri Z and Tilly JL (1998). When Cells Die. New York: John Wiley & Sons.

Lu BC, Gallo N and Kües U (2003). White-cap mutants and meiotic apoptosis in the basidiomycete *Coprinus cinereus*. Fungal Genet Biol 39:82-93.

Madeo F, Herker E, Maldener C, Wissing S, Lachelt S, Herlan M, Fehr M, Lauber K, Sigrist SJ, Wesselborg S and Frohlick KU (2002). A caspase related protease regulates apoptosis in yeast. Mol Cell 9:911-917.

Martinez Arias A, Zecchini V and Brennan K (2002). CSL-independent Notch signalling: a checkpoint in cell fate decisions during development? Curr Opin Genet Dev 12:524-533.

Massague J, Hata A and Liu F (1997). TGF-β signalling through the Smad pathway. Trends Cell Biol 7:187-192.

Meškauskas A, McNulty LJ and Moore D (2004). Concerted regulation of all hyphal tips generates fungal fruit body structures: experiments with computer visualisations produced by a new mathematical model of hyphal growth. Mycol Res 108:341-353.

Meyerowitz EM (1999). Plants, animals and the logic of development. Trends Cell Biol 9:M65-M68.

Miyata M, Miyata H and Johnson BF (1986). Asymmetric location of the septum in morphologically altered cells of the fission yeast *Schizosaccharomyces pombe*. J Gen Microbiol 132:883-891.

Money NP (2003). Suicidal mushroom cells. Nature 423:26.

Moore D (1998). Fungal Morphogenesis. New York: Cambridge University Press.

Moore D, Chiu SW, Umar MH and Sánchez C (1998). In the midst of death we are in life: further advances in the study of higher fungi. Bot J Scotland 50:121-135.

Mousavi SAA and Robson GD (2003). Entry into the stationary phase is associated with a rapid loss of viability and an apoptotic-like phenotype in the opportunistic pathogen *Aspergillus fumigatus*. Fungal Genet Biol 39:221-229.

Mulder NJ, Apweiler A, Attwood TK, Bairoch A, Barrell D, Bateman A, Binns D, Biswas M, Bradley P, Bork P, Bucher P, Copley RR, Courcelle E, Das U, Durbin R, Falquet L, Fleischmann W, Griffiths-Jones S, Haft D, Harte N, Hulo N, Kahn D, Kanapin A, Krestyaninova M, Lopez R, Letunic I, Lonsdale D, Silventoinen V, Orchard SE, Pagni M, Peyruc D, Ponting CP, Selengut JD, Servant F, Sigrist CJA, Vaughan R and Zdobnov EM (2003). The InterPro Database, 2003 brings increased coverage and new features. Nucl Acids Res 31:315-318.

Muller-Dieckmann HJ, Grantz AA and Kim SH (1999). The structure of the signal receiver domain of the *Arabidopsis thaliana* ethylene receptor ETR1. Structure 7:1547-1556.

Mullor JL, Sanchez P and i Altaba AR (2002). Pathways and consequences: Hedgehog signaling in human disease. Trends Cell Biol 12:562-569.

Novak Frazer L (1996). Control of growth and patterning in the fungal fruiting structure. A case for the involvement of hormones. In: S. W. Chiu & D. Moore, Ed. Patterns in Fungal Development. Cambridge, U.K: Cambridge University Press, pp 156-181.

Nybakken K and Perrimon N (2002). Hedgehog signal transduction: recent findings. Curr Opin Genet Dev 12:503-511.

Patterson DJ (1999). The diversity of eukaryotes. Amer Nat 154 (suppl.):S96-S124.

Patterson DJ and Sogin ML (2000). Tree of Life web project. http://tolweb.org/tree.

Pourquié O (1999). Notch around the clock. Curr Opin Genet Dev 9:559-565.

Prosser JI (1995). Kinetics of filamentous growth and branching. In: N. A. R. Gow & G. M. Gadd, Ed. The Growing Fungus. London: Chapman & Hall, pp 301-318.

Redhead SA, Vilgalys R, Moncalvo JM, Johnson J and Hopple JS (2001). *Coprinus* Pers. and the disposition of *Coprinus* species *sensu lato*. Taxon 50:203-241.

Reyes JC and Grossniklaus U (2003). Diverse functions of Polycomb group proteins during plant development. Seminars Cell Dev Biol 14:77-84.

Rocheleau CE, Downs WD, Lin R, Wittmann C, Bei Y, Cha YH, Ali M, Priess JR and Mello CC (1997). Wnt signalling and an APC-related gene specify endoderm in early *C. elegans* embryos. Cell 90:707-716.

Roger AJ (1999). Reconstructing early events in eukaryotic evolution. Amer Nat 154 (suppl.):S146-S163.

Saitou N and Nei M (1987). The Neighbour-Joining method: a new method for reconstructing phylogenetic trees. Mol Biol Evol 4:406-425.

Samuels AL, Giddings TH and Staehelin LA (1995). Cytokinesis in tobacco BY-2 and root tip cells - a new model of cell plate formation in higher plants. J Cell Biol 130:1345-1357.

Schulte U, Becker I, Mewes HW and Mannhaupt G (2002). Large scale analysis of sequences from *Neurospora crassa*. J Biotech 94:3-13.

Seto ES and Bellen HJ (2004). The ins and outs of Wingless signaling. Trends Cell Biol 14:45-53.

Shin TH, Yasuda J, Rocheleau CE, Lin RL, Soto M, Bei YX, Davis RJ and Mello CC (1999). MOM-4, a MAP kinase kinase kinase-related protein, activates WRM-1/LIT-1 kinase to transduce anterior/posterior polarity signals in *C. elegans*. Mol Cell 4:275-280.

Sogin ML, Morrison HG, Hinkle G and Silberman JD (1996). Ancestral relationships of the major eukaryotic lineages. Microbiologia 12:17-28.

Sogin ML and Silberman JD (1998). Evolution of the protists and protistan parasites from the perspective of molecular systematics. Int J Parasitol 28:11-20.

Souchelnytskyi S, Moustakas A and Heldin C-H (2002). TGF-β signaling from a three-dimensional perspective: insight into selection of partners. Trends Cell Biol 12:304-307.

Stark DR (2002). Hedgehog signalling: Pulling apart patched and smoothened. Curr Biol 12:R437-R439.

Stearns JC and Glick BR (2003). Transgenic plants with altered ethylene biosynthesis or perception. Biotechnol Adv 21:193-210.

Swafford DL (1998). PAUP*: Phylogenetic analysis using parsimony and other methods. Massachusetts: Sinauer Associates.

Tabin CJ and McMahon AP (1997). Recent advances in Hedgehog signalling. Trends Cell Biol 7:442-446.

Thorpe CJ, Schlesinger A, Carter JC and Bowerman B (1997). Wnt signaling polarizes an early *C. elegans* blastomere to distinguish endoderm from mesoderm. Cell 90:695-705.

Thorpe CJ, Schlesinger A and Bowerman B (2000). Wnt signalling in *Caenorhabditis elegans*: regulating repressors and polarizing the cytoskeleton. Trends Cell Biol 10:10-17.

Thrane C, Kaufmann U, Stumann BM and Olsson S (2004). Activation of caspase-like activity and poly(ADP-ribose) polymerase degradation during sporulation in *Aspergillus nidulans*. Fungal Genet Biol 41:361-368.

Thummler F, Kirchner M, Teuber R and Dittrich P (1995). Differential accumulation of the transcripts of 22 novel protein kinase genes in Arabidopsis thaliana. Plant Mol Biol 29:551-565.

Umar MH and van Griensven LJLD (1997). Morphogenetic cell death in developing primordia of *Agaricus bisporus*. Mycologia 89:274-277.

Umar MH and van Griensven LJLD (1998). The role of morphogenetic cell death in the histogenesis of the mycelial cord of *Agaricus bisporus* and in the development of macrofungi. Mycol Res 102:719-735.

van Es JH, Barker N and Clevers H (2003). You Wnt some, you lose some: oncogenes in the Wnt signaling pathway. Curr Opin Genet Dev 13:28-33.

Wainright PO, Hinkle G, Sogin ML and Stickel SK (1993). Monophyletic origin of the Metazoa: an evolutionary link with fungi. Science 260:340-342.

Walker JC (1994). Structure and function of the receptor-like protein kinases of higher plants. Plant Mol Biol 26:1599-1609.

Wang YX, Kauffman EJ, Duex JE and Weisman LS (2001). Fusion of docked membranes requires the armadillo repeat protein Vac8p. J Biol Chem 276:35133-35140.

Watanebe N and Lam E (2004). Recent advance in the study of caspase-like proteases and Bax inhibitor-1 in plants: their possible roles as regulator of programmed cell death. Mol Plant Pathol 5:65-70.

Weinmaster G (2000). Notch signal transduction: a real Rip and more. Curr Opin Genet Dev 10:363-369.

Wetmore C (2003). Sonic hedgehog in normal and neoplastic proliferation: insight gained from human tumors and animal models. Curr Opin Genet Dev 13:34-42.

Wu Y, Wang X, Liu X and Wang Y (2003). Data-Mining approaches reveal hidden families of proteases in the genome of malaria parasite. Genome Res 13:601-616.

Yu SW and Tang KX (2004). MAP kinase cascades responding to environmental stress in plants. Acta Bot Sin 46:127-136.

Zhao XC, Qu X, Mathews DE and Schaller GE (2002). Effect of ethylene pathway mutations upon expression of the ethylene receptor ETR1 from Arabidopsis. Plant Physiol 130:1983-1991.

Applied Mycology
and Biotechnology
An International Series
Volume 5. Genes and Genomics
© 2005 Elsevier B.V. All rights reserved

ELSEVIER

8

Applications of Fungal Site-specific Recombination as a Tool in Biotechnology and Basic Biology

Yuri Voziyanov, Ian Grainge and Makkuni Jayaram
Section of Molecular Genetics and Microbiology, University of Texas at Austin, Austin, TX
78712, USA (jayaram@icmb.utexas.edu).

Two classes of conservative site-specific recombinases, those belonging to the tyrosine and serine families, have been identified, and several of its members characterized in genetic and biochemical detail. These families are named after the active site amino acid, tyrosine or serine, that is utilized as the nucleophile during the strand breaking step of recombination. The Flp recombinase encoded by the 2 micron plasmid of *Saccharomyces cerevisiae* and related recombinases encoded by similar plasmids found in other yeast species belong to the tyrosine family. The Flp protein has provided several insights into the mechanism of target DNA recognition, strand cleavage and strand exchange during the recombination reaction. Here we describe how the Flp system has been used as a tool for tackling basic and applied problems in biology.

1. INTRODUCTION

Recombination is a universal mechanism that organisms employ for reshuffling their genetic deck and creating novel genome configurations. Recombination can be broadly divided into two classes: homologous and site-specific. During gamete production in diploid creatures, genetic exchange between homologous chromosomes and chiasmata formation are prerequisites for their proper reductional segregation. An equally important function of recombination in prokaryotes, and during mitotic cell division in eukaryotes, is the repair of stalled or collapsed replication forks resulting from DNA nicks or other types of DNA lesions (Cox et al. 2000; Lusetti and Cox, 2002). Unlike homologous recombination that utilizes rather long stretches of homology between partner DNA molecules, site-specific recombination targets relatively short DNA sites

---

Corresponding author: Makkuni Jayaram

with well-defined sequences. Whereas a large number of proteins with distinct biochemical activities cooperate to carry out homologous recombination, it is usually four subunits of a single protein or two subunits each of a pair of related proteins that carry out the catalytic steps of site-specific recombination. In some instances, the site-specific recombinase may be aided by one or two accessory proteins in synapsing the DNA partners and assembling the chemically competent recombination complex. Because of their overall simplicity, these systems have served as models for investigating the mechanisms of phosphoryl transfer reactions during recombination and related reactions.

Homologous recombination serves, at a global level, the dual purpose of maintaining genome integrity on the one hand, and of promoting genetic diversity on the other. By contrast, the action of site-specific recombinases is local and precise. Yet, the genetic switches they throw have profound functional effects on the genome as a whole. Site-specific recombination has been utilized in evolution to bring about a variety of physiological consequences: integration of phage genomes into or their excision from bacterial chromosomes, resolution of the cointegrate intermediates formed during DNA transposition, determination of host specificity in bacterial viruses, flagellar phase variation in *Salmonella*, dimer resolution and equal segregation of bacterial chromosomes as well as unit copy or low copy phage and plasmid genomes, activation of gene expression during development and differentiation in blue green algae and copy number control in yeast plasmids (reviewed in Jayaram et al. 2002; Jayaram et al. 2004a).

Through the centuries, plant and animal breeders have applied recombination to select for crops that are disease or drought resistant, farm animals that produce more milk or yield more meat and beasts of burden with improved strength and stamina. So too, the modern day biotechnologists utilize recombination for genome engineering, creation of transgenic plants and animals and the production of biofactories. These applications of recombination can bring enormous benefits to mankind. They also have the potential, in the wrong hands, of being employed for highly destructive purposes.

In this chapter, I wish to illustrate how the Flp site-specific recombinase encoded by the 2 micron plasmid of *Saccharomyces cerevisiae* and enzymes related to Flp may be exploited to create novel tools for genomic manipulations from bacteria to man as well as to advance our basic understanding of biological phenomena.

2. THE FLP RECOMBINASE: ITS NORMAL PHYSIOLOGICAL FUNCTION

The source of Flp protein is the 2 micron circle, a double stranded DNA plasmid that exists in the yeast nucleus as a high copy, extrachromosomal element (Jayaram et al. 2004a; Jayaram et al. 2004b). The plasmid is generally regarded as a 'selfish' DNA molecule, as it apparently confers no selective advantage to its host. It has also been characterized as a benign parasite genome. The plasmid imposes little or no burden on the host's metabolic machinery, provided the normal copy number does not rise significantly above the steady state value of approximately 60 per cell. The high stability of the 2 micron circle in yeast is accomplished through its special genomic architecture and its genetic coding capacity (Fig. 1, left) Since the plasmid replication origin is functionally similar to chromosomal origins, each plasmid molecule replicates once (and only once) per cell cycle. A plasmid partitioning system consisting of the Rep1 and Rep2 proteins and the *cis*-acting locus *STB* mediates the equal or roughly equal distribution of the replicated molecules to daughter cells. Although the plasmid is a multicopy element, it has a clustered organization in the nucleus (Fig. 1, right), suggesting an effective copy number of unity or close to unity (Scott-Drew et al. 2002; Velmurugan et al. 2000). Time-lapse microscopy of fluorescence tagged reporter plasmids suggests that this cluster also forms the unit in segregation (Velmurugan et al. 2003). Hence the existence of an active partitioning system is readily justified. Duplication of plasmid molecules during the S phase followed by even distribution to daughter cells represents the normal life style of the plasmid.

The site-specific recombination system consisting of the Flp protein and its target DNA sites (*FRT*; see Fig. 1) constitutes an amplification mechanism, which is normally kept quiescent. In case there is a decrease in copy number due to a rare missegregation event, the amplification reaction is triggered to quickly restore it to the normal value. The Raf1 protein is believed to positively control *FLP* expression, and thus accelerate the amplification response. The basis of amplification is a carefully timed recombination event during the bidirectional replication of the plasmid (Futcher, 1986; Reynolds et al. 1987; Volkert and Broach, 1986). The asymmetric location of the origin causes the

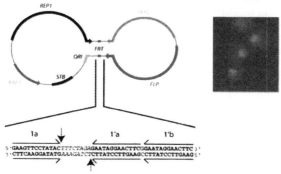

Fig. 1. Structural and functional organization of the 2 micron plasmid. *Left.* The double stranded circular plasmid is shown in the standard dumbbell form in which it is normally represented. The parallel lines (the handle of the dumbbell) indicate the inverted repeats (IRs) of the plasmid. The open reading frames are highlighted, with the arrowheads pointing in the direction of their transcription. The *cis*-acting DNA elements in the plasmid are the replication origin (*ORI*), the partitioning locus (*STB*) and the Flp recombination target sites (*FRT*). The *FRT* site consists of three 13 bp Flp binding elements, 1a, 1'a and 1'b whose orientations are denoted by the horizontal arrows. The elements 1a and 1'a together with the 8 bp spacer region included between them constitute the sequences directly relevant to the recombination reaction (the minimal *FRT* site). The points at which strand cleavage and exchange occur are indicated by the vertical arrows. *Right.* By fluorescence tagging, 2 micron circle derived reporter plasmids are found to localize in the yeast nucleus as tightly clustered and highly dynamic foci (real-time mobility of the foci within a plasmid cluster may be viewed at: http://www.sbs.utexas.edu/jayaram/Plasmid2004.htm).

proximal *FRT* site to be duplicated first, while the distal one is still in a single copy state. Flp mediated recombination between the unduplicated *FRT* and one of its duplicated partners causes the inversion of one replication fork with respect to the other (Futcher, 1986; see Fig. 2) As the forks chase each other around the circular template, multiple tandem copies of the plasmid can be spun out. Individual molecules may be resolved from the amplification product by Flp recombination itself or by homologous recombination.

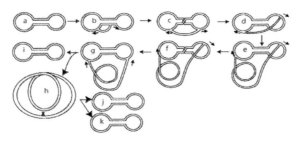

Fig. 2. The Futcher model for plasmid amplification. The schematic representation of the Futcher model (Futcher, 1986) is adapted from Broach and Volkert (1991). Bi-directional replication starting at the origin in a plasmid molecule duplicates the proximal *FRT* site before the distal one (a-b). A Flp mediated inversion (c) results in two replication forks oriented in the same direction (d). Movement of the two forks around the circular template amplifies copy number (e). A second recombination event (f) restores bi-directional fork movement (g). The products of replication are an amplified moiety containing multiple tandem copies of the plasmid (h) and a template copy (i). The tandem multimer can be resolved by Flp recombination into plasmid monomers (j, k).

3. FLP: A MEMBER OF THE TYROSINE FAMILY OF RECOMBINASES

Flp and Flp related recombinases encoded by 2 micron-like plasmids found in other yeast species related to *Saccharomyces cerevisiae* belong to the tyrosine family of site-specific recombinases (Jayaram et al. 2002; Van Duyne, 2002). In general, they utilize four subunits of a single recombinase protein to carry out recombination between two DNA partners of defined sequence. In rare cases, as exemplified by the XerC/XerD recombinase of *E. coli* and its homologues in other bacteria, the active recombinase is constituted by two subunits each of two distinct, though related, recombinases. The basic chemistry of recombination follows the type IB topoisomerase mechanism, in which an active site tyrosine is used as the nucleophile. As a result, strand cutting generates a covalent linkage between the 3'-phosphate and tyrosine, at the same time exposing a 5'-hydroxyl group. During strand exchange, the 5'-hydroxyl group from one DNA substrate attacks the 3'-phosphotyrosine bond formed within its partner in a reciprocal manner. The overall reaction is completed in two steps of single stranded exchanges. The exchange of the first pair of strands generates a Holliday junction intermediate, which is resolved into reciprocal recombinants by the exchange of the second pair of strands. The organization of the recombination target site, the pathway of tyrosine family recombination and the crystal structure of the Flp-*FRT* recombination complex (Chen et al. 2000) are outlined in Fig. 3.

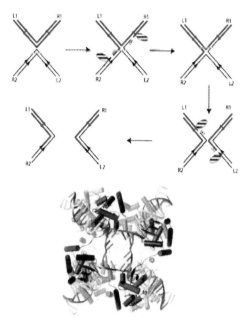

Fig. 3. Pathway of tyrosine family recombination. *Top.* Recombination between two substrates L1-R1 and L2-R2 to yield recombinant products L1-R2 and L2-R1, carried out by tyrosine recombinases in general, is schematically represented. Each substrate consists of a pair of recombinase binding elements (indicated by the parallel arrows) flanking the spacer sequence (shown as bent DNA) in a head to head orientation. The reaction consists of two steps of cleavage and exchange between single strands. The first step produces a Holliday junction intermediate; the second step resolves it. During strand cleavage, the recombinase gets covalently attached to the 3-phosphate via the tyrosine nucleophile (shown by the DNA linked ovals). L and R refer to the left and right recombinase binding arms. *Bottom.* The crystal structure of the Flp recombinase in association with its DNA substrate shown here (Chen et al. 2000) was provided by Dr. P. A. Rice, University of Chicago. In this nearly planar complex, the four Flp monomers establish a cyclic interaction network, responsible for mediating the physico-chemical steps of recombination.

The core DNA site at which the reaction occurs has a common architecture in the tyrosine family. There are two recombinase binding elements (11 to 13 bp long) arranged in inverted orientation on either side of a spacer or strand exchange region (6 to 8 bp long) (see Fig. 3). Some of the more complex members of the family, the lambda Int protein and XerC/XerD, for example, may utilize additional DNA sites and their cognate binding proteins as accessory factors in recombination (Azaro and Landy, 2002; Barre and Sherratt, 2002). The simpler members such as Cre and Flp have little or no constraints with respect to the topology of the DNA substrates on which they act, or the relative orientation of the target sites. They efficiently mediate intermolecular recombination (DNA integration or translocation), intramolecular DNA inversion (between inverted sites) and intramolecular DNA deletion (between directly oriented sites). Recombination sites present on supercoiled, relaxed or nicked circular molecules or linear molecules are readily acted upon by Flp and Cre. Because of the simplicity of their target DNA sites and their reaction conditions, Flp and Cre have been rather extensively used in biotechnological applications (Jayaram et al. 2002; Jayaram et al. 2004a; Sauer, 2002; Sauer and Henderson, 1990).

4. PHAGE INTEGRASES, TYROSINE AND SERINE RECOMBINASES: PRACTICAL APPLICATIONS

The prototype member of the tyrosine recombinases is the lambda integrase protein, and for this reason, until a few years ago, they were classified under the 'integrase' family. There are also phage integrases that belong to the second family of site-specific recombinases, namely, the serine family (or originally called the invertase/resolvase family) (Groth and Calos, 2004). As the names imply, the two families differ in the nucleophiles used for DNA cleavage, tyrosine in one case and serine in the other. They also differ in their mechanisms of strand cleavage and exchange. The serine family recombination is accomplished in one concerted step of double stranded cleavage of DNA partners, presumed relative rotation of the cleaved complex through 180 degrees and restoration of the double chains by strand joining. Furthermore, the active site serine gets covalently linked to the 5' phosphate during strand cleavage to generate a 3'-hydroxyl group that serves as the nucleophile for the joining step.

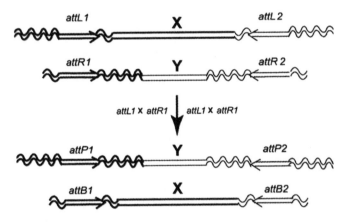

Fig. 4. Cassette replacement using lambda integrase recombination in the GATEWAY cloning system. The 'entry' cassette containing the gene of interest X flanked by *attL1* and *attL2* is recombined with the 'destination' target plasmid containing *attR1* and *attR2* flanking the sequence Y. Compatibility constrains restrict recombination to occur only between *attL1* and *attR1* and between *attL2* and *attR2*. The desired recombinant plasmid product harbors X flanked by *attB1* and *attB2*.

The highly specialized aspects of the lambda Int reaction, the large size of the *attP* DNA (phage attachment site). the requirement for this DNA to be negatively supercoiled and the role of bacterial host factors in recombination, have generally discouraged attempts to apply phage integrases for directing specific rearrangements in mammalian genomes. However, variants of Int that are IHF (integration host factor) independent have been demonstrated to carry out intramolecular recombination between *attP* and *attB* (bacterial attachment site) in HeLa cells as well as mouse embryonic stem cells (Lorbach et al. 2000). Kolot et al. have been successful in getting the closely related recombinase Int HK202 to perform both intermolecular and intramolecular recombination reactions (integration and excision) in COS-1, NIH3T3 and human cell lines (Kolot et al. 2003; Kolot et al. 1999).

Some phage integrases, those of C31, R4 and TP90-1, for example, follow the serine family mechanism of recombination (Breuner et al. 2001; Christiansen et al. 1996; Matsuura et al. 1996; Smith and Thorpe, 2002). The amino-terminal halves of these proteins show recognizable homologies to the canonical serine recombinases such as the Gin invertase or the Tn3/γδ resolvase, and contain the expected subunit interaction motifs and catalytic residues. Because these proteins act autonomously, have rather simple target *att* sites and are generally specific to *attB* x *attP* reactions, they are eminently suitable for bringing about unidirectional integration events in mammalian cells (Groth and Calos, 2004; Groth et al. 2000; Olivares et al. 2001; Stoll et al. 2002). As it turns out, the mouse and human genomes contain sequences resembling ϕC31 *attP* sites (pseudo *att* sites) at which ϕC31 Int mediated integrations occur with at least a ten fold higher frequency than random integrations (Thyagarajan et al. 2001). Recently, Tn3 resolvase fused to a DNA recognition domain from the mouse transcription factor Zif268 has been shown to catalyze recombination specifically between synthetic target sites containing two Zif268 recognition motifs (Akopian et al. 2003). The functional autonomy of the resolvase catalytic domain raises the possibility of creating custom-built recombinases that can act at chosen target sites within a genome.

The utility of phage integrases in biotechnology is illustrated by the GATEWAY™ cloning methodology, based on the lambda Int system, developed by Life Technologies (Invitrogen, Carlsbad, CA) (Fig. 4). The normal integration reaction mediated by Int involves the recombination between the large *attP* site and the relatively small *attB* site to generate two recombinant sites *attL* and *attR* flanking the integrated prophage at the left and right ends, respectively. The reverse reaction or excision involves the recombination between *attL* and *attR* to yield *attP* and *attB*. Because of the relative indifference of the tyrosine recombination systems to the actual DNA sequence of the strand exchange (or overlap) region, contrasted by the strict requirement for overlap homology between recombining partners, it is fairly straightforward to design pairs of compatible and incompatible target sites. In practice, a gene of interest located in an 'entry' cassette, and flanked by *attL1* and *attL2*, is recombined with a 'destination' plasmid containing a negative selection marker flanked by *attR1* and *attR2*. In this configuration of recombination partners, *attL1* can functionally interact with *attR1* and not *attR2*; and the opposite is true for *attL2*. As a result, the recombinant plasmid receives the gene with *attB1* and *attB2* marking its left and right borders. Based on the design of the destination plasmid, it is possible to generate expression vectors with cassettes programmed for transcription from the T7 promoter or the production of recombinant proteins fused to His-tag or to GST.

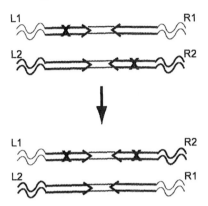

Fig. 5. Strategy for one-way recombination by a tyrosine recombinase. The substrates L1-R1 and L2-R2 are each debilitated by a mutation, marked X, in one of the two recombinase binding elements. Because of the recombinase binding cooperativity provided by the native binding element, these sites can undergo recombination, although the efficiency of the reaction is impaired. Following recombination, the site L1-R2 is mutated in both binding elements, whereas L2-R1 is wild type. Because L1-R2 is an extremely poor recombination substrate, the probability of the reverse reaction is reduced significantly.

One of the fundamental hurdles in attempting to perform an integration event at a local site within a large genome pertains to the kinetics and thermodynamics of the reaction. Since the reaction is bimolecular, the targeting vector has to cope with the enormous background of DNA noise while searching for its cognate site. Furthermore the excision reaction, being intramolecular is efficient, and is entropically favored over integration. Usually, one tries to overcome the kinetic barrier by using a large excess of the incoming DNA. To counter the unfavorable entropic situation, the following strategy is useful. The donor recombination site weakened by a point mutation in the recombinase binding element on one side of the strand exchange sequence, and the recipient site is also weakened by the same or a similar mutation, but in the binding element on the opposite side (Fig. 5). Following recombination, one of the sites flanking the integrated DNA contains two wild type binding elements whereas the second one contains two mutant binding elements. The extremely poor reactivity of the double mutant site significantly decreases the rate of excision.

A clever experimental design to promote the integration event is the recombinase mediated cassette exchange (RMCE) protocol (Baer and Bode, 2001). The logic employed here is an extension of that underlying the GATEWAY™ system described above (see Fig. 4) A pre-existing recipient cassette containing a reporter gene and flanked by two heterospecific target sites is replaced through recombination by a donor cassette that is also flanked by an identical pair of heterospecific sites. Provided the resident cassette harbors a marker that can be selected against, such as the TK (thymidine kinase) gene, it is possible to obtain large enrichment of the desired integrants over those established at non-target loci by promiscuous recombination.

As illustrated by the examples above, phage integrases of the tyrosine and serine families and Tn3 resolvase and related serine recombinases hold promise as potential tools for genomic manipulations. Nevertheless, the tyrosine recombinases Cre and Flp have clearly outperformed them so far in biotechnological applications. Both the Cre and Flp systems have been suitably adapted to perform efficiently in bacteria, fungi, flies and animals.

5. FLP SITE-SPECIFIC RECOMBINATION AS A TOOL IN GENETIC ENGINEERING

Although the expression of Flp in prokaryotes was established over twenty years ago (Cox, 1983; Vetter et al. 1983), it was not until much later that it became popular as an instrument for routine genetic manipulations. The procedure for efficiently mobilizing exogenous DNA to directed locations in the *E. coli* chromosome in a reversible manner through Flp recombination was first worked out in the Cox laboratory (Huang et al. 1997). Since then, the Flp/*FRT* system has been successfully deployed for engineering genomes in prokaryotes and eukaryotes: to facilitate site-specific DNA insertions, targeted DNA deletions, and expression of proteins from selected chromosomal locales (Reviewed in Jayaram et al. 2002). Related applications with strong implications in medical and agricultural technology include excision of large chromosomal DNA segments for sequencing, rescue of pathogenic islands, removal of exogenous sequences inserted into a chromosome and production of biofactories.

The Flp/*FRT* system has been employed as an accessory tool for enhancing the efficacy of general gene disruption strategies. Here recombination is used for 'recycling' a selectable marker, thus making it possible to perform multiple, iterative rounds of gene knockouts within a genome. A case in point is provided by the bacterial targeting vectors based on the type II mobile introns (targetrons). They can be programmed, on the grounds of base pairing, to insert into 'any' gene of choice (Zhong et al. 2003). Furthermore, the targetron design ensures that the expression of the selectable marker harbored by it is activated only following the retrotransposition event. Excision of this marker by recombination between a pair of head to tail *FRT* sites flanking it prepares the disruptant for a subsequent round of gene interruption at another locale. An analogous Flp/*FRT* based strategy has been employed for sequential gene disruptions of the pathogenic fungus, *Candida albicans*, whose genetic manipulation is complicated by its diploid genome, the lack of a defined sexual phase and its unusual codon usage (Morschhauser et al. 1999).

The high efficiency allele replacement by Flp/*FRT* recombination, combined with recyclable mutagenic cassettes (see above), affords high throughput genetic analyses of bacteria for basic as well as applied research. The ability to generate multiple unmarked mutations in a chromosome can expedite the development of live vaccine strains as well as 'food-safe' bacteria (Schweizer, 2003). The exquisite specificity with which extraneous sequences can be introduced into or excised from a genome makes Flp/FRT ideal for engineering bacterial strains designed for environmental release: for example, those used as biosensors or intended for bioremediation.

Helper dependent (HD) high capacity adenovirus vectors are one of the most efficient vehicles in vogue for gene therapy. One critical step in the production of HD vectors is to ensure that there is no significant contamination by the helper virus. With the help of an *in vitro* evolved robust Flp variant called Flpe (see below; Buchholz et al. 1998), it has been possible to obtain nearly quantitative elimination of the helper virus packaging signal by recombination in the producer cells (Umana et al. 2001). As a result, large scale production of HD vectors with helper virus contamination well below acceptable levels can be done rapidly by eliminating cumbersome and time consuming purification steps.

The native Flp protein has a rather limited degree of thermal tolerance, and tends to partially lose activity above 40°C. This difficulty has been largely overcome by the isolation of thermostable versions of the recombinase (Flpe, for example, as noted above) using methods of directed evolution and selection (Buchholz et al. 1998). Another refinement in technology has been the design of ligand dependent recombination for imposing rather tight control on the reaction. A hybrid recombinase with Flp fused to the steroid hormone ligand binding domain, for example, has been shown to mediate recombination in cell lines in response to hormone administration (Logie and Stewart, 1995). In a variation of this theme, a chimeric Cre recombinase has been engineered to contain the hormone binding domain from a mutated human progesterone receptor that responds to the synthetic ligand RU 486 but not progesterone (Kellendonk et al. 1996). Similarly, an ecdysone controlled Flp/FRT system has been developed that has the ability to channel recombination events to specific cell types (Sawicki et al. 1998). These regulated and cell specific recombination capabilities are suitable for cell lineage studies and for devising gene therapy protocols, especially in stem cells.

6. EXPANDING THE TARGET REPERTOIRE OF SITE-SPECIFIC RECOMBINATION: ALTERED SPECIFICITY FLP VARIANTS

One serious limitation in the use of site-specific recombination as a means for genome modification is that the recombination target site or sites must be first inserted, or must be fortuitously present (a highly unlikely situation), at the genomic locale of interest. The power of site-specific recombination in genetic engineering can indeed be revolutionized if one is able to pre-select a genomic site that resembles a native recombination target and coax the recombinase to acquire this new specificity. Work in the past six or seven years has made it feasible to obtain, by directed evolution, variants of Flp and Cre recombinases with new target specificities (Baldwin et al. 2003; Buchholz and Stewart, 2001; Konieczka et al. 2004; Santoro and Schultz, 2002; Voziyanov et al.

2003; Voziyanov et al. 2002). Since these recombinases are rather lax with respect to the spacer sequences of their target sites, new specificities refer to sequence alterations in the recombinase binding elements of these sites.

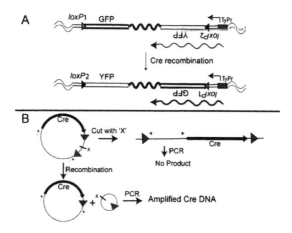

Fig. 6. Strategies for directed evolution of altered specificity recombinases. The successful methodologies employed for obtaining variants of the Cre recombinase that are active on new *loxP* target sites are diagrammed. **A.** In one approach, a library of recombinase mutants is screened for recombination on a pair of the altered target sites, denoted here as *loxP1* and *loxP2*. They are identical in sequence, but differ from native *loxP*, and are placed on either side of a dual reporter cassette in inverted orientation. In the parental configuration of the cassette, only YFP (yellow fluorescent protein) is expressed. DNA inversion by Cre recombination causes the expression of GFP (green fluorescent protein). **B.** In a second approach, a pool of recombinase mutants is subjected to a PCR based selection for a successful recombinase and against an unsuccessful one. The gene for a recombinase variant and the altered recombination target sites in direct orientation (shown as the two shaded triangles) are present on the same plasmid molecule. The primers (indicated by the thin arrows) cannot amplify the recombinase gene from the linearized plasmid obtained by cutting at site 'X' between the two targets. Since successful recombination unlinks site X from the plasmid, the same primers can amplify the recombinase gene.

In a structure based rational design employed for Cre, random mutations were made in pre-selected amino acid residues that are involved in contacting (or lie in close proximity to) certain base pairs (bp) of the Cre binding element (Santoro and Schultz, 2002). The library of mutants was screened in *E. coli* by flow cytometry for the recombination mediated switch from the expression of one fluorescent reporter protein to a second one (Fig. 6A). This basic experimental design can be readily modified to not only positively select for a new specificity but also negatively select against wild type specificity. In an alternative strategy devised by Buchholz and Stewart (2001), desired variants of Cre were selected from a vast pool of random mutants by a procedure called 'substrate-linked protein evolution (SLiPE)'. The rationale here is that the occurrence of recombination within a substrate placed adjacent to the recombinase coding region will physically differentiate a successful recombinase from an unsuccessful one by the rearranged DNA sequence (Fig. 6B). The winner can be retrieved from the background of losers by PCR amplification using appropriate primers.

In our lab we have developed a dual reporter screen in *E. coli* for identifying altered specificity variants of the Flp recombinase (Voziyanov et al. 2002; Fig. 7). In one reporter, the lacZ α gene segment is flanked by the altered target sites (*mFRTs*) in direct orientation; in the other, the red fluorescence protein (RFP) gene is flanked by native FRTs. The color of a colony on an X-gal indicator plate denotes the recombination potential of the variant Flp protein expressed in it: blue if no recombination or only *FRT* recombination occurs, red if only *mFRT* recombination occurs and white if both *FRT* and *mFRT* recombinations occur. The most interesting mutations from the altered specificity standpoint are represented by the red colony color, namely, those that accept a new target site and reject the old target site (native *FRT*).

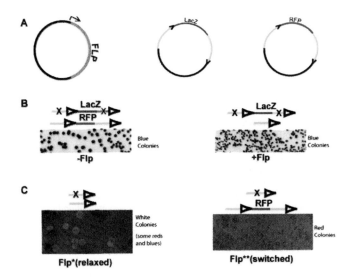

Fig. 7. The double reporter scheme for identifying Flp variants with relaxed or altered target specificities. A. The *in vivo* bacterial assay system contains three compatible plasmids: an expression plasmid for Flp variants and two reporter plasmids, one containing directly oriented native *FRT* sites flanking the gene for RFP (red fluorescent protein) and the other containing directly oriented altered *FRT* sites (or *mFRTs*) flanking *lacZα*. B. A Flp mutant that cannot recombine *FRT* or *mFRT* will leave both reporters intact, leading to RFP and *lacZα* expression. In the presence of the indicator X-gal, the color of RFP is masked by the blue color resulting from β-galactosidase activity. Wild type Flp will eliminate the RFP gene by *FRT* x *FRT* recombination (but not *lacZα*), giving rise to blue colonies as well. C. A relaxed specificity variant can yield white colonies by recombination of *FRT* x *FRT* (loss of *lacZα*) and *mFRT* x *mFRT* (loss of RFP). The interspersed red colonies result from *mFRT* recombination and the blue ones from *FRT* recombination (or a lack of recombination). A switched specificity Flp variant will remove *lacZα* by *FRT* x *FRT* recombination but will leave the RFP gene intact, producing red colonies.

The utility of the double reporter scheme has been validated by the identification and *in vivo* and *in vitro* characterization of Flp variants that show either relaxed specificity (active on *FRT* and *mFRT*) or shifted specificity toward *mFRT*. Subsequently, gene shuffling and further mutagenesis of Flp variants with defined *mFRT* specificities have yielded new variants that now functionally recognize *FRT* sites containing combinatorial mutations (Voziyanov et al. 2003). In addition, it has become possible to perform recombination on bi-specific hybrid *FRT* sites (*hmFRTs*) using a binary combination of Flp variants: one variant specific for one binding element of the *hmFRT* and the other specific for the second binding element (Konieczka et al. 2004). These results hold the promise that iterated rounds of gene shuffling and mutagenesis, along with powerful screening or selection procedures, will generate a library of evolved recombinases for target sites that are quite diverged from native *FRT*.

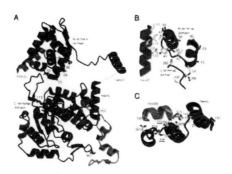

Fig. 8. Amino acid substitutions that alter or modulate recombination target specificity of Flp tend to cluster in certain locales. **A.** In the crystal structure of a Flp monomer, the amino acid positions whose alterations lead to changes in target specificity or modulate interprotomer collaboration during recombination are shown. The peptide regions housing these residues are in close proximity to those from the two neighboring Flp monomers that are part of the cyclical Flp tetramer in the recombination complex. **B** and **C** present isolated views of the interactions between the amino-terminal domains and the carboxyl-terminal domains of partner monomers, respectively. For more details, refer to Konieczka et al. (2004). Current genetic and biochemical evidence suggests that amino acid changes at the dimer interfaces play a role in coupling target DNA recognition to the chemical steps of recombination, and are important for relaxed or shifted target specificities.

7. APPLICATIONS OF FLP FOR SOLVING BASIC PROBLEMS IN BIOLOGY

We now present several instances where the Flp recombinase has provided a valuable tool for tackling a number of fundamental biological problems. Although the primary emphasis of this chapter has been the use of Flp in the field of biotechnology, it is well-nigh impossible to divorce the utilities of Flp recombination in basic biology from those in applied biology. For instance, in addition to their utility in genetic engineering, altered specificity variants of Flp are likely to advance our understanding of how DNA recognition is coupled to the chemical steps of recombination (Fig. 8).

Furthermore, as is clear from the discussion leading up to this section, the logic behind creating transgenic organisms and biofactories by site-specific recombination can also drive seminal investigations in chromosome dynamics and developmental biology. As an impressive example, programmed genetic rearrangements induced in *Drosophila* by P element imported Flp/*FRT* systems have opened new avenues in investigations on chromosome structure and function as well as developmental programs and their regulation (Golic and Golic, 1996; Golic and Lindquist, 1989). This technique permits the recovery of paracentric or pericentric chromosomal inversions when two P elements lie on the same chromosome in inverted orientation. Similarly, deficiencies and duplications can be recovered when the elements lie in direct orientation on the same chromosome or on homologous chromosomes. Construction of mosaic flies in *Drosophila* by site-specific recombination between homologues and the analyses of these mosaics have provided a simple and convenient method for tracking cell lineages during development (Theodosiou and Xu, 1998; Xu and Harrison, 1994).

Long before its emergence as a model biochemical system to study mechanisms of DNA recombination, the genetics of Flp recombination was exploited for mapping cloned yeast genes to individual chromosomes (Falco et al. 1982). When the inverted repeats of the 2 micron plasmid, containing the *FRT* sites, are integrated into a chromosome, they can readily undergo recombination in the presence of Flp. Flp mediated unequal sister chromatid exchange (recombination between *FRT* site-1 on one chromosome and the oppositely oriented *FRT* site-2 on its sister) will result in a dicentric chromosome, whose breakage is repaired by copying the allelic information from the homologue. The Flp induced instability of specific markers identifies the chromosome involved in the recombination and dicentric formation. Keep in mind that a cloned yeast gene, along with a pair of inverted *FRT* sites engineered alongside it, can be readily integrated by homologous recombination at its native chromosomal locus. This method has since become obsolete, the advent of genomics having simplified the mapping task enormously. Nevertheless, it exemplifies an elegant application of site-specific recombination for the purpose of genetic mapping. Furthermore, the generation of dicentrics by Flp/*FRT* recombination in *Drosophila* (Ahmad and Golic, 2004) and other organisms paves the way to studying a number of chromosome associated phenomena: segregation of acentric chromosomes, behavior and resolution of chromosome bridges and cellular repair responses to broken chromosomes.

A second application of Flp has been in curing yeast strains of the endogenous 2 micron circles to obtain what are called [cir⁰] derivatives. One effective way of doing this is to overexpress a Flp mutant that can cut the *FRT* site efficiently but cannot mediate strand joining, which is the step following strand cleavage during recombination (Tsalik and Gartenberg, 1998). Because of the accumulation of unrepaired strand breaks, [cir⁰] cells arise in the population at an easily detectable frequency. Overexpression of even wild type Flp can also generate [cir⁰] cells. Cells die or grow poorly under this condition, presumably because of the unregulated amplification in plasmid copy number. Rare [cir⁰] cells formed by missegregation of the plasmid become enriched in the population because of their relative growth advantage.

The [cir⁰] strains have proven to be extremely useful in dissecting the roles of plasmid encoded proteins as well as *cis*-acting loci of the plasmid in sustaining its stable propagation.

The most valuable contribution of Flp to basic biology has been as a template for understanding DNA-protein recognition, protein subunit interactions and mechanisms of strand cleavage and strand joining during tyrosine family recombination (Jayaram et al. 2002; Jayaram et al. 2004b). Biochemical and structural studies of Flp, together with similar analyses of Flp related tyrosine recombinases, have solved nearly all of the critical physicochemical and mechanistic questions pertaining to recombination. We now describe below how our intimate knowledge of the assembly of the Flp (or Cre) recombination complex and the orientation of strand exchange within it can be used to understand the organization of DNA in high-order DNA-protein machines whose structures are unknown.

8. THE GEOMETRY OF THE FLP RECOMBINATION COMPLEX

The Flp protein can act with equal efficiencies on a pair of *FRT* sites arranged in head to head orientation (inverted sites) or head to tail orientation (direct sites). The former reaction inverts the DNA between the two *FRT* sites, and the latter deletes (or excises) the included DNA as a circular molecule. The *FRT* sites come together in the synapse by random collision. In a negatively supercoiled circular substrate, therefore, a random number of DNA crossings can be trapped outside the synapse. As a result, the inversion reaction produces, along with the unknotted circle, a series of knots with

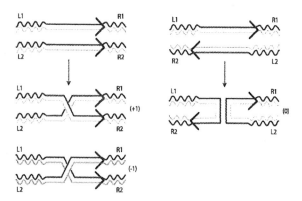

Fig. 9. Geometry of site alignment during tyrosine family recombination and the topological consequence on the recombination product. The two partner sites are named L1-R1 and L2-R2 to orient them left to right. In the reaction shown at the left, the two sites are arranged in a parallel fashion (left to right in both cases). If strand exchange occurs with a right handed rotation of the DNA, the recombinant products (L1-R2 and L2-R1) cross each other to introduce a +1 node. If the sense of rotation is left handed during exchange, the crossing between L1-R2 and L2-R1 will be -1. In the reaction shown at the right, the partner sites are arranged in the antiparallel orientation (left to right for L1-R1 and right to left for R2-L2). The act of recombination does not introduce a crossing between L1-R2 and L2-R1.

increasing complexity (3, 5, 7 etc. crossings). Similarly, the deletion reaction produces, along with two unlinked circles, a series of catenated (interlinked) circles (2, 4, 6 etc crossings). An important question is whether recombination occurs within a fixed local geometry of the partner *FRT* sites inside the synapse, regardless of their global orientation, head to head or head to tail, in the DNA substrate as a whole.

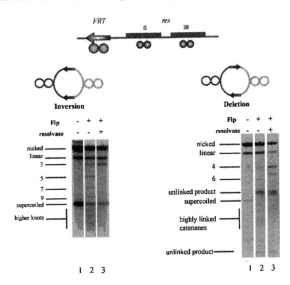

Fig. 10. Flp mediated recombination after the assembly of the resolvase synapse that specifically traps three negative crossings. The general organization of a hybrid *FRT/res* site is diagrammed schematically at the top. The *res* II and *res* III accessory sites together with the resolvase protein will establish a well characterized synapse with a three noded topology (see Fig. 11). The *res* I site, where the chemical steps of resolvase recombination occur, is replaced by an *FRT* site. In the two hybrid *FRT/res* sites present on a plasmid, the *res* II/III site are arranged in their native head to tail orientation. The *FRT* sites are present in a head to head orientation in the inversion substrate (left) and a head to tail orientation in the deletion substrate (right). The Flp alone reactions (lanes 2) or the resolvase assisted Flp reactions (lanes 3) are analyzed by electrophoresis after removing supercoils by DNase I nicking. Flp mediated inversion gives unknotted circles (which co-migrate with the nicked parental circles) as well as knotted circles with 3, 5, 7 etc. crossings (lane 2, left panel). The faint interleaved ladder of bands in this lane are the products of a second round of recombination (knots with 4, 6, 8 etc. crossings). Flp mediated deletion yields unlinked circles as well as linked ones with 2, 4, 6 etc. crossings (lane 2, right panel). Here also faint second round recombination bands (knots with 3, 5, 7 etc. crossings) are detected. Recombination after preincubation with resolvase (to establish the 3 noded synapse) enriches the 3 noded knot for inversion (lane 3, left panel) and the 4 noded catenane for deletion (lane 3, right panel). The slight enrichment of the 4 noded knot in lane 3 of the left panel and the 5 noded knot in lane 3 of the right panel are due to a second round of recombination.

In principle, the *FRT* sites may be arranged in a parallel or antiparallel orientation, provided the recombination complex is essentially planar. Crystal structures of Flp and

Cre in association with their DNA targets are in strong agreement with the notion of planarity. Recombination from the parallel geometry of FRT sites will result in a DNA crossing between the exchanged sites (Fig. 9). It may be denoted by +1 or -1, depending on whether the DNA rotation is right handed or left handed during the exchange event. By contrast, recombination from the antiparallel geometry of FRT sites will not introduce a DNA node (zero crossing). As briefly outlined below, Grainge et al. (2000) have used this difference to figure out that the actual geometry of the FRT sites during recombination is unique, and is antiparallel.

In their experiments, Grainge and colleagues (Grainge et al. 2000) used the accessory sites, res II and res III, of the topologically well characterized Tn3 resolvase system and ordered protein additions to precisely trap three extraneous negative supercoils before assembling the Flp synapse (Fig. 10). They then carried out Flp recombination on the FRT sites, which replace the res I recombination sites, in the plasmid substrates. They used a matched pair of plasmids, an inversion substrate containing the FRT sites in head to head orientation and a deletion substrate containing the sites in the head to tail orientation. For the inversion reaction from the resolvase/Flp hybrid synapse, the unique product was a three noded knot; and for the corresponding deletion reaction, the product was a four noded catenane. These results are most consistent with the antiparallel arrangement of the FRT sites (Fig. 11). When resolvase traps three outside DNA crossings, the head to head FRT sites will come together in an antiparallel manner. The inversion product will be a knot containing these three crossings. For the matched, deletion substrate, the three outside crossings will orient the FRT sites in a parallel fashion. By trapping a fourth crossing, provided by a supercoil from the plasmid, the sites can be reoriented as antiparallel. Recombination will then yield the four noded catenane, as observed experimentally. Furthermore, as illustrated in tabular form in Fig. 12, recombination in the parallel mode with a +1 crossing or a -1 crossing can be ruled out.

The top panel of the Table in Fig. 12 refers to a right handed rotation of DNA during strand exchange (+1 crossing), and the bottom panel refers to a left handed rotation (-1 crossing). Note that the 'additional' crossing represents the extra negative supercoil from the plasmid substrate that is trapped to bring the FRT sites in parallel geometry. Note also that the signs of all the crossings change during inversion (rows 1 and 3), and hence the knot nodes are '+'. For the +1 mode of strand exchange from the parallel sites (top panel in Fig. 12), the predicted products are a three noded knot for inversion and a two noded catenane for deletion. For the -1 mode of strand exchange (bottom panel), the corresponding products are a five noded knot and a four noded catenane. The experimentally observed pair of products, a three noded knot and a four noded catenane (see Fig. 10 and 11), disagree with the topological predictions for parallel site alignment, and strongly support antiparallel synapsis of the FRT sites. These conclusions from the topological analyses of the Flp system agree with those from similar analyses of the Cre system (Kilbride et al. 1999) and with the crystal structures of Flp-DNA and Cre-DNA complexes (Rice, 2002; Van Duyne, 2002).

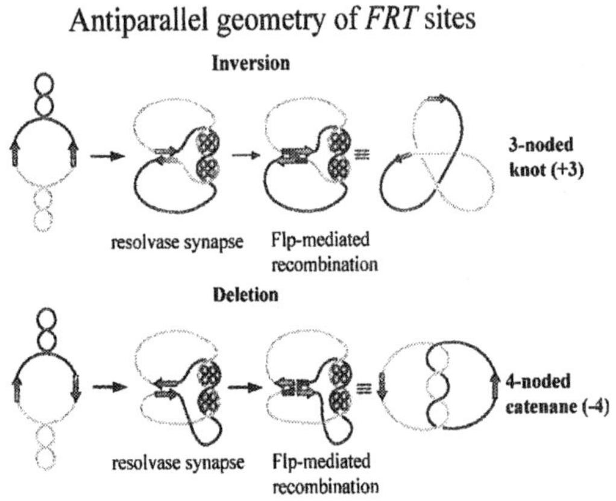

206

Antiparallel geometry of *FRT* sites

Fig. 11. The product topologies during Flp mediated recombination performed from a prearranged resolvase synapse are consistent with antiparallel alignment of the *FRT* sites. The three negative supercoils (odd number of DNA crossings) sequestered by the resolvase synapse will arrange the *FRT* sites in an antiparallel manner in the Flp synapse during the inversion reaction. The product, in which these DNA crossings are trapped, will be a 3 noded knot. With one more negative supercoil from the substrate (or a total of four crossings), the *FRT* sites can establish antiparallel synapse during the deletion reaction. The catenane product will contain four links between the two deletion circles. The signs of the DNA crossings in the knot and catenane products are opposite because the direction of the DNA axis is changed in the inverted DNA segment.

The unique geometric character of Flp/Cre recombination and the lack of any DNA crossing due to strand exchange *per se* permit one to apply a 'difference topology' method for deciphering the number of DNA crossings within an uncharacterized synapse. The logic of the analysis (as extended from the resolvase/Flp reaction) is that the inversion knot and the deletion catenane formed from the 'unknown/Flp' hybrid synapse, for a pair of matched inversion and deletion substrates, respectively, will differ by one crossing. The smaller number is contributed by the unknown synapse, and the additional crossing reflects the antiparallel constraint on the *FRT* sites in order for recombination to occur. The success of the difference topology method in unveiling the path of DNA within the phage Mu DNA transposition system is described below (Pathania et al. 2002; Pathania et al. 2003) .

9. DIFFERENCE TOPOLOGY REVEALS A THREE SITE, FIVE NODED DNA SYNAPSE FORMED DURING PHAGE MU TRANSPOSITION

The transposition of phage Mu requires negative supercoiling of the DNA substrate plus the interaction of three separate DNA sites, the left and right ends of Mu (L and R, respectively) and the enhancer element (E), mediated by the transposase protein MuA. Do these interactions sequester a fixed number of supercoils within the transpososome?

Parallel geometry of *FRT* sites: topological predictions

	Resolvase Synapse	Additional crossing	Crossing during recombiantion	Product topology
Inversion	-3	-1	+1	+3 Knot
Deletion	-3	0	+1	-2 Catenane
Inversion	-3	-1	-1	+5 Knot
Deletion	-3	0	-1	-4 Catenane

Fig. 12. Predicted topologies of inversion knots and deletion catenanes during resolvase assisted Flp recombination if the *FRT* sites have a parallel geometry. In the top panel, strand rotation during Flp mediated cross-over is assumed to be right handed (+1); in the bottom panel, the rotation is assumed to be left handed. For the inversion and deletion reactions, the external resolvase synapse has a fixed topology (-3 crossings). An additional negative supercoil (-1) is trapped in the inversion substrate (rows 1 and 3) for parallel alignment of the *FRT* sites. The signs of the DNA crossings in the substrate and product are reversed when recombination inverts DNA (+3 knot in row 1 and +5 knot in row 3). Flp recombination with *FRT* sites arranged in parallel fashion predicts either the +3 knot/-2 catenane (top panel) or the +5 knot/-4 catenane (bottom panel) pair of inversion and deletion products. The experimental results, yielding the +3 knot/-4 catenane combination, contradict these predictions.

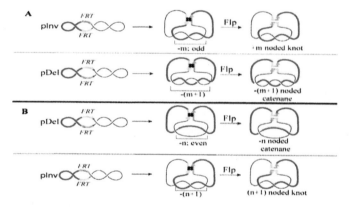

Fig. 13. The rationale for the application of difference topology for deciphering the DNA path within an uncharacterized synapse. Imagine that two distant DNA sites, resident on the thick and thin DNA domains are bridged by specific DNA-protein interactions to introduce a fixed number of thick x thin crossings. The DNA domains are bordered by the target sites for a tyrosine recombinase (*FRT* for Flp and *loxP* for Cre, for example). **A.** Imagine that an odd number 'm' of negative supercoils are sequestered in the exterior synapse. As a result, for the inversion substrate pInv, the *FRT* sites will come together in antiparallel mode. The product of Flp recombination will be a knot with +m crossings. For the deletion substrate pDel, one more negative supercoil, -(m+1), has to be utilized for antiparallel alignment of *FRT* sites. The catenane product therefore will contain one more DNA crossing than the inversion knot, but the crossings are of the opposite (minus) sign. **B.** When the two DNA sites trap an even

number 'n' of negative supercoils, the deletion catenane formed by Flp recombination will contain -n crossings. The corresponding inversion knot will contain one more crossing, n+1, with plus signs.

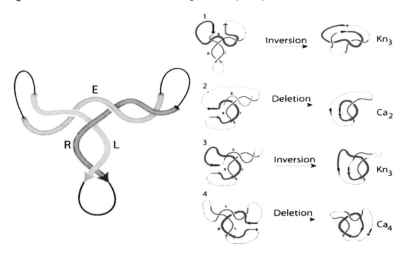

Fig. 14. The topology of the three site synapse during the phage Mu transposition reaction. The five DNA crossings trapped by the left (L) and right (R) ends of Mu and the transposition enhancer (E) when the transposase protein bridges these three sites are schematically depicted at the left. The crossing numbers between each of the three sites and its two partners were deduced by the difference topology method (see text for details). Some of the key experimental results supporting the topology of the transposition synapse are shown at the right. When the enhancer is sequestered into the thick DNA domain by the recombination target sites, the inversion product is a 3 noded knot (row 1). The corresponding deletion product is a 4 noded catenane. Hence the enhancer (thick line) crosses L and R (thin line) thrice. When L is isolated into the thick domain and R into the thin domain, and the enhancer is supplied in *trans*, the deletion product is a 2 noded catenane (row 2). L and R must therefore cross each other two times. The product for the matched inversion reaction is a 3 noded knot. When L is resident in the thick domain, and E and R in the thin domain, inversion produces a 3 noded knot (row 3) and deletion a 4 noded catenane. Hence L must cross E and R thrice (E once and R twice). Finally, when R is quarantined into the thick domain, the deletion product is a 4 noded catenane (row 4) and the inversion product is a 5 noded knot. Hence R must make four crossings with L and E (two with E and two with L).

Difference topology can only tell the number of crossings between two DNA domains (shown in thick and thin lines in Fig. 13): those separated by the *FRT* sites, or more precisely, the crossover points in recombination. Hence the larger question regarding the total number of L x E x R nodes has to be subdivided into three smaller ones: (a) how many times does E cross L and R (E x L,R); (b) how many times does L cross R and E (L x R,E); and (c) how many times does R cross E and L (R x E, L)? To answer a, the recombination sites (*FRT* sites if Flp is the recombinase of choice and *lox*P sites if Cre is the preferred recombinase) must be placed flanking E, isolating this site into one DNA domain while sequestering L and R into the second domain. To answer b and c, the recombination sites must flank L in the first case and R in the second.

Furthermore, as explained in the previous section for the resolvase assisted Flp reaction, a matched pair of inversion and deletion substrates must be constructed in each case. The experimental protocol involves the arrangement of the Mu synapse first, followed by the execution of recombination using Flp (or Cre), and finally the analysis of the knot and catenane products after all supercoils have been removed by DNase I nicking.

Imagine that in a given experiment, we find an $|m|$ noded knot as the inversion product and an $|m+1|$ noded catenane as the deletion product (Fig. 13A). We can immediately deduce that there are 'm' crossings between the thick and thin domains, and 'm' is odd. With one extra crossing, $|m+1|$, the *FRT* sites of the deletion substrate will become 'antiparallel', and will give the $|m+1|$ catenane as the recombination product. The minus signs for the DNA crossings in Fig. 13 indicate that the plasmid substrates are negatively supercoiled. The sign of the crossings in the knot products becomes plus because DNA is inverted during their formation. If, on the other hand, the deletion product is an $|n|$ noded catenane and the inversion product is an $|n+1|$ noded knot (Fig. 13B), the number of trapped crossings between the domains must be n, and n is even. The experimental results with the Mu transposition synapse and their interpretations are given below.

For (a), E x L, R, the products were found to be a three noded knot for inversion and a four noded catenane for deletion. Hence, there must be three crossings between E and L, R. For (b), L x R,E, the products were, again, a three noded knot and a four noded catenane. So there must be also three crossings between L and R,E. For (c), R x E,L, the products were a five noded knot and a four noded catenane. Or, there must be four crossings between R and E,L. The only mutually consistent DNA path that satisfies the results from a, b and c contains one crossing between E and L, two between L and R and two between R and E. In other words, the transposition synapse has the five noded configuration shown in Fig. 14. The number of crossings between L and R were further verified by an alternative strategy. Here, the enhancer was deleted from the substrate and provided as a linear DNA fragment in *trans*, thus topologically dissociating it from the synapse. In agreement with the two DNA crossings between L and R (Fig. 14), the inversion product in this case was a three noded knot, and the deletion product was a four noded catenane.

In principle, the difference topology analysis can be extended to complex interactions involving multiple DNA sites, for example, replication, transcription or repair complexes. By the strategy discussed for the transpososome, the number of crossings made by each site with the rest of the sites can be determined after isolating it into one DNA domain with the recombination target sites placed on either side of it. The subtopologies obtained by iterating this procedure for all of the sites can then be integrated to obtain the final composite topology of the entire DNA path.

10. CONCLUSION

Site-specific recombinases in general have served as excellent model systems for understanding the molecular details of DNA-protein interactions, protein-protein cooperativity and strand exchange mechanisms during precise information swap between double helical DNA partners. Studies on the fungal site-specific recombinase Flp have made significant conceptual advances in this field of research. Furthermore, as outlined in this chapter, the Flp system and related recombinase systems have been manipulated in a variety of ways in order to apply them for executing pre-meditated DNA rearrangements in large genomes. As this chapter also illustrates, no less impressive are the contributions of Flp and its kin recombinases as tools in advancing basic research in development, chromosome dynamics and DNA topology.

Acknowledgements: We acknowledge with gratitude the contributions of past and present colleagues whose contributions have sustained our efforts to understand the recombination and partitioning systems of the yeast plasmid. Work in the Jayaram laboratory has received funding from the National Institutes of Health, the National Science Foundation, the Human Frontiers in Science Program, the Robert F. Welch Foundation, the Council for Tobacco Research and the Texas Higher Education Coordinating Board.

REFERENCES

Ahmad K, and Golic KG (2004). Analyzing chromosome function by high frequency formation of dicentric chromosomes in vivo. Methods Mol Biol. 247:333-341.

Akopian A, He J, Boocock MR, and Stark WM (2003). Chimeric recombinases with designed DNA sequence recognition. Proc Natl Acad Sci USA. 100:8688-8691.

Azaro MA, and Landy A (2002). l integrase and the l int family. In Mobile DNA II. ASM Press, Washington DC. pp 118-148.

Baer A, and Bode J (2001). Coping with kinetic and thermodynamic barriers: RMCE, an efficient strategy for the targeted integration of transgenes. Curr Opin Biotechnol. 12:473-480.

Baldwin EP, Martin SS, Abel J, Gelato KA, Kim H, Schultz PG, and Santoro SW (2003). A specificity switch in selected Cre recombinase variants is mediated by macromolecular plasticity and water. Chem Biol. 10:1085-1094.

Barre F-X, and Sherratt DJ (2002). Xer site-specific recombination: promoting chromosome segregation. In Mobile DNA II. ASM Press, Washington DC. pp 149-161 .

Breuner A, Brondsted L, and Hammer K (2001). Resolvase-like recombination performed by the TP901-1 integrase. Microbiology. 147:2051-2063.

Broach JR, and Volkert FC (1991). Circular DNA plasmids of yeasts: In The Molecular Biology of the Yeast *Saccharomyces*. Genome Dynamics, Protein Synthesis and Energetics. Cold Spring Harbor Laboratory Press, Cold Spring harbor, New York. pp 297-331.

Buchholz F, Angrand PO, and Stewart AF (1998). Improved properties of FLP recombinase evolved by cycling mutagenesis. Nat Biotechnol. 16:657-662.

Buchholz F, and Stewart AF (2001). Alteration of Cre recombinase site specificity by substrate-linked protein evolution. Nat Biotechnol. 19:1047-1052.

Chen Y, Narendra U, Iype LE, Cox MM, and Rice PA (2000). Crystal structure of a Flp recombinase-Holliday junction complex. Assembly of an active oligomer by helix swapping. Mol Cell. 6:885-897.

Christiansen B, Brondsted L, Vogensen FK, and Hammer K (1996). A resolvase-like protein is required for the site-specific integration of the temperate lactococcal bacteriophage TP901-1. J Bacteriol. 178:5164-5173.

Cox MM (1983). The FLP protein of the yeast 2-micron plasmid: expression of a eukaryotic genetic recombination system in *Escherichia coli*. Proc Natl Acad Sci USA. 80:4223-4227.

Cox MM, Goodman MF, Kreuzer KN, Sherratt DJ, Sandler SJ, and Marians KJ (2000). The importance of repairing stalled replication forks. Nature. 404:37-41.

Falco SC, Li Y, Broach JR, and Botstein D (1982). Genetic properties of chromosomally integrated 2 micron plasmid DNA in yeast. Cell. 29:573-584.

**Applied Mycology
and Biotechnology**
An International Series
Volume 5. Genes and Genomics
© 2005 Elsevier B.V. All rights reserved

9

ELSEVIER

Heterologous Gene Expression in Filamentous Fungi: A Holistic View

Helena Nevalainen, Valentino Te'o, Merja Penttilä[1] and Tiina Pakula
Department of Biological Sciences, Macquarie University, Sydney, NSW 2109, Australia
(hnevalai@els.mq.edu.au); VTT Biotechnology, P.O. Box 1500, FIN-02044 VTT, Finland

As scavengers of recalcitrant polymers in the nature, filamentous fungi are excellent secretors of proteins outside the growing mycelium. This characteristic has been targeted and systematically improved in industrially-exploited production strains. Over the last five years there has been a significant shift from one-gene-at-a-time approaches to wider understanding of the organism, made possible, for example by gene array and proteome technologies that can now also be applied to filamentous fungi. This has presented novel opportunities for studies into gene regulation under specific conditions such as a particular carbon source or developmental stage with a view of advancing the basic knowledge and gaining information that can be applied for strain and process improvement. Filamentous fungi offer enormous potential for efficient and large scale production of heterologous gene products. Importantly, protein secretion provides a platform for the eukaryotic style post-translational modification of gene products. Fungi are cheap to cultivate and down-stream processing is made easy with no need to break cells open for product recovery. In order to capitalize on fungi as heterologous production hosts, research is now directed to revealing the cellular mechanisms for internal protein quality control, secretion stress, functional genomics of protein expression and secretion, protein modification and linking the physiology to productivity.

1. INTRODUCTION

As natural protein secretors and industrially exploited enzyme producers, filamentous fungi hold promise for becoming effective producers of gene products originating from other organisms, especially higher eukaryotes. The yields obtained, for example, for mammalian proteins produced in fungal systems are competitive to those achieved in yeast or animal cell cultures (Table 1). Several strategies including the use of strong fungal promoters, optimizing the codon usage of the foreign gene, producing the foreign protein as a fusion to a homologous highly secreted protein, using protease-minus strains as expression hosts and co-expression of proteins that assist in protein folding have been employed in order to successfully produce various foreign proteins in fungal hosts. There are several recent reviews that

---

Corresponding author: Helena Nevalainen

discuss these approaches in detail (Conesa et al. 2001; Nevalainen and Te'o 2003; Nevalainen et al. 2004). While the levels of some secreted foreign gene products from fungal systems have been pushed to gram per liter level (Ward et al. 1990, 2004). there is still room for improvement judging from the tens of grams of homologous protein that filamentous fungi are secreting into the external growth medium (Durand et al. 1988; Berka et al. 1991).

In addition to good yields, it is equally important that a given foreign protein is produced in an active form. Proteins secreted from filamentous fungi are modified in the secretory pathway by folding, potential proteolytic processing and addition of glycans or compounds such as phosphate and sulphate. Experimental evidence gained so far suggests that several foreign proteins expressed in filamentous fungi are lost in the secretory pathway because of incorrect processing or misfolding, resulting in their elimination by cellular quality control mechanisms (Archer and Peberdy 1997; Gouka et al. 1997). Genetic and proteomic studies into these

Table 1. Yields of some heterologous gene products produced in fungal, baculovirus and mammalian expression systems

Expression system	Protein	Yield	Reference
Aspergillus niger	Immunoglobulin G1 (γ),	0.9 g/L	Ward et al. 2004
	Fab antibody fragment	0.2 g/L	
Aspergillus niger	Human interleukin 6	150 mg/L	Punt et al. 2002
Saccharomyces cerevisiae	Human serum albumin	90 mg/L	Okabayashi et al. 1991
Saccharomyces cerevisiae	Human serum albumin	150 mg/L	Sleep et al. 1991
Pichia pastoris	Human serum albumin	10 g/L	Kobayashi et al. 2000
Trichoplusia	Human collagenase IV	300 mg/L	George et al. 1997
Spodoptera frugiperda	Human proapolipoprotein AI	80 mg/L	Pyle et al. 1995
CHO cells	Human apolipoprotein AI	80 mg/L	Schmidt et al. 1997
Human embryo kidney cells	Human laminin	1 mg/L	Kariya et al. 2002

mechanisms have revealed a multilayered system for protein quality control involving several genes and regulatory circuits (Chapman et al. 1998; Pakula et al. 2003; Al-Sheikh et al. 2004).

A typical microfungal lifecycle features vegetative growth of the organism as filamentous hyphae and formation of asexual conidia that germinate forming new hyphae or production of or sexual spores that undergo meiosis. The nature of the organism provides both advantages and disadvantages in terms of heterologous gene expression and product synthesis. The central dogma is that proteins are mainly secreted through the growing hyphal tip, therefore growth and secretion are intimately linked and difficult to study separately (Wessels 1993). Importantly, secretion provides a platform for complex post-translational processing of proteins discussed above. Alternation of the hyphal growth with the formation of uni- or multinuclear conidia makes it hard to maintain a population of autonomously replicating plasmids to boost the gene copy numbers and thereby product yields. There is a revived interest concerning the relationship of productivity and the physiological form of the growing fungus as well as the role of the largest organelle, the fungal cell wall in protein secretion, explored this time with genomic and proteomic tools. In this Chapter, we will discuss the role of protein quality control systems in expression of heterologous gene products and physiological aspects relating to protein excretion.

2. CELLULAR CONTROL POINTS IN PROTEIN EXPRESSION
2.1. Fungal Secretory Pathway in Brief

Proteins destined for secretion outside the fungal hyphae first enter the ER where they fold and undergo modifications such as proteolytic processing, core glycosylation and disulfide bridge formation. From the ER, proteins travel in secretory vesicles to the Golgi where further modifications such as trimming of glycosylation and proteolytic cleavage may take place. The final stages of secretion involve packing the proteins into secretory vesicles and fusion of the protein-containing vesicles to the plasma membrane followed by their externalization into the surrounding medium through the fungal cell wall (reviewed in Consea et al. 2001). Some proteins may settle in the ER or go to cytoplasmic organelles such as mitochondria and specific vesicles, depending on the address tag attached to the protein-encoding sequence.

2.2. Endoplasmic Reticulum (ER)

ER provides the compartment for the folding of proteins targeted to different destinations in the cell as well as for proteins that are secreted into the culture medium. Protein quality control, with subsequent elimination of incorrectly folded and/or unassembled multimeric proteins, is an essential function of the endoplasmic reticulum (ER). Without quality control, accumulation of aberrant proteins in the cell would soon reach toxic levels. This accumulation is elegantly prevented by two mechanisms that are activated by the presence of extensively misfolded proteins: unfolded protein response (UPR) (reviewed in Chapman et al. 1998) and ER-associated degradation (ERAD) (Brodsky and McCracken 1999; Ellgaard and Helenius 2001). UPR leads to transcriptional upregulation of a large set of genes to relieve the ER stress including genes encoding for chaperones such as Bip and genes involved in ER-to-Golgi trafficking amongst many others (summarized in Sitia and Braakman 2003). In the event that the UPR fails to relieve the stress in the ER caused

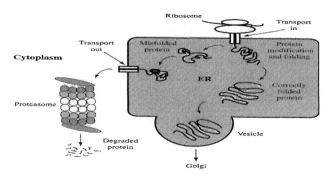

Fig. 1. Protein folding in the ER and disposal of misfolded proteins by ERAD through protein retrotranslocation into cytoplasm and degradation by proteasome (adapted from Dobson 2003).

by unfolded proteins, terminally misfolded polypeptides, inappropriately modified proteins and unassembled units of protein complexes will be eliminated by ER-associated degradation, ERAD. For ERAD to take place, the aberrant proteins will be retrotranslocated across the ER membrane to the cytoplasm, ubiquinylated and degraded by a giant protease, the 26S proteasome (Brodsky and Mc Cracken 1999; Bochtler et al. 1999; Glickman 2000). Thus, the cellular environments for protein folding and degradation are spatially separated, yet coordinated. Protein folding in the ER and its link to the ERAD are depicted in Figure 1.

2.3. Current Studies of ER Quality Control in Fungi

The unicellular yeast *Saccharomyces cerevisiae* is an acknowledged model for a eukaryotic cell and most of the information on ER quality control in fungi comes from yeast studies. With regard to the cellular quality control machinery, protein ubiquination has been resolved at the molecular level and the yeast proteasome has been characterised at the biochemical, structural and gene level (reviewed in Glickman et al. 1996; Hochstrasser 1996; Cagney et al. 2001). A recent study in *S. cerevisiae* has shown that impaired secretion of a heterologous hydrophobic cutinase enzyme resulted in the formation of protein aggregates, changes in ER membrane morphology, oxidative stress and activation of UPR and ERAD, therefore supporting the view that proteasome-mediated proteolysis has a role in the degradation of heterologous gene products produced in fungi (Sagt et al. 2002). There is a growing body of genetic and transcriptomic evidence that UPR and ERAD are tightly linked and act cooperatively in elimination of misfolded proteins from ER in fungal systems. A genome-wide expression analysis showed up-regulation of about 400 genes in yeast, induced by UPR, and that these genes also included those associated with ERAD (Travers et al. 2000). So far, comparative information is not available for filamentous fungi.

2.4. Unfolded Protein Response (UPR)

The ER houses several chaperones and folding catalysts that participate in efficient protein folding and protein subunit assembly. Accumulation of unfolded proteins in the ER triggers transcriptional up-regulation of the genes encoding ER-resident proteins by UPR (Chapman et al. 1998). Other phenomena interfering with the function of ER include inhibition of protein glycosylation or disulphide bond formation and depletion of ER of $Ca2^+$ with a Ca^+ ionophore. Genetic constituents of the signal transduction pathway leading to UPR have been studied in yeast, mammals and to some extent, filamentous fungi. For example, the unfolded protein response (UPR) pathway regulator gene, *hac*, has been cloned from *Trichoderma reesei*, *Aspergillus nidulans* and *A. niger* and the activation mechanism studied in detail (Saloheimo et al. 2003). The main features of UPR seem to have been well conserved across species (Patil and Walter 2001; Ma and Hendershot 2001).

Amongst the best studied proteins under the UPR control are the molecular chaperone BiP and protein sulfide isomerase (PDI) which is involved in the formation of disulfide bridges in the folding proteins. BiP, a resident protein in the ER and a member of the heat shock 70 protein family (van Gemeren et al. 1997; Punt et al. 1998). is induced, for example, by heat shock, glucose starvation and particular conditions that induce the unfolded protein response (see references in Conesa et al.

2001). A number of studies show that overproduction of homologous (e.g. glucoamylase) and recombinant proteins (e.g. tissue plasminogen activator t-PA) induces the expression of UPR-related gene products in *Aspergillus niger* (Punt et al. 1998; Ngiam 2000; Wiebe et al. 2001). However, success in studies starting from intentional overexpression of a single chaperone or foldase gene in a particular production host in order to increase the levels of secreted heterologous gene products made in that host have been somewhat inconsistent in filamentous fungi (Table 2; see also references in Conesa et al. 2002).

Overexpression of the induced form of *hacA* transcript in *A. niger* resulted in enhancement of production of the heterologous proteins *Trametes versicolor* laccase and bovine preprochymosin (Valkonen et al. 2003b; Table 2). Interestingly, overexpression of the *T. reesei hac1* gene in the yeast *S. cerevisiae* increased extracellular production of a *Bacillus* α-amylase and the endogenous yeast protein invertase (Valkonen et al. 2003a). while overexpression of the homologous *hac1* did not have a beneficial effect on expression of a heterologous laccase in *T. reesei* (Valkonen et al. personal communication). Studies addressing the effect of overexpression of *ire1* (gene involved in activation of translation of the transcription factor Hac1p) on growth and protein production in *T. reesei* transformants expressing *Phlebia radiata* laccase indicated that even though UPR was strongly induced, the transfomants secreted less total protein than the parental strain. Moreover, there were no major differences in laccase production levels between the transformants and the parental strain (Valkonen et al. 2004).

In summary, the effect of overexpression of UPR-induced gene products on the heterologous product yield has been inconsistent and seems to be product dependent in filamentous fungi. The portfolio of appropriate foldases and chaperones may be different for different proteins and overexpression of particular foldases may interfere with the cellular balance of the folding machinery. This situation demonstrates the difficulty of obtaining conclusive results by observing individual genes in a complex system, therefore calling for methods that enable taking a more holistic view into the changes in gene and protein expression in the fungal hyphae under particular circumstances.

Notwithstanding the not yet fully understood mechanisms of UPR, it is currently one of the best known mechanisms affecting the levels of heterologous gene products expressed in filamentous fungi. The mechanisms for protein secretion and UPR are currently studied across a range of filamentous fungi and new target genes are being isolated and characterized, including e.g. *bip1/bipA*, *pdi1/pdiA*, *prpA* (Wang and Ward, 2003), *cypB* (Derkx and Madrid 2001), *clxA*, and *tigA* (Jeenes et al. 1997). Known genes related to protein folding and secretion are shown in Figure 2.

2.5. Repression of Genes Encoding Extracellular Proteins During Secretion Stress in Filamentous Fungi (RESS)

A large set of genes are activated in the cells to enhance the protein folding and transport in response to accumulation of unfolded proteins within the ER. These genes include e.g. genes encoding foldases as well as proteins involved in quality control, discussed above, and protein transport. In mammalian cells, ER stress results also in attenuation of translation. The response is mediated by phosphorylation of the translation initiation factor eIF2α by the ER transmembrane kinase PERK. However, inhibition of translation in response to ER stress or identification of the counterpart of PERK has not been reported in fungi. Recently, a novel type of feedback mechanism has been shown to be activated in response to secretion stress with filamentous fungi leading to down-regulation of genes encoding extracellular proteins. The response provides the cells an efficient means to reduce the amount of protein to be transported under the stress conditions.

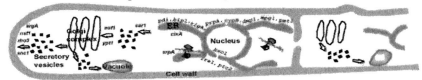

Fig. 2. Secretion-related genes in filamentous fungi. Highlighted are some of the known genes, location of the proteins they encode and their respective roles in relation to the maintenance of the secretory pathway: *dpm1*, *mpg1*, involved in protein core glycosylation (cytosolic); *bip1*, involved in protein folding (ER lumen); *pdi1*, *tigA*, *prpA*, involved in protein folding, disulphide bond oxidation and isomerization (ER lumen); *hac1*, transcription factor of unfolded protein response (nuclear); *ire1*, positive regulator of *hac1* splicing (ER membrane); *ptc2*, negative regulator of UPR (cytosolic); *clxA*, calnexin and a lectin-like chaperone (ER membrane); *srpA*, protein translocation (cytosol/ER membrane); *sar1*, GTPase and vesicle budding from ER (cytosolic face of ER); *ypt1*, GTPase and vesicle fusion at Golgi (cis-golgi membrane); *nsf1*, ATPase and disassembly of SNARE complexes (cis-Glogi membrane/plasma membrane); *srgA*, GTPase and vesicle fusion at plasma membrane (Golgi-plasma membrane); *snc1*, T-SNARE vesicle fusion at plasma membrane (Golgi-plasma membrane); *rho3*, small GTP-binding protein and cell signaling (Golgi-plasma membrane).

Table 2. Examples of effects of overexpression of UPR-related genes on heterologous gene expression in *Aspergillus* hosts.

Expression	Overexpressed gene of target protein	Effect on yield	References
Aspergillus awamori	*bipA*	Cutinase from *Fusarium solanum pisi*, no effect	van Gemeren et al. 1998
A. niger	*prpA*	chymosin, no effect	Wang and Ward, 2000
A. niger	cyclophilin	Human t-PA, no effect	Wiebe et al. 2001
A. niger	*clxA* (calnexin)	MnP of *Phanerochaete chrysogenum* up 5x	Conesa et al. 2002
A. awamori	*pdiA*	Plant thaumatin up 5x	Moralejo et al. 2001
A. niger	*hacA*, expressed constitutively	*Trametes versicolor* laccase up 7x Bovine preprochymosin up 2.8 x	Valkonen et al. 2003b

The RESS response (repression under secretion stress) was shown in cultures of *Trichoderma reesei* treated with chemical agents such as DTT or brefeldin A which inhibit protein folding and/or transport (Pakula et al. 2003). The group of genes subjected to repression included the cellulose genes *cbh1, cbh2, egl1* and *egl2*, as well as the hemicellulase gene *xyn1* encoding the major extracellular proteins produced by the fungus, and a hydrophobin gene *hfb2*. Repression of the genes took place soon after the drug treatment, approximately at the same time with activation of the *hac1*gene encoding the UPR transcription factor. The analysis also revealed a set of genes encoding intracellular proteins whose expression levels were not significantly affected during the treatments. Interestingly, the transcript level of *bgl2*, a gene encoding an intracellular β-glucosidase and displaying a similar type of carbon source dependent regulation as the major cellulase genes, was maintained at a constant level during the drug treatments. The result suggests that the RESS type of regulation is not common to all the cellulase genes, but targets the secreted representatives of the group. A similar type of stress response has been reported to be activated in *Aspergillus niger* and lead to reduced expression levels of glucoamylase and aspergillopepsin genes in cultures treated with DTT. RESS type response is also seen in a strain expressing an antisense construct of *pdiA* thus producing reduced amounts of the foldase PDIA (Al-Sheikh et al. 2004).

Analyses of reporter gene systems, *lacZ* expression under *cbh1* promoters of different length in *T. reesei* and *uidA* expression under *glaA* promoter constructs in *A. niger*, have shown that the feed-back regulation of the genes in DTT treated cultures takes place at transcriptional level rather than affecting mRNA stability and that particular promoter regions are involved in mediating the response (Pakula et al. 2003; Al-Sheikh et al. 2004). However, further studies are required for characterization of the molecular mechanism of RESS activation and the regulatory factors involved. Currently, it is not known whether components of the UPR signalling pathway are involved in the activation of RESS or whether the responses are controlled independently. Experimental work in both *T. reesei* and *A. niger* suggest that the transcription factor HACI/HacA is not directly involved in the RESS activation, or that only minor amount of the factor would be sufficient to activate the response. This view is supported by the fact that RESS has been observed also under conditions in which *hac1/hacA* activation was weak or negligible, e.g. in *T. reesei* cultures treated with tunicamycin or the ionophore A23187 or in *A. niger* strains expressing the antisense construct for *pdiA*. However, a more upstream component in the pathway could be involved, or a more diverse signaling pathway activated in response to limitations in folding and secretory capacity to result in RESS activation.

Even though many aspects in the cellular responses to secretion stress have been well conserved during evolution, there are also differences in the mechanisms and the variety of responses. Repression of genes encoding the secretory cargo proteins in response to secretion stress has not been described in the yeast *Saccharomyces cerevisiae* or in mammalian cells. On the other hand, transcriptional profiling data in *Arabidopsis thaliana* indicate that a large group of genes encoding proteins with putative signal sequences were down-regulated under DTT or tunicamycin stress (Martinez and Chrispeels, 2003). This type of repression mechanism induced upon secretion stress may be particularly beneficial for organisms such as filamentous fungi that are capable of producing very large quantities of extracellular proteins. Translational inhibition has not been shown to take place in fungi and transcriptional repression would thereby provide an alternative means to efficiently relieve the secretory load under stress conditions. The capacity of filamentous fungi to secrete large amounts of proteins may also set a special demand for the folding machinery of the cells as well as the quality control system. Induction of the UPR pathway has been reported to take place concomitantly with induction of endogenous secreted proteins in cultures of *T. reesei*. This was demonstrated by inducing the cellulase production by addition of lactose containing medium into cultures first grown under repressing conditions on glucose medium. Induction of cellulase and UPR genes followed a similar temporal expression pattern upon the shift in the culture conditions (Collén et al. 2004).

3. FUNCTIONAL GENOMICS OF PRODUCTIVITY

Genome data, when available, combined with microarray technology and proteomic profiling offers the opportunity for studying gene expression and identifying the genes involved on a large scale. The technology suits e.g. for metabolic mapping (Nakajima et al. 2000; Chambergo et al. 2002) expression profiling in pathogenicity (Allen and Nuss 2004) and identification of genes coordinately expressed with gene(s) or conditions of interest (Schmoll et al. 2004). Analysis of carbon source dependent expression of genes in *T. reesei* has revealed new previously undescribed DNA sequences encoding hydrolytic enzymes and in addition, information has been obtained on genes that are differentially expressed under conditions inducing or repressing cellulase gene expression (Foreman et al. 2003). A recent paper by Sims et al. (2004) describes the monitoring of gene expression profiles from a conidial germination library of a wild type *Aspergillus nidulans* consisting of 4092 ESTs in cultures grown to mid-exponential phase with ethanol as a sole carbon source followed by addition of glucose. The microarray analysis revealed genes regulated by the growth rate, for example genes of which the expression is affected by carbon catabolite repression. The analysis of gene expression after addition of glucose to the cultivation medium allowed the study of changes in the expression of genes representing various metabolic pathways such as glycolysis, TCA cycle, gluconeogenesis, glyoxylate cycle and ethanol utilization. Understanding the effects of carbon catabolite repression on gene expression on a broad scale is of primary importance for selecting promoters and devising cultivation strategies for the expression of heterologous gene products in filamentous fungi.

Transcriptional profiling using microarray technology and proteome analysis have been used for identification of differentially expressed genes and gene products in response to production of heterologous proteins in *Saccharomyces cerevisiae* (Casagrande et al. 2000, Sagt et al. 2002) and in bacterial species whose genome sequences are known, e.g. *Escherichia coli* (Choi et al. 2003; Baglioni et al. 2003; Champion et al. 2003). and *Bacillus subtilis* (Jürgen et al. 2001). The increasing amount of fungal genome sequences becoming available enables efficient analysis of heterologous protein production at genome-wide level in filamentous fungi as well.

3.1. Gene Expression Profiling Methods Not Based on DNA Arrays

In addition to array based technologies, also novel methods for efficient multiplexed analysis of gene expression have been established. VTT TRAC (TRanscriptional profiling with the aid of Affinity Capture; Söderlund et al. 2003a,b,c; Satokari et al. 2004) is based on hybridization of affinity-labelled sample RNA with pools of probes of distinct sizes, affinity capture of the hybrids and subsequent size fractionation and quantification of the hybridized probes in the pool using capillary electrophoresis. A similar type of approach is used in the MLPA method (Multiplex Ligation-dependent Probe Amplification) which is based on PCR amplification of hybridized probes of unique sizes and subsequent analysis of the PCR products using capillary electrophoresis (Eldering et al. 2003). The methods are suitable for fast analysis of sets of genes relevant for the application, e.g. for monitoring of bioprocesses. This type of methods can potentially be applied also to wider transcriptional analysis. Computational algorithms have been developed for the setting up of pools of probes on the genomic scale using VTT TRAC (Kivioja et al. 2002). Computational methods have also been used for optimization of cDNA-AFLP (Amplified Fragment Length Polymorphism) analysis, a method also applicable for transcriptional profiling in organisms with unknown genome sequences (Kivioja et al. manuscript in preparation). In cDNA-AFLP, the cDNAs derived from the sample mRNA are arranged as pools using restriction enzymes and selective PCR of the fragments and the resulting fragment pools are analyzed using gel or capillary electrophoresis (Breyne and Zabeau 2001). Partial sequence information or genomic sequences of related organisms can be used to aid selection of the suitable restriction enzymes and primers for the PCR in order to obtain optimal pools for the analysis.

3.2. Proteomic Approaches to Study Production of Secreted Proteins in Filamentous Fungi

Proteome analysis provides an efficient tool to study cellular responses to various internal and external conditions. The methodology can be used to analyze quantitative changes in intracellular or extracellular protein patterns including changes in posttranslational modifications. The methods have been applied to study responses related to protein production in *Trichoderma reesei*. In addition, methods for measuring the efficiency of protein synthesis and transport itself have been set up.

3.2.1 Methodology for monitoring protein transport

Metabolic labeling and subsequent 2D gel analysis of the nascent proteins have been applied to study protein synthesis and transport under different conditions in cultures of *Trichoderma reesei*. Specifically, kinetics of synthesis and secretion of cellobiohydrolase I (CBHI; Cel7A), the major protein secreted into the culture medium by the fungus, has been measured and used as a model to elucidate protein secretion (Pakula et al. 2000, 2003, 2005). Quantification of intracellular and extracellular protein in samples collected at frequent intervals from the labeled cultures enables determination of parameters describing the efficiency of protein synthesis and secretion, such as the average synthesis and secretion time of the protein as well as the specific synthesis and secretion rates of the protein.

The method can also be applied for detailed monitoring of maturation of proteins during transport. Posttranslational modifications taking place during the transport along the secretory pathway may also affect the pI of the protein and can therefore be detected using 2D gel analysis. In the case of CBHI (Pakula et al. 2000) the nascent protein in the cell extracts is first detected as one major pI isoform followed by formation of more acidic forms. Finally, the protein is secreted into the culture medium as 7-8 isoforms in the pI range of 3.5-4.3 (Figure 3). Parallel labeling of the cultures with [³⁵S]methionine and [¹⁴C]mannose produced identical pI patterns of CBHI, indicating that all the forms detected were glycosylated with mannose-containing structures. The result is in accordance with glycosylation taking place concomitantly with synthesis and translocation of the protein in the ER. However, comparison of the synthesis kinetics of the labeled protein revealed a two-minute delay in the [¹⁴C]mannose labeled cultures compared to [³⁵S]methionine labeled cultures, most probably due to the time taken by incorporation of the mannose units in the glycan precursors. The nascent pI form first detected in the labeling experiments is likely to represent the ER form of the protein. The maturation of the pI pattern of CBHI was prevented in cultures treated with DTT, a reducing agent known to inhibit disulphide bridge formation taking place in the ER. In the DTT treated cultures, only the early forms of CBHI were detectable even at the late time points of labeling and the protein was not secreted into the culture medium.

Heterogeneity in the pI of the protein in the later stages of transportation can be explained, at least in part, by differences in glycan structures on the protein. Removal of N-glycans from the protein using EndoH or PNGaseF resulted in a loss of approximately 10 kDa in the apparent molecular weight of all the isoforms. In addition, the pI pattern was slightly shifted to the basic direction and the relative amount of the acidic forms was reduced. Furthermore, chemical removal of both N- and O-glycans caused a more drastic change in the mobility of CBHI in 2D gel analysis: the apparent molecular weight was reduced by 15 kDa and the protein appeared as a broad spot with a more basic pI compared to untreated CBHI samples.

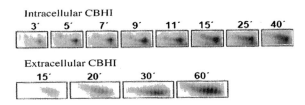

Fig. 3. 2DE analysis of CBHI in intracellular and extracellular fractions during metabolic labeling of *T. reesei* cultures. The time point (min) in the labeling experiment is indicated above each panel.

Apparently, the glycan structures on the protein are modified to gain negative charge as the protein proceeds along the secretory pathway, most probably at post-ER state. However, it is possible that in addition to glycan structures, also other types of posttranslational modifications would contribute to the pI of the protein.

Modification of N-glycan structures in the ER (reviewed in Goochee et al. 1991). as well as their role in the quality control of folding of the glycoprotein (Helenius and Aebi 2004) are rather well conserved between species. However, further modifications taking place in Golgi compartment show more variation and species specific features. Structural evidence on N-glycans on fungal proteins indicate that this is the case also in filamentous fungi. The N-glycans of secreted CBHI produced by *T. reesei* have been reported to be of high-mannose type, consisting of a mammalian type of core glycan extended with a limited number of mannose and glucose units (Salovuori et al. 1987; Maras et al. 1997, 1999). However, also presence of charged components in the glycans have been shown, e.g. atypical terminal mannose units linked by phosphodiester bridges (Maras et al. 1997) which could affect the pI of the protein. Also single non-substituted GlcNAc attached to N-glycosylation sites has been detected (Klarskov et al. 1997; Harrison et al, 1998). The O-glycans of the protein consist mainly of neutral mannose units (Salovuori at al. 1987; Harrison et al. 1998) but the O-glycosylated linker region of the protein has been shown to be sulfated (Harrison et al. 1998). which could also give rise to more acidic pI isoforms. However, it is also possible that yet unidentified modifications in the glycan structures would be involved in formation of the different pI isoforms or that other type of posttranslational modifications would also contribute to the pI of the protein. Whichever the case, the methodology discussed here can be applied to follow transport of a particular protein along the secretory pathway by analyzing changes in the posttranslational modifications of the protein and to compare protein transport under different conditions.

3.2.2 Kinetics of protein synthesis and secretion - the effect of specific growth rate on protein production by *T. reesei*

The capacity of fungal cells to synthesize and transport proteins under different physiological conditions has been studied in chemostat cultures of *T. reesei* Rut-C30 on lactose-containing medium (Pakula et al. 2005). The cultures were analyzed for growth characteristics and protein production, as well as for kinetics of synthesis and secretion of CBHI and expression levels of cellulase and UPR marker genes. Analysis of the cultures showed that production of secreted proteins, consisting mainly of cellulases and hemicellulases, was most efficient at low specific growth rates, 0.022/h-0.033/h. The specific production rate of the extracellular proteins first increased with increasing specific growth rate reaching the maximum value at the specific growth rate of 0.03/h, after which the production of extracellular proteins was significantly reduced. At the low specific growth rates, the extracellular proteins formed up to 29 % of the total proteins synthesized by the cells, whereas at the high specific growth rates, 0.045/h-0.066/h, only 6-8 % of the proteins synthesized were extracellular.

To study production of secreted proteins more in detail the kinetics of synthesis and secretion of CBHI was measured in cultures that were metabolically labeled with ^{35}S-methionine using the methodology described in Chapter 3.2.1 (Pakula et al. 2005). In accordance with the other data on protein production, both the specific synthesis and secretion rate of labeled CBHI were maximal at the specific growth rate of 0.03/h. Also the mRNA levels of the cellulase genes *cbh1* and *egl1* were the highest at the growth rate 0.03/h indicating that the high CBHI synthesis rate under that condition can, at least in part, be explained by the expression level of the corresponding gene. The *in vivo* labeling studies also showed that under the conditions in which the synthesis of CBHI as well as production of extracellular proteins into the culture medium were the highest, the ratio secretion rate/synthesis rate of CBHI was significantly lower compared to the ratio at higher specific growth rates. At 0.031/h, the secretion rate of CBHI was only 37 % of the synthesis rate, whereas at the high growth rate 0.066/h the secretion rate was 62 % of the synthesis rate of the protein. The result suggests that, at the low specific growth rates, the secretion capacity of the cells cannot adjust to the level of synthesis of secreted proteins which leads to production of reduced amount of extracellular protein compared to the amount originally synthesized. A slight induction in the expression level of the UPR target genes *bip1* and *pdi1*, as well as the induced form of *hac1*, were also detected under these conditions which could indicate a limitation in the secretion capacity of the cells.

The transport machinery in actively growing hyphae is expected to function efficiently to ensure that cell wall material needed for growth is available in hyphal tips, which could also enhance secretion of extracellular proteins. Many examples of growth rate-associated production of secreted proteins are known in filamentous fungi, e.g. production of α-amylase by *Aspergillus oryzae* (Carlsen and Nielsen 2001; Spohr et al. 1998) and glucoamylase production by *A. niger* (Pedersen et al. 2000; Schrickx et al. 1993; Withers et al. 1998). The production of (hemi)cellulases by *T. reesei* is not directly growth rate associated. Efficient production of the enzymes under low nutrient conditions or during slow growth is beneficial for the organism as the hydrolytic enzymes are required for extracting the energy and carbon source from large polymeric substances. The specific consumption rate of the carbon source increased with increasing specific growth rate in the cultures. Interestingly, the most efficient expression of the cellulase genes and production of the enzymes took place under conditions in which the specific lactose consumption rate in the cultures was relatively low. It has been previously shown that cellulase gene expression is induced in batch cultures after the carbon source is exhausted (Ilmén et al. 1997). One possibility is that a similar type of starvation signalling could be involved in this case. At the very low growth rates in the chemostat cultures also a decrease in the production of the extracellular proteins was observed. Under these conditions a rather large part of the carbon source consumed is estimated to be used for maintenance and thus probably limiting production of both biomass and extracellular protein.

3.2.3 Analysis of secretion stress

Inability of the cells to efficiently fold or transport heterologous proteins, or endogenous secreted proteins in large amounts, is likely to induce stress responses in the cells to reduce the load in the secretory pathway and to alleviate the deleterious effects of unfolded proteins accumulating within the cells. Secretion stress may also be induced in the cells by inhibiting protein folding or transport by treatment with chemical agents. The drugs, such as dithiothreitol (DTT), brefeldin A (BFA), tunicamycin or the ionophore A23187 have been widely used in studies on protein secretion in model organisms of lower and higher eukaryotes, and in addition shown to induce UPR target genes in filamentous fungi (Jeenes et al. 1997;

Ngiam et al. 1997, 2000; Al-Sheikh et al. 2004; Mulder et al. 2004; Saloheimo et al. 2003, 2004; Pakula et al. 2003; Derkx and Madrid 2001).

Comparison of the cellular responses to stress conditions induced by different means can be used for identification of key components and limiting factors in protein production and to revealing potential targets for modifications to enhance protein production. This strategy has been applied to analyze secretion stress conditions in *T. reesei*. *T. reesei* Rut-C30 and its transformant producing human tissue plasminogen activator (tPA) were cultivated in similar conditions in chemostat cultures and subjected to proteome analysis (Pakula et al. manuscript in preparation). The steady-state protein levels were analyzed using 2D gel electrophoresis and additional information was obtained from Western analysis using phosphoprotein specific antibodies and analysis of separate cultures labelled *in vivo* with ^{35}S-methionine. The results were then compared to those obtained in cultures treated with DTT. Some of the responses were induced both in cultures producing the heterologous protein as well as in cultures in which protein folding and transport was inhibited by the treatment with DTT, whereas some of the responses were unique to the particular culture conditions. The analysis revealed a group of novel proteins, not previously identified from *Trichoderma*, that were up-regulated under secretion stress conditions. These included e.g. several proteins involved in protein folding as well as a group of proteins involved in more general stress responses. The same culture conditions were analyzed also using subtraction libraries and cDNA-AFLP analysis in order to identify differentially expressed genes under these conditions and to compare proteome and transcriptome level changes under secretion stress (Arvas et al. manuscript in preparation). Altogether, the analysis gives a broader view to the potential UPR target genes up-regulated under secretion stress conditions, and also reveals new features in the response in filamentous fungi to the conditions applied.

3.3. Transcriptional and Proteomic Responses in *Trichoderma reesei* Expressing a Bacterial Thermophilic Xylanase and a Heterologous Fungal Laccase

While research into heterologous expression of gene products in filamentous fungi has largely targeted proteins originating from higher organisms, there is also considerable interest in producing high levels of oxidative (see examples in Table 1) and thermostable enzymes (Bergquist et al. 2003). that are not produced at commercially viable amounts in the original host organisms. The effect of (over)expression of the *Dictyoglomus thermophilum xynB* gene encoding a thermophilic xylanase enzyme in the heterologous fungal host *T. reesei* was studied in a transformant strain grown in batch culture in a bench top laboratory fermenter. Samples taken from the fermenter cultivations were subjected to enzyme activity and yield assays, transcriptional analysis of UPR induction and comparison of the proteomes of the parental host strain with the transformant expressing the bacterial XynB enzyme (Nevalainen et al. 2004). Transcription of the *xynB gene* under the *T. reesei cbh1* (cellobiohydrolase 1) promoter was comparable to that of the native *cbh1*, however, the yield of the heterologous xylanase produced in a particular transformant studied was of the order of 100 mg/l. Transcription analysis of selected genes involved in UPR indicated a slight increase in the expression of *pdi1* but otherwise the UPR pathway seemed not to have been induced. Comparative proteomic analysis of the culture supernatant from the non-transformed host and the transformant expressing thermophilic xylanase showed differential expression of 23 proteins one of which was identified as an aspartic protease. This protein had a molecular weight of about 43 kDa and pI 4.6. Tryptic peptides sequenced from the protein spot gave a good match with a number of fungal aspartic proteinases. Aspartic proteinases have reported to have a considerable role in the degradation of heterologous proteins produced in e.g. Aspergillus which has led to mutagenesis and screening for protease negative/low protease mutant strains (Mattern et al. 1992) and disruption of the gene encoding aspartic proteinase enzyme (Berka et al. 1990). The amino acid sequence of the heterologous XynB protein has cleavage sites for an aspartic proteinase, therefore it is possible that the protein is degraded in the culture medium.

A Melanocarpus albomyces laccase was expressed in T. reesei under the cbh1 promoter as a non-fused laccase and a hydrophobin-laccase fusion protein (Kiiskinen et al. 2004). In a yield comparison using shake flask cultures, the non-fused laccase was produced at about 230 mg/L which was significantly higher to the yield obtained with the fusion protein. While Northern blot analyses showed relatively similar mRNA levels from both expression constructs, Western analysis revealed intracellular accumulation and degradation of the hydrophobin-laccase fusion protein. Therefore, the fusion product was misbehaving at the post-transcriptional level, yet there was no induction of UPR.

The two examples provide very little additional clues as to what may be triggering the UPR response in heterologous protein production and what are the other possible cellular mechanism affecting the yields. In the case of XynB production, there is a clear discrepancy between the xynB specific mRNA levels and the amount of secreted gene product: xynB specific mRNA was present at about the same amounts as the endogenous cbh1 mRNA, yet the secreted levels of the foreign xylanase were about 100 mg/L instead of grams. There seemed to be very little accumulation of the XynB inside the T. reesei hyphae (Hekelaar 2002) which would point either to degradation of the foreign product after secretion or mediating the degradation of XynB by a mechanism other than UPR. The third explanation for the low yield, yet unexplored, would involve, for example, the localization and stability of xynB specific mRNA and its translability. A new way to look into protein-related factors affecting secretion of the small family 11 xylanases from Trichoderma, made possible by the availability of 3D structures for these enzymes from different organisms, would be a comparative study of xylanases of T. reesei, XynB from D. thermophilum and XYN2 of Humicola insolens. All these enzymes exhibit a very similar structure, yet their secreted yields in T. reesei differ considerably (Hekelaar 2002; Faria et al. 2000).

In the work with M. albomyces laccase, the comparison of mRNA levels was carried out between the non-fusion and hydrophobin-laccase fusion proteins. This time it was clearly shown that the fusion product was degraded in the cell without induction of UPR whereas the non-fusion protein could be produced up to 920 mg/L in a fermenter culture. The experimental data emphasizes the fact that there are other so far less well characterized quality control mechanisms such as ERAD that operate in the cell in parallel with or maybe independently from UPR. These mechanisms are perfect targets for research using the technologies bridging genomes and proteomes.

4. YIELDS AND AUTHENTICITY OF RECOMBINANT PROTEINS MADE IN FILAMENTOUS FUNGI

Heterologous expression of genes encoding secreted fungal hydrolases has proven straightforward enough resulting in secretion of gram(s) per liter levels of enzymes into the cultivation medium of the heterologous fungal host (reviewed in Paloheimo et al. 1993; Schmidt 2004). The increasing importance of recombinant low volume-high value products of pharmaceutical importance, typically of mammalian origin, has further strengthened the interest in using filamentous fungi as expression hosts for these products. Filamentous fungi fulfill most of the preferred criteria for such organisms being easy to cultivate and able to secrete the gene products therefore simplifying the down-stream processing procedures. The challenge is in making "the same in another organism". From the recombinantly produced pharmaceuticals commanding a market of about $60 billion in the U.S alone, some of the products currently on the market are produced using yeasts such as *Saccharomyces cerevisiae* (e.g. Hepatitis B vaccine, insulin, platelet-derived growth factor PDGF and its agonist) and *Hansenula polymorpha* (Hepatitis B vaccine) but none in a filamentous fungus (reviewed in Schmidt 2004). The leading production systems are largely based on the use of Chinese hamster ovary cells (CHO) and *E. coli*.

There have been attempts to secrete pharmaceutically relevant gene products from industrially exploited filamentous fungi such as *Aspergillus niger* (var. *awamori*) and *Trichoderma reesei*. The examples include human interleukin 6 at 150 mg/L in *Aspergillus* (Punt et al. 2002). and tissue plasminogen activator t-PA at 30 mg/L in *T. reesei* (Uusitalo et al. manuscript in preparation).

4.1. Production of Antibodies

Current studies addressing heterologous expression of antibodies in yeasts and filamentous fungi have been reviewed recently (Joosten et al. 2003; Gerngross 2004). Production strategies have involved targeting the antibodies for secretion and their expression as fusion proteins either to an endogenously produced carrier protein or as fusions to an enzyme with a role in a particular application such as in laundry cleaning. The latter approach provides an example of a 'magic bullet' strategy where the fusion partner is guiding the antibody to a location where it will be needed.

Amongst the different types of antibodies trialed are murine Fab fragments in *T. reesei* (Nyyssönen et al. 1993). llama heavy chain antibody fragments (V_{HH}) in *A. awamori* (Joosten et al. 2004). functionalized single chain antibody fragments (scFV) in *A. awamori* (Frenken et al. 1998). and humanized full-length humanized immunoglobulin G1 (κ) antibodies in *A. niger* (Ward et al. 2004). Recently, we have also expressed a cDNA encoding a V_{NAR} fragment of IgNAR antibody (Immunoglobulin new shark antigen receptor) from a Wobbegong shark in *Trichoderma reesei* (Macdonald et al. unpublished). Similarly to llama V_{HH}, V_{NAR} antibodies consist of a single variable heavy chain domain of an antibody made from heavy chains only. However, the general layout of the IgNAR antibodies differ from heavy chains only. However, the general layout of the IgNAR antibodies differ from that of *Camelidae* antibodies. The heavy chain arrangement in the *Camelidae* antibodies is similar to the conventional IgG antibodies whereas the IgNARs have a less hydrophobic structure and therefore may be easier to produce in a soluble form in a heterologous host.

The IgG antibody was expressed in *A. awamori* at the level of 0.9 g/L with correct folding and efficient antigen binding despite of fungal type of N-glycosylation ranging from $Hex_6GlcNAc_2-Hex_{15}GlcNAc_2$ (Ward et al. 2004). In this particular case, glycosylation did not seem to be of primary importance in terms of affinity, avidity, pharmacokinetics and cellular cytotoxicity. However, there are a number of other cases such as with human erythropoietin (EPO) where glycosylation has an effect on the therapeutic activity of the compound (Takeuchi et al. 1998). Therefore, glycosylation is clearly an issue that needs to be addressed in filamentous fungi when developing them for hosts for functional biopharmaceuticals.

4.2. Protein Glycosylation in Fungi

While the glycosylation pathways in mammalian cells and in yeast such as *Pichia pastoris* have been worked out in detail, corresponding information for filamentous fungi is largely missing. Based on the glycans characterized from filamentous fungi and glycosylation pathway genes isolated so far it appears that N-glycans in filamentous fungi resemble more the high-mannose mammalian glycans than the oligosaccharides added on proteins by yeast that may consist up to 200 mannoses.

4.2.1 N-glycosylation

The complex process by which N-glycans are assembled and attached is orchestrated by numerous cytosolic and ER-associated enzymes (Kornfeld and Kornfeld 1985). Glycosylation pattern of N-glycans is significantly diverse according to their origin among eukaryotes, however, the initial steps, such as attachment of an activated precursor, dolichol-$GlcNAc_2Man_9Glc_3$ to the nascent protein and its subsequent trimming to $Man_8GlcNAc_2$ in the ER seem to be similar in mammals and fungi (reviewed in Vervecken et al. 2004; Gerngross 2004). After this initial glycosylation, further processing steps of N-glycans show substantial differences: in mammalian Golgi, the structure will be transformed to $Man_5GlcNAc_2$ followed by several trimming steps, such as addition of a GlcNAc residue, which step is catalyzed by a β-N-acetylglucosaminyl-transferase. Further processing steps are controlled by several glycan processing enzymes. These include trimming of two mannose units by the mannosidase II enzyme, addition of GlcNAc by means of GlcNAc transferases, addition of sialic acid by sialyl transferase and galactose by galactosyl transferase. In contrast to sugar trimming in mammalian cells, yeast cells continue adding mannose through the activity of mannosyltransferases (Dean 1991). Yeast enzymes involved in the N-glycosylation of proteins and their counterparts in *A. nidulans* are listed in Table 3. Significant matches were found to all other yeast enzymes/gene loci except for Mnn1p which adds the last mannose units to the protein. This may partly contribute to the less-mannosylated nature of proteins produced in filamentous fungi. While the enzymes and their genes participating to the above reactions have been isolated and characterized in mammalian cells and yeast, reports addressing filamentous fungi have addressed only some genes and enzymes, among these the *T. reesei* α-1-2-mannosidase (Maras et al. 2000; Van Petegem et al. 2001). Differently to the *S. cerevisiae* α-1-2-mannosidase located in the ER, the *T. reesei* enzyme can sequentially cleave four 1,2-linked mannose sugars from Man₉GlcNAc₂which is characteristic for Golgi-based mannosidases. Cellular location of the *T. reesei* α-1-2-mannosidase is yet to be solved. The ability of the *T. reesei* enzyme to readily covert Man₈GlcNAc₂ and a mixture of Man₆₋₉GlcNAc₂ substrates to Man₅GlcNAc2, a core glycan structure in mammalians has inspired studies into converting fungal glycans to the mammalian hybrid glycan structures by in vitro processing with mammalian enzymes (Maras et al. 1997) or cloning selected mammalian genes into the fungus. For example, expression of the human N-acetylglucosaminyltransferase I gene under the *T. reesei* cellobiohydrolase 1 (*cbh1*)

Table 3. Enzymes involved in N-linked glycosylation of the proteins in the yeast *Saccharomyces cerevisiae* and their homologues in the filamentous fungus *A. nidulans*

	Gene Locus / Enzyme	Accession No. / Reference	Definition / Function	Homologs found in *Aspergillus nidulans* [#]
1.	CWH41 or GLS1, (α–1,2-glucosidase I)	EMBL:Z36098	ER glucosidase I, removes the terminal α-1,2 linked glucose	AN6606.2
2.	ROT2 or GLS2, (α–1,3-glucosidase II)	EMBL:Z36098	ER Glucosidase II, removes the two α-1,3 linked glucose residues after the glucosidase I enzyme activity	AN8217.2
3.	MNS1, (α-1,2-mannosidase)	EMBL:Z49631	ER α-1,2-mannosidase, specific for removal of one mannose from Man₉GlcNAc₂ to Man₈GlcNAc₂	AN0551.2
4.	OCH1p	EMBL:Z72560	α-1,6 mannosyltransferase	AN4716.2
5.	MNN9p	EMBL:U44030	has α-1,6 mannosyltransferase activity	
6.	Van1p	EMBL: Z49210	involved in regulation of the phosphorylation of a number of proteins, some of which appear to be important in cell growth control	AN2159.2
7.	Mnn9p	EMBL:U44030	α-1,6 mannosyltransferase	AN4716.2
8.	Anp1p	EMBL:U18779	required to maintain a functional golgi apparatus and for glycosylation in the Golgi, role in retention of glycosyltransferases in the Golgi	AN2159.2
9.	Mnn10p	EMBL:Z49701	subunit of mannosyltransferase complex	AN7562.2
10.	Mnn11p	EMBL:Z49458	related to Mnn10p	AN1969.2
11.	Hoc1p	EMBL:U62942	forms a complex with Anp1p and Mnn9p in the Golgi membrane	AN4716.2
12.	Mnn2p	EMBL:Z35884	has α-1,2 mannosyltransferase activity	AN6571.2
13.	Mnn5p	EMBL:Z49461	has α-1,2 mannosyltransferase activity	AN6571.2
14.	Mnn4p	EMBL:Z28201	regulates the mannophosphorylation	AN1268.2
15.	Mnn6p	EMBL:U39205	mannosylphotransferase	AN2752.2
16.	Mnn1p	EMBL:U18778	α-1,3 mannosyltransferase, required for complex glycosylation of both N- and O-linked oligosaccharides	No significant match

[#]http://www.broad.mit.edu/annotation/fungi/aspergillus/

promoter for the addition of a GlcNAc residue on Man₅GlcNAc2, resulted in formation of GlcNAcMan5GlcNAc2, confirmed by NMR analysis (Maras et al. 1999). The first, corresponding effort by Kalsner et al. (1995) to introduce a mammalian glycosylation step into the genetically well chacterized *A. nidulans*, the gene product was not active and therefore had no impact on the fungal N-glycans. There has also been attempts to increase the amount of GDP-mannose in *T. reesei* by overexpression of the *mpg1* gene encoding mannose-1-phosphate guanytransferase in a quest for ensuring mannose supply in the protein secreting hyphae (Zakrewska et al. 2003b).

A considerable amount of work addressing genetic modification of the fungal glycosylation machinery to produce human type hybrid and complex glucans has been carried out with the yeasts *S. cerevisiae* and as *P. pastoris* (reviewed in Gerngross 2004; Vervecken et al. 2004). The approach has advanced to the last step towards humanizing the glycans, addition of sialic acid to the sugar by introduction of sialyltransferase activity into the *P. pastroris* cells (Gerngross 2004). As for filamentous fungi, the work has only begun and would require thorough mapping of the different stages and genes in the glycosylation pathway.

4.2.2 O-glycosylation

The majority of proteins produced by filamentous fungi for which the glycosylation pattern has been analyzed in some detail are highly secreted hydrolytic enzymes such as *Aspergillus niger* glucoamylase (GLA; Neustroev at al. 1993; Gunnarsson et al. 1984) and cellobiohydrolase I (CBHI or Cel7A) from *Trichoderma reesei* (Klarskov et al. 1997; Maras et al. 1997; Harrison et al.1998). In these enzymes, the O-glycosylation is concentrated on the linker region separating the catalytic core from the substrate binding domain, therefore performing a structural function. In *Aspergillus* as well as *Trichoderma*, the O-linked sugars are mainly composed of one to three mannose units, linear or branched (Pazur et al. 1980; Gunnarsson et al. 1984) attached to threonine or serine residues on the linker peptide. In addition, these oligosaccharides can contain glucose and galactose units as well (Gunnarsson et al. 1984; Wallis et al. 1999: Harrison et al. 2002) all features which make them different to O-linked oligosaccharides on yeast proteins composed of linear structures of mannose units only (Strahl-Bolsinger et al. 1999).

It has been suggested that O-glycosylation of proteins has a role in the secretion (Kubicek et al. 1987; Zakrewska et al. 2003a), stability (Goto et al. 1999; Harty et al. 2001) and localization of proteins to the fungal cell wall (Bourdineaud et al. 1998). which provides an interesting and potentially an important link of O-glycosylation to the synthesis and integrity of the cell wall, the last obstacle for the proteins secreted outside the cell into the culture medium. However, most of the O-glycosylated proteins localized in the cell wall or plasma membrane of filamentous fungi remain to be identified. A number of current studies are addressing this task by the application of e.g. proteomics, discussed below in section 5.

There has been no systematic mapping of the genes involved in O-glycosylation in filamentous fungi, however, yeast gene equivalents such as *pmtA* and *pmt1* encoding protein O-mannosyltransferase have been isolated from *Aspergillus* (Oka et al. 2004). and *Trichoderma* (Zakrewska et al. 2003a) respectively.

4.2.3 Effect of cultivation conditions on protein glycosylation

It has been evident for a long time that the culture medium somehow influences protein glycosylation in filamentous fungi. There are also published reports on differential glycosylation between different high protein-secreting mutant strains of the same fungal species (see references in Stals et al. 2004). On top of these, there is microheterogeneity in the glycosylation on a particular protein such as the *T. reesei* Cel7A which originates largely from the varying ratio of single N-acetylglucosamine residues to higher mannose structures on three of the four potential N-glycosylation sites on the Cel7A catalytic core and the presence or absence of phosphate residues. In their systematic study into glycosylation of Cel7A produced by *T. reesei* Rut-C30, Stals et al. (2004) showed that complete N- and O-glycosylation of the enzyme was achieved only on minimal medium lactose as a carbon source. Both uncharged $GlcMan_{7-8}GlcNAc_2$ and corresponding phosphorylated glycans were detected in these cultures. On the other hand, the glycans attached to Cel7A purified from the medium containing corn steep liquor featured high amount of $Man_5GlcNAc_2$ but a vanishing amount of phosphorylated oligosaccharides. As already indicated in some previous work, glycosylation can also be modified after secretion of the protein into the growth medium by extracellular enzymes such as α-1-2-mannosidase, α-1-3-glucosidase and an EndoH type activity. Adding the effect of the pH of the culture medium and the mode of cultivation (shake flask/fermentor) into this equation, setting the parameters for a prescribed glycosylation of proteins produced in fungal cell factories seems a huge task (see also sections in 3.2). However, heterogeneity in protein glycosylation is more a norm than an exception.

5. CELL WALL AND PHYSIOLOGICAL ASPECTS OF PRODUCTIVITY

The fungal cell wall is a dynamic organelle that comprises some 15-25 % of dry weight of the fungal cell (van der Vaart and Verrips 1998). In addition to electron microscopic studies carried out in the early 1970's (reviewed in Nykänen 2002) and chemical characterization of the cell wall featuring the main skeletal polysaccharides chitin and β-1,3- and and β-1,6 glucans, there are surprisingly few studies addressing the functional biochemistry of the wall, especially, when considering the importance of filamentous fungi as industrial production organisms. The development of novel technologies such as proteomics and the availability of genome sequence data, have clearly revived the interest into the cellular and molecular mechanisms that allow the secretion and release of gram levels of proteins from the fungal mycelium.

It has been well established that the majority of secreted proteins including heterologous gene products are secreted through the growing hyphal tip (Wessels 1993). This default pathway is functional, for example, in the secretion of the glucoamylase enzyme in *Aspergillus niger* (Wösten et al. 1991) and cellobiohydrolase I in *Trichoderma reesei*. The ability of *T. reesei* hyphae to synthesize and secrete several grams of CBHI per liter of the cultivation medium could perhaps be explained further by assuming another mechanism operating in the hyphae in addition to the default secretion via hyphal apices. Indeed, it has been shown that the CBHI enzyme is secreted also from the more mature parts of hyphae (Nykänen 2002). Besides CBHI, *T. reesei* endoglucanase I (EGI) and some heterologous enzymes such as *Hormoconis resinae* glucoamylase P and calf chymosin also occurred from mature parts of the *T. reesei* hyphae. Contrary to these observations, secretion of the heterologous barley cysteine proteinase EPB seemed to occur solely at the hyphal apex thereby following the default pathway in *T. reesei* (Nykänen et al. 1997). These findings would imply that there are other, yet unidentified factors in play regarding protein externalization from fungal hyphae. Some of these may relate to the physical subcellular localization of the specific mRNA, localization information printed in the amino acid sequence of a protein and protein glycosylation.

The link between protein glycosylation and secretion and the cell wall is provided by the involvement of protein O-mannosyltransferase enzyme in the cell wall biosynthesis, studied in *A. nidulans* (Oka et al. 2004). Disruption of the *pmtA* gene encoding this enzyme resulted in prominent reduction of enzyme activity, inhibition of extension of hyphae, reduced formation of conidia, increased sensitivity to heat, antifungal reagents and the fluorescent brightener Calcofluor White, hypersensitivity to Congo Red and altered carbohydrate composition, especially involving the skeletal wall polysaccharides. Oka et al. (2004) also studied the effect of the *pmtA* deletion on the glycosylation of *A. awamori* glucoamylase produced in *A. nidulans*. A difference of about 11 mannose units was detected, however, there is no information on the potential effect of the altered O-glycosylation on the secretion of this protein.

5.1 Proteomics of the Fungal Cell Wall

There are some published proteome-based studies on the fungal cell wall addressing the proteins located in the cell envelope (Lim et al. 2001; Vierula 2004). Even though these studies are currently at the level of mapping out the cell wall proteins, this information can later be applied to answer questions such as are there proteins that can be linked to high level secretion or what proteins play a significant role in the physiology and productivity of the growing hyphae. Amongst proteins identified so far in the studies of the *Neurospora crassa* and *Trichoderma reesei* cell walls are enolase, vacuolar protease A, glyceraldehyde-3-phosphate dehydrogenase, transaldolase, alcohol dehydrogenase III, protein disulfide isomerase, triose phosphate isomerase, mitochondrial outer membrane porin, mitochondrial ATP synthase β, diphosphate kinase, translation elongation factor beta and HEX1, the major protein in Woronin body, a structure unique to filamentous fungi. Protein identification has been pending on the throughput of mass spectrometry data and availability of genome sequence data both of which are improving exponentially. Therefore, a large part of the future work will involve bringing together and processing the information available from different approaches.

6. CONCLUSION

Recent developments in areas such as molecular biology, fluorescent microscopy, genomics, proteomics and bioinformatics are facilitating research beyond one-gene-at-a-time-approach steering it towards systems biology, the way how biological systems work. The novel global approach into protein localization using fluorescent tagging of the genes in the entire *Saccharomyces cerevisiae* and subcellular localization of the expressed products from a library of haploid yeast strains each expressing an individual protein has already been carried out (Huh et al. 2003). Envisaging a similar approach with filamentous fungi for which the genome has been sequenced is possible even though there are some technical drawbacks presented, for example, by the relative low transformation frequencies obtained compared to those with yeast. Therefore, a fair amount of raw work is required to acquire corresponding information from filamentous fungi.

In terms of protein production and secretion, there have been publications tracing proteins through the secretory pathway in filamentous fungi such as *Aspergillus niger* (Gordon et al. 2000). However, due to the diffraction barrier of light, the resolution of the general microscopy techniques has been insufficient to reveal a detailed map of protein localization. Recent advances in optical bioimaging in microscopy instrumentation and sophisticated microscopy techniques, such as Fluorescence Resonance Energy Transfer (FRET) and fluorescent lifetime imaging (FLIM) are now making possible localization of proteins of interest with respect to cellular compartments providing novel insights into protein secretion at the whole organism level. This is particularly important considering the often encountered situation where recent developments in computing applied to large-scale biological imaging datasets and cutting edge bioinformatics approaches are critical to efficiently utilize the data obtained from microscopic studies. Stepping up to the next level calls for novel type of collaboration and developing a common language across the different disciplines contributing to the future research.

REFERENCES

Allen TD and Nuss DL (2004). Specific and common alterations in host gene transcript accumulation following infection of the chestnut blight fungus by mild to severe hypoviruses. J Virol 78:4145-4155.

Al-Sheikh H, Watson AJ, Lacey GA, Punt PJ, MacKenzie DA, Jeenes DJ, Pakula T, Penttilä M, Alcocer MJC and Archer DB (2004). Endoplasmic reticulum stress leads to the selective transcriptional downregulation of the glucoamylase gene in *Aspergillus niger*. Mol Microbiol 53:1731-1742.

Archer DB and Peberdy JF (1997). The molecular biology of secreted enzyme production by filamentous fungi. Crit Rev Biotechnol 17:273-306.

Baglioni P, Bini L, Liberatori S, Pallini V and Marri L (2003). Proteome analysis of *Escherichia coli* W3110 expressing a heterologous sigma factor. Proteomics 3:1060-1065.

Bergquist P, Te'o V, Gibbs M, Curach N and Nevalainen K (2003). Recombinant bleaching enzymes from thermophiles expressed in fungal hosts. In: Applications of Enzymes to Lignocellulosics. Mansfield SD and Saddler JN, eds. Amer Chem Soc Symp Ser 855: 435-445.

Berka RM, Bayliss FT, Bleobaum P, Cullen D, Dunn-Coleman N, Kodama KH, Hayenga KJ, Hizteman RA, Lamsa MH, Przetak MM, Rey MW, Wilson LJ and Ward M (1991). *Aspergillus niger* var *awamori* as a host for the expression of heterologouse genes. In: JW Kelly and TO Baldwin, eds. NewYork: Plenum Press, pp 273-292.

Berka R, Hayenga K, Lawlis VB and Ward M (1990). Aspartic proteinase deficient filamentous fungi. WO 90/00192.

Bochtler M, Ditzel L, Groll M, Hartman C and Huber R (1999). The proteasome. Annu Rev Biophys Biomol Struct 28:295-317.

Bourdineaud JP, van den Vaart JM, Donzeau M, de Sampalo G, Verrips CT and Lauquin GJ (1998). Pmt1 mannosyl transferase is involved in cell wall incorporation of several proteins in *Saccharomyces cerevisiae*. Mol Microbiol 27:85-98.

Breyne P and Zabeau M (2001). Genome-wide expression analysis of plant cell cycle modulated genes. Curr Opin Plant Biol 4: 136-142.

Brodsky JL and McCracken AA (1999). ER protein quality control and proteasome-mediated protein degradation. Cell Devel Biol 19:507-513.

Cagney G, Uetz P and Fields S (2001). Two-hybrid analysis of the *Saccharomyces cerevisiae* 26S proteasome. Physiol Genomics 7:27-34.

Carlsen M and Nielsen J (2001). Influence of carbon source on alpha-amylase production by *Aspergillus oryzae*. Appl Microbiol Biotechnol. 57:346-349.

Casagrande R, Stern P, Diehn M, Shamu C, Osario M, Zuniga M, Brown PO and Ploegh H (2000). Degradation of proteins from the ER of *S. cerevisiae* requires an intact unfolded protein response pathway. Mol Cell 5: 729-735.

Chambergo FS, Bonaccorsi ED, Ferreira AJS, Ramos, ASP, Ferreira Jnr JR, Abrahão-Netos J, Simon Fraha JP and El-Dorry H (2002). Elucidation of the metabolic fate of glucose in the filamentous fungus *Trichoderma reesei* using expressed sequence tag analysis (EST) and cDNA microarrays. J Biol Chem 277:13983-13988.

Champion KM, Nishihara JC, Aldor IS, Moreno GT, Andersen D, Stults KL and Vanderlaan M (2003). Comparison of the *Escherichia coli* proteomes for recombinant human growth hormone producing and nonproducing fermentations. Proteomics 3:1365-1373.

Chapman R, Sidrauski RS and Walter P (1998). Intracellular signaling from the endoplasmic reticulum to the nucleus. Annu Rev Cell Dev Biol 14:459-485.

Choi JH, Lee SJ, Lee SJ and Lee SY (2003). Enhanced production of insulin-like growth factor I fusion protein in *Escherichia coli* by coexpression of the down-regulated genes identified by transcriptome profiling. Appl Environ Microbiol. 69:4737-4742.

Collén, A, Saloheimo M, Bailey M, Penttilä M and Pakula, TM (2004). Protein production and induction of the unfolded protein response in *Trichoderma reesei* strain Rut-C30 and its transformant expressing endoglucanase I with a hydrophobic tag. Biotech Bioeng, in press.

Conesa A, Jeenes D, Archer DB, van den Hondel CAMJJ and Punt P (2002). Calnexin overexpression increases manganese peroxidase production in *Aspergillus niger*. Appl Environ Microbiol 68:846-851.

Conesa A, Punt PJ, van Luijk N and van den Hondel CAMJJ (2001). The secretion pathway in filamentous fungi: a biotechnological view. Fung Genet Biol 33:155-171.

Dean N (1991). Asparagine-linked glycosylation in the yeast Golgi. Biochim Biophys Acta 1426:309-322.

Derkx PM and Madrid SM. (2001). The foldase CYPB is a component of the secretory pathway of *Aspergillus niger* and contains the endoplasmic reticulum retention signal HEEL. Mol Genet Genomics 266:537-545.

Dobson CM (2003). Protein folding and misfolding. Nature 426:884-890.

Durand H, Clanet M and Tiraby G (1988). Genetic improvement of *Trichoderma reesei* for large scale cellulase production. Enzyme Microb Technol 10:341-346.

Eldering E, Spek CA, Aberson HL, Grummels A, Derks IA, de Vos AF, McElgunn CJ and Schouten JP (2003). Expression profiling via novel multiplex assay allows rapid assessment of gene regulation in defined signalling pathways. Nucleic Acids Res 31:e153.

Ellgaard L and Helenius A (2001). ER quality control: towards an understanding at the molecular level. Curr Opin Cell Biol 13:431-437.

Faria FP, Te'o VSJ, Bergquist PL, Azevedo MO and Nevalainen KMH (2002). Expression and processing of a major xylanase (XYN2) from the thermophilic fungus *Humicola grisea* var *thermoidea* in *Trichoderma reesei*. Lett Appl Microbiol 34:119-123.

Foreman P, Brown D, Dankmeyer L, Dean R, Diener S, Dunn-Coleman N, Goedebuur F, Houfec T, England G, Kelley A, Meerman H, Mitchell T, Mitchinson C, Olivares H, Teunissen P, Yao J and Ward M (2003). Transcriptional regulation of biomass-degrading enzymes in the filamentous fungus *Trichoderma reesei*. J Biol Chem 287:31988-31997.

Frenken LG, Hessing JG, Van den Hondel CA and Verrips CT (1998). Recent advances in the large-scale production of antibody fragments using lower eukaryotic microorganisms. Res Immunol 149:589-599.

George HJ, Marchand P, Murphy K, Wiwall BH, Dowling RT, Giannara J, Hollis GF, Trzaskos JM and Copeland RA (1997). Recombinant human 92-kDa type IV collagenase/gelatinase from baculovirus-infected insect cells: expression, purification, and characterization. Protein Expr Purif 10:154-161.

Gerngross TU (2004). Advances in the production of human therapeutic proteins in yeasts and filamentous fungi. Nature Biotech 22:1409-1414.

Glickman MH (2000). Getting in and out of the proteasome. Cell Devel Biol 11:149-158.

Glickman M, Rubin DM, Coux O, Wefes I, Pfeiffer G, Cjeka Z, Baumeister W, Fried VA and Finley D (1998). Cell 94:615-623.

Goochee CF, Gramer MJ, Andersen DC, Bahr JB and Rasmussen JR (1991). The oligosaccharides of glycoproteins: bioprocess factors affecting oligosaccharide structure and their effect on glycoprotein properties. Biotechnology (NY) 9:1347-1355.

Gordon CL, Archer DB, Jeenes DJ, Doonan JH, Wells B, Trinci APJ and Robson GD (2000). A glucoamylase:GFP gene fusion to study protein secretion by individual hyphae of Aspergillus niger. J Microbiol Methods 42:39-48.

Goto M, Tsukamoto M, Kwon I, Ekino K and Furukawa K (1999). Functional analysis of O-linked oligosaccharides in threonine/serine rich region of Aspergillus glucoamylase by expression in mannosyl-transferase disruptants of yeast. Eur J Biochem 260:596-602.

Gouka RJ, Punt PJ and van den Hondel CA (1997). Efficient production of secreted proteins by Aspergillus: progress, limitations and prospects. Appl Microbiol Biotechnol 47:1-11.

Gunnarsson A, Svensson B, Nilson B and Svensson S (1984). Structural studies on the O-glycosidically linked carbohydrate chains of glucoamylase G1 from Aspergillus niger. Eur J Biochem 145:463-468.

Harrison MJ, Nouwens AS, Jardine DR, Zachara NE, Gooley AA, Nevalainen H and Packer NH (1998). Glycosylation of cellobiohydrolase I from Trichoderma reesei. Eur J Biochem 256:119-127.

Harrison MJ, Wathugala IM, Tenkanen M, Packer NH and Nevalainen KMH (2002). Glycosylation of acetylxylan esterase from Trichoderma reesei. Glycobiology 12:291-298.

Harty C, Strahl S and Romish K (2001). O-mannosylation protects mutant alpha-factor precursor from endoplasmic reticulum associated degradation. Mol Biol Cell 12:1093-1101.

Hekelaar J (2002). Production of the heterologous thermostable XynB from Dictyoglomus thermophilum by the filamentous fungus Trichoderma reesei. MSc Thesis, Noordelijke Hogeschool Leeuwarden and Van Hall Instituut Universities for Professional Education, Department of Biotechnology, Leeuwarden, The Netherlands.

Helenius A and Aebi M (2004). Roles of N-linked glycans in the endoplasmic reticulum. Annu Rev Biochem 73:1019-1049.

Hochstrasser M (1996). Ubiquitin-dependent protein degradation. Annu Rev Genet 30:405-439.

Huh WK, Falvo JV, Gerke LC, Carroll AS, Howson RW, Weissman JS and O'Shea EK (2003). Global analysis of protein localization in budding yeast. Nature 425:686-691.

Ilmén M, Saloheimo A, Onnela ML and Penttilä ME (1997). Regulation of cellulase gene expression in the filamentous fungus Trichoderma reesei. Appl Environ Microbiol 63:1298-1306.

Jeenes DJ, Pfaller R and Archer DB (1997). Isolation and characterisation of a novel stress-inducible PDI-family gene from Aspergillus niger. Gene 193:151-156.

Joosten V, Gouka RJ, Van Den Hondel CA, Verrips CT and Lokman BC (2004). Expression and production of llama variable heavy-chain antibody fragments (V(HH)s) by Aspergillus awamori. Appl Microbiol Biotechnol, in press.

Joosten V, Lokman C, van den Hondel CAMJJ and Punt PJ (2003). The production of antibody fragments and antibody fusion proteins by yeasts and filamentous fungi. Microb Cell Fact 2:1-15.

Jürgen B, Hanschke R, Sarvas M, Hecker M and Schweder T (2001). Proteome and transcriptome based analysis of Bacillus subtilis cells overproducing an insoluble heterologous protein. Appl Microbiol Biotechnol 55:326-332.

Kalsner I, Hintz W, Reid LS and Schachter H (1995). Insertion into Aspergillus nidulans of functional UDP-GlcNAc: alpha 3-D- mannoside beta-1,2-N-acetylglucosaminyl-transferase I, the enzyme catalysing the first committed step from oligomannose to hybrid and complex N-glycans. Glycoconj J 12:360-370.

Kariya Y, Ishida K, Tsubota Y, Nakashima Y, Hirosaki T, Ogawa T and Miyazaki K (2002). Efficient expression system of human recombinant laminin-5. J Biochem 132:607-612.

Kiiskinen LL, Kruus K, Bailey M, Ylösmaki E, Siika-Aho M and Saloheimo M (2004). Expression of Melanocarpus albomyces laccase in Trichoderma reesei and characterization of the purified enzyme. Microbiology 150:3065-3074.

Kivioja T, Arvas M, Kataja K, Penttilä M, Söderlund H and Ukkonen E (2002). Assigning probes into a small number of pools separable by electrophoresis. Bioinformatics 18 (suppl 1):199-206.

Klarskov K, Piens K, Ståhlberg J, Høj PB, Beeumen JV and Claeyssens M (1997). Cellobiohydrolase I from Trichoderma reesei:identification of an active-site nucelophile and additional information on sequence including the glycosylation pattern of the core protein. Carbohydr Res 304:143-154.

Kobayashi K, Kuwae S, Ohya T, Ohda T, Ohyama M, Ohi H, Tomomitsu K and Ohmura T (2000). High-level expression of recombinant human serum albumin from the methylotrophic yeast *Pichia pastoris* with minimal protease production and activation. J Biosci Bioeng 89:55-61.

Kornfeld R and Kornfeld S (1985). Assembly of asparagines linked oligosaccharides. Annu Rev Biochem 54:631-664.

Kubicek CP, Panda T, Schreferl-Kunar G, Gruber F and Messner R (1987). O-linked but not N-linked glycosylation is necessary for the secretion of endoglucanases I and II by *Trichoderma reesei*. Can J Microbiol 33: 698-703.

Lim D, Hains P, Walsh B, Bergquist P and Nevalainen H (2001). Proteins associated with the cell envelope of *Trichoderma reesei*: A proteomic approach. Proteomics 1:899-910.

Ma Y and Hendershot L (2001). The unfolding tale of the unfolded protein response. Cell 107:827-830.

Maras M, de Bruyn A, Schraml J, Herdewijn P, Claeyssens M, Fiers W and Conteras W (1997). Structural characterization of N-linked oligosaccharides from cellobiohydrolase I secreted by the filamentous fungus *Trichoderma reesei* RUTC 30. Eur J Biochem 245:617-625.

Maras M, de Bruyn A, Vervecken W, Uusitalo J, Penttilä M, Busson R, Herdewijn P and Contreras R (1999). *In vivo* synthesis of complex N-glycans by expression of human N-acetylglucosaminyltransferase I in the filamentous fungus *Trichoderma reesei*. FEBS Lett 452:3653-3670.

Maras M, Callewaert N, Piens K, Claeyssens M, Martinet W, Dewaele S, Contreras H, Dewerte I, Penttilä M and Contreras R (2000). Molecular cloning and enzymatic characterization of a *Trichoderma reesei* α-1-2-mannosidase. J Biotechnol 77:255-263.

Martinez IM and Chrispeels MJ (2003). Genomic analysis of the unfolded protein response in *Arabidopsis* shows its connection to important cellular processes. Plant Cell 15:561-576.

Mattern IE, van Noort JM, van den Berg P, Archer DB, Roberts IN and van den Hondel CA (1992). Isolation and characterization of mutants of *Aspergillus niger* deficient in extracellular proteases. Mol Gen Genet 234:332-336.

Meerman HJ, Wang H, Ward M, Meneses R, Baliu E, Rashid H, Wong S and Schellenberger V (2004). Abstracts of Society for Industrial Microbiology Annual Meeting, Anaheim CA, p 55.

Moralejo FJ, Watson AJ, Jeenes DJ, Archer DB and Martin JF (2001). A defined level of protein disulfide isomerase expression is required for optimal secretion of thaumatin by *Aspergillus awamori*. Mol Genet Genomics 266:246-253.

Mulder HJ, Saloheimo M, Penttilä M and Madrid SM (2004). The transcription factor HACA mediates the unfolded protein response in *Aspergillus niger*, and up-regulates its own transcription. Mol Genet Genomics 27:130-140.

Nakajima K, Kunihiro S, Sano M, Zhang Y, Eto S, Zhangh YC, Suzuki T, Jigami Y and Machida M (2000). Comprehensive cloning and expression analysis of glycolytic genes from the filamentous fungus *Aspergillus oryzae*. Curr Genet 37:322-327.

Neustroev KN, Golubev AM, Firsov LM, Ibatullin FM, Potaevich I and Makarov A (1993). Effect of modification of carbohydrate component on properties of glucoamylase. FEBS Lett 2:157-160.

Nevalainen H, Hekelaar J, Uusitalo J, Te'o J, Jonkers I, Bergquist P and Penttilä M (2004). Transcriptional and proteomic changes in *Trichoderma reesei* expressing a heterologous bacterial *xynB* gene from the thermophile *Dictyoglomus thermophilum*. Poster Abstracts of the 7th European Conference on Fungal Genetics, Copenhagen, p 201.

Nevalainen H, Penttilä M and Te'o VJS (2004). Application of genetic engineering for strain development in filamentous fungi. In: DK Arora, PD Bridge and D Bhatnagar, eds. Handbook of Fungal Biotechnology, Mycology Series 20. New York: Marcel Dekker, pp 193-208.

Nevalainen H and Te'o VJS (2003). Enzyme production in industrial fungi-molecular genetic strategies for integrated strain improvement. In: DK Arora and GG Kchachatourians, eds. Applied Mycology & Biotechnology. Volume 3, Fungal Genomics. The Netherlands: Elsevier Science BV, pp 241-259.

Ngiam C, Jeenes DJ and Archer DB (1997). Isolation and characterisation of a gene encoding protein disulphide isomerase, *pdiA*, from *Aspergillus niger*. Curr Genet. 31:133-138.

Ngiam C, Jeenes DJ, Punt PJ, van den Hondel CAMJJ and Archer DB (2000). Characterization of a foldase, protein disulfide isomerase A, in the secretory pathway of *Aspergillus niger*. Appl Environ Microbiol 66:775-782.

Nykänen M (2002b). Protein secretion in *Trichoderma reesei*. Expression, secretion and maturation of cellobiohydrolase I, barley cysteine endoproteinase and calf chymosin in Rut-C30. PhD Dissertation, University of Jyväskylä, Finland.

Nykänen, M, Saarelainen R, Raudaskoski M, Nevalainen KMH and Mikkonen A (1997). Expression and Secretion of Barley Cysteine Endopeptidase B and Cellobiohydrolase I in *Trichoderma reesei*. Appl Environ Microbiol 63:4929-4937.

Nyyssönen E, Penttilä M, Harkki A, Saloheimo A, Knowles JKC and Keränen S (1993). Efficient production of antibody fragments by the filamentous fungus *Trichoderma reesei*. Bio/Technology 11:591-595.

Oka T, Hamaguchi T, Sameshima Y, Goto M and Furukawa K (2004). Molecular caharcterization of protein O-mannosyltransferase and its involvement in cell-wall synthesis in *Aspergillus nidulans*. Microbiology 150: 1973-1982.

Okabayashi K, Nakagawa Y, Hayasuke N, Ohi H, Miura M, Ishida Y, Shimizu M, Murakami K, Hirabayashi K and Minamino H (1991). Secretory expression of the human serum albumin gene in the yeast *Saccharomyces cerevisiae*. J Biochem 110:103-110.

Pakula TM, Laxell M, Huuskonen A, Uusitalo J, Saloheimo M and Penttilä M (2003). The effects of drugs inhibiting protein secretion in the filamentous fungus *Trichoderma reesei*: Evidence for down-regulation of genes that encode secreted proteins in the stressed cells. J Biol Chem 278:45011-45020.

Pakula T M, Salonen K, Uusitalo J and Penttilä M (2005). The effect of specific growth rate on protein synthesis and secretion in the filamentous fungus *Trichoderma reesei*. Microbiology 151:135-143.

Pakula TM, Uusitalo J, Saloheimo M, Salonen K, Aarts RJ and Penttilä M (2000). Monitoring the kinetics of glycoprotein synthesis and secretion in the filamentous fungus *Trichoderma reesei*: cellobiohydrolase I (CBHI) as a model protein. Microbiology 146:223-232.

Patil C and Walter P (2001). Intracellular signaling from the endoplasmic reticulum to the nucleus:the unfolded protein response in yeast and mammals. Curr Opin Cell Biol 13:349-355.

Paloheimo M, Mäntylä A, Kallio J and Suominen P (2003). High-Yield Production of ac Bacterila Xylanase in the Filamentous Fungus *Trichoderma reesei* Requires a Carrier Polypeptide with an Intact Domain Structure. Appl Environ Microbiol 69:7073-7082.

Pazur JH, Tominaga Y, Forsberg LS and Simpson DL (1980). Glycoenzymes: an unusual type of glycoprotein structure for a glucoamylase. Carbohydr Res 84:103-114.

Pedersen H, Beyer M and Nielsen J (2000). Glucoamylase production in batch, chemostat and fed-batch cultivations by an industrial strain of *Aspergillus niger*. Appl Microbiol Biotechnol 53:272-277.

Punt PJ, van Biezen N, Conesa A, Albers A, Mangnus J and van den Hondel C (2002). Filamentous fungi as cell factories for heterologous protein production. Trends Biotechnol 20:200-206.

Punt PJ, van Gemeren IA, Drint-Kuijvenhoven J, Hessing JG, van Muijlwijk-Harteveld GM, Beijersbergen A, Verrips CT and van den Hondel CAMJJ (1998). Analysis of the role of the gene *bipA*, encoding the major endoplasmic reticulum chaperone protein in the secretion of homologous and heterologous proteins in black *Aspergilli*. Appl Microbiol Biotechnol 50: 447-454.

Pyle LE, Barton P, Fujiwara Y, Mitchell A and Fidge N (1995). Secretion of biologically active human proapolipoprotein A-I in a baculovirus-insect cell system: protection from degradation by protease inhibitors. J Lipid Res 36:2355-2361.

Sagt CMJ, Müller WH, van der Heide L, Boonstra J, Verkleij AJ and Verrips CT (2002). Impaired cutinase secretion on *Saccharomyces cerevisiae* induces irregular endoplasmic reticulum (ER) membrane proliferation, oxidative stress and ER-associated degradation. Appl Environ Microbiol 68:2155-2160.

Saloheimo M, Valkonen M and Penttilä M (2003). Activation mechanisms of the HACI-mediated unfolded protein response in filamentous fungi. Mol Microbiol 47: 1149-1161.

Saloheimo M, Wang H, Valkonen M, Vasara T, Huuskonen A, Riikonen M, Pakula T, Ward M and Penttilä M (2004). Characterization of secretory genes *ypt1/yptA* and *nsf1/nsfA* from two filamentous fungi: induction of secretory pathway genes of Trichoderma reesei under secretion stress conditions. Appl Environ Microbiol 70:459-467.

Salovuori I, Makarow M, Rauvala H, Knowles J and Kääriäinen L (1987). Low molecular mass high-mannose type glycans in a secreted protein of the filamentous fungus *Trichoderma reesei*. Biotechnology 5:153-157.

236

Satokari RM, Kataja K and Söderlund H (2004). Multiplexed quantification of bacterial 16S rRNA by solution hybridization with oligonucleotide probes and affinity capture. Microbial Ecology, in press.

Schmidt FR (2004). Recombinant expression systems in the pharmaceutical industry. Appl Microbiol Biotechnol 65:363-372.

Schmidt HHJ, Genschel J, Haas R, Buttner C and Manns MP (1997). Expression and purification of recombinant human apolipoprotein A-I in chinese hamster ovary cells. Protein Expr Purif 10:226-236.

Schmoll M, Zeilinger S, Mach RL and Kubicek CP (2004). Cloning of genes expressed early during cellulase induction in *Hypocrea jecorina* by a rapid subtraction hybridization approach. Fungal Genet Biol. 41:877-887.

Schrickx JM, Krave AS, Verdoes JC, van den Hondel CA, Stouthamer AH and van Verseveld HW (1993). Growth and product formation in chemostat and recycling cultures by *Aspergillus niger* N402 and a glucoamylase overproducing transformant, provided with multiple copies of the *glaA* gene. J Gen Microbiol 139:2801-2810.

Sims AH, Gent ME, Robson GD, Dunn-Coleman NS and Oliver SG (2004). Combining transcriptome data with genomic and cDNA sequence alignments to make confident functional assignments for *Aspergillus nidulans* genes. Mycol Res 108:853-857.

Sitia R and Braakman I (2003). Quality control in the endoplasmic reticulum protein factory. Nature 426: 891-894.

Sleep D, Belfield GP, Balance DJ, Steven J, Jones S, Evans LR, Moir PD and Goodey AR (1991). *Saccharomyces cerevisiae* strains that overexpress heterologous proteins. Bio/Technology 9:183-187.

Spohr A, Dam-Mikkelsen C, Carlsen M, Nielsen J and Villadsen J (1998). On-line study of fungal morphology during submerged growth in a small flow-through cell. Biotechnol Bioeng 58:541-553.

Stals I, Sandra K, Geysens S, Contreras R, Van Beeumen J and Claeyssens M (2004). Factors influencing glycosylation of *Trichoderma reesei* cellulases. I: Postsecretorial changes of the O- and N-glycosylation pattern of Cel7A. Glycobiology 14:713-724.

Strahl-Bolsinger S, Gentzsh M and Tanner W (1999). Protein O-mannosylation. Biochim Biophys Acta 1426:297-307.

Söderlund H, Kataja K, Paloheimo M, Ilmén M and Takkinen K (2003a). Method and test kit for quantitative and/or comparative assessment of variations in polynucleotide amounts in cell or tissue samples. FI20010041.

Söderlund H, Kataja K, Paloheimo M, Ilmén M and Takkinen K A (2003b). A Method and Test Kit for Quantitative and/or Comparative Assessment of Variations in Polynucleotide Amounts in Cell or Tissue Samples. PCT/FI02/00023.

Söderlund H, Satokari R, Kataja K and Takkinen K (2003c). Method and Test Kit for Determining the Amounts of Individual Polynucleotides in a Mixture. PCT/FI03/00544.

Takeuchi M, Inoue N, Strickland TW, Kubota M, Wada M, Shimizu R, Hoshi S, Kozutsumi H, Takasaki S and Kobata A (1998). Relationship between sugar chain structure and biological activity of recombinant erythropoietin produced in Chinese hamster ovary cells. Proc Natl Acad Sci USA 86:7817-7822.

Travers KJ, Patil CK, Wodicka L, Lockhart DJ, Weissman JS and Walter P (2000). Functional and genomic analyses reveal an essential coordination between the unfolded protein response and ER-associated degradation. Cell 101:249-258.

Valkonen M, Penttilä M and Saloheimo M (2003a). Effects of inactivation and constitutive expression of the unfolded- protein response pathway on protein production in the yeast *Saccharomyces cerevisiae*. Appl Environ Microbiol. 69:2065-2072.

Valkonen M, Penttilä M and Saloheimo M (2004). The *ire1* and *ptc2* genes involved in the unfolded protein response pathway in the filamentous fungus *Trichoderma reesei*. Mol Genet Genomics 272:443-451.

Valkonen M, Ward M, Wang H, Penttilä M and Saloheimo M (2003b). Improvement of foreign-protein production in *Aspergillus niger* var *awamori* by constitutive induction of the unfolded-protein response. Appl Environ Microbiol. 69:6979-6986.

van Gemeren IA, Beijersbergen A, van den Hondel CAMJJ and Verrips CT (1998). Expression and secretion of defined cutinase variants by *Aspergillus awamori*. Appl Environ Microbiol 64: 2794-2799.

van Gemeren IA, Punt PJ, Drint-Kuyvenhoven A, Broekhuijsen MP, van't Hoog A, Beijersbergen A, Verrips CT and van den Hondel CAMJJ (1997). The ER chaperone encoding *bipA* gene of black *Aspergilli* is induced by heat shock and unfolded proteins. Gene 198:43-52.

Van Petegem F, Contreras H, Contreras R and Van Beeum J (2001). *Trichoderma reesei* a-1-2-mannosidase: structural basis for the cleavage of four consecutive mannose residues. J Mol Biol 312:157-165.

van der Vaart JM and Verrips CT (1998). Cell wall proteins of *Saccharomyces cerevisiae*. Biotechnol Gen Engin Rev 15:387-411.

Vervecken W, Kaigorodov V, Callewaert N, Geysens S, De Vusser K and Contreras R (2004). *In Vivo* Synthesis of Mammalian-Like, Hybrid-Type N-Glycans in Pichia pastoris. Appl Environ Microbiol 70: 2639-2646.

Vierula PJ, Valencia CA, Tessier I and Kelly FJ (2004). Analysis of the *Neurospora* cell wall proteome by mass spectrometry. Abstracts of Society for Industrial Microbiology Annual Meeting, Anaheim CA, p 86.

Wallis GL, Swift RJ, Hemming FW, Trinci AP and Peberdy JF (1999). Glucoamylase overexpression and secretion in *Aspergillus niger*. Biochim Biophys Acta 1472:576-586.

Wang H and Ward M (2003). Molecular characterization of a PDI-related gene *prpA* in *Aspergillus niger* var *awamori*. Curr Genet 37:57-64.

Ward M, Lin C, Victoria D.C, Fox, J.A, Wong DL, Meerman HJ, Pucci JP, Fong RB, Heng MH, Tsurushita N, Gieswein C and Wang H (2004). Characterization of Human Antibodies Secreted by *Aspergillus niger*. Appl Environ Microbiol 70: 2567-2576.

Ward M, Wilson LL, Kodama KH, Rey MW and Berka RM (1990). Improved production of chymosin in *Aspergillus* by expression as a glucoamylase-chymosin fusion. Bio/Technology 8:435-440.

Wessels, JGH (1993). Transley review No. 45, Wall growth, protein excretion and morphogenesis in fungi. New Phytol 123:397-413.

Wiebe, MG, Karandikar A., Robson GD, Trinci AP, Cansia JL, Trappe S, Wallis G, Rinas U, Derkx, PM, Madrid SM, Sisniega H, Faus I, Montijn R, van Hondel CA and Punt PJ (2001). Production of tissue plasminogen activator (t-PA) in *Aspergillus niger*. Biotechnol Bioeng 76:164-174.

Wiebe MG, Robson GD, Schuster JR and Trinci AJP (1999). Growth rate independent production of recombinant glucoamylase by *Fusarium venenatum* JeRS 325. Biotechnol Bioeng 68:245-251.

Withers JM, Swift RJ, Wiebe MG, Robson GD, Punt PJ, van den Hondel CA and Trinci AP (1998). Optimization and stability of glucoamylase production by recombinant strains of *Aspergillus niger* in chemostat culture. Biotechnol Bioeng 59:407-418.

Wösten H, Moukha S, Sietsma J and Wessels J (1991). Localization of growth and secretion of proteins in *Aspergillus niger*. J Gen Microbiol 137:2017-2023.

Zakrewska A, Migdalski A, Saloheimo M, Penttilä ME, Palamarczyk G and Kruszewska JS (2003a). cDNA encoding protein O-mannosyltransferase from the filamentous fungus *Trichoderma reesei*; functional equivalence to *Saccharomyces cerevisiae* PMT2. Curr Genet 43:11-16.

Zakrewska A, Palamarczyk G, Krotkiewski H, Zbedska E, Saloheomo M, Penttilä M and Kruszewska J (2003b). Overexpression of the gene encoding GTP:mannose-1-phosphate guanyltransferase, *mpg1*, increases cellular GDP-mannose levels and protein mannosylation in *Trichoderma reesei*. Appl Environ Microbiol 69:4383-4389.

**Applied Mycology
and Biotechnology**
An International Series
Volume 5. Genes and Genomics
© 2005 Elsevier B.V. All rights reserved

10

ELSEVIER

Expression and Engineering of Fungal Hydrophobins

Karin Scholtmeijer[1], Rick Rink[1], Harm J Hektor[1] and Han AB Wösten[2]
[1] BiOMaDe Technology, Nijenborgh 4, 9747 AG Groningen, The Netherlands.
(Scholtmeijer@biomade.nl); [2] Department of Microbiology, Institute of Biomembranes.
University of Utrecht, Padualaan 8, 3584 CH Utrecht, The Netherlands.

Filamentous fungi secrete unique proteins called hydrophobins. Upon contact with a hydrophilic-hydrophobic interface these proteins self-assemble into an amphipathic membrane. Differences in the solubility of the assemblages divides hydrophobins into two groups. The class I hydrophobins form highly insoluble membranes that can only be dissolved in trifluoroacetic acid (TFA) and formic acid, while assemblies of class II hydrophobins can be readily dissolved in ethanol or SDS. Self-assembly allows hydrophobins to change the nature of a surface; hydrophobic surfaces turn hydrophilic and hydrophilic surfaces become hydrophobic. These properties make hydrophobins interesting candidates for use in technical and medical applications. Class I hydrophobins seem to be particularly interesting to coat solid surfaces, while class II seem to be the molecules of choice for use in liquid systems. Application of hydrophobins would benefit from the availability of a library of hydrophobin variants. Moreover, production should be increased. Nature provides a fast amount of hydrophobins with slightly different characteristics. In addition to these hydrophobins, new variants may be obtained via random mutagenesis or by rational design. Functional class I hydrophobins (i.e. capable of self-assembling into an amphipathic membrane) could only be produced by filamentous fungi that by nature secrete hydrophobins into the culture medium. However, yields are still relatively low. On the other hand, class II hydrophobins can already be produced at high levels using *Trichoderma reesei* as a host.

1. INTRODUCTION

Filamentous fungi belonging to the ascomycetes and the basidiomycetes (and possibly zygomycetes as well) produce hydrophobins (de Vries et al. 1993; Wessels, 1997). These small secreted proteins with unique properties are classified as class I or class II (Wessels, 1994). All hydrophobins have eight conserved cysteine residues but the hydropathy pattern of class I and class II are different. They do not share extensive sequence homology, not even within a class (Wessels 1994, 1997). Hydrophobins fulfill several functions in fungal growth and development. They are

---

Corresponding author: Karin Scholtmeijer

involved in formation of hydrophobic aerial structures (Wösten and Wessels, 1997; Wösten, 2001) and mediate attachment of hyphae to hydrophobic surfaces (Wösten et al. 1994b; Talbot et al. 1996). Hydrophobins fulfill these functions by self-assembling at hydrophilic–hydrophobic interfaces (Wösten et al. 1993; 1994b; 1995). Hydrophobins produced by submerged hyphae diffuse into the aqueous environment and self-assemble upon contact with the interface between medium and air. This self-assembly results in a decrease of the water surface tension, allowing hyphae to breach the interface and to grow into the air (Wösten et al. 1999). Hydrophobins secreted by hyphae in contact with air or a hydrophobic solid self-assemble at the surface of the cell wall. This results in the exposure of the hydrophobic side of the membrane, while the hydrophilic side interacts with the polysaccharides of the cell wall. As a consequence, aerial hyphae and spores become hydrophobic, while hyphae that grow over a hydrophobic substrate firmly attach.

SC3 of *Schizophyllum commune* is the best studied class I hydrophobin. Upon contact of an aqueous solution of SC3 with a hydrophobic surface, SC3 self-assembles into a 10 nm thick amphipathic film (Wösten et al. 1993, 1994a, 1995). The water contact angles of the hydrophilic and hydrophobic sides of the SC3 membrane are 36° (moderately hydrophilic) and 110° (highly hydrophobic), respectively (Wösten et al. 1993, 1994a). Assembled SC3 is highly insoluble and can only be dissociated by formic acid and trifluoroacetic acid (TFA) (Wessels et al. 1991; de Vries et al. 1993). Self-assembly is accompanied by an increase in lectin and surface activity (van Wetter et al. 2000; van der Vegt et al. 1996; Wösten et al. 1999). In fact, with a maximal lowering of the water surface tension from 72 to 24 mJ m^{-2} SC3 (50 µg ml^{-1}) is the most surface-active protein identified. Although the gross properties of other class I hydrophobins are similar to those of SC3, they do have slightly different properties (Wösten and de Vocht, 2000). The water contact angle of the hydrophobic side of the assemblage is always approximately 110° and their surface activity is similar, ranging between 32 and 37 mJ m^2 at 100 µg ml^{-1} (van der Vegt et al. 1996; Wösten and Wessels, 1997; Lugones et al. 1999; Wösten and de Vocht, 2000). However, the wettability at the hydrophilic side of the assemblage differs, ranging between 36° and 63°. Moreover, the diameter of the rodlets varies between 5 and 12 nm and their lectin activity seems to be directed to different sugars (Wösten and de Vocht, 2000). Solubility of the monomers in aqueous organic solvents also differs (Scholtmeijer, et al. 2002).

Class II hydrophobins also assemble at hydrophilic-hydrophobic interfaces. Most is known about the class II hydrophobins HFBI and HFBII of *Trichoderma reesei* (Linder et al. 2002 Askolin et al. 2004). CU of *Ophiostoma Ulmin* (Russo et al. 1982) and CRP of *Cryphonectria parasitica* (Carpenter et al. 1992; Wösten and de Vocht, 2000). In contrast to assemblages formed by class I hydrophobins, assemblages of class II hydrophobins are less stable. Those of CU, CRP, HFBI and HFBII can be dissociated in 60% ethanol and 2% SDS (Russo et al., 1982; Carpenter et al. 1992; Wösten and de Vocht, 2000; Askolin et al. 2004) and assembled CU, HFBI and HFBII also dissociate by applying pressure or cooling (Russo et al. 1982; Askolin et al. 2004). In addition, the water contact angle of the hydrophobic side of assembled HFBI and HFBII seems to be lower than that of class I hydrophobins (60-70° and 100-130°, respectively)(Askolin et al. 2004). Unlike class I hydrophobins, the assemby properties of HFBI and HFBII were shown to differ (Askolin et al. 2004). HFBI and HFBII were both shown to decrease the water surface tension. However, HFBI lowered the water surface tension to a minimum of 28 mJ m^2 at 20 µg ml^{-1}, while HFBII only reached a maximal reduction to 37 mJ m^2 at 200 µg ml^{-1}. Furthermore, assembly of HFBII at Teflon does not change the wettability of the solid, while HFBI and CRP were shown to lower the water contact angle to 59° and 22°, respectively (Wösten and de Vocht, 2000; Askolin et al. 2004). Finally, both HFBI and HFBII adsorb to hydrophobic alkylated gold and silanized glass, however, binding of HFBII was shown to be somewhat weaker (Linder et al. 2002).

1.1. Structure of Class I Hydrophobins

Little is known about the tertiary structure of soluble or assembled class I hydrophobins. Multidimensional NMR spectroscopy of the soluble state of the class I hydrophobin EAS from *Neurospora crassa* showed that the water soluble protein is mainly unstructured with a small core region containing antiparallel β-sheet structure (Mackay et al. 2001). The latter was proposed to be stabilized by the four intramolecular disulphide bonds. Most of the protein has a significant degree of local motion, while only a few smaller regions, corresponding to cysteine containing sequences, appear rigid. NMR and molecular dynamic (MD) simulation also indicate that water soluble SC3 is largely unstructured (Wösten and de Vocht, 2000; Zangi et al. 2002). This would explain why H/D exchange experiments did not show any highly protected regions in the protein (Wang, 2004).

More is known about the secondary and ultra-structure of class I hydrophobins. Self-assembly of SC3 is accompanied by several conformational changes (de Vocht et al. 1998; 2002; Wösten and de Vocht, 2000; Wang et al. 2002; 2004; Wang, 2004). Upon contact with the water-air or water-oil interface, β-sheet rich water soluble dimers and tetramers (Wang et al. 2002; 2004) proceed via a conformation with increased α-helix (α-helix state) to a conformation with increased β-sheet (de Vocht et al. 1998; 2002). Initially, the β-sheet state has no clear ultrastructure (β-sheet I state). However, after a few hours a mosaic of bundles of 10 nm wide rodlets can be observed (β-sheet II state). These rodlets consist of amyloid fibrils as evidenced by the fact that they interact with the amyloid specific dye Thioflavine T (ThT) as well as with Congo Red (Wösten and de Vocht, 2000; Butko et al. 2001; Mackay et al. 2001). Rodlets are composed of two tracks of 2-3 protofilaments, each having a diameter of about 2.5 nm. At the water-Teflon interface SC3 is arrested in the α-helical conformation. Assembly into the β-sheet II state can be induced by treatment with detergent at increased temperature (Wösten and de Vocht 2000) or low pH (R. Rink, unpublished data). The β-sheet state appears to be more stable than the α-helical form. Both strongly adhere to hydrophobic surfaces but only the α-helical form can be dissociated by treatment with cold diluted detergents (Wösten and de Vocht, 2000).

The region of SC3 between amino acids Thr52-Ser76 (numbering starting at the first Gly residue of mature SC3) could be fitted into an amphipathic helix (de Vocht et al. 1998). These amino acids are located between the third and fourth cysteine residue encompassing 32 amino acids. In class II hydrophobins this region is much smaller (Wösten and Wessels, 1997). Since class II hydrophobins exert less strong binding to Teflon surfaces (see below), this region was expected to play a role in the strong binding of class I hydrophobins to hydrophobic surfaces. Indeed, in contrast to other peptides of SC3, the peptide corresponding to amino acids Thr52-Ser76 strongly adsorbed to Teflon, at which it adopted α-helical structure (Wang, 2004). Hot detergent treatment, however, did not result in the β-sheet state. Molecular dynamic simulation of SC3 assembled at a water-hexane interface predicted β-sheet structure throughout the hydrophobin molecule (Zangi et al. 2002). Possibly, the region in between the third and fourth cysteine residue binds SC3 to the hydrophobic solid, inducing β-sheet structure in the rest of the molecule.

1.2. Structure of Class II Hydrophobins

The crystal structure of the class II hydrophobin HFBII was determined at 1 Å resolution (Hakanpää et al. 2004). HFBII contains two β-hairpin motifs linked via an α-helix which form a barrel consisting of four antiparallel β-strands. Several aliphathic amino acid residues within the two β-hairpin motifs are conserved in class II hydrophobins. From this it was suggested that other class II hydrophobins have a similar structure. The conserved residues form a hydrophobic patch with a relatively flat surface in an otherwise globular protein (Hakanpää et al. 2004). This patch was suggested to confer the amphiphilic character of the protein.

It is not expected that class I hydrophobins have an identical 3D structure as class II hydrophobins. Structural analysis of HFBII indicated that the protein is rather rigid and contains a disulphide pairing that may hamper conformational changes (Hakanpää et al. 2004). Disulphide bridges in HFBII are formed between cysteine 1 and 6, cysteine 2 and 5, 3 and 4 and 7 and 8. This came as a surprise since these disulphide bridges differ from those previously determined for CU (Yaguchi et al. 1993). In CU the disulphide bridges were shown to connect cysteine 1 with 2, 3 with 4, 5 with 6 and 7 with 8. Structural analysis of other hydrophobins should reveal whether or not disulphide bridges are conserved in hydrophobins.

Small angle X-ray scattering and size-exclusion chromatography of HFBI and HFBII solutions showed that the proteins occur as tetramers at high concentrations (10-100 mg/ml; Torkelli et al. 2002; Panaanen et al. 2003). Modelling indicated that the HFBI tetramers have a torus-shape (Panaanen et al. 2003) while those of HFBII are dumb-bell shaped (Torkelli et al. 2002). AFM analysis of Langmuir-Blodgett films of HFBI and HFBII showed a highly ordered 2D crystalline structure. The structure consists of repetitive units with dimensions and features similar to those of the tetramers found in solution (Panaanen et al. 2003). CD-analysis showed that, unlike class I hydrophobins, assembly of the class II hydrophobins HFBI and HFBII at the water-air interface does not result in structural changes (Askolin et al. 2004). Assembly of CRP, HFBI and HFBII at the water-Teflon interface did result in an increase of α-helical structure similar to that in class I hydrophobins. However, hot SDS treatment resulted in dissociation of the class II hydrophobins from the surface (Wösten and de Vocht, 2000; Askolin et al. 2004) and an increase in β-sheet structure was not detected. From the fact that class II hydrophobins do not form rodlets (Wösten and de Vocht, 2000) and do not increase ThT fluorescence during assembly (Torkelli et al. 2002) it is concluded that these hydrophobins do not assemble into amyloid fibrils.

2. MEDICAL AND TECHNICAL APPLICATION OF HYDROPHOBINS

By self-assembly at hydrophilic-hydrophobic surfaces hydrophobins change the nature of a surface. Hydrophobic surfaces (solid or liquid) become hydrophilic (Wösten et al. 1994a, 1995; Lugones et al. 1996) while hydrophilic surfaces become hydrophobic (Wösten et al., 1993; Lugones et al., 1996). This makes hydrophobins interesting candidates for the use in technical applications. Moreover, since hydrophobins do not seem to be toxic (they are consumed via mushrooms and fungus-fermented foods) nor cytotoxic (Janssen et al. 2002) or immunogenic, they may also be used in medical applications (Wessels 1997). Apart from changing the biophysical properties of a surface, hydrophobins can be used to immobilize cells, proteins or other molecules at solids or liquids. Immobilization can be achieved simply by adsorption to the hydrophobin (Bilewicz et al. 2001) or by chemical cross-linking (Wessels 1997). Proteins and peptides may also be immobilized at a surface by fusing them to a hydrophobin via genetic engineering, after which the fusion protein is assembled at the surface of interest. Self-assembled hydrophobin films are about 10 nm thick and monomers are predicted to have a diameter of about 2 nm. Therefore, these proteins may be used to pattern molecules on a surface with nanometer accuracy (Wessels, 1997). Preliminary results indicate that hydrophobins can form mixed membranes (X. Wang, unpublished). This implies that membranes can be made of hydrophobins contributing a certain biophysical property and hydrophobins exposing functional groups.

The fact that class I hydrophobins strongly adsorb to solid surfaces and form highly stable membranes make these proteins most suitable to coat solid surfaces. On the other hand, class II hydrophobins would be favored to use in liquids especially because these membranes dissociate more easily.

2.1. Use of Hydrophobins on Solid Surfaces

Hydrophobins have been used to promote growth of cells on a hydrophobic surface. The SC4 hydrophobin was shown to be superior to SC3. This may be due to the fact that assembled SC4 is less hydrophilic or that it is not glycosylated (Janssen et al. 2002). Improved growth of human and mouse fibroblasts was also obtained by engineering of SC3 (Janssen et al. 2002; Scholtmeijer et al. 2002; Janssen, 2004). Cell growth was promoted by deleting part of the glycosylated N-terminus of the mature hydrophobin and/or by fusing the RGD tripeptide to the N-terminus. *Saccharomyces cerevisiae* cells were immobilized on a hydrophobic surface by expressing a fusion protein consisting of the class II hydrophobin HFBI and the Flo1 protein (Nakari-Setälä et al. 2002). The yeast cells that were more apolar, and slightly less negatively charged had a two-fold higher binding affinity to hydrophobic materials.

Lipases (Palomo et al. 2003) and endoglucanase I (EGI) (Linder et al. 2002) have also been immobilized at solid surfaces using hydrophobins. The lipases were adsorbed to the hydrophobic side of the class I hydrophobins Vmh1, 2 and 3 of *Pleurotus ostreatus* that had been adhered to hydrophilic glyoxyl-agarose. The stability and the catalytic properties of the immobilized lipases were similar to that of the lipases immobilized on conventional hydrophobic supports (Palomo et al. 2003). EGI was genetically fused to the N-terminus of the class II hydrophobin HFBI.

Hydrophobin mediated attachment to hydrophobic surfaces did not affect activity of the endoglucanase (Linder et al. 2002).

Not only peptides or proteins can be immobilized on a surface via hydrophobins. The electro-active compounds ubiquinone, quinone and azobenezene adsorbed to surfaces of electrodes that had been coated with the hydrophobin HydPt-1 of *Pisolithus tinctorius* (Bilewicz et al. 2001). This resulted in functional electrodes that remained active for weeks. Moreover, the hydrophobin layer protected the electrode surfaces from oxidation and blocked access of small electro-active compounds to the electrode surface.

2.2. Use of Hydrophobins on Liquids

Self-assembled hydrophobins exhibit a surface activity similar to that of traditional biosurfactants such as glycolipids, lipoproteins, phospholipids, neutral lipids, substituted fatty acids and lipopolysaccharides (Desai and Banat, 1997; Wösten and de Vocht, 2000). Hydrophobins may substitute these surfactants in a wide range of applications (e.g. in emulsions and dispersions). In contrast to the traditional surfactants (Desai and Banat, 1997) the surface activity of hydrophobins is solely caused by the amino acid sequence and they are encoded by small (± 400 bp) simply organized genes. This implies that surfactant properties can be easily modified by genetic engineering of the hydrophobin.

Previously, peptide tags were used for purification of EGI (Köhler et al. 1991; Berggren et al. 1999; Collén et al. 2001a; 2001b; 2002a; 2002b; Rodenbrock et al. 2001). However, production of these fusion proteins was shown to be difficult (Yang et al. 1994; Berggren et al. 1999; Collén et al. 2002b). In contrast, a fusion between EGI and the HFBI hydrophobin was produced in high quantity and was efficiently purified from culture media using an aqueous two-phase extraction system (Collén et al. 2002). To this end, culture medium was mixed with the non-ionic detergent Triton X-114 and the polymer hydroxypropyl starch. Under defined conditions, this mixture separates into two phases. The EGI-HFBI fusion partitioned mainly into the micelle-rich phase (99%) while other, hydrophilic, secreted proteins ended up in the polymer-rich phase. The EGI-HFBI fusion protein was subsequently back-extracted into the water-phase by addition of ethylene oxide-propylene oxide (EOPO). As a result, 90% of the protein was recovered in one sixth of the original volume.

3. PRODUCTION OF HYDROPHOBIN VARIANTS

Each application is expected to require a hydrophobin with specific properties. For example, aspecific binding of molecules to a biosensor surface may depend on the wettability of the hydrophobin coating. A library of hydrophobins would allow screening of the best performing hydrophobin for a given application. Nature provides a fast diversity of hydrophobins with the occurrence of several hydrophobin genes in most of the 1.5 million species that make up the fungal kingdom (Hawksworth, 1991). These natural variants may be supplemented with hydrophobins obtained via random mutagenesis or by rational design. The latter requires structural information and knowledge of the effects of modification. The region preceding the first cysteine residue of hydrophobins is highly variable (Wösten, 2001). For example, it consists of 31 amino acids and 16-22 mannose residues in case of SC3 (de Vocht et al., 1998) while only 8-12 amino acids preceed the first cysteine residue in hydrophobins such as SC4 and ABH3 (Lugones et al. 1996; 1998; 1999). This variation was predicted to cause differences in the biophysical properties. Indeed, the water contact angle of the hydrophilic side of assembled SC3 increased when the protein was deglycosylated (de Vocht et al, 1998) or when 25 amino acids of the N-terminal region of mature SC3 (TrSC3) were deleted (Scholtmeijer et al. 2002). Yet, other properties were not affected. Addition of the cell-binding domain of human fibronectin (RGD) at the N-terminus of SC3 and TrSC3 (resulting in RGD-SC3 and RGD-TrSC3, respectively) also only affected the properties of the hydrophilic side of the assemblage (Janssen et al., 2002; Scholtmeijer et al. 2002).

The C-terminus of class I hydrophobins is also expected to determine the properties of the hydrophilic side of the assemblage, at least when the protein is in its β-sheet state. H/D exchange rates of the C-terminal part increased dramatically when SC3 converted from the α-helical to the β–sheet state (Wang, 2004). In agreement, peptides from this part of the molecule were more easily digested by pepsin when SC3 was in the latter conformation. Genetic engineering should confirm whether the C-terminus of class I hydrophobins can be used to expose proteins to the water phase after assembly. It should be noted that the variation in the number of amino acids following the 8th cysteine residue is restricted to only 2-13 amino acids (Wösten and de Vocht, 2000) which contrasts the observed variability at the N-terminal part. It may therefore be difficult to obtain functional fusion proteins at the C-terminus of hydrophobins.

Apart from changing the biophysical properties at either side of the assemblage it would be interesting to create hybrids of class I and class II hydrophobins. For instance, it may be desirable to create hydrophobins that strongly bind to a hydrophobic solid in an aqueous environment but that easily dissociate in ethanol. Such hybrids could be obtained by exchanging regions between the third and the fourth cysteine residue (see above).

4. OVER-EXPRESSION OF HYDROPHOBINS IN FILAMENTOUS FUNGI

Production levels determine whether hydrophobins can be used in bulk applications or whether only small scale applications should be considered. Filamentous fungi are known for their capacity to secrete large amounts of proteins. For instance, 30 g L^{-1} glucoamylase and cellobiohydrolase can be produced in the industrial fungal production systems *Aspergillus niger* and *Trichoderma reesei*, respectively (Finkelstein et al. 1989; Durand et al. 1998). Yeast and bacterial production systems have also been evaluated to produce hydrophobin, especially because they would allow easy screening of improved hydrophobin variants.

4.1. Production of Class I Hydrophobins in Filamentous Fungi

Initially, production of class I hydrophobins in the industrial production systems *Aspergillus niger* and *Trichoderma reesei* was examined. *A. niger* seemed to be attractive because wild-type strains do not secrete hydrophobins in their culture medium (O.M.H. de Vries, unpublished), allowing easy purification of the heterologous produced protein. Expression of *SC3* behind the *GPD* promoter of

Aspergillus nidulans resulted in low amounts of the hydrophobin in the culture medium (less than 1% compared to *S. commune*; Fig.1; Table 1) (Scholtmeijer, 2000). Since production did not correlate with mRNA levels (20-40% compared to *S. commune*), it seems that limitation occurrs for a large part at the post-transcriptional level.

Degradation of SC3 by extracellular proteases seems not to be a limiting factor since expression in protease-deficient *A. niger* strains did not result in increased production of the hydrophobin. SC3 was then fused to the 3′-end of the homologous *Gla* gene encoding the highly secreted glucoamylase form G2. In this strategy, the homologous protein would serve as a carrier to improve translocation of SC3 into the endoplasmatic reticulum (ER). In addition, the carrier could allow proper folding, thereby protecting the heterologous protein from degradation (Gouka et al. 1996). However, this fusion strategy did not result in increased levels of SC3 (Table 1). Possibly, certain foldases and/or chaperones necessary for the folding and secretion of class I hydrophobins are absent in the submerged mycelium of *A. niger*. *Aspergillus* species do produce hydrophobins in aerial structures. This suggests that foldases and/or chaperones involved in hydrophobin production are differentially expressed. Future experiments should identify these proteins and their corresponding genes, which should then be expressed in the submerged mycelium to improve hydrophobin production.

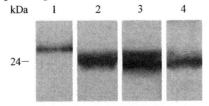

Fig. 1: Western blot analysis of SC3 in the culture medium of *A. niger* and *T. reesei* using antiserum against SC3. (1) *A. niger* strain AB4.1 expressing *SC3* behind the *GPD* promoter of *A. nidulans* using glucose as carbon source. (2) *T. reesei* strain Rut-C30 expressing *SC3* behind the *HFBI* promoter using glucose as a carbon source. (3) *T. reesei* strain Rut-C30 expressing *SC3* behind the *HFBII* promoter using lactose as a carbon source. (4) SC3 produced by *S. commune* serving as control. The equivalent of 4.5 ml (1) and 0.45 ml (2-4) of culture medium was loaded on gel.

T. reesei secretes high amounts of HFBI and HFBII in its culture medium (Nakari-Setälä et al. 1996; 1997; Askolin et al. 2001) and may therefore contain the correct machinery to produce correctly folded hydrophobin in liquid cultures. Expression of *SC3* behind the *HFBI* or *HFBII* promoter resulted in production levels of 60 mg L^{-1}, which is similar to that of wild-type strains of *S. commune* (Fig. 1; Table 1)(Scholtmeijer, 2000). Northern analysis revealed that *SC3* mRNA levels were also similar in *T. reesei* and *S. commune*, indicating that post-transcriptional limitations, as observed in *A. niger*, do not occur in *T. reesei*. SC3 was purified from the culture medium adopting the procedure used for purification of the hydrophobin from culture media of *S. commune*. In this way, SC3 and the class II hydrophobins were seperated (Fig. 2). Heterologously produced SC3 showed self-assembly properties similar to SC3 produced by *S. commune*. Production of the homologous class II hydrophobin HFBI in *T. reesei* was 10-fold higher than the production of SC3 (see below). The reason for this is not yet clear. As production and mRNA levels of SC3 were similar to those of *S. commune* it is expected that the machinery for production of hydrophobin is functional. Possibly, mRNA levels of class II hydrophobins are higher and if so, transcription and mRNA stability of *SC3* should be analysed. Moreover, production levels of *SC3* may be increased by (over)expressing foldases and/or chaperones in *T. reesei*.

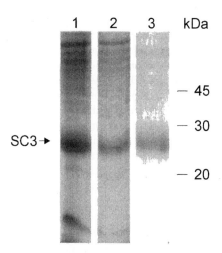

Fig. 2. SC3 could be purified from other proteins contained in the medium of a *T. reesei* strain expressing *SC3* behind the *HFBII* promoter by spinning down the assembled class I hydrophobin followed by dissolving the protein in TFA. TCA precipitate of the culture medium before (1) and after (2) centrifugation of the medium. Subsequent treatment of the pellet with TFA solubilizes SC3 (3). Proteins were separated by SDS-PAGE and visualized by coomassie brilliant blue staining. Amounts corresponding to 1 ml of culture medium were loaded in each lane.

In contrast to ascomycetes (e.g. *A. niger* and *T. reesei*), basidiomycetes secrete class I hydrophobins in liquid shaken cultures. Wild-type *S. commune* strains secrete up to 60 mg per liter of SC3 in their culture medium (Table 1)(Wösten et al. 1999). To improve production of SC3, extra copies of the gene were introduced in *S. commune*. However, introduction of more than one copy silenced both the introduced and the endogenous genes (Schuurs et al. 1997) resulting in the absence of SC3 in the culture medium. Silencing occurs at the transcriptional level by cytosin-methylation in the coding part of the gene (Schuurs et al. 1997). It was therefore not surprising that expression of *SC3* behind the *SC15* promoter also resulted in silencing in a wild-type background (Scholtmeijer, data not shown). Notably, SC3 production could be temporarily restored by inoculation of mycelial homogenates (Schuurs et al. 1997) but silencing re-occurred after a few days at the moment high levels of SC3 were produced. This indicates that silencing occurs upon (over)expression of the *SC3*-gene. In a dikaryon resulting from a cross of a wild-type and a silenced strain, the wild-type nucleus was not silenced. Therefore, the signal inducing silencing seems not to be diffusible. A model (Schuurs, 1998) involving the formation of DNA-RNA hybrids upon (over)expression of *SC3* was proposed. In this model the DNA-RNA hybrid serves as a template for methylation of the DNA. Identification of the underlying mechanism may result in strategies to overcome the silencing phenomenon. Alternatively, a mutagenesis approach can be taken to obtain overproducing strains.

Use of *S. commune* for the (over)production of hydrophobins is also hampered by the genetic instability of the strains used. The *thn* mutation occurs frequently, resulting in the elimination of expression of SC3 and other genes (Raper and Miles, 1958; Schwalb and Miles, 1967; Wessels et al. 1991b). The high frequency of this mutation was explained by the transposition of the *Scooter* transposable element in the *thn1*-gene, which encodes a putative regulator of a G-protein signalling pathway (Fowler and Mitton, 2000). Multiple copies of *Scooter* were identified in all strains analysed, which hampers elimination of this transposable element from the genomic DNA. We will examine production of SC3 in a *thn* background using the constitutive GPD promoter. Protease activity is severely reduced in this strain (L.G. Lugones, personal communication) which may overcome the observed degradation of hydrophobins like ABH1 and SC4 in the medium of wild-type *S. commune* (Lugones, 1998; van Wetter, 2000; R. Rink, unpublished).

Pycnoporus cinnabarinus seems to be an attractive protein production system. Fermentation of this basidiomycete is well established (Eggert et al. 1996; Falconnier et al. 1994; Oddou et al. 1999) High levels of the homologous lac1 laccase were produced when a laccase deficient strain of *P. cinnabarinus* was transformed with the *lac1* gene under control of its own promoter or that of GPD of *S. commune* (46 and 38 mg L^{-1}, respectively; Alves et al. 2004). Remarkably, production levels increased to 1 and 1.2 g L^{-1}, respectively, by adding 40 g L^{-1} of ethanol to the culture medium. Ethanol inhibits growth and can thus be considered a stress factor (Lomascolo et al. 2003). The promoters of *lac1* and GPD contrain stress elements, which would explain why mRNA levels increased dramatically (Alves et al. 2004). Currently, we are analyzing SC3 production in *P. cinnabarinus* using the *lac1* and GPD promoters (Table 1).

Table 1. Filamentous fungi examined as host for homologous and heterologous production of class I and class II hydrophobins.

Fungus	Gene / Fused to	Hydrophobin Class	Promoter / SS	Production [mg L^{-1}]
A. niger	SC3	I	GPD / SC3	<1
	SC3 / Gla (C)	I	Gla / Gla	<1
T. reesei	SC3	I	HFBI / SC3	60
	SC3	I	HFBII / SC3	60
	HFBI	II	HFBI / HFBI	600
S. commune	SC3	I	SC3 / SC3	60
P. cinnabarinus	SC3	I	Lac1 / SC3	nd
	SC3	I	GPD / SC3	nd
	SC3	I	SC3/ SC3	nd

(C) indicates that the homologous (fusion) gene is present at the C-terminus of the hydrophobin gene. Abbreviations: SS; Signal sequence for secretion, GPD; glyceraldehyde-3-phosphate dehydrogenase, Gla; glucoamylase, Lac1; Laccase 1. The hydrophobin genes SC3 and HFBI are from the filamentous fungi *S. commune* and *T. reesei*, respectively. nd: Not determined.

4.2. Production of Class I Hydrophobins in Yeast and Bacteria

Transformation procedures and culturing of filamentous fungi usually are time-consuming. To allow faster screening of hydrophobin variants, expression in yeasts and bacteria was examined. *ABH1* was expressed in the yeasts *S. cerevisae*, *Pichia methanolica* and *Pichia pastoris*. Production of the encoded hydrophobin was monitored by Western blot analysis of medium and intracellular fractions. Inducable and constitutive yeast promoters were used to express *ABH1* without signal sequence or with the α-factor signal sequence of *S. cerevisiae*. In none of the yeast strains a significant amount of ABH1 was produced (Table 2). Similarly, no production of the class I hydrophobin HydPt-1 of *Pisolithus tinctorius* could be detected upon expression in *S. cerevisiae* (Table 2)(Tagu et al. 2002) or when SC3 was expressed in *Hansenula polymorpha* (E. de Bruin and K. Scholtmeijer, unpublished). It should be noted that hydrophobin genes are absent in the genomes of yeasts. Therefore, it is reasonable to expect that foldases or chaperones dedicated for production of class I hydrophobins are absent as well.

Expression of hydrophobins was also examined in the bacteria *Bacillus megaterium*, *Lactococcus lactis*, *Caulobacter cresentus* and *Escherischia coli* (Table 2). Mature forms of TrSC3 (see above) and ABH1 were expressed in *B. megaterium* using the α-amylase signal sequence and the xylose inducable promoter of *xylA*. Both hydrophobins were produced at 1-2 mg L^{-1}. Circular dichroism indicated that ABH1, but not TrSC3, was correctly folded. By treating TrSC3 with bovine protein disulphide isomerase (PDI) a conformation was obtained that was similar to that of soluble state TrSC3 produced in *S. commune*. Both hydrophobins adsorbed to Teflon. However, they did not adopt the stable β-sheet state and did not resist extraction with hot SDS. Similar results were obtained when hydrophobins were expressed in *L. lactis* behind the *nisA* promoter. The different variants that were examined (Table 2) were produced at low level (< 1 mg L^{-1}) or were not produced at all. No β-sheet formation was observed for the produced variants. Expression of TrSC3 and ABH1 fused to the homologous RsaA protein and regulated by the lactose promoter in *C. cresentus* did not result in hydrophobin production at all (Table 2).

The highest production level of class I hydrophobins in a bacterium was obtained in *E. coli* (80 mg L^{-1}). A chimeric gene encoding a mature SC3 flanked by the bacteriophage fd signal sequence and a C-terminal His-tag and *Myc* epitope was placed under regulation of the arabinose inducible araBAD promoter. However, the protein was not secreted and the signal sequence remained attached. The fusion protein was able to adsorb to Teflon, but was removed by hot detergent. TrSC3 could not be produced using the same system. A fusion of TrSC3 to the C-terminus of dehaloalkane dehalogenase (*DhlA*) under regulation of the T7 promoter was constructed. Production of DhlA at 30 °C results in a soluble protein, while at 37 °C the protein accumulates in inclusion bodies (Schanstra et al. 1993). Expression of the fusion product at 30 °C resulted in the formation of inclusion bodies, while at 37 °C the cells were unable to grow. To increase solubility of the protein, TrSC3 was fused to the C-terminus of thioredoxin. Many proteins that are produced in an insoluble form in *E. coli* become more soluble when fused to this protein (Prinz et al. 1997; Bessette et al. 1999). In addition, thioredoxin catalyzes the formation of disulphide bonds in the cytoplasm of *E. coli* strains carrying the thioredoxin reductase (*trxB*) mutation. The *E. coli* strain origami (Novagen), containing this *trxB* mutation, was transformed with the thioredoxin-TrSC3 fusion construct. However, this approach did not result in expression of TrSC3 even though the origami strain contains an additional mutation in the glutathion reductase gene (*gor*), encoding another central enzyme involved in cytoplasmic disulphide formation. In addition, expression of the class I hydrophobin POH1 from *Pleurotus ostreatus* in *E. coli* only resulted in µg quantities per litre (Peñas et al. 1998).

The results show that class I hydrophobins can be produced in bacteria. However, these hydrophobins do not fold properly and can therefore not self-assemble. Co-expression of hydrophobins with the corresponding fungal foldases (PDI from the basidiomycete *S. commune* or the ascomycete *T. reesei*) may lead to functional hydrophobins in bacterial expression systems.

Table 2. Yeasts and bacteria examined as host for heterologous production of class I and class II hydrophobins.

Host	Gene / Fused to	Hydrophobin class	Promoter / SS	Production [mg l⁻¹]
S. cerevisiae	ABH1	I	GAL1	0
	ABH1	I	GAL1 / α	0
	ABH1 / AGA2 (C)	I	AGA2 / AGA2	0
	HydPtI	I	GPD / HydPtI	0
	HFBI / Flo1 (N)	II	GAL1 / FLO1	nd
P. methanolica	ABH1	I	AUG1	0
	ABH1	I	AUG1 / α	0
P. pastoris	ABH1	I	GAP / α-factor	<1
B. megaterium	TrSC3	I	xylA / α-amylase	1-2
	ABHI	I	XylA / α-amylase	1-2
L. lactis	SC3	I	Nis A	<1
	SC3	I	NisA / Nis A	0
	SC3 / PA (N)	I	Nis A	<1
	SC3 / PA (C)	I	Nis A	<1
	TrSC3 / PA (N)	I	Nis A	<1
	TrSC3 / PA (C)	I	Nis A	<1
C. cresentus	TrSC3 / RsaA (C)	I	Lac / RsaA	0
	ABH1 / RsaA (C)	I	Lac / RsaA	0
E. coli	SC3 / HisMyc (C)	I	araBAD / fd	80
	TrSC3 / HisMyc (C)	I	araBAD / fd	0
	TrSC3 / DhlA (N)	I	T7	nd
	TrSC3 / Thio (N)	I	T7	0
	POH1 / His (N)	I	T7	<1
	CU / GST (N)	II	Lac	<1

(C) or (N) indicates that the homologous (fusion) gene is present at the C- or N-terminus of the hydrophobin gene, respectively. Abbreviations: SS; Signal sequence for secretion, GAL1; Galactose 1, α; α-factor (pheromone), AGA2; a-agglutinin, GPD; glyceraldehyde-3-phosphate dehydrogenase, Flo1; Flocculin 1, AUG1; Alcohol oxidase 1, GAP; glyceraldehyde-3-phosphate dehydrogenase, xylA; Xylose A, Nis A; Nisin A, PA; *L. lactis* protein anchor, RsaA; S-layer protein *C. cresentus*, Lac; Lactose, His; His-tag for purification, Myc; epitope for antibody detection, araBAD; Arabinose BAD, fd; Bacteriphage fd, DhlA; Dehaloalkane dehalogenase, T7; T7 DNA polymerase *E. coli*, Thio; Thioredoxin, GST; Glutathion-S-transferase. ABH1, HydPt-1, HFBI, (Tr)SC3, POH1 and CU are hydrophobin genes, or derivatives thereof, from the filamentous fungi *A. bisporus*, *P. tinctorius*, *T. reesei*, *S. commune*, *P. ostreatus* and *O. Ulmin*, respectively. nd: Not determined.

4.3. Production of Class II Hydrophobins

Production of class II hydrophobins in filamentous fungi and yeasts seems to be less difficult compared to class I hydrophobins. Homologous expression of the class II hydrophobin gene *HFBI* under control of its own regulatory sequences in *T. reesei* resulted in a production level of 600 mg L⁻¹ (Table 1)(Askolin et al. 2001). Most of the hydrophobin was cell-bound (80%) but could be purified via extraction of the mycelium with SDS followed by KCl precipitation and hydrophobic interaction chromatography. The yield of the purified HFBI was 120 mg L⁻¹. HFBI was also produced in *S. cerevisiae* but no data are available on the amount of hydrophobin produced in this system (Nakari-Setälä et al. 2002). Expression of the class II hydrophobin CU in *E. coli* only resulted in µg L⁻¹ quantities of the protein (Table 2)(Bolyard and Sticklen, 1992).

5. CONCLUSIONS

High production of functional class I hydrophobins has been shown to be difficult. Functional class I hydrophobins could only be produced in filamentous fungi that naturally secrete hydrophobins into their culture medium, such as *S. commune* and *T. reesei*. These filamentous fungi may express specific chaperones or foldases in their submerged mycelium that aid in folding and disulphide bridge formation. The absence of these proteins in the submerged mycelium of *A. niger* and in yeasts and bacteria would result in misfolding of hydrophobins either or not followed by proteolytic degradation. Future studies should identify chaperones and foldases that are involved in folding of hydrophobins. These proteins could be (over)expressed in submerged mycelia of fungi or in yeasts and/or bacteria. Homologous production of the class II hydrophobin HFBI was shown to be very efficient. Whether class II hydrophobins can be efficiently produced heterologously is not yet known, but *T. reesei* seems to be a promising host.

Acknowledgements: We would like to thank Esther de Boef, Rolf Kanninga and Barbara Lussenburg for the construction of some of the expression clones and performing the subsequent expression experiments.

REFERENCES

Alves AMCR, Record E, Lomascolo A, Scholtmeijer K, Asther M, Wessels JGH and Wösten HAB (2004). Highly efficient production of laccase in the basidiomycete Pycnoporus cinnabarinus. Appl and Environ Microbiol. In press.

Askolin S, Nakari-Setälä T and Tenkanen M (2001). Overproduction, purification, and characterization of the Trichoderma reesei hydrophobin HFBI. Appl Microbiol Biotechnol 57:124-130.

Askolin S, Linder M, Scholtmeijer K, Tenkanen M, Penttilä M, de Vocht ML and Wösten HAB (2004). The class II hydrophobins HFBI and HFBII of Trichoderma reesei have distinct assembly characteristics at hydrophilic-hydrophobic interfaces. Submitted.

Berggren K, Veide A, Nygren P-Å and Tjerneld F (1999). Genetic engineering of protein-peptide fusions for control of protein partitioning in thermoseparating aqueous two-phase systems. Biotechnol Bioeng 62:135-144.

Bessette PH, Åslund F, Beckwith J and Georgiou G (1999). Efficient folding of proteins with multiple disulphide bonds in the Escherichia coli cytoplasm. PNAS 96:13703-13708.

Bilewicz R, Witomski J, van der Heyden A, Tagu D, Palin B and Rogalska E (2001). Modification of electrodes with self-assembled hydrophobin layers. J Phys Chem B 105:9772-9777.

Bolyard MG and Sticklen MB (1992). Expression of a modified dutch elm disease toxin in Escherichia coli. Mol Plant-Microbe Interact 5:520-524.

Butko P, Buford JP, Goodwin JS, Stroud PA, McCormick CL and Cannon GC (2001). Spectroscopic evidence for amyloid-like interfacial assembly of hydrophobin SC3. Biochem Biophys Res Commun 280:212-215.

Carpenter CE, Mueller RJ, Kazmierczak P, Zhang L, Villalon DK and van Alfen NK (1992). Effect of a virus on accumulation of a tissue-specific cell-surface protein of the fungus Cryphonectria (Endothia) parasitica. Mol Plant-Microbe Interact 4:55-61.

Collén A, Ward M, Tjerneld F and Stålbrand (2001a). Genetically engineered peptide fusions for improved protein partitioning in aqueous two-phase systems: Effect of fusion localisation on endoglucanase I (Cel7B) of Trichoderma reesei. J Chromatogr A 910:275-284.

Collén A, Ward M, Tjerneld F and Stålbrand (2001b). Genetic engineering of the Trichoderma reesei endoglucanase I (Cel7B) for enhanced partitioning in aqueous two-phase systems containing thermoseparating ethylene oxide-propylene oxide copolymers. J Biotechnol 87:179-191.

Collén A, Penttilä M, Tjerneld F and Veide A (2002a). Characterizing partitioning of endoglucanase I (Cel7B) fusion proteins in PEG-phosphate aqueous two-phase systems. J Chromatogr A 943:55-62.

Collén A, Selber K, Hyytia T, Persson J, Nakari-Setla T, Bailey M, Fagerstrom R, Kula MR, Penttila M, Stalbrand H, Tjerneld F (2002b). Primary recovery of a genetically engineered Trichoderma reesei endoglucanase I (Cel 7B) fusion protein in cloud point extraction systems. Biotechnol Bioeng 78:385-394.

Desai JD and Banat IM (1997). Microbial production of surfactants and their commercial potential. Microbiol Mol Biol Rev 61:47-64.

De Vocht ML, Scholtmeijer K, van der Vegte EW, de Vries OMH, Sonveaux N, Wösten HAB, Ruysschaert J- M, Hadziioannou G, Wessels JGH and Robillard GT (1998). Structural characterization of the hydrophobin SC3, as a monomer and after self-assembly at hydrophobic/hydrophilic interfaces. Biophys J 74:1-10.

De Vocht ML, Reviakine IR, Ulrich W-P, Bergsma-Schutter W, Wösten HAB, Vogel H, Brisson A, Wessels JGH and Robillard GT (2002). Self-assembly of the hydrophobin SC3 proceeds via two structural intermediates. Protein Sci 11:1199-1205.

De Vries OMH, Fekkes MP, Wösten HAB. and Wessels JGH (1993). Insoluble hydrophobin complexes in the walls of Schizophyllum commune and other filamentous fungi. Arch Microbiol 159:330-335.

Durand H, Clanet M and Tiraby G (1998). Genetic improvement of Trichoderma reesei for large scale cellulase production. Enzyme Microb Technol 10:341-345.

Eggert C, Temp U and Eriksson K-EL (1996). The ligninolytic system of the white rot fungus Pycnoporus cinnabarinus: Purification and characterization of the laccase. Appl Microbiol Biotechnol 62:1151-1158.

Falconnier B, Lapierre C, Lesage-Meesen L, Yonnet G, Brunerie P, Colonna BC, Corrieu G and Asther M (1994). Vannilin as a product of ferulic acid biotransformation by the white rot fungus Pycnoporus cinnabarinus I-937. J Biotechnol 37:123-132.

Finkelstein DB, Rambosek J, Crawford MS, Soliday CL, McAda PC and Leach J (1989). Protein secretion in Aspergillus niger. In: CL Hershberger, SW Queener and G Hegeman eds. Genetics and Molecular Biology of Industrial Microorganisms. Washington DC, American Society for Microbiology, pp 295-300.

Fowler TJ and Mitton MF (2000). Scooter, a new active transposon in Schizophyllum commune, has disrupted two genes regulating signal transduction. Genetics 156:1585-1594.

Giacomelli CE and Norde W (2003). influence of hydrophobic Teflon particles on the structure of amyloid β- peptide. Biomacromol 4:1719-1726.

Gouka RJ, Punt PJ, Hessing JGM and van den Hondel CAMJJ (1996). Analysis of heterologous protein production in defined recombinant Aspergillus awamori strains. Appl Environ Microbiol 62:1951-1957.

Hakanpää J, Paananen A, Askolin S, Nakari-Setälä T, Parkkinen T, Penttilä M, Linder MB and Rouvinen J (2004). Atomic resolution structure of the HFBII hydrophobin, a self-assembling amphiphile. J Biol Chem 279:534-539.

Hawksworth DL (1991). The fungal dimension of biodiversity: magnitude, significance and conservation. Mycol Res 95: 641-655.

Janssen MI (2004). Modification of (bio)material surfaces using hydrophobins. PhD dissertation, University of Groningen, Groningen, The Netherlands.

Janssen MI, van Leeuwen MBM, Scholtmeijer K, van Kooten TG, Dijkhuizen L and Wösten HAB (2002). Coating with natural and genetic engineered hydrophobins promotes growth of fibroblasts on a hydrophobic solid. Biomaterials 2:4847-4854.

Köhler K, Ljungqvist C, Kondo A, Veide A and Nilsson B (1991). Engineering proteins to enhance their partitioning coefficients in aqueous two-phase systems. Bio/Technology 9:642-646.

Linder M, Szilvay GR, Nakari-Setala T, Soderlund H, Penttila M (2002). Surface adhesion of fusion proteins containing the hydrophobins HFBI and HFBII from Trichoderma reesei. Protein Sci 11:2257-66.

Lomascolo A, Record E, Herpoël-Gimbert I, Delattre M, Robert JL, Georis J, Dauvrin T, Sigoillot JC and Asther M (2003). Overproduction of laccase by a monokaryotic strain of Pycnoporus cinnabarinus using ethanol as an inducer. J Appl Microbiol 94:618-624.

Lugones LG (1998). Function and expression of hydrophobins in the basidiomycetes Schizophyllum commune and Agaricus bisporus. PhD Thesis, University of Groningen, Groningen, The Netherlands.

Lugones LG, Bosscher JS, Scholtmeijer K, de Vries OMH and Wessels JGH (1996). An abundant hydrophobin (ABH1) forms hydrophobic rodlet layers in Agaricus bisporus fruiting bodies. Microbiol 142:1321-1329.

Lugones LG, Wösten HAB and Wessels JGH (1998). A hydrophobin (ABH3) specifically secreted by growing hyphae of Agaricus bisporus (common white button mushroom). Microbiol 144:2345-2353.

Lugones LG, Wösten HAB, Birkenkamp KU, Sjollema KA, Zagers J and Wessels JGH (1999). Hydrophobins line air channels in fruiting bodies of Schizophyllum commune and Agaricus bisporus. Mycol Res 103:635- 640.

Mackay JP, Matthews JM, Winefield RD, Mackay LG, Haverkamp RG and Templeton MD (2001). The hydrophobin EAS is largely unstructured in solution and functions by forming amyloid-like structures. Structure 9:83-91.

Nakari-Setälä T, Aro N, Kalkkinen N, Alatalo E and Penttilä M (1996). Genetic and biochemical characterization of the Trichoderma reesei hydrophobin HFBI. Eur J Biochem 235:248-255.

Nakari-Setälä T, Aro N, Ilmén M, Muñoz G, Kalkkinen N and Penttilä M (1997). Differential expression of the vegetative and sporebound hydrophobins of Trichoderma reesei. Eur J Biochem 248:415-423.

Nakari-Setälä T, Azeredo J, Henriques M, Oliveira R, Texteira J, Linder M and Penttilä M (2002). Expression of a fungal hydrophobin in the Saccharomyces cerevisiae cell wall: Effect on cell surface properties and immobilization. Appl Environ Microbiol 68:3385-3391.

Oddou JC, Stentelaire C, Lesage-Meesen L, Asther M and Ceccaldi BC (1999). Improvement of ferrulic acid bioconversion into vannilin by use of high-density cultures of Pycnoporus cinnabarinus. Appl Microbiol _ Biotechnol 1:1-16.

Palomo JM, Peñas MM, Fernández-Lorente LG, Mateo C, Pisabarro AG, Fernández-Lafuente LR, Ramírez L _ and Guisán JM (2003). Solid-phase handling of hydrophobins: Immobilized hydrophobins as a new tool to study lipases. Biomacromolecules 4:204-210.

Panaanen A, Vuorimaa E, Torkelli M, Penttilä M, Kauranen M, Ikkala O, Lemmetyinen H, Serimaa R and Linder MB (2003). Structural hierarchy in molecular films of two class II hydrophobins. Biochemistry 42:5253-5258.

Peñas MM, Ásgeirsdóttir SA, Lasa I, Culianez-Marcia FA, Pisabarro AG, Wessels JGH and Ramirez, L (1998). Identification, characterization and in situ detection of a fruit-body-specific hydrophobin of Pleurotus ostreastus. Appl Environ Microbiol 64:4028-4034.

Prinz WA, Åslund F, Holmgren A and Beckwith J (1997). The role of the thioredoxin and glutaredoxin pathways in reducing protein disulphide bonds in the Escherichia coli cytoplasm. J Biol Chem 272:15661- 15667.

Raper JR and Miles PG (1958). The genetics of Schizophyllum commune. Genetics 43:530-546.

Rodenbrock A, Selber K, Egmond M and Kula M-R (2001). Cutinase partitioning in detergent-based aqueous two-phase systems. Bioseparation 9:269-276.

Russo PS, Blum FD, Ipsen JD, Miller WG and Abul-Hajj YJ (1982). The surface activity of the phytotoxin cerato-ulmin. Can J Bot 60:1414-1422.

Schanstra JP, Rink R, Pries F and Janssen DB (1993). Construction of an expression and site-directed mutagenesis system of haloalkane dehalogenase in Escherichia coli. Protein Expression Purif 4:479-489.

Scholtmeijer K (2000). Expression and engineering of hydrophobin genes. PhD Thesis, University of Groningen, Groningen, The Netherlands.

Scholtmeijer K, Janssen MI, Gerssen B, de Vocht ML, van Leeuwen MBM, van Kooten TG, Wösten HAB and Wessels JGH (2002). Modification of surfaces using engineered hydrophobins. Appl Environ Microbiol 68:1367-1373.

Schuurs TA, Schaeffer EAM and Wessels JGH (1997). Homology-dependent silencing of the SC3 gene in Schizophyllum commune. Genetics 147:589-596.

Schuurs TA (1998). Regulation of hydrophobin genes in Schizophyllum commune. PhD Thesis, University of Groningen, Groningen, The Netherlands.

Schwalb MN and Miles PG (1967). Morphogenesis of Schizophyllum commune. I. Morphological variation and mating behaviour of the thin mutation. Am J Bot 54: 440-446.

Tagu D, Marmeisse R, Baillet Y, Rivière S, Palin B, Bernandini F, Méreau A, Gay G, Balestrini R, Bonfante P and Martin F (2002). Hydrophobins in ectomycorrhizas: heterologous transcription of the Pisolitus HydPt-1 gene in yeast and Hebeloma cylindrosporum. Eur J Histochem 46:23-29.

Talbot NJ, Kershaw M, Wakley GE, de Vries OMH, Wessels JGH and Hamer JE (1996). MPG1encodes a fungal hydrophobin involved in surface interactions during infection related development of the rice blast fungus Magnaporthe grisea. Plant Cell 8:985-999.

Torkelli M, Serimaa R, Ikkala O and Linder M (2002). Aggregation and self-assembly of hydrophobins from Trichoderma reesei: Low-Resolution Structural models. Biophys J 83:2240-2247.

Van der Vegt W, van der Mei HC, Wösten HAB, Wessels JGH and Busscher HJ (1996). A comparison of the surface activity of the fungal hydrophobin SC3p with those of other proteins. Biophys Chem 57:253-260.

Van Wetter M-A (2000). Functions of hydrophobins in Schizophyllum commune. PhD Thesis, University of Groningen, Groningen, The Netherlands.

Van Wetter M-A, Wösten HAB and Wessels JGH (2000). SC3 and SC4 hydrophobins have distinct roles in formation of aerial structures in dikaryons of Schizophyllum commune. Mol Microbiol 36:201-210.

Wang X, de Vocht ML, de Jonge J, Poolman B and Robillard GT (2002). Structural changes and molecular interactions of hydrophobin SC3 in solution and on a hydrophobic surface. Protein Science 11:1172-1181.

Wang X, Graveland-Bikker JF, de Kruif CG and Robillard GT (2004). Oligomerization of hydrophobin SC3 in solution: From soluble state to self-assembly. Protein Science 13:810-821.

Wang X (2004). Insight into the interfacial self-assembly and structural changes of hydrophobins. PhD Thesis, University of Groningen, Groningen, The Netherlands.

Wessels JGH, de Vries OMH, Ásgeirsdóttir SA and Schuren FHJ (1991). Hydrophobin genes involved in formation of aerial hyphae and fruit bodies in Schizophyllum commune. Plant Cell 3:793-799.

Wessels JGH (1994). Developmental regulation of fungal cell wall formation. Ann Rev Phytopathol 32:413- 437.

Wessels JGH (1997). Hydrophobins: proteins that change the nature of a fungal surface. Adv Microb Phys 38:1- 45.

Wösten HAB (2001). Hydrophobins: Multipurpose proteins. Annu Rev Microbiol 55:625-646.

Wösten HAB, de Vries OMH and Wessels JGH (1993). Interfacial self-assembly of a fungal hydrophobin into a hydrophobic rodlet layer. Plant Cell 5:1567-1574.

Wösten HAB, Schuren FHJ and Wessels JGH (1994a). Interfacial self-assembly of a hydrophobin into an amphipathic membrane mediates fungal attachment to hydrophobic surfaces. EMBO J 13:5848-5854.

Wösten HAB, Ásgeirsdóttir SA, Krook JH, Drenth JHH and Wessels JGH (1994b). The SC3p hydrophobin self- assembles at the surface of aerial hyphae as a protein membrane constituting the hydrophobic rodlet layer. Eur J Cell Biol 63:122-129.

Wösten HAB, Ruardy TG, van der Mei HC, Busscher HJ and Wessels JGH (1995). Interfacial self-assembly of a Schizophyllum commune hydrophobin into an insoluble amphipathic protein membrane depends on surface hydrophobicity. Coll Surf B: Biointer 5:189-195.

Wösten HAB and Wessels JGH (1997). Hydrophobins, from molecular structure to multiple functions in fungal development. Mycoscience 38:363-374.

Wösten HAB, van Wetter M-A, Lugones LG, van der Mei HC, Busscher HJ and Wessels JGH (1999). How a fungus escapes the water to grow into the air. Curr Biol 9:85-88.

Wösten HAB and de Vocht, ML (2000). Hydrophobins, the fungal coat unraveled. Biochim Biophys Acta 85524:1-8.

Yaguchi M, Pusztai-Carey M, Roy C, Surewicz WK, Carey PR, Stevenson KJ, Richards WC and Takai S (1993). Amino-acid sequence and spectroscopic studies of Dutch elm disease toxin cerato-ulmin. In: MB Sticklen and J Sherald eds. Dutch elm disease research, cellular and molecular approaches. New York: Springer Verlag, pp 152-170.

Yang S, Bergman T, Veide A and Enfors S-O (1994). Effects of amino acid insertions on the proteolysis of a staphylococcal protein A derivative in Escherichia coli. Eur J Biochem 226:847-852.

Zangi R, de Vocht ML, Robillard GT and Mark AE (2002). Molecular dynamics study of the folding of hydrophobin SC3 at a hydrophilic/hydrophobic interface. Biophys J 83:112-124.

**Applied Mycology
and Biotechnology**
An International Series
Volume 5. Genes and Genomics
© 2005 Elsevier B.V. All rights reserved

11

ELSEVIER

Genetic Regulation of Carotenoid Biosynthesis in Fungi

Johannes Wöstemeyer, Anke Grünler, Christine Schimek and Kerstin Voigt
Institute for Microbiology, Chair of General Microbiology and Microbial Genetics and Fungal Reference Centre, Friedrich-Schiller-University Jena, Neugasse 24, D-07743 Jena, Germany (b5wojo@rz.uni-uni-jena-de).

Fungal carotenoids are synthesized by the isoprenoid pathway with isopentenyl pyrophosphate as the general precursor. They are found in all divisions of the fungal realm, and several are at the edge of being exploited at an industrial scale for satisfying an increasing demand for carotenoid pigments, food and feed additives, and components of cosmetics and pharmaceuticals. Fungi as carotenoid source are highly appealing. At least prospectively, they should be easier amenable to genetic manipulation than plants, and thus will allow tailoring of specially designed substances. Genes for carotenoid synthesis were cloned from many different fungi. In order to stimulate further functional studies on genetic pathways for internal and environmental regulation of carotene synthesis, modification and degradation, an overview on the situation in the most thoroughly studied model organisms is presented. The role of carotenoids as antioxidants, light protective substances and as signalling compounds is discussed.

1. INTRODUCTION

Carotenoids are ubiquitously occurring coloured C_{40} isoprenoids, formed by archaebacteria, bacteria, algae, plants, and fungi. C_{40} hydrocarbons with a conjugated double bond polyene structure are usually termed carotenes; their oxidised derivatives containing hydroxyl, keto or carboxyl functions are traditionally termed xanthophylls. Metazoa are not able to synthesize carotenoids, but need them and their derivatives, the retinoids, as vitamins or cellular regulators and thus depend on carotenoid uptake by nutrition. As carotenoids are frequently found already in the prokaryotic world, the metazoan branch in the tree of life seems to be the only one to have lost this ability. Many fungi from all major phylogenetic groups produce carotenoids, especially β-carotene, sometimes in enormous amounts. A phylogenetic tree of the Fifth Kingdom, Eumycota, is presented in Fig. 1; important carotene producers are indicated. Technically exploited is *Blakeslea trispora* (Zygomycota, Mucorales, Choanephoraceae) and

---

Corresponding author: Johannes Wöstemeyer

258

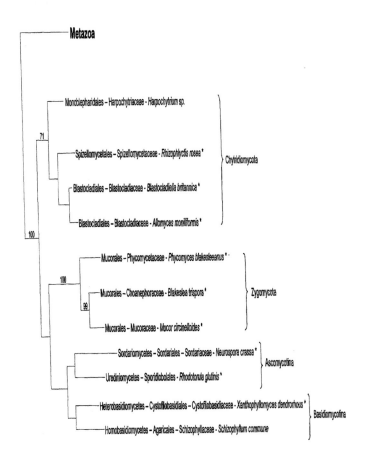

Fig. 1. Phylogenetic tree of Eumycota (neighbour joining algorithm), based on 1033 amino acids of the protein coding regions of actin, β-tubulin, and elongation factor EF1α (Voigt and Wöstemeyer 2000, 2001; Einax and Voigt 2003). The tree was rooted to *Bombyx mori* and *Caenorhabditis elegans* as typical Metazoa, the immediate monophyletic fungal sister group. Numbers above branches indicate bootstrap values above 70%. Carotene producers occur in all four fungal divisions; asterisks mark typical examples.

possibly on the edge of industrial usage are *Phycomyces blakesleeanus* (Zygomycota,Mucorales, Phycomycetaceae) and a basidiomycetous yeast, *Xanthophyllomyces dendrorhous* (Basidiomycotina, Cystofilobasidiales; imperfect form: *Phaffia rhodozyma*). Others are of considerable importance for understanding biochemical and especially regulatory pathways of carotenoid formation. Among those are predominantly *Neurospora crassa* (Ascomycotina, Sordariales, Sordariaceae) and *Mucor circinelloides* (Zygomycota, Mucorales, Mucoraceae).

Carotenoids in prokaryotes are always bicyclic, usually with β-ionone rings at both ends of the molecule; however, fungi may produce acyclic compounds or the monocyclic β-carotene instead of or in addition to the bicyclic β-carotene. A choice of frequently occurring structures are given in Fig. 2.

The phylogenetically basal fungal group Chytridiomycota produces β-carotene and sometimes β-carotene. Chytridiales seem to prefer β-carotene; Blastocladiales with their prominent genera *Allomyces* and *Blastocladiella* have a tendency towards formation of β-carotene (Cantino and Hyatt 1953). Zygomycota have a strong preference for β-carotene, but apparently unique in the fungal world, give rise to a wealth of different carotenoid induction and maintenance of sexual differentiation towards the formation of zygospores. In the largest fungal division, Ascomycotina, additional carotene forms are frequently found. *Neurospora crassa* forms the monocyclic Neurosporaxanthin. Acyclic carotenes are also found; an example is the oxidised carotene derivative phillipsiaxanthin in *Phillipsia carminea*. Among the basidiomycetes, especially *Xanthophyllomyces dendrorhous* is highly interesting, because it accumulates astaxanthin (3,3′-dihydroxy-β,β-carotene-4,4′-dione) in considerable amounts. A comprehensive survey on natural carotenoid diversity is compiled in Sandmann and Misawa (2002) and in Goodwin (1980).

2. TECHNICAL IMPORTANCE

Presently, only a single fungus, the zygomycete *Blakeslea trispora* is fermented at large scale for producing β-carotene. Carotenes, especially astaxanthin and β-carotene, are predominantly produced by chemical synthesis. However, the requirements of the market are changing, and especially in the European community there is a strong tendency towards natural products from biological sources. Thus, considerable efforts are made towards increasing carotene yields from fungi by fermentation. β-carotene or provitamin A is used as food additive for improving the colour of fat-containing food and as a source of vitamin A or retinol. Astaxanthin is added in huge and still increasing amounts as colorant to animal feed; the most prominent field of application is certainly its addition to the diet of aquaculture animals.

At the level of medical prophylaxis, carotenoids are discussed for preventing cardio-vascular disease and cancer, the latter possibly by apoptosis induction in tumour cells (Vershinin 1999). The predominant importance of carotenoids is probably their role as antioxidants in protection against singlet oxygen ($^1O_2$) and other reactive oxygen species (ROS). Experimental results on biological action of reactive oxygen derivatives *in vivo* are sometimes difficult to interpret and not always unequivocal. Under some conditions, reactions between ROS and carotenoids may even lead to products with prooxidant properties (Young and Lowe 2001; Echavarri-Erasun and Johnson 2002). Therefore, employment of carotenoids in food additives, cosmetics and pharmaceuticals need careful analysis and possibly control of subsequent chemical reactions.

Important but difficult projects in fungal genetic engineering concern channelling isoprenoid units from ergosterol and other terpenoid biochemical pathways into carotenoid synthesis and switching production profiles from β-carotene to other technically attractive carotenoids like astaxanthin. In the long run, the increase in knowledge on fungal genomic sequences will open the way towards manipulation of carotenoid channelling in the cell derivatives that are formed by degradation of β-carotene to trisporic acids. These compounds occur in many different isoforms and are highly important as pheromones by mediating

3. BIOLOGICAL ROLE

Carotenoids are believed to form the chemical principle of protection against light and UV irradiation. These effects can easily be measured by comparing growth of carotene producers with strains that have lost this ability by mutations. Experiments with the smut fungus *Ustilago violacea* support the light-protective role of carotene (Will et al. 1984). *Neurospora crassa* seems to behave similarly (Blanc et al. 1976) and in the basidiomycetous yeast *Xanthophyllomyces dendrorhous*, carotenoid production is positively regulated by ROS (Schroeder and Johnson 1995).

Experiments with the carotene producing zygomycete *Phycomyces blakesleeanus* lead to contradictory results. Comparison of survival rates between mutants producing different amounts of phytoene, lycopene, and β-carotene did not support the idea that these carotenoids provide protection against UV-B and UV-C irradiation (Cerdá-Olmedo et al. 1996). Carotenes are also believed to protect from ionising radiation by scavenging free radicals and other highly reactive products. This protective role seems to be supported for mammalia supplemented with carotene (Endinkeau and Woodward 1993) but was not substantiated for fungi. This was tested experimentally in *Phycomyces blakesleeanus* with mutants accumulating different amounts and sorts of carotenoids. Again, after exposing the cells to γ-irradiation from a ^{60}Co-source, no effect of carotenoids on survival was verified (Martín-Rojas et al. 1996). Until now, there is no generally accepted model for protective carotene action in the fungal producers. Surprisingly, genetic manipulating *Escherichia coli* for producing neurosporene, β-carotene or zeaxanthin improves resistance against UV-B light (Sandmann and Misawa 2002).

The role of carotenoid derivatives as mediators of sexual communication in the phylogenetically basal fungal division Zygomycota is clear. Mating type recognition and induction of the primary mating reactions at least in Mucorales and possibly in all orders of Zygomycota depends on trisporic acid, a degradation product of β-carotene, or on its immediate precursors (Schimek et al. 2003). Trisporic acid is formed by co-operative action of the complementary mating types. Individual mating types form precursors that are converted to trisporic acid by the sexual partner (Fig. 3). The consequence of this elaborate chemical communication pathway is induction of sexual differentiation in both mating types. Trisporic acid or very closely related derivatives are also responsible for recognition between the mucoralean mycoparasite *Parasitella parasitica* and its many different hosts (Wöstemeyer et al. 2002).

An important derivative of β-carotene, synthesized by oxidative polymerisation, is the crosslinked carbohydrate sporopollenin. Sporopollenin is a major compound of the outer wall layers in mucoralean zygospores, rendering them highly hydrophobic (empirical formula, arbitrarily normalised to C_{90} units: $C_{90}H_{130}O_{39}$). It is synthesized by oxidative polymerisation of β-carotene (Gooday 1973). Sporopollenin is also a major structural component in ascospore walls of the *Neurospora* species *crassa* and *tetrasperma* (Gooday et al. 1974). In these organisms, elementary analysis led to the empirical formula $C_{90}H_{128}O_{63}$. Sporopollenin is concentrated in perisporal ribs at the ascospore surface. It seems to be formed bound to particles, the rib-forming bodies that are found underneath the ribs near the inner spore wall.

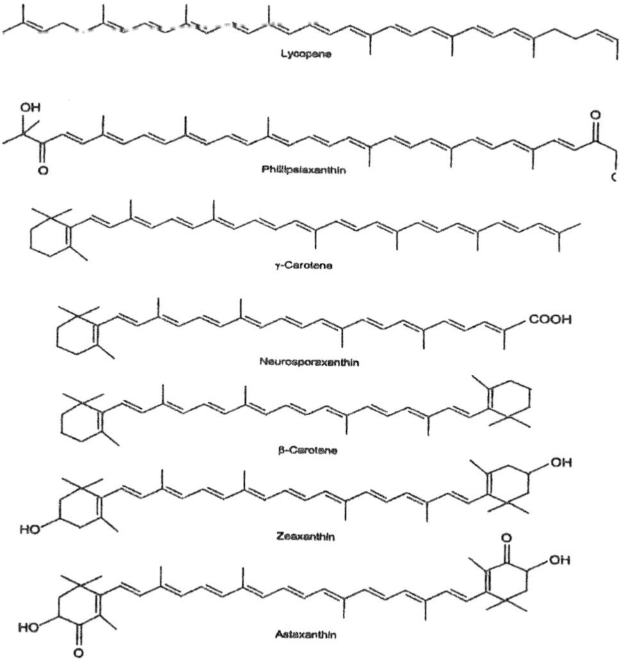

Fig. 2. Frequently occurring fungal carotenoids. Phillipsiaxanthin is derived from lycopene by oxidation in *Phillipsia carminea*. γ-carotene, the product of the first lycopene cyclase reaction, is partly degraded and oxidised to Neurosporaxanthin in *Neurospora crassa*, and alternatively converted to β-carotene by the second cyclase reaction. In *Xanthophyllomyces dendrorhous*, β-carotene is converted to astaxanthin by oxidizing the β-ionone rings via zeaxanthin as one of the major intermediates obtained by two successive hydroxylase reactions.

4. THE BIOCHEMICAL PATHWAY

Labelling experiments with ^{13}C-glucose followed by NMR analysis of mevalonate precursors in *Aschersonia aleyrodis* and *Rhodotorula glutinis* (Disch and Rohmer 1998) as well as in *Botrytis cinerea* and *Cercospora pini-densiflorae* (Hirai et al. 2000) documented that synthesis of isoprenoids in fungi depends exclusively on the mevalonate pathway. The alternative biosynthetic route from glyceraldehyde 3-phosphate via the primary C_5 intermediate, 1-deoxy-*threo*-pentulose 5-phosphate, was first detected in Bacteria (Rohmer et al. 1993) and later found to be typical for photosynthetic Eucarya. The fungal pathway from acetyl-CoA to the energetically activated biosynthetic unit of all terpenoids, isopentenyl- pyrophosphate, is depicted in Figure 4a. In principle, three acetyl groups are used to form mevalonate, which, after being phosphorylated by two subsequent kinase reactions, is decarboxylated to isopentenyl pyrophosphate. A series of condensation reactions leads to the first C_{40} carotenoid, phytoene (Fig. 4b). β-carotene is synthesized from phytoene via four successive desaturase reactions, leading to lycopene with 11 conjugated double bonds, and two subsequent cyclase reactions that form first the monocyclic β-carotene followed by β-carotene (Fig. 4c).

5. GENETIC REGULATION OF CAROTENOID BIOSYNTHESIS
5.1 *Mucor circinelloides* and other zygomycetes

While the most prominent mould for carotene fermentation is the zygomycete *Blakeslea trispora* (Mucorales, Choanephoraceae). the genetic analysis of this system and of the organism is severely limited. The fungus grows readily in culture and sporulates deliberately, mitotically by forming sporangiospores and sexually by developing zygospores after mating between (+) and (-) types. It is not possible to germinate the zygospores in the laboratory, and artificial methods for obtaining genetic recombinants are not available. However, heterokarya between mating types can be obtained by co-culturing complementary strains on potatodextrose agar followed by subculturing mycelial fragments from the confrontation line (Mehta et al. 2003) a technique that was initially developed for *Phycomyces blakesleeanus* (Gauger et al. 1980). Intersexual heterokaryons produce higher amounts of β-carotene than mated cultures, but have a tendency towards instability. In *Blakeslea trispora*, it is not   yet possible to produce stable diploids or aneuploids after parasexual fusion events, a technique that has been established for the mucoralean non-producer *Absidia glauca* (Wöstemeyer and Brockhausen-Rohdemann 1987; Wöstemeyer et al. 1990).

Quantitative analyses of *Blakeslea trispora* mutants with defects in the regulation of carotene biosynthesis have allowed the conclusion that the biosynthetic pathway of carotene formation is identical or very similar to the situation in *Phycomyces blakesleeanus*. The regulation of the process is however different. Whereas in *Phycomyces blakesleeanus* carotene synthesis is down-regulated by β-carotene itself, end product regulation does not play a major role in *Blakeslea trispora* (Mehta and Cerdá-Olmedo 1995). The genetic regulation of carotenoid synthesis has been studied in more detail in the genus *Mucor*, especially in *Mucor circinelloides* (Mucorales, Mucoraceae) and in *Phycomyces blakesleeanus* (Mucorales, Phycomycetaceae).

Geranylgeranyl pyrophosphate (GGPP) is the general precursor of all carotenoids. In *Mucor circinelloides*, the corresponding synthase is encoded by the *carG* gene (Velayos et al. 2003). It catalyses the successive head-to-tail condensations of the C_5 monomers, isopentenyl pyrophosphate and dimethylallyl pyrophosphate, via geranyl pyrophosphate (C_{10}) and farnesyl pyrophosphate (C_{15}) to the C_{20} compound geranylgeranyl pyrophosphate (Fig. 4b). The gene was functionally identified by its ability to complement mutations in the corresponding *crtE* gene from *Erwinia uredovora* in a recombinant *Escherichia coli* background. The gene product has a length of 303 amino acids and is thus the shortest of all known fungal GGPP synthases. In *Mucor circinelloides*, carotenogenesis is stimulated by blue light. This effect is reflected in the transcriptional regulation of *carG*. Following a   blue light pulse, the carG mRNA level increased during the first 10 minutes and, after a phase

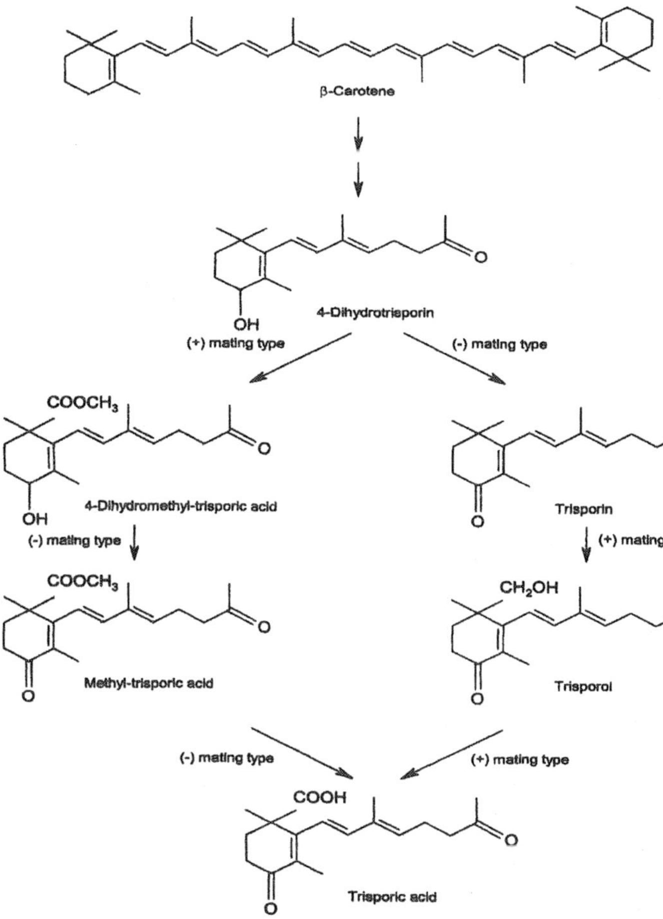

Fig. 3. The complementary mating types co-operatively perform the biosynthesis of trisporic acid, the general sexual pheromone of zygomycetes. Neither of the sexual partners is able to synthesize trisporic acid on its own. The degradation of β-carotene to the common precursor of both mating types, 4-dihydrotrisporin, is still hypothetical and may proceed via a C18 ketone. 4-Dihydrotrisporin is converted to trisporic acid by a series of mating type-specific reactions. The intermediates are exchanged between mating types in crossed cultures. Trisporic acid switches on sexual differentiation and induces carotenogenesis. The biosynthetic pathway follows the scheme by van den Ende (1978). The gene for 4-dihydromethyl-trisporic acid dehydrogenase has been cloned from the (-) mating type of *Mucor mucedo* (Czempinski et al. 1996). its regulation has been studied at the transcriptional and posttranscriptional levels in *Mucor mucedo* and *Parasitella parasitica* (Petzold et al. 2004; Schultze et al. 2004).

of decrease, raised again after approximately 20 min. The mRNA level depends on light intensity and ranges between 6-fold at 1 J/m² and 40-fold at 960 J/m² (Velayos et al. 2003). It is suspected that sequence motives with strong similarity to the *Neurospora crassa* APE (*al-3* proximal element) may be correlated to blue light regulation. *al-3* is the GGPP synthase gene homologue in *Neurospora crassa*.

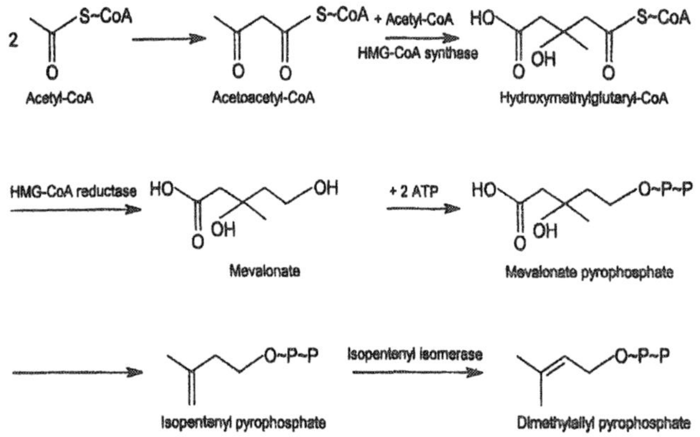

Fig. 4 A. Biosynthesis of carotene. From acetyl-CoA to isopentenyl pyrophosphate and dimethylallyl pyrophosphate.

Phytoene is formed from geranylgeranyl pyrophosphate by tail-to-tail condensation. The corresponding enzyme, phytoene synthase, is the gene product of the gene *carRP* in *Mucor circinelloides* (Velayos et al. 2000b). *carRP* codes for a bifunctional protein comprising in its C-terminal P-domain the phytoene synthase activity and in the N-terminal R-domain the lycopene cyclase activity. For activity, P relies on presence of the R-domain, whereas R is active even in truncated proteins without P activity. These observations strongly support that the *carRP* product is indeed a single bifunctional protein that is not posttranslationally processed to individually active catalytic subunits. The protein seems to be membrane-associated; it carries a transmembrane domain in the aminoterminal region. Phytoene, the product of the enzymatic action of carP, is converted to lycopene by phytoene dehydrogenase or phytoene desaturase, the product of *carB*. *carB* is closely linked to *carRP*; the promoter regions of both transcripts map within 450 bp. The successive dehydrogenation steps are catalysed by the same enzyme (Velayos et al. 2000a). The gene was functionally identified by complementing a genetically manipulated derivative of *Escherichia coli* with a defect in phytoene dehydrogenase (*crtB* from *Erwinia uredovora*). The enzyme has a calculated molecular weight of approximately 65 kb and is 73 % identical to the corresponding protein from *Phycomyces blakesleeanus*. Similar to the *carRP* situation, *carB* is inducible by blue light. The induction kinetics is biphasic also in this case, with a first maximum 10 minutes after the light pulse and a second one after approximately 20 minutes. The mRNA increase reaches 30-fold over controls without irradiation. The promoter region of the gene contains several motives that strongly resemble the APE consensus sequence from *Neurospora crassa al-3* and from *Mucor circinelloides carRP*. Thus, blue light regulation might follow comparable regulatory routes. The final step in β-carotene synthesis in *Mucor circinelloides* is the isomerisation of lycopene by introducing two β-ionone rings at the molecule's ends. The carR domain of the *carRP* gene product catalyses the reaction. At the sequence level, the R domain of the bifunctional enzyme differs markedly from cyclases of bacterial origin. This is completely different from the situation with the second part of the protein, the *carP* domain for phytoene synthase, where the gene is highly conserved between bacteria, fungi and plants (Velayos et al. 2000b).

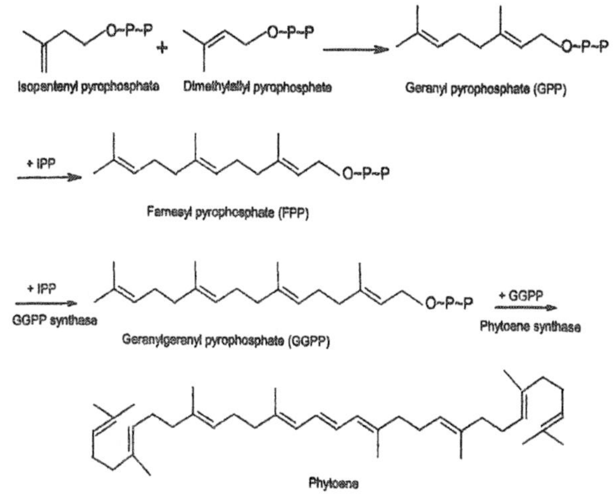

Fig. 4 B. Biosynthesis of carotene. From isopentenyl pyrophosphate to phytoene.

The requirement of blue light for carotenogenesis seems, at least in part, be caused by repression of carotenogenic genes by the product of *crgA* (Navarro et al. 2001) a protein, which was previously believed to be an activator of carotene synthesis (Navarro et al. 2000). Null mutants, constructed by gene replacement, over-accumulate carotenes independent of light conditions, which clearly proves the inhibitory mode of action for the *crgA* product. At the molecular level, the effect was ascribed to increased mRNA levels for *carB* expression. The protein carries a putative nuclear targeting signal, the cysteine-rich, zinc-binding RING-finger motive (Saurin et al. 1996). typical for many DNA binding proteins with regulatory character and also for proteins involved in larger aggregates. It also carries a site for cAMP dependent phosphorylation and an isoprenylation signal (Navarro et al. 2000). In addition to carotene synthesis, *crgA* seems to regulate processes involved in regulating vegetative growth (Quiles-Rosillo et al. 2003).

In *Phycomyces blakesleeanus*, the biochemical and regulatory routes towards β-carotene are very similar to the *Mucor* situation. Two genes for the early steps of isoprenoid biosynthesis were cloned, *hmgS* for HMG-CoA synthase and *hmgR* for HMG-CoA reductase, the enzyme leading to mevalonate (Ruiz-Albert et al. 2002). In contrast to the *hmgR* gene from *Absidia glauca*, which is repeated twice in the genome (Burmester and Czempinski, 1994). *Phycomyces blakesleeanus* harbours only a single copy. According to sequence motives in the promoter region of *hmgS*, the gene seems to be regulated by sterol compounds. HMG-CoA reductase contains amino acid sequences that anchor the enzyme within membranes and bear similarity to sterol sensing proteins from mammalia. Thus, the first specific steps towards terpenoid formation are possibly regulated by sterols at the protein level in case of HMG-CoA reductase and at the transcriptional level in case of HMG-CoA synthase. As mevalonate is the precursor for steroids as well as for carotenoids, the steroid regulation affects both routes, although these pathways are believed to be essentially independent and are associated with different compartments of the cell. Labelling experiments with ^{14}C-mevalonate lead to different specific radioactivities in β-carotene and ergosterol, thus suggesting different mevalonate pools for the two alternative routes towards isoprenoids (Bejarano and Cerdá-Olmedo 1992).

As in *Mucor circinelloides*, the enzymatic functions for phytoene synthase and lycopene cyclase are associated with a single protein, encoded by the *carRA* gene (Arrach et al. 2001) a molecular confirmation of earlier genetic deductions from recombination and complementation analyses (Ootaki et al. 1973; Roncero and Cerdá-Olmedo 1982). The aminoterminal carR domain catalyses the introduction of β-ionone rings into lycopene, whereas the A domain is responsible for converting two geranylgeranyl pyrophosphate moieties to phytoene. In addition and specific for *Phycomyces*, the A region seems to carry a regulatory domain. The mutation *carA113* allows mapping of the binding site for a regulatory protein, the gene product of *carI*. This gene mediates many of the regulatory stimuli affecting carotenogenesis in *Phycomyces blakesleeanus*, blue light (Bejarano et al. 1990) the chemicals retinol (Eslava et al. 1974) and dimethylphthalate (Cerdá-Olmedo and Hüttermann, 1986) and sexual activity (Govind and Cerdá-Olmedo 1986). Another regulatory gene of carotenogenesis, *carS*, is responsible only for stimulation by vitamin A and its derivatives. The corresponding protein is believed to interact with the *carI* product; alternative interpretations are, however, possible.

Phytoene dehydrogenase is a membrane-bound enzyme that was enriched 250-fold to apparent homogeneity from a lycopene-accumulating *carR* mutant after solubilization with Tween-60 (Fraser and Bramley 1994). A single protein catalyses all dehydrogenase steps from phytoene to lycopene; the reactions require two cofactors, NADP and FAD. Previous genetic complementation analyses led to the interpretation that carotenogenesis in fungi might be catalysed by an enzyme complex consisting of four molecules of phytoene dehydrogenase, one for each of the steps from phytoene via phytofluene, β-carotene and neurosporene to lycopene (Aragón et al. 1976). Possibly, this molecular arrangement is completed by two lycopene cyclase molecules, converting successively lycopene first to β-carotene with one and β-carotene with two β-ionone rings. A detailed complementation study in artificial heterokaryons between different *carB* mutants, completed by sequence analyses of the alleles, supports the idea of interallelic complementation via the formation of hybrid enzyme monomers (Sanz et al. 2002). However, direct evidence for the suggested polymeric and multifunctional complexes is still missing.

The corresponding gene, *carB*, was cloned and shown to encode a protein of 583 amino acids. The gene was expressed in recombinant *Escherichia coli* cells. Enzymatic studies in this heterologous system confirmed the previous view that a single enzyme performs all four dehydrogenation steps to lycopene. The start point of transcription is separated from the beginning of the *carRA* gene by 1381 nucleotides. The genes are transcribed in opposite direction, a situation very similar to that in *Mucor circinelloides*. Such arrangements offer the advantage that two genes can be regulated co-ordinately by binding transcription factors in the region of overlapping control elements.

The transcription of this gene, too, is stimulated by blue light, exhibiting a similar biphasic kinetic behaviour as in *Mucor circinelloides* (Ruiz-Hidalgo et al., 1997). The *Phycomyces carB* gene contains an APE box (consensus: GAANNTTGCC), similar to the situations in *Neurospora crassa* and *Mucor circinelloides*. Vitamin A stimulates the β-carotene content of *Phycomyces blakesleeanus* 20-fold. At the transcriptional level of *carB*, only a very slight effect can be measured. Thus, the predominant action of the compound seems to be exerted at the translational or post-translational level.

In its aminoterminal region, the protein carries a sequence that matches the consensus sequence for binding of the dehydrogenation cosubstrates FAD and NAD (G-G-GG-A-A- -L- - -G; Linden et al. 1994). It also carries the sequence VGA-THPG-G-P, which is believed to be responsible for carotenoid binding (Armstrong et al. 1989). Like other membrane-associated proteins, it has an explicit hydrophobic region near the carboxyterminus.

Among zygomycetes, *Phycomyces blakesleeanus* has probably the most explicit regulation of carotenogenesis by light. In order to be effective, blue light pulses must be applied at the defined developmental stage preceding the stop of mycelial growth by nutritional depletion (Bejarano et al. 1990). Inhibitor experiments support the assumption that the phototransduction process of carotenogenesis depends on protein kinases and phosphatases (Tsolakis et al. 1999). The positive control of carotene synthesis is, however, not a general phenomenon, neither in zygomycetes nor in the other fungal groups. Carotenogenesis of *Blakeslea trispora* is not activated by light (Sutter 1970). The ascomycete *Neurospora crassa* answers to light by increased carotenoid formation, the basidiomycetous yeast *Xanthophyllomyces dendrorhous* reacts on high light intensities by growth retardation, a decrease of most carotenoids and an increase of some (An and Johnson 1990).

5.2 Carotenoid Biosynthesis in *Xanthophyllomyces dendrorhous*

From a technical point of view, the heterobasidiomycetous yeast Xanthophyllomyces dendrorhous (imperfect form: Phaffia rhodozyma) is a highly interesting organism, because it produces the oxidised β-carotene derivative astaxanthin, a widely used additive in aquacultures of salmon, trout and shrimps. Compared with the other microbial producer of this industrially relevant compound, the unicellular green alga *Haematococcus pluvialis*, *Xanthophyllomyces* offers the advantages of convenient cultivation and of better genetic procedures for strain improvement. The yeast allows stable multicopy integration of heterologous plasmids into its rDNA clusters (Wery et al. 1997). cDNAs from *Xanthophyllomyces dendrorhous* allowed complementation of corresponding mutations in the *Erwinia uredovora crt* gene cluster, harboured by recombinant *Escherichia coli* cells. Again, as in other fungi, the functions for converting geranylgeranyl pyrophosphate to phytoene and for formation of the β-ionone rings from lycopene to β-carotene are encoded by a single gene, *crtYB* (Verdoes et al. 1999a). The gene has remarkably low sequence similarity with corresponding genes from other fungi. The most similar gene is found in Neurospora crassa with 17% identity. With 18% identity, essentially the same degree of sequence similarity is found between the aminoterminal part of *crtYB*, coding for the lycopene cyclase activity, and ORF7 in the carotene gene cluster of the myxobacterium *Myxococcus xanthus*.

The gene for phytoene dehydrogenase, *crtI*, was also cloned by complementation in *Escherichia coli* (Verdoes et al. 1999b). Complementation analysis in the heterologous system shows that a single gene product is sufficient for catalysing all four successive oxidation steps from phytoene to lycopene. Several enzymatic routes towards the formation of astaxanthin from β-carotene are realised, and some of the genes involved were cloned from bacteria, algae and plants (Misawa et al. 1995; Lotan and Hirschberg 1995; Cunningham and Gantt 1998). For *Xanthophyllomyces dendrorhous*, a biochemical pathway has been proposed (Andrewes et al. 1976) that introduces first a keto group into one of the β-carotene rings, leading to echinenone, which is then converted to 3-hydroxyechinenone. The equivalent course of reactions is then performed at the second ring. The model is still a hypothetical one for fungi; neither the enzymes nor the corresponding genes have been isolated, although appropriate *Escherichia coli* genotypes are available that should allow efficient screening for astaxanthin producers after introducing the genes as cDNA copies. The corresponding genes from the marine *Agrobacterium aurantiacum*, *crtW* and *crtZ*, have however been cloned (Misawa et al. 1995) thus rendering the model a reasonable degree of probability.

Analyses of unexpected cDNAs for *crtI* and *crtYB* at the sequence level suggest alternative splicing of primary transcripts for both mRNAs that would not lead to functional carotenogenic enzymes (Lodato et al. 2003). Although not proven at the biochemical or genetic level, it might be hypothesised that alternative splicing to non-functional transcripts could represent a regulatory mechanism allowing adaptation to environmental parameters.

Astaxanthin seems to fulfill similar functions in *Xanthophyllomyces dendrorhous* as β-carotene in mucoralean fungi. It protects the yeast against killing by singlet oxygen. In accordance with this observation, $^1O_2$ and peroxyl radicals positively regulate carotenoid biosynthesis. There is also preliminary evidence that carotenoid biosynthesis is feedback-regulated by astaxanthin, a situation that is similar to the end product inhibition of β-carotene synthesis in *Phycomyces blakesleeanus* (Schroeder and Johnson 1995).

5.3 Regulation of Carotenoid Biosynthesis in *Neurospora crassa*

In *Neurospora crassa*, genes for carotenoid synthesis are designated *albino* (*al*). due to the white phenotype they exhibit. *al-3*, the gene for condensing two geranyl pyrophosphate moieties, was the first geranylgeranyl pyrophosphate synthase to be cloned and identified by complementing an *al-3* mutant to the carotenoid producing orange phenotype (Nelson et al. 1989; Carattoli et al. 1990). The identity of the cloned gene was further shown by complementation experiments in recombinant *Escherichia coli* cells that lacked the corresponding *crtE* function (Sandmann et al. 1993). The gene encodes a protein of 428 amino acids and has no introns. According to computer modelling of the tertiary structure and comparing the protein with other prenyltransferases, the pocket for putative substrate binding and the catalytic site were mapped to a conserved aspartate-rich region with associated basic amino acid residues (Quondam et al. 1997).

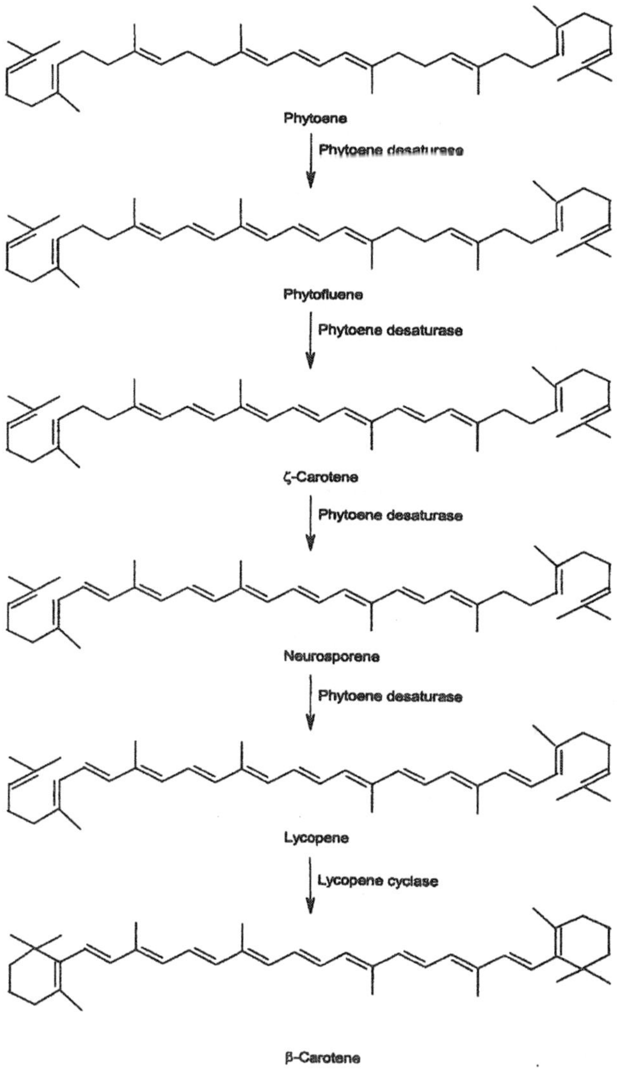

Fig. 4 C. Biosynthesis of carotene. From phytoene to β-carotene.

In growing mycelia, the gene is not transcribed in the dark and needs short blue light pulses for being activated. This is consistent with the general light regulation of carotenogenesis in this mould. *Neurospora crassa* needs light for developing orange hyphae, but expresses the carotenogenesis genes constitutively in macroconidia. A study based on site-directed mutagenesis and progressive 5′-deletions maps the genetic elements involved in blue-light regulation to the region between positions −226 and −55. This minimal promoter fragment still confers photoregulation to the reporter gene *mtr*, a gene coding for a membrane-associated amino acid transporter. Two copies of the *al-3* proximal element or APE play the key role for light inducibility. This element is 10-12 bp long and contains the sequence GAA- -TTGCC. Very similar motives are also found in other light-regulated genes in *Neurospora crassa*. Deletion and base substitution mutations especially in the first one of these elements are sufficient for abolishing light-dependent transcription of *al-3* (Carattoli et al. 1994). This simple model explains control by light during mycelial growth. For understanding the necessary constitutive expression of *al-3* during conidiogenesis, an additional regulatory mechanism is required. The *al-3* locus codes for two overlapping transcripts, al-3(m) and al-3(c). al-3(m) is a 1.6 kb mRNA that is regulated by light and is expressed in the mycelium after blue light induction. The promoter of al-3(m) maps in the transcribed region of the 2.2 kb al-3(c) transcript. This mRNA is not transcribed in mycelia, but is developmentally controlled and exclusively expressed in differentiating macroconidia. al-3(c) mRNA has a constitutive level during conidiogenesis, but is additionally stimulated by blue light and subject to regulation by the circadian cycle (Arpaia et al. 1995).

As in all other fungi analysed, a single *Neurospora crassa* gene, *al-2*, encodes the two functions phytoene synthase and carotene cyclase (Arrach et al. 2002). The gene has a length of 603 amino acids and carries the cyclase function at its amino terminus. As for *al-3*, the gene is photoinducible by blue light during mycelial growth Schmidhauser et al. 1994). *al-2* transcripts accumulate in germinating conidia, too, and this developmentally regulated transcription of *al-2* seems to be the only one to be completely independent from control by light (Li and Schmidhauser 1995). *al-1*, the gene for phytoene desaturase, maps within one map unit near the selectable genetic marker *hom*, conferring dependency on homoserine. The gene was cloned by concomitant complementation of a *hom al-1* double mutant to homoserine independent growth and restoring orange colour (Schmidhauser et al. 1990). In accordance to the other carotenogenesis genes in this organism, al-1 is photoinduced at the transcriptional level 70-fold by blue light. By expression in *Escherichia coli*, the gene product was shown to be sufficient for catalysing five successive NAD-dependent dehydrogenation steps, the usual four ones from phytoene to lycopene, and an additional one from lycopene to 3,4-didehydrolycopene. The latter two are the main products of the reaction chain. The enzyme is membrane-associated and, like in zygomycetes, is believed to be assembled to a loose complex from identical subunits that pass the intermediates on to the subsequent protein moieties (Hausmann and Sandmann 2000).

Carotenogenesis and asexual sporulation are tightly associated at the regulatory level. This reminds of the situation in zygomycetes with their intimate correlation between carotene synthesis and sexual development and the induction of sexual differentiation by the degradation products of β-carotene, the trisporoids. In *Neurospora crassa*, carotene synthesis during mycelial growth is induced by light and in addition by developmental signals. al-1, al-2, and al-3 are transcribed in response to light and nitrogen limitation. During conidiation, al-1 and al-2 respond to light as well as to developmental stage. Especially the process of conidial septation coincides with a light-independent increase of transcription for *al-1* and *al-2*. Light seems to add to the developmental effect via a different regulatory pathway (Li and

Schmidhauser 1995). The three *albino* transcripts 1-3 are turned on early during formation of macroconidia (Li et al. 1997). Photoinduction of carotenogenesis requires functional white-collar genes, *wc-1* and *wc-2*, coding for the heterodimeric subunits of a functional transcription factor, required among other things for the circadian rhythm of conidiation. *al-1* and *al-2* need *wc*-action for light-dependent expression during mycelial growth, but are also transcribed *wc*-independently during conidiation.

6. CONCLUSIONS

Carotenoids represent a large and manifold group of natural products. They are biologically important by providing organisms with colour, by protecting against light and oxidative stress, and by acting as signal molecules. Carotene degradation products, especially retinoids and the chemically very similar trisporoids, play crucial roles in signal transduction pathways of all organismic groups. Although being synthesized from a single universal chemical module, isopentenyl pyrophosphate, many different isomers and enantiomers exist, thus providing organisms with a fascinating tool kit for intra- and intercellular com-munication. The most elaborate system for making versatile use of carotenoid derivatives is probably found in zygomycetes. They regulate partner recognition as well as the differentiation pathway towards sexual structures by many different trisporoids, degradation products of β-carotene. With respect to the many functions in biology, carotenoids are certainly one of the most rewarding systems for exploring and understanding evolution of metabolic diversity.

Acknowledgements: We thank 'Deutsche Forschungsgemeinschaft' and 'Fonds der Chemischen Industrie' for grants allowing us to study the effects of carotenoid derivatives on zygomycetous differentiation and the evolution of carotenoid diversity.

REFERENCES

An G-H and Johnson EA (1990). Influence of light on growth and pigmentation of the yeast *Phaffia rhodozyma*. Antonie van Leeuwenhoek 57: 191-203.

Andrewes AG, Phaff HJ and Starr MP (1976). Carotenoids of *Phaffia rhodozyma*, a red pigmented fermenting yeast. Phytochem 15: 1003-1007.

Aragón CMG, Murillo FJ, De la Guardia MD and Cerdá-Olmedo E (1976). An enzyme complex for the dehydrogenation of phytoene in *Phycomyces*. Eur J Biochem 63: 71-75.

Armstrong GA, Alberti M, Leach F, and Hearst JE (1989). Nucleotide sequence, organization and nature of the protein products of the carotenoid biosynthesis gene cluster of *Rhodobacter capsulatus*. Mol Gen Genet 216: 254-268.

Arpaia G, Carattoli A and Macino G (1995). Light and development regulate the expression of the *Albino-3* gene in *Neurospora crassa*. Development Biol 170: 626-635.

Arrach N, Fernández-Martín R, Cerdá-Olmedo E and Avalos J (2001). A single gene for lycopene cyclase, phytoene synthase, and regulation of carotene biosynthesis in *Phycomyces*. Proc Natl Acad Sci. USA 98: 1687-1692.

Arrach N., Schmidhauser TJ and Avalos J (2002). Mutants of the carotene cyclase domain of *al-2* from *Neurospora crassa*. Mol Genet Genomics 266: 914-921.

Bejarano ER, Avalos J, Lipson ED and Cerdá-Olmedo E (1990). Photoinduced accumulation of carotene in *Phycomyces*. Planta 183: 1-9.

Bejarano ER and Cerdá-Olmedo E (1992). Independence of the carotene and sterol pathways of *Phycomyces*. FEBS-Lett 306: 209-212.

Blanc PL, Tuveson RW and Sargent ML (1976). Inactivation of carotenoid-producing and albino strains of *Neurospora crassa* by visible light, black light and ultraviolet radiation. J Bacteriol 125:

616-625.

Burmester A and Czempinski K (1994). Sequence comparison of a segment of the gene for 3-hydroxy-3-methylglutaryl-coenzyme A reductase in zygomycetes. Eur J Biochem 220: 403-408.

Cantino EC and Hyatt MT (1953). Carotenoids and oxidative enzymes in the aquatic Phycomycetes *Blastocladiella* and *Rhizophlyctis*. Am J Bot 40: 688-694.

Carattoli A, Romano N, Ballario P, Morelli G and Macino G (1991). The *Neurospora crassa* carotenoid biosynthetic gene (*albino 3*) reveals highly conserved regions among prenyltransferases. J Biol Chem 266: 5854-5859.

Carattoli A., Cogoni C, Morelli G and Macino G (1994). Molecular characterization of upstream regulatory sequences controlling the photoinduced expression of the *albino-3* gene of *Neurospora crassa*. Mol Microbiol 13: 787-795.

Cerdá-Olmedo E and Hüttermann A (1986). Förderung und Hemmung der Carotinsynthese bei *Phycomyces* durch Aromaten. Angew Botanik 60: 59-70.

Cerdá-Olmedo E and Martín-Rojas V, Cubero B (1996). Causes of cell death following ultraviolet B and C exposures and the role of carotenes. Photochem Photobiol 64: 547-551.

Cunningham Jr FX and Gantt E (1998). Genes and enzymes of carotenoid biosynthesis in plants. Annu Rev Plant Physiol Mol Biol 49: 557-583.

Disch A and Rohmer M (1998). On the absence of the glyceraldehyde 3-phosphate/pyruvate pathway for isoprenoid biosynthesis in fungi and yeasts. FEMS Microbiol Lett 168: 201-208.

Echavarri-Erasun C and Johnson EA (2002). Fungal carotenoids. Appl Mycol Biotechnol 2: 45-85.

Einax E and Voigt K (2003). Oligonucleotide primers for the universal amplification of β-tubulin genes facilitate phylogenetic analysis among fungi (Eumycota). Org Divers Evol 3: 185-194.

Endinkeau K and Woodward TW (1993). Protective effects of chlorogenic acid, curcumin and □-carotene against gamma-radiation-induced *in vivo* chromosomal damage. Mutat Res 303: 109-112.

Eslava AP, Álvarez MI and Cerdá-Olmedo E (1974). Regulation of carotene biosynthesis in *Phycomyces* by vitamin A and β-ionone. Eur J Biochem 48: 617-623.

Fraser PD and Bramley PM (1994). The purification of phytoene dehydrogenase from *Phycomyces blakesleeanus*. Biochim Biophys Acta 1212: 59-66.

Gauger W, Peláez MI, Álvarez MI and Eslava AP (1980). Mating type heterokaryons in *Phycomyces blakesleeanus*. Exp Mycol 4: 56-64.

Gooday GW (1973). The formation of fungal sporopollenin in the zygospores wall of *Mucor mucedo*: a role for the sexual carotenogenesis in the Mucorales. J Gen Microbiol 74: 233-239.

Gooday GW, Green, D Fawcett P and Shaw G (1974). Sporopollenin formation in the ascospore wall of *Neurospora crassa*. Arch Microbiol 101: 145-151.

Goodwin TW (1980). The biochemistry of the carotenoids. Vol. I: Plants, 2nd edition, Chapman and Hall, London.

Govind NS and Cerdá-Olmedo E (1986). Sexual activation of carotenogenesis in *Phycomyces blakesleeanus*. J Gen Microbiol 132: 2775-2780.

Hausmann A and Sandmann G (2000). A single five-step desaturase is involved in the carotenoid biosynthesis pathway to β-carotene and torulene in *Neurospora crassa*. Fung Genet Biol 30: 147-153.

Hirai N, Yoshida R, Todoroki Y and Ohigashi H (2000). Biosynthesis of abscisic acid by the non-mevalonate pathway in plants, and by the mevalonate pathway in fungi. Biosci Biotechnol Biochem 64: 1448-1458.

Li C and Schmidhauser TJ (1995). Developmental and photoregulation of *al-1* and *al-2*, structural genes for two enzymes essential for carotenoid biosynthesis in *Neurospora*. Dev Biol 169: 90-95.

Li C, Sachs MS and Schmidhauser TJ (1997). Developmental and photoregulation of three *Neurospora crassa* carotenogenic genes during conidiation induced by desiccation. Fung Genet Biol 21: 101-108.

Linden H, Misawa N, Saito T and Sandmann G (1994). A novel carotenoid biosynthesis gene coding for β-carotene desaturase – functional expression, sequence and phylogenetic origin. Plant Mol Biol 24: 369-379.

Lodato P, Alcaino J, Barahona S, Retamales P and Cifuentes V (2003). Alternative splicing of transcripts from *crtI* and *crtYB* genes of *Xanthophyllomyces dendrorhous*. Appl Environ Microbiol 69 : 4676-4682.

Lotan T and Hirschberg J (1995). Cloning and expression in *Escherichia coli* of the gene encoding β-carotene-4-oxygenase that converts β-carotene to the ketocarotenoid canthaxanthin in

Haematococcus pluvialis. FEBS Lett 364: 125-128.

Martín-Rojas V, Gómez-Puerto A and Cerdá-Olmedo E (1996). Lack of protection by carotenes against gamma-radiation damage in *Phycomyces*. Radiat Environ Biophys 35 : 193-197.

Mehta BJ and Cerdá-Olmedo E (1995) Mutants of carotene production in *Blakeslea trispora*. Appl Microbiol Biotechnol 42: 836-838.

Mehta BJ, Obraztsova IN and Cerdá-Olmedo E (2003). Mutants and intersexual heterokaryons of *Blakeslea trispora* for production of β-carotene and lycopene. Appl Environ Microbiol 69 : 4043-4048.

Misawa N, Satomi Y, Kondo K, Yokohama A, Kajiwara S, Saito T, Ohtani T and Miki W (1995). Structure and functional analysis of a marine bacterial carotenoid biosynthesis gene cluster and astaxanthin biosynthetic pathway proposed at the gene level. J Bacteriol 177: 6575-6584.

Navarro E, Lorca-Pascual JM, Quiles-Rosillo MD, Nicolás FE, Garre V, Torres-Martínez S and Ruiz-Vásquez RM (2001). A negative regulator of light-inducible carotenogenesis in *Mucor circinelloides*. Mol Genet Genomics 266: 463-470.

Navarro E, Ruiz-Pérez L and Torres-Martínez S (2000). Overexpression of the *crgA* gene abolishes light requirement for carotenoid biosynthesis in *Mucor circinelloides*. Eur J Biochem 267: 800-807.

Nelson MA, Morelli G, Carattoli A, Romano N and Macino G (1989) Molecular cloning of a *Neurospora crassa* carotenoid biosynthetic gene (*Albino-3*) regulated by blue light and the products of the white collar genes. Mol Cell Biol 9: 1271-1276.

Ootaki T, Lighty AC, Delbrück M and Hsu WJ (1973). Complementation between mutants of *Phycomyces* deficient with respect to carotenogenesis. Molec Gen Genet 121: 57-70.

Petzold, A, Burmester, A, Schultze, K, Wolschendorf, F, Wöstemeyer, J, Schimek and C (2004). 4-Dihydro-methyl trisporate dehydrogenase, an enzyme of the sex hormone pathway, is posttranscriptionally regulated in *Mucor mucedo*. Fung Genet Biol, submitted.

Quiles-Rosillo MD, Torres-Martínez S and Garre V (2003). *cigA*, a light-inducible gene involved in vegetative growth in *Mucor circinelloides* is regulated by the carotenogenic repressor *crgA*. Fung Genet Biol 38: 122-132.

Quondam M, Barbato C, Pickford A, Helmer-Citterich M and Macino G (1997). Homology modelling of *Neurospora crassa* geranylgeranyl pyrophosphate synthase: structural interpretation of mutant phenotypes. Protein Engineering 10: 1047-1055.

Rohmer M, Knani M, Simonin P, Sutte, B and Sahm H (1993). Isoprenoid biosynthesis in bacteria: a novel pathway for the early steps leading to isopentenyl diphosphate. Biochem J 295: 517-524.

Roncero MIG. and Cerdá-Olmedo E (1982). Genetics of carotene biosynthesis in *Phycomyces*. Curr Genet 5: 5-8.

Ruiz-Albert J., Cerdá-Olmedo E and Corrochano LM (2002). Genes for mevalonate biosynthesis in *Phycomyces*. Mol Genet Genomics 266: 768-777.

Ruiz-Hidalgo MJ, Benito EP, Sandmann G and Eslava AP (1997). The phytoene dehydrogenase gene of *Phycomyces*: regulation of its expression by blue light and vitamin A. Mol Gen Genet 253: 734-744.

Sandmann G and Misawa N (2002). Fungal carotenoids. The Mycota X, Industrial applications, Osiewacz HD, ed, Springer, Heidelberg.

Sandmann G, Misawa N, Wiedemann N, Vittorioso P, Carattoli A, Morelli G and Macino G (1993) Functional identification of *al-3* from *Neurospora crassa* as the gene for geranylgeranyl pyrophosphate synthase by complementation with *crt* genes, *in vitro* characterization of the gene product and mutant analysis. J Photochem Photobiol 18: 245-251.

Sanz C, Álvarez MI, Orejas M, Velayos A and Eslava AP (2002). Interallelic complementation provides genetic evidence for the multimeric organization of the *Phycomyces blakesleeanus* phytoene dehydrogenase. Eur J Biochem 269: 902-908.

Saurin AJ, Borden KLB, Boddy MN and Freemont PS (1996). Does this have a familiar RING? Trends Biochem Sci 21: 208-214.

Schimek C, Kleppe K, Saleem A-R, Voigt K, Burmester A and Wöstemeyer J (2003). Sexual reactions in Mortierellales are mediated by the trisporic acid system. Mycol Res 107: 736-747

Schmidhauser TJ, Lauter F-R, Russo VEA and Yanofsky C (1990). Cloning, sequence, and photoregulation of *al-1*, a carotenoid biosynthetic gene of *Neurospora crassa*. Mol Cell Biol 10: 5064-5070.

Schmidhauser, TJ, Lauter F-R, Schumacher M, Zhou W, Russo VEA and Yanofsky C (1994).

Characterization of *al-2*, the phytoene synthase gene of *Neurospora crassa*. J Biol Chem 269: 12060-12066.

Schroeder WA and Johnson EA (1995). Singlet oxygen and peroxyl radicals regulate carotenoid biosynthesis in *Phaffia rhodozyma*. J Biol Chem 270: 18374-18379.

Schultze K, Schimek C, Wöstemeyer J and Burmester A (2004) Sexuality and parasitism share common regulatory pathways in the fungus *Parasitella parasitica*. Gene, in press.

Sutter RP (1970) Effect of light on β-carotene accumulation in *Blakeslea trispora*. J Gen Microbiol 64: 215-221.

Tsolakis G, Parashi E, Galland P and Kotzabasis K (1999). Blue light signaling chains in *Phycomyces*: Phototransduction of carotenogenesis and morphogenesis involves distinct protein kinase/phosphatases elements. Fung Genet Biol 28: 201-213.

Van den Ende H (1978). Sexual morphogenesis in the phycomycetes. In: The filamentous fungi. (Smith JE and Berry DR eds), Vol. III, pp 257-274, Edward Arnold, London.

Velayos A and Blasco JL, Álvarez MI, Iturriaga EA and Eslava AP (2000a) Blue-light regulation of phytoene dehydrogenase (*carB*) gene expression in *Mucor circinelloides*. Planta 210: 938-946.

Velayos A and Eslava AP, Iturriaga EA (2000b). A bifunctional enzyme with lycopene cyclase and phytoene synthase activities is encoded by the *carRP* gene of *Mucor circinelloides*. Eur J Biochem 267: 5509-5519.

Velayos A and Papp T, Aguilar-Elena R, Fuentes-Vicente M, Eslava AP, Iturriaga EA, and Álvarez MI (2003) Expression of the *carG* gene, encoding geranylgeranyl pyrophosphate synthase, is up-regulated by blue light im *Mucor circinelloides*. Curr Genet 43: 112-120.

Verdoes JC, Krubasik P, Sandmann G, and van Ooyen AJJ (1999a) Isolation and functional characterisation of a novel type of carotenoid biosynthetic gene from *Xanthophyllomyces dendrorhous*. Mol Gen Genet 262: 453-461.

Verdoes JC, Misawa N and van Oyen AJJ (1999b) Cloning and characterization of the astaxanthin biosynthetic gene encoding phytoene desaturase of *Xanthophyllomyces dendrorhous*. Biotechnol Bioeng 63: 155-755.

Vershinin A (1999). Biological functions of carotenoids – diversity and evolution. BioFactors 10, 99-104.

Voigt K and Wöstemeyer J. (2000). Reliable amplification of actin genes facilitates deep-level phylogeny. Microbiol Res 155: 179-195.

Voigt K and Wöstemeyer J (2001). Phylogeny and origin of 82 zygomycetes from all 54 genera of the Mucorales and Mortierellales based on combined analysis of actin and translation elongation factor EF-genes. Gene 270: 113-120.

Wery J, Gutker D and Renniers ACHM, Verdoes JC, van Ooyen AJJ (1997) High-copy-number integration into the ribosomal DNA of the yeast *Phaffia rhodozyma*. Gene 184, 89-97.

Will OH, Newland NA and Repee CR (1984). The photosensitivity of pigmented and non-pigmented strains of *Ustilago violacea*. Curr Microbiol 10, 295-302.

Wöstemeyer A and Teepe H and Wöstemeyer J (1990). Genetic interactions in somatic inter-mating type hybrids of the zygomycete *Absidia glauca*. Curr Genet 17, 163-168.

Wöstemeyer J and Brockhausen-Rohdemann E (1987). Inter-mating type protoplast fusion in the zygomycete *Absidia glauca*. Curr Genet 12, 435-441.

Wöstemeyer J, Burmester A, Wöstemeyer A, Schultze K and Voigt K. (2002). Gene transfer in the fungal host-parasite system *Absidia glauca - Parasitella parasitica* depends on infection. In: Horizontal gene transfer, 2nd edition (M Syvanen and CI Kado, eds) Academic Press, pp 237-247

Young AJ and Lowe GM (2001). Antioxidant and prooxidant properties of carotenoids. Arch Biochem Biophys 385, 20-27.

**Applied Mycology
and Biotechnology**
An International Series
Volume 5. Genes and Genomics
© 2005 Published by Elsevier B.V.

ELSEVIER

12

Combination of Suppression Subtractive Hybridization and Microarray Technologies to Enumerate Biomass-Induced Genes in the Cellulolytic Fungus *Trichoderma reesei*

Elena V. Bashkirova, Michael W. Rey, and Randy M. Berka
Novozymes Biotech, Inc., 1445 Drew Avenue, Davis, California 95616-4880, USA
(ebas@novozymes.com)

The concerted action of many enzymes is required for the hydrolysis of cellulosic biomass. However, only a few have been identified and characterized in detail. Mixtures of isolated cellulase components have been shown to be less efficient than crude culture filtrates for the conversion of biomass to fermentable sugars. This may suggest that additional components are required for complete saccharification of biomass. *Trichoderma reesei* is the best-studied cellulolytic fungus. To identify new *T. reesei* genes involved in biomass conversion, we have employed the high throughput analysis of expression of subtractive cDNA libraries by DNA microarray technology. The cDNA libraries have been generated by suppression subtractive hybridization (SSH), which allowed not only the selection of differentially expressed mRNAs, but also the enrichment for rare mRNAs and equalization of cDNA in a pool. Messenger RNA pools from *T. reesei* cells grown on glucose, cellulose, or acid-pretreated corn stover (PCS) was isolated and used for construction of the SSH cDNA libraries. Three SSH libraries representing cellulose-induced, PCS-induced and PCS minus cellulose-induced transcripts were constructed. Approximately 3600 cDNA clones from three SSH libraries were amplified by high throughput rolling circle amplification (RCA) to produce DNA for microarray printing. Microarray hybridization with tester and driver probes revealed728. DNA sequence analysis and bioinformatics were used to assemble these clones into approximately 90 previously unrecognized genes/proteins. Among them we have identified a number of novel enzymes/proteins with potential direct benefit for improving biomass degradation. Thus, the combination of SSH and cDNA microarray technologies has proven to be a powerful tool for discovery of new differentially expressed genes involved in biomass utilization.

Corresponding author: Randy.M.Berka

1. INTRODUCTION

The identification and efficient large-scale production of improved cellulase and hemicellulase mixtures is critical for reducing the costs of industrial conversion of biomass to ethanol. While such enzymes are produced among a diversity of organisms, fungal cellulases from only a few species have been applied extensively for industrial uses. Naturally high levels of secretion and specific activity have been primary motivators for such interest. It seems likely that improvements in specific activity, thermostability, or yields of the enzyme mixtures used for saccharification of plant cell wall polysaccharides may prove to be the most direct and cost effective routes toward efficient biomass conversion.

A complex mixture of biomass-degrading enzymes is naturally secreted by cellulolytic fungi. Among them the mesophile *Trichoderma reesei* (syn. *Hypocrea jecorina*) has perhaps been the most widely studied, and this species provides the source of several commercial cellulase and hemicellulose preparations. *T. reesei* produces these enzymes primarily when grown on biomass-derived substrates, and the corresponding genes are regulated by a sophisticated scheme involving both repression and transcriptional activation, depending on the carbon source available (reviewed by Schmoll and Kubicek, 2003). Some of these regulatory mechanisms may be shared by genes encoding cellulases and hemicellulases (Margolles-Clark et al. 1997). When *T. reesei* is grown on glucose, the genes are tightly controlled by carbon catabolite repression mediated by the *cre1* gene encoding a transcriptional repressor of the Cys_2His_2 zinc finger protein class (Ilmén et al. 1996). Homologues of *cre1* are present in several other fungal species (Bailey and Arst, 1975; Arst et al. 1990; Takashima *et al.* 1998; De la Serna et al. 1999; Vautard-Mey et al. 1999; Jekosch and Kuck, 2000), suggesting that the essential mechanisms mediating glucose catabolite repression may be similar among many fungal species. In *T. reesei* the interaction between the zinc finger domains and upstream regions of several cellulase genes revealed a specific Cre1 binding motif (5'-SYGGRG-3') that is similar to the consensus CreA binding site in *Aspergillus nidulans* (Takashima et al. 1996). In addition to *cre1*, two transcriptional modulators, termed *ace1* and *ace2* (Saloheimo et al. 2000; Aro et al. 2001) have been described in *T. reesei*. Like Cre1, the *ace1* gene product is a Cys_2His_2 zinc finger protein that presumably binds to specific sequences in the promoters of target genes. Deletion of the *ace1* gene results in an increase in the expression of the major cellulases and two xylanases, suggesting that Ace1 acts as a repressor (Penttilä, 2003). In contrast, the *ace2* gene product is a transcriptional activator belonging to the class of zinc binuclear cluster proteins that are found exclusively in fungi (Penttila, 2001). Ace2 binds to the DNA motif 5'-GGCTAATAA-3' that is similar to the proposed binding site of the xylanase regulator XlnR in *Aspergillus niger* (van Peij et al. 1998). Deletion of *ace2* results in reduced rates of transcription of the major cellulase and xylanase genes in *T. reesei*, but cellulase induction *per se* is not affected. Despite the elegant molecular biology that has been done to discover the functions of Cre1, Ace1 and Ace2, the complete subset of genes that responds to these regulatory proteins has not been determined.

The DNA sequences encoding several of the major cellulases of *T. reesei* have been identified, including several endoglucanases, cellobiohydrolases, and β-glucosidases (Shoemaker et al. 1983; Penttilä et al. 1986; Teeri et al. 1987; Saloheimo et al. 1988; Barnett et al. 1991; Saloheimo et al. 1994; Saloheimo et al. 1997; Okada et al. 1998; Saloheimo et al. 1998; Foreman et al. 2003). However, a better understanding of the genes that encode less abundant proteins and their respective contributions toward cellulose/hemicellulose hydrolysis may be required to provide effective targets for improvement of industrial conversion of biomass to ethanol. Despite the intense interest in degradation and utilization of plant biomass material among environmental biologists and industrial biotechnology companies, surprisingly little is known about the molecular mechanisms that regulate the genes encoding tens or even hundreds of enzymes necessary for efficient and complete decomposition of cellulose and hemicellulose. Only a handful of enzymes in this symphony of catalytic activities have been enumerated and characterized. Undoubtedly new enzymes and synergistic relationships could be revealed by a systematic effort to identify and catalog these components and/or their corresponding genes. An appealing approach for this assignment involves comparing patterns of differential gene expression in cells growing on biomass material versus non-cellulosic substrates. There has recently been tremendous recent interest in employing powerful genome-wide expression analysis techniques for discovery of new enzymes and regulatory schemes affecting fungal cellulose degradation and biomass conversion. Foreman et al. (2003) recently described the sequencing of 5100 random cDNA clones (ESTs) from *T. reesei* cells that were cultured on several carbon sources including cellulose and sophorose. They identified 23 cDNAs that corresponded to known proteins and 12 clones that encoded previously unknown enzymes that were likely to function in biomass degradation. Furthermore, they employed DNA microarrays to show that most of the genes encoding known and putative biomass-degrading enzymes were transcriptionally co-regulated.

In addition to sequencing of random cDNA clones from induced cells, several alternative techniques for detecting differentially expressed genes have been described in the literature including differential display (Liang and Pardee 1992), subtractive hybridization (Hedrick 1984), representational difference analysis (RDA) (Lisitsyn et al. 1993; Hubank et al. 1994), serial analysis of gene expression (SAGE) (Velculescu et al. 1995; Zhang et al. 1997), and DNA microarrays (Schena et al. 1995). DNA microarray hybridization techniques have recently emerged as the preferred methods for determining differential gene expression on a genome-wide scale. First, the use of different fluorescent tags allows a direct comparison of the relative mRNA abundance in two RNA populations. Second, the small size of array allows hybridizations in decreased volumes with a small quantity of probe material. Computerized scanning of arrays provides a quantitative assessment of the relative abundance of specific mRNAs two cell types or culture conditions (based on color). However, differential hybridization itself has inherent difficulties in detection of rare mRNA species. To overcome this limitation, a combination of cDNA fractionation and microarray hybridization may be used to increase the sensitivity for rare mRNAs that are differentially expressed (Yang et al. 1999; Sakai et al. 2000; Beck et al. 2001; Porkka and Visakorpi, 2001; Seta et al. 2001).

Subtractive hybridization is a powerful technique for identification of differentially expressed genes, and numerous versions of this approach have been reported. In general, they all involve hybridization of cDNA from one population (tester) to an excess of cDNA from another population (driver) followed by separation of the unhybridized cDNA fraction (target) from hybridized sequences that are common to both populations. Despite the successes reported in the literature, these methods are usually ineffective for obtaining low-abundance cDNAs, and they require large amounts of starting material (typically 20 μg of poly(A)$^+$ RNA), involve multiple subtraction/hybridization steps, and are labor intensive. Newer PCR-based methods, such representational difference analysis (RDA) and rapid subtraction hybridization (RaSH) (Lisitsyn et al. 1993; Hubank et al. 1994; Jiang et al. 2000) simplify the procedure and reduce costs, but neither of these methods resolves the problem of wide differences in abundance of individual mRNA species.

Among various techniques used to enrich for rare transcripts, suppression subtractive hybridization (SSH) is a powerful technique allowing both subtraction and normalization of transcripts and subsequent efficient generation of cDNA libraries (Diatchenko et al. 1996). We deployed a combination of SSH-mediated subtraction and normalization of cDNA populations with DNA microarray technology. In this chapter we describe the generation of SSH libraries and detail the nucleotide sequence analysis of the resulting cDNA libraries in conjunction with DNA microarrays to reveal more than 90 new biomass-induced genes that are likely to encode transcriptional regulators, transport proteins, and enzymes involved in biomass degradation. We have generated a catalogue of biomass-induced *T. reesei* genes that can be used as a starting point for identification of components to improve the production or activity of cellulase/hemicellulase enzymes in this fungus.

2. CONSTRUCTION OF SUBTRACTED AND NORMALIZED cDNA LIBRARIES

To identify new *T. reesei* genes involved in biomass conversion, we employed DNA microarray technology to analyze the expression profiles of individual clones from subtractive cDNA libraries followed by DNA sequencing and bioinformatics to annotate possible gene function. The cDNA libraries were generated by suppression subtractive hybridization (SSH), which allows not only the selection of differentially expressed mRNAs, but also the enrichment for rare mRNAs and equalization of cDNA in a pool. The following sections outline the methods that were used.

2.1 Cell Growth

T. reesei strain RutC30 was grown in two-liter Applikon laboratory fermentors using basal salts media and a batch fermentation protocol. The carbon sources included glucose, cellulose, or PCS. All were loaded based on the carbon equivalent of 52 g/L glucose. The fermentation was done at 28°C, pH 4.5, and a growth time of approximately 120 hours. In addition to the carbon source, all fermentations contained the following components: glucose (5 g/L), corn steep solids (10 g/L), CaCl₂ (2.08 g/L), (NH₄)₂SO₄ (3.87 g/L), KH₂PO₄ (2.8 g/L), MgSO₄•7H₂O (1.63 g/L), trace metals (0.75 ml/L), and pluronic (1.8 ml/L). Samples of mycelia were harvested one, two, three, four, and five days post-inoculum. These tissue samples were quickly separated from the culture medium by filtration through Miracloth, frozen in liquid nitrogen, and stored at -80°C.

2.2 RNA Isolation

Total cellular RNA was isolated from frozen cells using slight modifications to the method of Timberlake and Barnard (1981). Briefly, RNA extraction buffer was prepared by adding a freshly prepared solution of p-aminosalicylic acid (9.6 g in 80 ml of DEPC-treated water) to a solution of triisopropylnaphthalene sulfonic acid (1.6 g in 80 ml of DEPC-treated water). This mixture was added to 40 ml of 5×RNB solution (1 M tris-HCl, pH 8.5, 1.25 M NaCl, 0.25 M EGTA) with stirring. Frozen mycelia were ground to a fine powder in an electric coffee grinder with a few chips of dry ice to prevent thawing. The ground mycelia were poured directly into 20 ml of RNA extraction buffer on ice, and an equal volume of TE-saturated phenol was added. After vigorous agitation, the samples were centrifuged at 2500 rpm for 10 min to separate phases. The aqueous phase was transferred to a new tube that contained 10 ml of phenol and 10 ml of chloroform-isoamyl alcohol (24:1), while an additional 5 ml of extraction buffer was added to the phenol phase. The latter mixture was incubated at 68°C for 5 min to liberate RNA trapped in polysomes and in the interface material. Following the incubation, the tubes were centrifuged, and the aqueous phase was combined with that obtained from the first extraction. These mixtures were subjected to repeated extraction with phenol-chloroform until there was no longer protein at the interface (usually five or six times). The RNA was recovered by centrifugation following precipitation with 0.3 M sodium acetate, pH 5.2, and 50% isopropanol. From each sample consisting of approximately 1-2 g of frozen mycelia we obtained 0.4-1.8 mg of total cellular RNA.

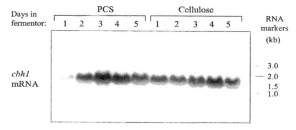

Fig. 1. Northern blot hybridization analysis of total cellular RNA extracted from *T. reesei* RutC30 grown on pretreated corn stover and cellulose (Avicel). The blot was probed with and HRP-labeled fragment of the *T. reesei* cbh1 gene. Lanes 1 through 5 in each bracket denote the time of growth in fermentors using corn stover or cellulose as the principal carbon sources (days 1-5). The positions and sizes of RNA markers are indicated to the right.

Total RNA fractions derived from cells grown on corn stover had a slight brown color, presumably due to corn pigments and/or lignin material that co-purified with the RNA. As shown in Figure 1 below we assessed the integrity of total RNA from *T. reesei* cells grown on cellulose and corn stover on a Northern blot (Thomas et al. 1980) that was probed with an HRP-labeled DNA fragment comprising the coding region of the *T. reesei cbh1* gene. The hybridization was done at 55°C using the buffers and protocols provided in the North2South Direct HRP labeling and detection kit purchased from Pierce (Rockford, IL). The blots were washed three times in 2×SSC with 0.1% SDS at 55°C for five minutes each, followed by three additional washes in 2×SSC (no SDS) for five minutes each. Virtually all of the hybridization signal in each lane was contained in a 1.8 kb *cbh1* mRNA species that migrated to a position just slightly above (perhaps partially occluded by) the 18S ribosomal RNA band. There was no evidence of significant mRNA degradation on either the autoradiogram (Fig. 1) or on the ethidium bromide stained gel (not shown).

A polyadenylated mRNA fraction was purified from total RNA by chromatography on oligo-dT cellulose (Qiagen Oligotex™ kit). Yields of poly(A)+ mRNA from each of these samples ranged from 2 µg to 25 µg. Each of the mRNA fractions was subsequently analyzed by Northern blot hybridization using HRP-labeled probes derived from the *T. reesei* γ-actin and *cbh1* genes (Fig. 2). As expected, virtually all the hybridization signals were localized in bands that corresponded to the γ-actin and *cbh1* mRNAs (ca. 1.2 kb and 1.8 kb, respectively) in each lane, indicating that the mRNA samples were of high quality and suitable for cDNA synthesis.

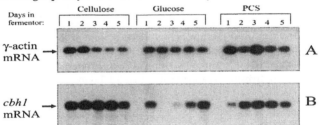

Fig. 2. Northern blot hybridization analysis of polyA+ mRNA purified from *T. reesei* RutC30 grown on corn stover, glucose, and cellulose (Avicel). Lanes 1 through 5 in each bracket denote the time of growth in fermentors (days 1-5). Panel A depicts the blot probed with γ-actin cDNA, and Panel B shows the same blot probed with *cbh1*.

Surprisingly, the *cbh1*-specific hybridization signal was observed in several of the mRNA samples extracted from glucose-grown *T. reesei* cells (Fig. 2B). Presumably this signal reflects a partial derepression of cellulases in *T. reesei* RutC30, which has undergone several rounds of mutagenesis for yield improvement. Furthermore, it appears that a high concentration of glucose is required to fully repress *cbh1* transcription in this strain. Conversely, when the glucose level falls below a critical threshold level, transcription of the *cbh1* gene occurs. This hypothesis is supported by the increasing intensity of the *cbh1* mRNA band from days 2 through 5 in the glucose supplemented fermentation. The notable exception to this premise is the observation of a prominent *cbh1* band in the day 1 sample. It is possible that this band could reflect the presence of *cbh1* mRNA in cells from the inoculum culture that was largely glucose-depleted and therefore induced.

2.3 cDNA Synthesis, Normalization and Subtraction

The procedure of suppression subtractive hybridization (SSH) (Fig. 3) described by Diatchenko et al. (1996) was used to generate a cDNA pool from *T. reesei* RutC30 that was both enriched for cellulose- and PCS-induced sequences and normalized to aid in recovery of rare transcripts. Synthesis and subtraction of cDNA was done using the PCR-Select™ kit obtained from BD Biosciences Clontech (Palo Alto, CA). First, we converted mRNA from three separate fermentations of *T. reesei* RutC30 grown on glucose, cellulose, and PCS into double-stranded cDNA. For synthesis of cDNA we combined 400 ng of poly(A)$^+$ mRNA derived from each time point (1-5 days) for a total of 2 µg of template. Table 1 below lists the combinations of driver and tester cDNAs used for these experiments.

Table 1. Driver and tester cDNA pools used for SSH.

SSH Reaction	Driver cDNA source	Tester cDNA source
1	Glucose-grown cells	PCS-grown cells
2	Glucose-grown cells	Cellulose-grown cells
3	Cellulose-grown cells	PCS-grown cells

The differentially expressed cDNAs are present in both the "tester" cDNA pool (*i.e.*, from cells grown on cellulose or PCS) and the "driver" cDNA, but are present at much lower levels in the "driver" pool (Table 1). Both of these cDNA pools were digested with the restriction enzyme *Rsa*I which recognizes a four-base pair palindrome (GT↓AC) and yields blunt-end fragments. The tester cDNA pool was then divided into two samples and ligated with two different adaptor oligonucleotides, resulting in two populations of tester cDNA. The adaptors were designed without 5'-phosphate groups such that only the longer strand of each adaptor could be covalently linked to the 5'-ends of the cDNA.

In the first of two hybridizations, an excess of driver cDNA was added to each portion of tester cDNA. The mixtures were denatured by heating to 95°C then allowed to anneal. Figure 3 shows the four types of molecules generated by this annealing (designated as *a, b, c,* and *d* molecules). Type *a* molecules include equal concentrations of high- and low-abundance cDNAs, because the second-order kinetics of hybridization are faster for more abundant molecules in the pool which will preferentially form *b* type molecules. At the same time, type *a* molecules are significantly enriched for differentially expressed (*e.g.*, cellulose or PCS induced) sequences, since common non-target cDNAs form type *c* molecules with the driver. In a second hybridization, the two pools of primary hybridized products are combined so that the type *a* molecules from each tester sample can associate and form new type *e* hybrids (Fig. 3). These are double-stranded tester molecules with different adaptor sequences on each end. Fresh denatured driver cDNA is also added to further enrich the pool for differentially expressed sequences.

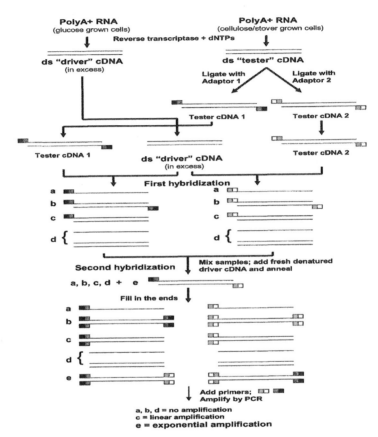

Fig. 3. SSH method for generation of subtractive and normalized cDNA libraries (Diatchenko et al. 1996). The components for creating SSH libraries are available commercially from BD Biosciences Clontech (Palo Alto , CA).

In the final step of the SSH procedure, the differentially expressed cDNAs are selectively amplified by PCR. As shown in Figure 3, only type *e* molecules that have two different primer-annealing sites can be amplified exponentially. These differentially expressed sequences are greatly enriched in the final subtracted cDNA pool, and they may be used to create a subtractive library.

2.4 Library Construction and Validation

Subtracted and normalized cDNA fractions generated by the SSH procedure were ligated with pCRII-TOPO (Invitrogen, Carlsbad, CA) and the ligation mixtures were used to transform electrocompetent E. coli TOP10 cells (Invitrogen). Transformants were selected on LB agar plates that contained 250 μg/ml X-Gal (no IPTG) and ampicillin at a final concentration of 100 μg/ml. Approximately 3600 white and light blue colonies were picked, grown overnight in selective medium in 96-well plates, and frozen for future use.

In order to evaluate the efficiency of subtraction and normalization in SSH cDNA libraries, we used two approaches: colony hybridization and sequencing of random clones from each SSH library. Colony-hybridization analysis included approximately 700 independent clones from each subtracted [PCS minus glucose (SG), cellulose minus glucose (CG), PCS minus cellulose (SC)] and un-subtracted (cellulose, PCS) cDNA libraries with DIG-labeled cbh1 probe (abundant transcript), and γ-actin probe, a moderately abundant transcript representing a house-keeping gene (Table 2).

Table 2. Colony hybridization of SSH libraries probed with cbh1 and γ-actin cDNA fragments.

Library	Frequency cbh1	Frequency γ-actin
Cellulose (no SSH)	3.3%	<0.17%
PCS (no SSH)	3.7%	0.13%
PCS minus glucose	0.4%	ND
Cellulose minus glucose	0.5%	ND
PCS minus cellulose	0.1%	ND

ND, not detected.

While the cbh1 is rather abundant in the non-subtracted cellulose and PCS libraries (3.3% and 3.6% correspondingly), the subtracted SG and CG libraries contain almost 10 times less cbh1 clones, which indicates that the abundant transcript was successfully normalized. Colony-hybridization of the SC library shows very low occurrence of cbh1 (only 0.1% of cbh1 clones) indicating an efficient subtraction of this abundant transcript when performing SSH with cell populations both expressing high levels of cbh1.

The cDNA clones from approximately 360 randomly picked colonies were purified by rolling circle amplification (RCA) (Dean et al., 2001) and analyzed by DNA sequencing (70 from Reaction 1, 96 from Reaction 2, and 192 from Reaction 3). Clustering of the sequences using Transcript Assembler™ software (Paracel, Inc.) showed that each pool contained a high percentage of non-redundant clones; 76% for reaction 1, 90% for reaction 2, and 67% for reaction 3. In addition, the contigs (overlapping sequences of the same cDNA) identified in this analysis contained on the average only two sequences. Collectively, these observations suggested that efficient normalization of the libraries was achieved during the SSH reactions, yielding a low level of redundancy in the corresponding cDNA libraries.

3. IDENTIFICATION OF DIFFERENTIALLY EXPRESSED cDNAs USING MICROARRAYS

3.1. Rolling Circle Amplification of Plasmid DNA and Microarray Construction

Plasmid DNA samples from 3608 colonies representing the three SSH cDNA libraries (1152 clones each from SG and SC libraries, and 1304 clones from the CG library) were prepared by rolling-circle amplification (RCA). RCA (Dean et al. 2001) of plasmid DNA from frozen cells was done using TempiPhi™ reagents (Amersham, Arlington Heights, IL). A method was created that allowed high-throughput processing of these reactions using a Beckman Biomek FX robot (Bashkirova, in preparation). The amplified cDNA clones were diluted to a concentration of 100-400 ng/µl in 3×SSC and spotted from 384-well plates onto poly-L-lysine coated glass microscope slides using equipment and methods that were described previously (Eisen and Brown 1999). Several control DNAs were included as well: *cbh1, cbh2, egl1, egl2*, serine hydroxymethyl transferase cDNA, γ-actin, and 28S rDNA.

3.2. Synthesis of Fluorescent cDNA Probes

Fluorescent probes were prepared by reverse transcription of poly(A)$^+$ RNA, incorporating aminoallyl-dUTP into first strand cDNA. The amino-cDNA products were subsequently labeled by direct coupling to either Cy3 or Cy5 monofunctional reactive dyes (Amersham, Arlington Heights, IL). The details of this protocol are available on the internet at http://cmgm.stanford.edu/pbrown/protocols. In all cases, cDNA from cells grown on glucose was used as the control (Cy3 label), and cDNA from cells grown on cellulose or PCS was labeled with Cy5. Cy3 and Cy5 labeled probes were combined, purified using a QIAquick PCR purification kit (Qiagen, Valencia, CA) and dried under a vacuum, resuspended in 15.5 µl of water and combined with the following: 3.6 µl of 20×SSC, 2.5 µl of 250 mM HEPES (pH 7.0), 1.8 µl of poly-dA (500 µg/ml), and 0.54 µl of 10% SDS. Before hybridization, the solution was filtered with a 0.22 µm filter, heated to 95°C for 2 min and cooled to room temperature.

3.3. Hybridization, Microarray Imaging, and Data Analysis

The fluorescently labeled cDNAs were applied to microarrays under cover glasses, placed in a humidified chamber, and incubated at 63°C overnight (15 – 16 hr). Before scanning, the arrays were washed consecutively in 1×SSC with 0.03% SDS, 0.2×SSC, and 0.05×SSC and centrifuged for 2 min at 500 rpm to remove excess liquid. Microarray slides were imaged using an Axon GenePix® 4000B scanner (Axon Instruments, Union City, CA), and the fluorescence signals were quantified using GenePix® Pro 5.0 software (Axon Instruments). PMT voltages were adjusted during image collection such that the average ratio of fluorescence intensities for the entire array was approximately 1.0. The intensity values were normalized using the loess function provided with S+ ArrayAnalyzer software (Insightful Corporation, Seattle, WA), and a statistical filtering for differentially expressed transcripts was made using the Local Pooled Error (LPE) analysis function in S+ArrayAnalyzer. After filtering for statistically significant changes, those spots for which the Cy5:Cy3 normalized intensity ratios were ≥ 2.0 on replicate arrays were chosen for DNA sequencing analysis.

From the 3608 clones that were screened, 728 cDNAs were found to be differentially expressed in technical replicates using threshold Cy5:Cy3 ratios ≥ 2.0. This represents a substantial fraction of differentially expressed genes in the SSH libraries (19%) compared to 0.7% found by microarray-based screening of 25,000 random clones from a *T. reesei* genomic DNA library (not shown). The distribution of biomass-induced genes among three SSH cDNA libraries is shown in Table 3.

4. ANALYSIS OF DIFFERENTIALLY EXPRESSED SSH-cDNAs
4.1 DNA Sequence Analysis

To enumerate and catalogue the *T. reesei* genes that are induced on cellulose and PCS, we performed nucleotide sequencing analysis of the differentially expressed cDNAs that were identified in microarray experiments. DNA sequencing was done using an

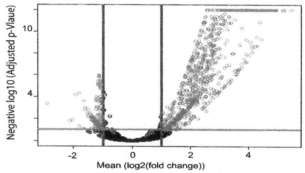

Fig. 4. Volcano plot from LPE test summarizing statistics for differential expression of biomass-induced genes. The horizontal line marks the threshold of cutoff Bonferroni-adjusted *p*-value=0.1. The two vertical lines mark the 2-fold change threshold (*i.e.*, $\log_2(2) = 1$). Genes that correspond to points in the upper left and upper right sextants were differentially expressed (significantly up- or down-regulated, respectively).

Table 3. Biomass-induced cDNAs from subtractive (SSH) libraries

Library	Number screened	Cellulose/PCS-Induced cDNAs
PCS minus glucose (SG)	1152	209
Cellulose minus glucose (CG)	1332	429
PCS minus cellulose (SC)	1152	90
TOTAL	**3840**	**728**

ABI3700 capillary sequencer with methods and reagents prescribed by the manufacturer (PE Applied Biosystems, Foster City, CA). The nucleotide sequences were edited using Phred/Phrap and Consed algorithms (Ewing and Green, 1998;

Gordon et al. 1998) and assembled into contigs using the Paracel Genome Assembler (Paracel, Inc., Pasadena, CA). Only sequences with high quality values (less than 1 error per 10^3-10^4 bp) were assembled and used to query the following nucleotide and protein databases using BLASTX and TBLASTX algorithms: (a) Novozymes Biotech proprietary *T. reesei* EST database, (b) a database consisting of all publicly available *T. reesei* gene/cDNA sequences collected from Genbank and EMBL, (c) publicly-available databases such as SwissProt, TREMBL, NR (NCBI), and PIR (Wu et al. 2003), (d) the *Phanerochaete chrysosporium* genome (protein-coding gene models obtained from the Joint Genome Institute, Walnut Creek, CA), and (e) the *Aspergillus fumigatus* genome (protein-coding gene models; licensed from The Institute for Genome Research, Rockville, MD). The search results were compiled in a FileMakerPro database (FileMaker, Inc., Santa Clara, CA).

4.2 Catalogue of *T. reesei* Genes Induced on Cellulose and PCS

BLAST analyses of the differentially expressed cDNA clones from the SSH libraries are summarized in Tables 4 and 5. Among these cDNAs, we identified nearly all of the genes that were previously known to be induced by growth on cellulosic substrates (Table 4). We also detected several of the new cellulose-induced genes reported by Foreman et al. (2003) including a family GH61 glycosyl hydrolase (numbering system of Henrissat (Coutinho and Henrissat, 1999a and 1999b), a β-glucosidase (GH1), an acetyl xylan esterase (CE5), and two previously unrecognized genes, termed *cip1* and *cip2*, whose functions are unknown (Table 5). In addition, we found a number of additional novel cDNAs whose expression was not known to be up-regulated by growth on cellulosic substrates (Table 5). Some of these novel cDNAs represented close relatives of genes that were known in the literature, and putative functions could be assigned. However, 102 of these novel gene sequences had no clear homology to any previously-known gene/protein in the publicly available databases or matched only hypothetical genes in other fungi and thus, they could only be designated as "unknowns." A minority (20%) of the unknowns could be assembled into eight contigs that each contained two to four sequences. The majority (80%) of the unknowns were represented by a single cDNA sequence. Among these unknowns we noted several interesting observations regarding the locations of the corresponding genes in a draft assembly of the *T. reesei* genome (Joint Genome Institute, Walnut Creek, CA). One of the unknowns is located next to a novel glycosyl hydrolase belonging to family GH5, and another is located adjacent to a gene that encodes a probable β-glucosidase of family GH1. Another previously discovered biomass-induced gene lies just upstream of the *xyn1* gene, and still another is located next to the gene encoding EGI.

It should be noted that several mitochondrial genes appeared to be induced by growth on cellulose/PCS. The apparent up-regulation of mitochondrial transcripts during growth on cellulose/PCS initially seemed troublesome, since mitochondrial RNA should not be present in the poly(A)$^+$ fraction of cellular mRNA. However, given that some low level of contamination with mitochondrial RNA should be anticipated in mRNA preparations, it should not be surprising that differences in metabolic rates resulting from utilization of these different carbon sources would lead to differences in respiration, energy requirement, lipid metabolism, growth rate, and the number of mitochondria. If similar levels of mitochondrial RNA contamination can be assumed for mRNA preparations derived from glucose, cellulose, and PCS, then the observed differential expression of several mitochondrial genes may represent a true physiological phenomenon that was revealed by the powerful SSH method employed in this study.

Table 4. List of biomass-induced transcripts detected in microarray experiments that encode previously-known (expected) enzymes involved in hydrolysis of cellulose and hemicellulose.

Enzyme	Family	Number of clones	Reference
CBH I	GH7	41	Shoemaker et al. 1983
CBH II	GH6	72	Teeri et al. 1987
EG I	GH7	3	Penttilä et al. 1986
EGII	GH5	55	Saloheimo et al. 1988
EGIII	GH12	3	Okada et al. 1998
EGIV	GH61	4	Saloheimo et al. 1997
EGV	GH45	6	SWALL accession no. Q7Z7X0
BGL1	GH3	1	Barnett et al. 1991
BGL2	GH1	3	Takashima et al. 1999
Swollenin	--	39	EMBL accession no. TRE245918
Xylanase I	GH11	28	Toerroenen et al. 1992
Xylanase II	GH11	7	Toerroenen et al. 1992
Xylanase III	GH10	0	EMBL accession no. AB036796
Xylanase IV	GH5	8	Parkkinen et al. 2004
β-xylosidase	GH3	5	Margolles-Clark et al. 1996a
Acetyl xylan esterase	CE5	21	Margolles-Clark et al. 1996b
α-Glucuronidase	GH67	1	Margolles-Clark et al. 1996c
β-mannanase	GH5	3	SWALL accession no. Q99036
Ribosomal proteins	--	2	SWALL accession nos. P26783 and Q12087
Mitochondrial	--	5	N.A.

N.A., not applicable

Foreman et al. (2003) observed that transcription of genes involved with protein synthesis, secretion, ribosome assembly, ER-Golgi trafficking, chaperonins, elements of the protein glycosylation machinery or components of the unfolded protein response were not significantly elevated by growth on cellulose. Our results were similar in that genes involved in protein secretion and glycosylation were not found to be differentially expressed in cellulose/PCS compared to glucose. However, we noticed several genes encoding ribosomal proteins (S5, L30, and a possible *RIC1* homologue) that showed increased transcript levels when cells were grown in cellulose or PCS, hinting that translation increases slightly under growth conditions in which high levels of extracellular protein are produced. Nevertheless, it is surprising that the secretion, protein folding, and glycosylation pathways do not appear to be induced or challenged even when the cells are producing high levels (greater than 10 g/L) of extracellular cellulases. Therefore, it seems likely that additional cellulase/hemicellulose yield improvements are not only possible, but likely with a traditional mutagenesis and screening program.

A comparison of the frequency distribution of selected cDNA clones among the three SSH libraries suggests that there may be significant differences between libraries. For example, the distribution of 13 different cDNA species between each of the SSH libraries (Table 6) was compared using a χ^2 test (3×13 contingency table with 24

Table 5. List of transcripts detected in microarray experiments that encode new enzymes/proteins that are up-regulated during growth on cellulosic substrates.

Enzyme	Number of clones
Glycosyl hydrolase (GH61)	8 †
Endoxylanase (GH11)	6
Sugar transporters; represents multiple genes/loci	77
Chitinase (GH18)	9
β-glucosidase (GH1)	3 †
Laminarinase/β-glucanase (GH16)	3
Endoxylanase/ β-1,6-glucanase	1
Acetyl xylan esterase (CE5)	7 †
Multicopper oxidoreductase	1
Possible ferric reductase transmembrane component	5
Plasma membrane ATPase	3
Probable RIC1 homologue	1
Putative flavohemoglobin-like protein	1
Possible Zn(2)-Cys(6) transcription factor	1
Similar to clock-controlled gene ccg-9 of N. crassa; has GT1 Interpro domain	2
Similar to MDR (ABC transporter) of A. nidulans	1
Possible transcription factor; similar to P. chrysosporium 27.4.1	1
Possible DNA repair exonuclease	2
Putative β-1,6-glucanase (GH30)	1
Possible glycosyl hydrolase, α-1,6-mannanase (GH76)	1
Similar to α -aminoadipate reductase	2
Predicted protein with similarity to coiled-coil protein in Asperrgillus fumigatus	1
cip1	6 †
cip2	11 †
No significant similarity to known proteins or similar only to hypothetical proteins in other fungi	92

†Also detected as a new cellulose-induced protein in the study of Foreman et al. (2003).

degrees of freedom). Differences for the frequencies of these 13 cDNAs were calculated with the following probabilities: CG versus SG: $P = 2.99 \times 10^{-14}$, CG versus SC: $P = 5.78 \times 10^{-6}$, and SG versus SC: $P = 1.22 \times 10^{-9}$. These calculations suggest that there might be different subsets of genes induced or different induction ratios when cells are grown on cellulose versus PCS. Clearly there are differences in the frequencies of xylanases (xyn1 and xyn2), acetylxylan esterase (axe1), as well as cel61b compared to the observed frequencies of cbh1, cbh2, and egl2 shown in Table 6.

4.3 Selected Genes Identified by Screening SSH Clones on DNA microarrays

Our collection of cellulose/PCS-induced cDNAs includes a number of novel enzymes and proteins that might contribute to efficient hydrolysis of biomass material and utilization of the resulting carbohydrates (Table 5). They include glycosyl hydrolases from families GH1, GH11, GH16, GH18, GH30, GH61, GH76, and putative sugar transport proteins. We also uncovered a new acetyl xylan esterase and a multi-copper oxidase (laccase) that may oxidize phenolic material in plant biomass. Several possible transcription factors were also found among the genes that are induced during growth on cellulosic substrates. The role of these proteins in modulating the choreography of cellulase/hemicellulase gene expression awaits further analysis by gene knockout and/or regulated expression. By combining the nucleotide sequence data from our SSH clones with that derived from sequencing of ESTs we were able to construct several full-length cDNAs *in silico*. Several of these are described in the following sections.

Table 6. Frequency distribution of 13 selected cDNA species among the SSH libraries constructed from T. reesei. Sample sizes (n) are as follows: CG (266), SG (88), and SC (204).

Transcript	# in CG	# in SG	# in SC	SUM
cbh1	8	24	9	41
cbh2	15	42	15	72
egl1	1	1	1	3
egl2	20	22	13	55
xyn1	24	4	0	28
xyn2	6	1	0	7
cel61a	1	1	2	4
cel61b	6	2	0	8
axe1	16	4	1	21
swo1	16	12	10	39
bgl (all)	0	3	1	4
bmn1	2	1	0	3
hxt (all)	31	32	14	77
SUM	146	149	66	361

4.3.1 Family GH61 Glycosyl Hydrolase

Several SSH clones were identified in microarray experiments that encode a previously undiscovered glyosyl hydrolase belonging to family GH61. Members of this family reportedly exhibit weak endoglucanase activity (Saloheimo et al. 1997), and they have been identified in the following fungal species: *Agaricus bisporus* (CEL1), *Aspergillus kawachii* (endoglucanase B), *Filobasidiella* (*Cryptococcus*) *neoformans* (CEL1), *Neurospora crassa* (genomic ORFs B24P7.180 and G15G9.090), *Phanerochaete chrysosporium* (genomic ORFs 38.73.1, 61.18.1, 10.35.1, 47.15.1, 48.13.1, 60.56.1, 88.21.1, 92.53.1, and 92.56.1), as well as in T. *reesei* (endoglucanase IV, cel61A). It should be noted that the genomes of most filamentous fungi examined to date encode multiple GH61 enzymes. This observation suggests a vital role for these enzymes in either biomass degradation and/or cell wall metabolism. Like Foreman *et al.* (2003) we refer to this newly discovered glycosyl hydrolase as *cel61B*, and its corresponding protein is designated as Cel61B. The nucleotide sequence and deduced translation of *cel61B* cDNA

```
GGATCTAAGCCCCATCGATATGAAGTCCTGCGCCATTCTTGCAGCCCTTGGCTGTCTTGCCGGGAGCGTTCTCGGCCA
                     M  K  S  C  A  I  L  A  A  L  G  C  L  A  G  S  V  L  G  H

TGGACAAGTCCAAAACTTCACGATCAATGGACAATACAATCAGGGTTTCATTCTCGATTACTACTATCAGAAGCAGAA
  G  Q  V  Q  N  F  T  I  N  G  Q  Y  N  Q  G  F  I  L  D  Y  Y  Y  Q  K  Q  N

TACTGGTCACTTCCCCAACGTTGCTGGCTGGTACGCCGAGGACCTAGACCTGGGCTTCATCTCCCCTGACCAATACAC
  T  G  H  F  P  N  V  A  G  W  Y  A  E  D  L  D  L  G  F  I  S  P  D  Q  Y  T

CACGCCCGACATTGTCTGTCACAAGAACGCGGCCCCAGGTGCCATTTCTGCCACTGCAGCGGCCGGCAGCAACATCGT
  T  P  D  I  V  C  H  K  N  A  A  P  G  A  I  S  A  T  A  A  A  G  S  N  I  V

CTTCCAATGGGGCCCTGGCGTCTGGCCTCACCCCTACGGTCCCATCGTTACCTACGTGGCTGAGTGCAGCGGATCGTG
  F  Q  W  G  P  G  V  W  P  H  P  Y  G  P  I  V  T  Y  V  A  E  C  S  G  S  C

CACGACCGTGAACAAGAACAACCTGCGCTGGGTCAAGATTCAGGAGGCCGGCATCAACTATAACACCCAAGTCTGGGC
  T  T  V  N  K  N  N  L  R  W  V  K  I  Q  E  A  G  I  N  Y  N  T  Q  V  W  A

GCAGCAGGATCTGATCAACCAGGGCAACAAGTGGACTGTGAAGATCCCGTCGAGCCTCAGGCCCGGAAACTATGTCTT
  Q  Q  D  L  I  N  Q  G  N  K  W  T  V  K  I  P  S  S  L  R  P  G  N  Y  V  F

CCGCCATGAACTTCTTGCTGCCCATGGTGCCTCTAGTGCGAACGGCATGCAGAACTATCCTCAGTGCGTGAACATCGC
  R  H  E  L  L  A  A  H  G  A  S  S  A  N  G  M  Q  N  Y  P  Q  C  V  N  I  A

CGTCACAGGCTCGGGCACGAAAGCGCTCCCTGCCGGAACTCCTGCAACTCAGCTCTACAAGCCCACTGACCCTGGCAT
  V  T  G  S  G  T  K  A  L  P  A  G  T  P  A  T  Q  L  Y  K  P  T  D  P  G  I

CTTGTTCAACCCTTACACAACAATCACGAGCTACACCATCCCTGGCCCAGCCCTGTGGCAAGGCTAGATCCAGGGGTAC
  L  F  N  P  Y  T  T  I  T  S  Y  T  I  P  G  P  A  L  W  Q  G  *

GGTGTTGGCGTTCGTGAAGTCGGAGCTGTTGACAAGGATATCTGATGATGAACGGAGAGGACTGATGGGCGTGACTGAG

TGTATATATTTTTGATGACCAAATTGTATACGAAATCCGAACGCATGGTGATCATTGTTTATCCCTGTAGTATATTGTC

TCCAGGCTGCTAAGAGCCCACCGGGTGTATTACGGCAACAAAGTCAGGAATTTGGGTGGCAATGAACGCAGGTCTCCAT

GAATGTATATGTGAAGAGGCATCGGCTGGCATGGGCATTACCAGATATAGGCCCTGTGAAACATATAGTACTTGAACGT

GCTACTGGAACGGATCATAAGCAAGTCATCAACATGTGAAAAAACACTACATGTAAAAAAAAAAAAAAAAAAAAAAA
```

Fig. 5. Nucleotide sequence and deduced amino acid sequence of a new family GH61 glycosyl hydrolase (*cel61B*) cDNA from *T. reesei*. The predicted signal peptide encoded by residues 1-19 is shown in italics (gray shading), and the mature enzyme is denoted by bold text.

identified in this study is shown in Figure 5. The predicted protein sequence contains an amino-terminal signal peptide (Nielsen et al. 1997) with a likely cleavage site between residues 19 and 20 (VLG-HG), suggesting that Cel61B is secreted. The mature protein has a predicted molecular weight of 25 kdal, however, a potential site for N-linked glycosylation (NFT) may affect the actual mass of the enzyme. Unlike Cel61A (EGIV), the deduced amino acid sequence of Cel61B does not include a cellulose binding module (CBM1). Gene expression studies with DNA microarrays indicate that *cel61B* is coordinately expressed with the previously-known complement of cellulases and hemicellulases, hinting that Cel61B plays a significant role in degradation of these substrates.

4.3.2 "Unknown core" (*cip1*)

In addition to *cel61B* we detected several cDNA clones that matched sequences we previously observed among our proprietary EST collection from *T. reesei*. The nucleotide and deduced amino acid sequences from one of these clones are shown in fig 6. This cDNA was also observed by Foreman et al. (2003), and they referred to the transcript as *cip1*. Their microarray experiments as well as ours suggest that the *cip1* transcript is induced when *T. reesei* is grown on cellulosic substrates. Using the method of von Heijne (1984) we noticed the presence of a presecretory signal peptide comprising the first nineteen residues of the predicted translation product. A consensus N-linked glycosylation site was also noted. In addition, the presence of a putative C-terminal cellulose binding domain (CBM1) and serine-threonine rich linker peptide were also revealed (Fig. 6), suggesting a modular architecture, not unlike the organization of domains in cellulolytic enzymes such as CBHI. The remaining portion of the protein presumably comprises the catalytic core. However, this region contains no motifs or signatures that have previously been associated with carbohydrase activities (based on Pfam and InterPro searches). Hence, we refer to this molecule as the "unknown core." The closest homologue is a "putative secreted hydrolase" annotated in the *Streptomyces coelicolor* genome (SPTREMBL: O69962, $E = 4 \times 10^{-37}$). Apparently the unknown core represents a unique class of biomass-induced extracellular proteins, and it is the first representative molecule of this type detected in filamentous fungi. At present the catalytic activity, substrate specificity, and contribution of the enzyme toward the collective activity of extracellular cellulases are unknown.

4.3.3 Novel Sugar Transport Proteins

As noted in Table 5, we observed biomass-induced transcription of several genes encoding putative sugar transport proteins in our microarray experiments. In total, we found that these genes accounted for an unexpectedly high number of clones compared to other cDNA species that were found to be up-regulated during growth on cellulose or PCS. This prompted us to investigate whether they represented a single gene or several functionally related genes. Interestingly, these cDNAs could be assembled into seven contigs and three singletons; however, since the cDNA was cleaved with RsaI during construction of the libraries, it is likely that these contigs represent fragments of fewer genes. Using BLAST to compare the contigs to the draft assembly of the *T. reesei* genome, it appears that the seven cDNA contigs and three singletons comprise four genes that encode sugar/ oligosaccharide transport proteins (Table 7). The predicted

```
ATGGTTCGCCGGACTGCTCTGCTGGCCCTTGGGGCTCTCTCAACGCTCTCTATGGCCCAAATCTCAGACGA
 M  V  R  R  T  A  L  L  A  L  G  A  L  S  T  L  S  M  A  Q  I  S  D  D
CTTCGAGTCGGGCTGGGATCAGACTAAATGGCCCATTTCGGCACCAGACTGTAACCAGGGCGGCACCGTCA
 F  E  S  G  W  D  Q  T  K  W  P  I  S  A  P  D  C  N  Q  G  G  T  V
GCCTCGACACCACAGTAGCCCACAGCGGCAGCAACTCCATGAAGGTCGTTGGTGGCCCCAATGGCTACTGT
 S  L  D  T  T  V  A  H  S  G  S  N  S  M  K  V  V  G  G  P  N  G  Y  C
GGACACATCTTCTTCGGCACTACCCAGGTGCCAACTGGGGATGTATATGTCAGAGCTTGGATTCGGCTTCA
 G  H  I  F  F  G  T  T  Q  V  P  T  G  D  V  Y  V  R  A  W  I  R  L  Q
GACTGCTCTCGGCAGCAACCACGTCACATTCATCATCATGCCAGACACCGCTCAGGGAGGGAAGCACCTCC
 T  A  L  G  S  N  H  V  T  F  I  I  M  P  D  T  A  Q  G  G  K  H  L
GAATTGGTGGCCAAAGCCAAGTTCTCGACTACAACCGCGAGTCCGACGATGCCACTCTTCCGGACCTGTCT
 R  I  G  G  Q  S  Q  V  L  D  Y  N  R  E  S  D  D  A  T  L  P  D  L  S
CCCAACGGCATTGCCTCCACCGTCACTCTGCCTACCGGCGCGTTCCAGTGCTTCGAGTACCACCTGGGCAC
 P  N  G  I  A  S  T  V  T  L  P  T  G  A  F  Q  C  F  E  Y  H  L  G  T
TGACGGAACCATCGAGACGTGGCTCAACGGCAGCCTCATCCCGGGCATGACCGTGGGCCCTGGCGTCGACA
 D  G  T  I  E  T  W  L  N  G  S  L  I  P  G  M  T  V  G  P  G  V  D
ATCCAAACGACGCTGGCTGGACGAGGGCCAGCTATATTCCGGAGATCACCGGTGTCAACTTTGGCTGGGAG
 N  P  N  D  A  G  W  T  R  A  S  Y  I  P  E  I  T  G  V  N  F  G  W  E
GCCTACAGCGGAGACGTCAACACCGTCTGGTTCGACGACATCTCGATTGCGTCGACCCGCGTGGGATGCGG
 A  Y  S  G  D  V  N  T  V  W  F  D  D  I  S  I  A  S  T  R  V  G  C  G
CCCCGGCAGCCCCGGCGGTCCTGGAAGCTCGACGACTGGGCGTAGCAGCACCTCGGGCCCGACGAGCACTT
 P  G  S  P  G  G  P  G  S  S  T  T  G  R  S  S  T  S  G  P  T  S  T
CGAGGCCAAGCACCACCATTCCGCCACCGACTTCCAGGACAACGACCGCCACGGGTCCGACTCAGACACAC
 S  R  P  S  T  T  I  P  P  P  T  S  R  T  T  T  A  T  G  P  T  Q  T  H
TATGGCCAGTGCGGAGGGATTGGTTACAGCGGGCCTACGGTCTGCGCGAGCGGCACGACCTGCCAGGTCCT
 Y  G  Q  C  G  G  I  G  Y  S  G  P  T  V  C  A  S  G  T  T  C  Q  V  L
GAACCCATACTACTCCCAGTGCTTATAA
 N  P  Y  Y  S  Q  C  L  *
```

Fig. 6. Nucleotide sequence and deduced amino acid sequence of the "unknown core" enzyme (*cip1*). The signal peptide region is denoted by gray shading encompassing residues 1-19. The putative catalytic core is underlined (residues 20-245); the linker region is in italics (residues 246-282); and the cellulose-binding domain (CBD) is in bold text (residues 283-316).

amino acid sequences from the contigs and singletons possess Interpro domains IPR003663, IPR005828, and IPR005829 corresponding to sugar transporter and major facilitator super family motifs. Contigs 3 and 7 (Table 7) were found to overlap portions of an EST clone (designated Tr0889) and yielded the nucleotide sequence of the entire coding region in the cDNA. The nucleotide sequence and deduced amino acid sequence are shown in Figure 7. An unusually long signal peptide was predicted, comprising residues 1 to 63, however, two additional ATG codons in this region may indicate alternate translation initiation sites that might yield signal peptides that are shorter and have a more conventional amino acid composition. A hidden-Markov model (PFAM) comparison of the deduced amino acid sequence to those in the publicly available databases revealed a consensus sugar transporter domain spanning residues 51-513

Table 7. Contigs and singletons encoding putative cellulose/PCS-induced sugar transport proteins.

Contig/singleton	Number of sequences	Comment
1	2	Similar to hypothetical genes in *G. zeae* ($E = 1 \times 10^{-55}$) and *N. crassa* ($E = 1 \times 10^{-53}$); Best BLASTX hit to known gene is a monosaccharide transporter from *A. niger* ($E = 2 \times 10^{-15}$); Contains Interpro domains IPR007114 (major facilitator superfamily) and IPR003663 (sugar transporter).
2	6	Low identity to a hypothetical gene in *N. crassa* ($E = 5 \times 10^{-3}$); Contains Interpro domains IPR003663 (sugar transporter), IPR005828 (metabolite transport), IPR005829 (sugar transport), IPR000683 (oxidoreductase N-terminal), and IPR004104 (oxidoreductase C-terminal).
3	31	Similar to hypothetical gene in *A. nidulans* ($E = 6 \times 10^{-59}$); Best BLASTX hits to known genes are hexose transporters from *A. parasiticus* ($E = 6 \times 10^{-18}$), *A. oryzae* ($E = 3 \times 10^{-17}$) and lactose permease of *K. lactis* (1×10^{-19}); Contains Interpro domains IPR003663 (sugar transporter), IPR005828 (metabolite transport), and IPR005829 (sugar transport).
4	4	Similar to hypothetical genes in *A. nidulans* ($E = 1 \times 10^{-20}$), *G. zeae* ($E = 6 \times 10^{-18}$), and *N. crassa* ($E = 4 \times 10^{-17}$); Best BLASTX hit to known genes are *K lactis* lactose permease ($E = 7 \times 10^{-11}$) and *Uromyces viciae-fabae* hexose transporter ($E = 2 \times 10^{-9}$); Contains Interpro domains IPR003663 (sugar transporter), IPR005828 (metabolite transport), and IPR005829 (sugar transport).
5	11	Similar to hypothetical genes in *N. crassa* ($E = 2 \times 10^{-17}$), *A. nidulans* ($E = 4 \times 10^{-18}$), and *G. zeae* ($E = 1 \times 10^{-15}$); Contains Interpro domains IPR003663 (sugar transporter), IPR005828 (metabolite transport), and IPR005829 (sugar transport).
6	7	Similar to hypothetical genes in *A. nidulans* ($E = 5 \times 10^{-15}$) and *G. zeae* ($E = 3 \times 10^{-12}$); Contains Interpro domains IPR003663 (sugar transporter), IPR005828 (metabolite transport), and IPR005829 (sugar transport).
7	12	Similar to hypothetical genes in *A. nidulans* ($E = 4 \times 10^{-41}$), *G.*

zeae (E = 2×10^{-41}), *M. grisea* (E = 3×10^{-29}), and *N. crassa* (E = 2×10^{-24}); Best BLASTX hits to known genes are *N. crassa* hexose transporter (E = 2×10^{-10}) and *K. lactis* lactose permease (E = 7×10^{-10}). Contains Interpro domains IPR003663 (sugar transporter), IPR005828 (metabolite transport), and IPR005829 (sugar transport).

SG0857	1	Similar to hypothetical genes in *N. crassa* (E = 0) and *M. grisea* (E = 1×10^{-180}); Best BLASTX hits to known genes are *A. niger* monosaccharide transporter (E = 1×10^{-50}), *P. angusta* hexose transporter (E = 1×10^{-48}), and *T. reesei* glucose transporter *hxt1* (E = 4×10^{-47}); Contains Interpro domains IPR003663 (sugar transporter), IPR005828 (metabolite transport), IPR002086 (aldehyde dehydrogenase), and IPR005829 (sugar transport).
CG0395	1	Similar to hypothetical genes in *A. nidulans* (E = 5×10^{-44}), *M. grisea* (E = 4×10^{-45}), and *G. zeae* (E = 2×10^{-40}); Best BLASTX hits to known genes are *K. lactis* lactose permease (3x10^{-33}), *A. oryzae* hexose transporter (E = 9×10^{-18}), and *A. parasiticus* hexose transporter (E = 3×10^{-18}); Contains Interpro domains IPR003663 (sugar transporter), IPR005828 (metabolite transport), IPR005829 (sugar transport), IPR000683 (oxidoreductase N-terminal), and IPR004104 (oxidoreductase C-terminal).
CG0557	1	Similar to unnamed product in *Debarymyces hansenii* (E = 1×10^{-104}); Best BLASTX hits to known genes are *A. oryzae* maltose permease (E = 1×10^{-77}), *N. crassa* α-glucoside transporter (E= 2×10^{-76}), *Leptoshpaeria maculans* maltose permease (E = 1×10^{-35}), and *S. cerevisiae* maltose permease (E = 3×10^{-35}); Contains Interpro domains IPR003663 (sugar transporter), IPR005828 (metabolite transport), IPR001687 (ATP/GTP-binding site motif A), and IPR005829 (sugar transport

(PFAM score = 278.1; E = 1.2×10^{-79}). The closest homologue in the publicly-available databases is the lactose permease from the yeast *Kluyveromyces lactis*. Considering that lactose and cellobiose are both β-linked disaccharides, it is tempting to speculate that the newly-identified transporter functions as a cellobiose or cello-oligosaccharide permease. This hypothesis is somewhat strengthened by the fact that intracellular β-glucosidases are also induced in the same cultures (see Table 4 and 5). These β-glucosidases would be required to release glucose from the imported cellooligosaccharides. However, it seems equally plausible that this gene product may function to transport glucose that has been generated by the concerted action of the cellulase enzymes. Given that this transporter is specifically transcribed in response to growth on cellulosic material, it is feasible that a disruption in the corresponding chromosomal sequence might lead to increased derepression of the cellulases that are

```
ATTACCCGTAACAATGGGCGAGAAAGAAGACATTCACGCTCACGAGGAGCTCGACCATGGAGAGATCAGGACCAAGGTCG
                M  G  E  K  E  D  I  H  A  H  E  E  L  D  H  G  E  I  R  T  K  V
TGACCGGACACGAGGCCTTTGAGGAGGCCATGATGAAGGAGCCGCCCAAGGCCTGGACCAAGGCTCAGGTCCTCGTCTAC
  V  T  G  H  E  A  F  E  E  A  M  M  K  E  P  P  K  A  W  T  K  A  Q  V  L  V  Y
AGCTTCTCCATCATTGCCTTCTTCTGCAGCACCATGAACGGCTACGACGGCTCGCTCATCAACAACCTGCTGCAGAACCC
  S  F  S  I  I  A  F  F  C  S  T  M  N  G  Y  D  G  S  L  I  N  N  L  L  Q  N  P
CTGGTTCAAGGCCAAGTACACTGTGGGAAACGACGGCATCTGGGCCGGCATTGTGTCTTCCATGTACCAGATTGGTGGTG
  W  F  K  A  K  Y  T  V  G  N  D  C  I  W  A  G  I  V  S  S  M  Y  Q  I  G  G
TCGTCGGCCCTTCCCTTTGTCGGCCCTGCCATTGACGGCTTTGGCCGCCGAATCGGCATGCTGTTGGGTGCCATCCTCATT
  V  V  A  L  P  F  V  G  P  A  I  D  G  F  G  R  R  I  G  M  L  L  G  A  I  L  I
GTCGTCGGCACCATCATCCAGGGTCTGTCAAACTCGCAGGGCCAGTTCATGGGCGGCCGCTTTCTGCTTGGATTCGGCGT
  V  V  G  T  I  I  Q  G  L  S  N  S  Q  G  Q  F  M  G  G  R  F  L  L  G  F  G  V

CTCCATTGCAGCGGCAGCGGGCCCCATGTACGTGGTTGAGATTAACCACCCTGCATACCGTGGACGCGTTGGCGCCATGT
  S  I  A  A  A  A  G  P  M  Y  V  V  E  I  N  H  P  A  Y  R  G  R  V  G  A  M
ACAACACTCTCTGGTTCTCGGGTGCCATCATCTCGGCCGGTGCCGCTCGAGGCGGCCTCAACGTCGGAGGCGACTACTCG
  Y  N  T  L  W  F  S  G  A  I  I  S  A  G  A  A  R  G  G  L  N  V  G  G  D  Y  S
TGGCGACTCATCACCTGGCTCCAGGCCCTCTTCTCCGGCCTCATCATCATCTTCTGCATGTTCCTGCCCGAGTCCCCCCG
  W  R  L  I  T  W  L  Q  A  L  F  S  G  L  I  I  I  F  C  M  F  L  P  E  S  P  R
CTGGCTCTACGTGCACCACAAGAAGGACGCCGCCAAGGCTGTGCTCACCAAGTATCATGGCAACGGAAACCCCGACTCCG
  W  L  Y  V  H  H  K  K  D  A  A  K  A  V  L  T  K  Y  H  G  N  G  N  P  D  S
TCTGGGTCCAGCTCCAGCTCTTCGAGTATGAGCAGCTCCTCAACATGGACGGCGCCGATAAGCGCTGGTGGGATTACCGG
  V  W  V  Q  L  Q  L  F  E  Y  E  Q  L  L  N  M  D  G  A  D  K  R  W  W  D  Y  R
GCGCTCTTCCGCTCGCGCGACGCCGTCTACCGTCTGTTGTGCAACGTCACCATCACCATTTTTGGCCAGTGGGCTGGCAA
  A  L  F  R  S  R  A  A  V  Y  R  L  L  C  N  V  T  I  T  I  F  G  Q  W  A  G  N
TGCGGTTCTTTCCTACTTCCTCGGCTCCGTCCTCGATACGGCCGGCTACACGGGCACCATTGCGCAGGCCAACATCACGC
  A  V  L  S  Y  F  L  G  S  V  L  D  T  A  G  Y  T  G  T  I  A  Q  A  N  I  T
TCATCAACAACTGCCAGCAGTTCGCCTGGGCCATTCTGGGCGCCTTCCTGGTCGACCGCGTTGGTCGTCGCCCCTTGCTG
  L  I  N  N  C  Q  Q  F  A  W  A  I  L  G  A  F  L  V  D  R  V  G  R  R  P  L  L
CTCTTCTCCTTTGCTGCCTGCACCGTGGTCTGGCTGGGCATGACGGTTGCCTCATCCGAATTTGCGCAGTCGTTCATCGG
  L  F  S  F  A  A  C  T  V  V  W  L  G  M  T  V  A  S  S  E  F  A  Q  S  F  I  G
AAATGACGCCAACGGCGATCCCATCTACAGCAACCCCAGCGCTTCCAAGGCTGCCCTGGCCATGATCTTCATCTTTGGTG
  N  D  A  N  G  D  P  I  Y  S  N  P  S  A  S  K  A  A  L  A  M  I  F  I  F  G
CCGTCTACTCTGTGGGCATCACTCCTCTGCAGGCCCTGTATCCCGTCGAGGTGCTCTCCTTTGAGATGCGCGCCAAGGGC
  A  V  Y  S  V  G  I  T  P  L  Q  A  L  Y  P  V  E  V  L  S  F  E  M  R  A  K  G
ATGGCCTTTTCCAGCTTTGCCACCAACGCTGCTGGACTCCTGAACCAGTTTGCATGGCCCGTGTCCATGGACAAGATTGG
  M  A  F  S  S  F  A  T  N  A  A  G  L  L  N  Q  F  A  W  P  V  S  M  D  K  I  G
CTGGAAGACGTACATTATCTTTACCATCTGGGATCTCGTCCAGACGGTTGTCGTCTACTTTTTCATTCCCGAGACCAAGG
  W  K  T  Y  I  I  F  T  I  W  D  L  V  Q  T  V  V  V  Y  F  F  I  P  E  T  K
GACGCACTTTGGAAGAGCTTGACGAAATCTTCGAGGCCAAGAACCCGGTCAAGACGTCGACGACGAAGAAGGCCGTGGCC
  G  R  T  L  E  E  L  D  E  I  F  E  A  K  N  P  V  K  T  S  T  T  K  K  A  V  A
GTGGACAGCCACGGCGACATTGTCAATATCGAGAAGGCTTAATGCCACGGACTTTTACTTGCGGTCACGATACTATACCA
  V  D  S  H  G  D  I  V  N  I  E  K  A  *
CTATATCAAGAATATCTGGGCAGTTGTGCGCAGGGCTTGGGGCTGTGAGCTGATGTTTTGTTTCGATGGTTCCTTGTCAG
GGCAGGAGGAAACAACTTTGGTTGCTATTTTAGCTGTTACTTTTGTTTCCGCTGATAATTGTGAAATATGAGGGTGAGGG
GAGAGCCAAGAGGAGAGCCCTAAGGATGCAGCATGAATCTGCAACTGCCACACAGCCTGCTATTCTCAATGGATGCGTGC
CTCTTTCGCAAAAAAAAAAAAAAAAAAAAAAAAAAAAAAAAAAAAA
```

Fig. 7. Nucleotide sequence and deduced amino acid sequence from the cDNA encoding the cellulose-induced oligosaccharide transport protein. The signal peptide region is in italics text by a dashed line encompassing residues 1-63.

controlled by carbon catabolite repression. Similar strategies have been successfully implemented for deregulating expression of amylotic enzymes in fungi (Fiedurek et al. 1987; Allen et al. 1989; Gamo et al. 1994). In one case the yield of glucoamylase increased 200% in a transport-deficient mutant (Gamo et al. 1994).

5. CONCLUSIONS

We have demonstrated that the combination of SSH and cDNA microarray technologies is an effective approach for the discovery of differentially expressed genes involved in biomass utilization. Using this approach in a high-throughput manner we were able to identify and catalogue a diverse assortment of biomass-induced genes from the cellulolytic fungus *T. reesei*. In addition to the genes that encode previously-known (expected) enzymes involved in hydrolysis of cellulose and hemicellulose the catalogue of biomass-induced genes elucidated in this study includes a number of novel genes that encode hydrolytic enzymes (glycosyl hydrolases and carbohydrate esterases) that may contribute to the symphony of gene products that are involved in degradation of plant biomass materials.

Perhaps one of the most exciting results emanating from this work was the revelation that a significant number of the cellulose-induced cDNA clones (90) encode proteins that are "unknown" or hypothetical in that they have no known functions, although there may be homologues in other microorganisms. These clones represented the largest fraction of those that we found to be cellulose/PCS induced on DNA microarrays. Only 20 of these clones could be assembled into contigs, and most of the clones were represented only once. A few of these unknowns have signal peptides and/or cellulose-binding modules, suggesting that they may participate in the breakdown of cellulose/hemicellulose. Future investigations may allow for the expression and biochemical characterization of these novel genes/proteins in order to elucidate their activities and substrate specificities. The present study will complement the genome sequencing and annotation that are ongoing at the time of this writing.

It has not been overlooked that in the course of this work we also identified a number of clones corresponding to genes that were either down-regulated during growth on cellulosic substrates or up-regulated when cells were grown on glucose medium. At present we have not investigated these genes in detail, however, the completed *T. reesei* genome sequence is expected soon, and it expected to provide an extensive catalogue of all CDSs and allow for a thorough elucidation of various regulatory networks.

Acknowledgments: We gratefully acknowledge the assistance of Michael Lamsa and Brian Gorre-Clancy for help with robotic colony picking and automated liquid-handling. We also gratefully recognize Beth Nelson, and Preethi Ramaiya for their help with nucleotide sequence analysis, and Dave Steuer for *T. reesei* fermentations.

REFERENCES

Allen KE, McNally MT, Lowendorf HS, Slayman CW, and Free SJ (1989). Deoxyglucose-resistant mutants of *Neurospora crassa*: isolation, mapping, and biochemical characterization. J Bacteriol 171: 53-58.

Aro N, Saloheimo A, Ilmén M, and Penttilä M (2001). ACEII, a novel transcriptional activator involved in regulation of cellulase and xylanase genes of *Trichoderma reesei*. J Biol Chem 276: 24309-24314.

Arst HN, Tollervey D, Dowzer CEA, and Kelly JM (1990). An inversion truncating the *creA* gene of *Aspergillus nidulans* results in carbon catabolite derepression. Mol Microbiol 4: 851-854.

297

Bailey CR, and Arst HN (1975). Carbon catabolite repression in *Aspergillus nidulans*. Eur J Biochem 51: 573-577.

Barnett CC, Berka RM, and Fowler T (1991). Cloning and amplification of the gene encoding an extracellular β-glucosidase from *Trichoderma reesei*: Evidence for improved rates of saccharification of cellulosic substrates. Bio/Technol 9: 562-567.

Beck MT, Holle L, and Chen WY (2001). Combination of PCR subtraction and cDNA microarray for differential gene expression profiling. Biotechniques 31: 782-786.

Coutinho PM and Henrissat B (1999a). Carbohydrate-active enzymes: an integrated database approach. In *Recent Advances in Carbohydrate Bioengineering*, HJ Gilbert, G Davies, B Henrissat and B Svensson eds., The Royal Society of Chemistry, Cambridge, pp. 3-12.

Coutinho, PM and Henrissat B (1999b). The modular structure of cellulases and other carbohydrate-active enzymes: an integrated database approach. In *Genetics, Biochemistry and Ecology of Cellulose Degradation*, K Ohmiya, K Hayashi, K Sakka, Y Kobayashi, S Karita and T Kimura eds., Uni Publishers Co., Tokyo, pp. 15-23.

Dean FB, Nelson JR, Giesler TL, and Laskin RS (2001). Rapid amplification of plasmid and phage DNA using Phi29 DNA polymerase and multiply-primed rolling circle amplification. Genome Res 11: 1095-1099.

De la Serna L, Ng D, and Tyler BM (1999). Carbon regulation of ribosomal genes in *Neurospora crassa* occurs by a mechanism which does not require Cre-1, the homologue of the *Aspergillus* carbon catabolite repressor, CreA. Fungal Genet Biol 26: 253-69.

Diatchenko L, Lau Y-FC, Campbell AP, Chenchik A, Moqadam F, Huang B, Lukyanov S, Lukyanov N, Sverdlov ED, and Siebert PD (1996). Suppression subtractive hybridization: A method for generating differentially regulated or tissue-specific cDNA probes and libraries. Proc Nat Acad Sci USA 93: 6025-6030.

Eisen MB, and Brown PO (1999). DNA arrays for analysis of gene expression. Methods Enzymol 303: 179-205.

Ewing B and Green P (1998). Base-calling of automated sequencer traces using phred. II. Error probabilities. Genome Res 8: 196-194.

Fiedurek J, Paszczynski A, Ginalska G, and Ilczuk Z (1987). Selection of amylolytically active *Aspergillus niger* mutants to 2-deoxy-D-glucose. Zentralbl Mikrobiol 142: 407-412.

Foreman PK, Brown D, Dankmeyer L, Dean R, Diener S, Dunn-Coleman NS, Goedegebuur F, Houfek TD, England GJ, Kelley AS, Meerman HJ, Mitchell T, Mitchinson C, Olivares HA, Teunissen PJM, Yao J, and Ward M (2003). Transcriptional Regulation of Biomass-Degrading Enzymes in the Filamentous Fungus *Trichoderma reesei*. J Biol Chem 278: 31988-31997.

Gamo FJ, Lafuente MJ, and Gancedo C (1994). The mutation DGT1-1 decreases glucose transport and alleviates carbon catabolite repression in *Saccharomyces cerevisiae*. J Bacteriol 176: 7423-7429.

Gordon D, Abajian C, and Green P (1998). Consed: A graphical tool for sequence finishing. Genome Res 8: 195-202.

Hedrick SM, Cohen DI, Nielsen EA, and Davis MM (1984). Isolation of cDNA clones encoding T-cell specific membrane-associated proteins. Nature 308: 149-153.

Hubank M and Schatz DG (1994). Identifying differences in mRNA expression by representational difference analysis of cDNA. Nucl Acids Res 22: 5640-5648.

Ilmén M, Thrane C, and Penttilä M (1996). The glucose repressor gene *cre1* of *Trichoderma*: isolation and expression of a full-length and a truncated mutant form. Mol Gen Genet 251: 451-460.

Jekosch K and Kuck U (2000). Glucose dependent transcriptional expression of the *cre1* gene in *Acremonium chrysogenum* strains showing different levels of cephalosporin C production. Curr Genet 37: 388-95.

Jiang H, Kang D, Alexandre D, and Fisher PB (2000). RaSH, a rapid subtraction hybridization approach for identifying and cloning differentially expressed genes. Proc Nat Acad Sci USA 97: 12684-12689.

Liang, P., and Pardee, A.B. 1992. Differential display of eukaryotic messenger RNA by means of the polymerase chain reaction. Science 257: 967-971.

Lisitsyn N, Lisitsyn N, and Wigler M (1993). Cloning the differences between two complex genomes. Science 259: 946-951.

Margolles-Clark E, Tenkanen M, Nakari-Setälä T, and Penttilä M (1996a). Cloning of genes encoding α-L-arabinofuranosidase and β-xylosidase from *Trichoderma reesei* by expression in *Saccharomyces cerevisiae*. Appl Environ Microbiol 62: 3840-3846.

Margolles-Clark E, Tenkanen M, Soderlund H, and Penttilä M (1996b). Acetyl xylan esterase from *Trichoderma reesei* contains an active-site serine residue and a cellulose-binding domain. Eur J Biochem 237: 553-560.

Margolles-Clark E, Saloheimo M, Siika-aho M, and Penttilä M (1996c). The α-glucuronidase-encoding gene of *Trichoderma reesei*. Gene 172: 171-172.

Margolles-Clark M, Ilmén M, and Penttilä M (1997). Expression patterns of ten hemicellulase genes of the filamentous fungus *Trichoderma reesei*. J Biotechnol. 57: 167-179.

Nielsen H, Engelbrecht J, Brunak S, and von Heijne G (1997). A neural network method for identification of prokaryotic and eukaryotic signal peptides and prediction of their cleavage sites. Int J Neural Sys 8: 581-599.

Okada H, Tada K, Sekiya T, Yokoyama K, Takahashi A, Tohda H, Kumagai H, and Morikawa Y (1998). Molecular characterization and heterologous expression of the gene encoding a low-molecular-mass endoglucanase from *Trichoderma reesei* QM9414. Appl Environ Microbiol 64: 555-563.

Parkkinen T, Hakulinen N, Tenkanen M, Siika-aho M, and Rouvinen J (2004). Crystallization and preliminary X-ray analysis of a novel *Trichoderma reesei* xylanase IV belonging to glycoside hydrolase family 5. Acta Cryst D60: 542-544.

Penttilä M, Lehtovaara P, Nevalainen H, Bhikhabhai R, and Knowles J. (1986). Homology between cellulase genes of *Trichoderma reesei*: Complete nucleotide sequence of the endoglucanase I gene. Gene 45: 253-264.

Penttilä M (2001). ACEII, a novel transcriptional activator involved in regulation of cellulase and xylanase genes of *Trichoderma reesei*. J Biol Chem 276: 24309-24314.

Penttilä M (2003). ACEI of *Trichoderma reesei* is a repressor of cellulase and xylanase expression. Appl Environ Microbiol 69: 56-65.

Porkka KP, and Visakorpi T (2001). Detection of differentially expressed genes in prostate cancer by combining suppression subtractive hybridization and cDNA library array. J Pathol 193: 73-79.

Sakai K, Higuchi H, Matsubara K, and Kato K (2000). Microarray hybridization with fractionated cDNA: enhanced identification of differentially expressed genes. Anal Biochem 287: 32-37.

Saloheimo M, Lehtovaara P, Penttilä M, Teeri TT, Stahlberg J, Johansson G, Pettersson G, Claeyssens M, Tomme P, and Knowles JKC (1988). EGIII, a new endoglucanase from *Trichoderma reesei*: the characterization of both gene and enzyme. Gene 63: 11-22.

Saloheimo A, Henrissat B, Hoffren A, Teleman O, and Penttilä M (1994). A novel small endoglucanase gene *egl5* from *Trichoderma reesei* isolated by expression in yeast. Mol Microbiol 13: 219-228.

Saloheimo M, Nakari-Setälä T, Tenkanen M, and Penttilä M (1997). cDNA cloning of a *Trichoderma reesei* cellulase and demonstration of endoglucanase activity by expression in yeast. Eur J Biochem 249: 584-591.

Saloheimo A, Aro N, Ilmén M, and Penttilä M (2000). Isolation of the *ace1* gene encoding a Cys2-His2 transcription factor involved in regulation of activity of the cellulase promoter *cbh1* of *Trichoderma reesei*. J Biol Chem 275: 5817-5825.

Schena M, Shalon D, Davis RW, and Brown PO (1995). Quantitative monitoring of gene expression patterns with a complementary DNA microarray. Science 270: 467-470.

Schmoll M and Kubicek CP (2003). Regulation of *Trichoderma reesei* cellulase formation: lessons in molecular biology from an industrial fungus. Acta Microbiol Immunol Hungarica 50: 125-145.

Shoemaker S, Schweickart V, Ladner M, Gelfand D, Kwok S, Myambo K, and Innis M (1983). Molecular cloning of exo-cellobiohydrolase I derived from *Trichoderma reesei* strain L27. Bio/Technol 1:691-696.

Seta KA, Kim R, Kim H-W, Millhorn DE, and Beitner-Johnson D (2001). Hypoxia-induced regulation of MKP-1: Identification by subtractive suppression hybridization and cDNA microarray analysis. J Biol Chem 276: 44405-44412.

Takashima S, Iikura H, Nakamura A, Masaki H, and Uozumi T (1996). Analysis of Cre1 binding sites in the *Trichoderma reesei cbh1* upstream region. FEMS Microbiol Lett 145: 361-366.

Takashima S, Nakamura A, Hidaka M, Masaki H, and Uozumi H (1998). Isolation of the *creA* gene from the cellulolytic fungus *Humicola grisea* and analysis of CreA binding sites upstream from the cellulase genes. Biosci Biotechnol Biochem 62: 2364-70.

Takashima S, Nakamura A, Hidaka M, Masaki H, and Uozumi T (1999). Molecular cloning and expression of the novel fungal β glucosidase genes from *Humicola grisea* and *Trichoderma reesei*. J Biochem 125: 728-736.

Teeri TT, Lehtovaara P, Kauppinen S, Salovuori I, and Knowles J (1987). Homologous domains in *Trichoderma reesei* cellulolytic enzymes: gene sequence and expression of cellobiohydrolase II. Gene 51:43-52.

Thomas PS (1980). Hybridization of denatured RNA and small DNA fragments transferred to nitrocellulose. Proc Nat Acad Sci USA 77: 5201-5205.

Timberlake WE and Barnard EC (1981). Organization of a gene cluster expressed specifically in the asexual spores of *Aspergillus nidulans*. Cell 26: 29-37.

Törrönen A, Mach RL, Messner R, Gonzalez R, Kalkkinen N, Harkki A, and Kubicek CP (1992). The two major xylanases from *Trichoderma reesei*: characterization of both enzymes and genes. Bio/Technol 10: 1461-1465.

van Peij NNME, Visser J, and de Graaff LH (1998). Isolation and analysis of *xlnR*, encoding a transcriptional activator co-ordinating xylanolytic expression in *Aspergillus niger*. Mol Microbiol 27: 131-142.

Vautard-Mey G, Cotton P, and Fevre M (1999). The glucose repressor *CRE1* from *Sclerotinia sclerotiorum* is functionally related to *CREA* from *Aspergillus nidulans* but not to the *MIG* proteins from *Saccharomyces cerevisiae*. FEBS Lett 453: 54-8.

Velculescu VE, Zhang L, Vogelstain B, and Kinzler KW (1995). Serial analysis of gene expression. Science 270: 484-487.

von Heijne G (1984). How signal sequences maintain cleavage specificity. J Mol Biol 173:243-251.

Wu CH, Yeh LL, Huang H, Arminski L, Castro-Alvear J, Chen Y, Hu ZZ, Ledley RS, Kourtesis P, Suzek BE, Vinayaka CR, Zhang J and Barker WC (2003). The Protein Information Resource. Nucl Acids Res 31: 345-347.

Yang GP, Ross DT, Kuang KW, Brown PO, and Weigel R.J (1999). Combining SSH and cDNA microarrays for rapid identification of differentially expressed genes. Nucl Acids Res 27: 1517-1523.

Zhang L, Zhou W, Velculescu VE, Kern SE, Hruban RH, Hamilton SR, Vogelstain B, and Kinzler KW (1997). Gene expression profiles in normal and cancer cells. Science 276: 1268-1272.

**Applied Mycology
and Biotechnology**
An International Series
Volume 5. Genes and Genomics
© 2005 Elsevier B.V. All rights reserved

ELSEVIER

13

Genomics of Some Human Dimorphic Fungus

André Rodrigues Lopes, Luciano Angelo, Everaldo dos Reis Marques, Maria Helena de S. Goldman†, and Gustavo H. Goldman**

Faculdade de Ciências Farmacêuticas de Ribeirão Preto and †Faculdade de Filosofia, Ciências e Letras de Ribeirão Preto, Universidade de São Paulo; Av. do Cafe S/N, CEP 14040-903, Ribeirão Preto, São Paulo, Brazil (ggoldman@usp.br).

The behavior and mode of infection of the dimorphic pathogenic fungi *B. dematitidis*, *H. capsulatum*, and *P. brasiliensis* display many similarities. Thermally regulated dimorphism is their main defining genetical trait. Based on their phylogenetical origin and similar biological behavior, a comparative genomic approach could illuminate the mechanisms responsible for phase transition and virulence/pathogenicity in these fungi. Here, we will present some preliminary evidences about the usefulness of this approach. The occurrence of *P. brasiliensis* genes that do not show any similarity with the NCBI non-redundant databank, but high similarity with putative *H. capsulatum* genes, and their concomitant expression at the same phase, demonstrate the power of comparative genomics in these species to understand their biology. Furthermore, promoters from different virulence and yeast phase-specific genes from *B. dermatitidis* and *H. capsulatum* demonstrate reciprocal functionality. This indicates the existence of common regulatory networks for virulence/pathogenicity and phase specificity among these three fungi.

1. INTRODUCTION

Most systemic infections of imunocompetent humans are caused by dimorphic fungi. This group of pathogens includes *Blastomyces dermatitidis*, *Histoplasma capsulatum*, and *Paracoccidioides brasiliensis* (Bd-Hc-Pb). Thermally regulated dimorphism is their main defining genetical trait. The behavior and mode of infection of these three pathogens display many similarities. The human host is generally infected by inhalation of airborne microconidia, which reach the pulmonary alveolar epithelium and transform into the parasitic yeast form (Borges-Walmsley et al. 2002; Magrini et al. 2001; Restrepo et al. 2001; Brandhorst et al. 2002; Woods 2001; 2002). There are still many blanks in our knowledge about the epidemiology of these pathogens. Cladistic analysis of partial 26S rRNA sequences from these three fungi showed that these species were closely related, with about 5 % base differences (Gueho et al. 1997; Bialek et al. 2000) by using 18S

Corresponding author: Gustavo H. Goldman

rRNA gene showed that *P. brasiliensis* and *B. dermatitidis* are more closely related than *H. capsulatum* and *B. dermatitidis*. It has been suggested by Medoff et al.(1986) by treating *H. capsulatum* mycelia with the sulfhydryl inhibitor *p*-chloromercuriphenylsulfonic acid (PCMS) which prevents conversion to the yeast form, that the transition from mold to yeast morphology is a requirement for pathogenesis. The dimorphic transition occurs simultaneously with changes in the cell wall composition, such as migration and reorganization of membrane lipids, especially glycosphingolipids (GLS) (Levery et al. 1998, Toledo et al. 1999; Vigh et al. 1998) and structural alterations in the carbohydrate polymers. As the fungus adopts the yeast form, an increase in the chitin content is observed in the cell wall, followed by a change in the glucan anomeric structure from a β-1,3-linked polymer to an α-1,3-glucan (San-Blas and Niño-Veja 2001). The surface α-glucan may have a role as a protective layer against the host defense mechanisms because of the incapacity of phagocytic cells to digest α-1,3-glucan (San-Blas 1982). Medoff et al. (1987) demonstrated that the temperature-induced transition is accompanied by a shared and well-characterized sequence of biochemical events: after an increase in the temperature from 25 °C to 37 °C, there is a rapid decline in intracellular ATP that follows the uncoupling of oxidative phosphorylation. This is followed by a progressive decrease in respiration rate (stage 1). After 24-40 hours, cells enter a dormant period (stage 2) that can last as long as 4-6 days. In stage 3, cytochrome components are restored, normal respiration is resumed, yeast-phase-specific cysteine oxidase is induced, and the transition to yeast morphology is completed.

Based on their phylogenetical origin and similar biological behavior, a comparative genomic approach could illuminate the mechanisms responsible for phase transition and virulence/pathogenicity in these fungi. Here, we will present some preliminary evidences about the usefulness of this approach. Initially, we will discuss about the genomics of *C. albicans* and if it is appropriate to use this organism as a model system for Bd-Hc-Pb. Since there are only ongoing genome projects for *H. capsulatum* (http://genome.wustl.edu/projects/hcapsulatum/) and *P. brasiliensis* (Goldman et al. 2003; Marques et al. 2004; Felipe et al. 2003; Magrini et al. 2004) most of our comparisons will be made with these two species. However, we will also present interesting data about heterologous reciprocous expression of promoters of genes encoding a phase-specific and virulence factor proteins from *H. capsulatum* and *B. dermatitidis*, respectively.

2. *Candida albicans*: THE MODEL DIMORPHIC FUNGUS ?

The predominant cause of fungal infections in hospitalized patients remains *C. albicans*, a pathogenic yeast that causes oral, vaginal, and systemic infections (De Backer et al. 2000). *C. albicans* is a comensal of the gastrointestinal and genitourinary tract, but it can also be an important pathogen that cause a range of conditions including painful superficial infections, such as vaginitis in healthy women, surface infections of the mouth and esophagus in human immunodeficiency virus (HIV) patients, and life-threatening blood stream infections among vulnerable intensive care patients, such as those undergoing cancer chemotherapy or immunosuppressive therapy following organ or bone marrow transplant procedures (Kao et al. 1999; Sudbery et al. 2004). The ability to switch between yeast, hyphal and pseudohyphal morphogenesis is often considered to be necessary for virulence, although formal proof remains lacking (Sudbery et al. 2004). There are an impressive number of molecular tools to work on the biology of *C. albicans* (Magee et al. 2003). Surely, it can be said that *C. albicans* is the model system for fungal pathogens. However, how close is *C. albicans* from Bd-Hc-Pb and how related are the virulence determinants and the dimorphic transition among these organisms ? Can *C. albicans* be used as a model dimorphic fungus to understand the virulence and/or pathogenicity of Bd-Hc-Pb ? Certainly, many conceptual aspects of virulence/pathogenicity of this organism can be applied to Bd-Hc-Pb, such as adhesion, penetration through the tissues via protease production, mechanisms of macrophage survival etc. (for a review, see van Burik and Magee 2001). These concepts will evolve as soon as other true pathogen genomes are sequenced, e.g. *Cryptococcus neoformans*. In this sense, the comparison among all the fungal pathogen genomes is advisable and must be done in order to identify genes that could have a role in virulence/pathogenesis in these species.

The diploid genome sequence of the fungal pathogen *C. albicans* was published by the group headed by Ronald Davis at the Stanford DNA Sequencing and Technology Center (Jones et al. 2004). Because *C. albicans* has no known haploid or homozygous form, sequencing was performed as a whole-genome shotgun of the heterozygous diploid genome in strain SC5314, a clinical isolate that is the parent of strains widely used for molecular analysis. Heterozygosity at numerous alleles originally resulted in single-copy genes being assigned to two distinct contigs (Odds et al. 2004). To solve this problem, this group has developed computational methods to eliminate the problems of aligning sequence contigs for an organism with no known haploid state.

C. albicans has eight distinct chromosomes that constitute a haploid genome size of 6,419 open reading frames (ORF) longer than 100 codons, of which some 20 % have no known counterpart in other available genome sequences. The codon CUG, which is translated abnormally by *C. albicans* as serine rather than leucine, is found at least once in approximately two-thirds of ORFs. *C. albicans* has a very high frequency of polymorphisms, i.e., one polymorphism per 237 bases (Jones et al. 2004). These heterozygosities are distributed unevenly across the *C. albicans* genome, with the highest prevalence on chromosomes 5 and 6. Highly polymorphic loci include the mating type-like (*MTL*) locus and a region of chromosome 6 that encodes several genes in the agglutinin-like sequence (*ALS*) gene family, that was described as involved in adhesion to and interaction with host surfaces (Zhao et al. 2003; Odds et al. 2004). Over half of the *C. albicans* genes contain allelic differences, and two-thirds of these polymorphisms are predicted to alter the protein sequence (Jones et al. 2004). The significance of the extensive allelic differences in *C. albicans* is unknown but may function to increase genetic diversity and contribute to the evolution of drug resistance (Jones et al. 2004).

Another interesting aspect revealed by the *C. albicans* genome is the fact that sulfur metabolism appears to differ between *C. albicans* and *S. cerevisiae* (Jones et al. 2004). *C. albicans* has genes likely to encode a direct pathway to cysteine in addition to a transsulfuration pathway from homocysteine. Genes encoding cysteine catabolic enzymes may also be present. These additional cysteine pathways might reflect an increased significance for glutathione metabolism in *C. albicans*. This information could be highly relevant to Bd-Hc-Pb considering the essentiality of organic sulfur for the yeast phase growth. Early reports on the mycelial-to-yeast transition of *H. capsulatum* attributed an important role to sulfhydryl groups, mainly in the form of cysteine (Maresca et al. 1977; 1978). Maresca et al. (1977) showed that a sulphydryl blocking drug, *p*-chloromercuriphenylsulfonic acid (PCMS) blocks the mycelium to yeast transition at 37 °C. Maresca et al. (1978) demonstrated the presence, in the yeast phase only, of a cystine reductase enzyme. Additionally, a cysteine dioxigenase activity was previously identified as specific to the yeast phase of *Histoplasma capsulatum* (Maresca and Kobaysashi 2000). It has been previously shown that *P. brasiliensis* mycelial phase can grown in the presence of both organic and inorganic sources of sulfur while yeast cells can only grow in the presence of a sulfur-containing amino acid (Paris et al. 1985).

Hwang et al. (2003) extended these results by showing that the gene that encodes this protein is more expressed in *H. capsulatum* yeast than mycelia. These same authors by using a microarray approach have shown that several yeast-expressed genes which share sequence similarity to genes involved in sulfur metabolism in other organisms: choline sulfatase; ATP sulfurylase (the first enzyme in the sulfate-assimilation arm of the methionine/cysteine biosynthetic pathway).; glutamate-cysteine ligase (which affects glutathione and glutamate metabolism).; and methionine permease (which mediate both methionine and cysteine uptake in *S. cerevisiae*; Kosugi et al. 2002). Marques et al. (2004) have identified by suppression subtraction hybridization (SSH). screening, the *CDO1* gene that encodes a cysteine dioxygenase in *P. brasiliensis*. Earlier, Goldman et al. (2003). have recognized by sequencing *P. brasiliensis* ESTs, *APS1*, *CHS1*, *GCL1* genes and more recently the same group also identified the *MEP1* gene in a *P. brasiliensis* random sequencing tag project (Nobrega and Goldman, unpublished data).. that encode adenylsulfate kinase, choline sulfatase, glutamate-cysteine ligase, and methionine permease, respectively. Marques et al. (2004) verified that these genes are more expressed in the yeast than in the mycelia, suggesting a possible role, like in *H. capsulatum* for the sulfur metabolism in the maintenance of the yeast state.

Another aspect that could display a synergistic effect in terms of oxidative protection and the sulfur organic source is the high temperature that is necessary for the yeast phase growth. Lethal heat stress generates oxidative stress in *S. cerevisiae*, and anaerobic cells are several orders of magnitude more resistant than aerobic cells to a 50 °C heat shock (Davidson and Schiestl 2001). As above mentioned, the transition from 25 ° to 37 °C in Bd-Hc-Pb is accompanied by marked changes in respiration and rapid decline in intracellular ATP levels, due to uncoupling of oxidative phosphorylation (Maresca and Kobayashi 2000). This is possible, because the mitochondria of fungi possess an alternative form of respiration in addition to the cytochrome pathway, which is resistant to cyanide and is mediated by alternative oxidase, an enzyme that transfers electrons from reduced ubiquinone to molecular oxygen (for a review, see Vanlerbeghe and McIntosh L 1997). When electrons produced from the oxidation of NADH flow through the alternative pathway, two of the three sites of energy conservation are by-passed and the level of ATP is decreased. The alternative oxidase expression in *P. brasiliensis* was shown to parallel the behavior of the enzyme in *H. capsulatum*, where the cytochrome system and the alternative oxidase decrease in parallel over the first 24-40 hours during the M-Y transition (Goldman et al. 2003). Even with the ATP decrease, growth in the yeast phase at 37 ºC should give birth to increased oxygen free radical production due to the growth at such elevated temperature. These observations suggest that antioxidant systems in Bd-Hc-Pb could be extremely important not only for survival in macrophages but also for detoxification of the oxygen free radicals. Marques et al. (2004) showed that *TSA1* (thiol-specific antioxidant) and *GST1* genes, encoding a thioredoxin peroxidase and a glutathione-S-transferase, respectively, are more expressed in the *P. brasiliensis* yeast phase than in the mycelial phase. Accordingly, these genes could be used as antioxidants to reduce the reactive oxygen species during the yeast growth. Curiously, Choi et al. (2003) identified Tsa1p as highly expressed in the hyphae of *C. albicans* by two-dimensional gel electrophoresis.

Whether organic sulfur is needed for satisfaction of a nutritional requirement of the yeastlike cells or for regulation of the oxidation-reduction potential of the medium at levels required for induction of conversion from the mycelial to the yeast phase or for both is not known. However, *C. albicans* might help in the elucidation of this interesting problem.

3. FEW PHASE-SPECIFIC GENES HAVE BEEN IDENTIFIED IN DIMORPHIC FUNGI.

There are very few genetic determinants isolated from Bd-Hc-Pb which are involved either in phase transition or virulence/pathogenicity (Table 1). Curiously, each of these genetic determinants seems to be unique, since it is only present in each respective species. The products of genes induced or upregulated upon this conversion may play a role in the structural or metabolic changes required for morphogenesis or may be necessary for survival in and colonization of the host (Rooney et al. 2001). The number of genetic determinants involved in these processes in dimorphic fungi will surely increase due to genomic and pos-genomic strategies that are already in progress or being established (http://genome.wustl.edu/projects/hcapsulatum/; Venancio et al. 2002; Hwang et al. 2003; Goldman et al. 2003; Marques et al. 2004).

Venancio et al. (2002) used differential display (DD) to isolate and identify differentially expressed genes of *P. brasiliensis* in the two cell types yeast (Y) and mycelium (M) as well as at different time intervals during temperature induced M to Y transition. The authors were able to detect the presence of at least 20 differentially transcribed cDNA fragments. Three of these DNA fragments (M32, M51, and M73) were found to be specific for the mycelial form while two cDNA fragments (M-Y1 and M-Y2) were upregulated during M to Y transition.

To identify phase-regulated genes of *H. capsulatum*, Hwang et al. (2003) carried out expression analysis by using a genomic shotgun microarray representing approximately one-third of the genome, and identified 500 clones that were differentially expressed. Genes induced in the mycelial phase included several involved in conidiation, cell polarity, and melanin production in other organisms. Genes induced in the yeast phase included several involved in sulfur metabolism. Interestingly, differential regulation of the site of transcript initiation was also observed in the two phases.

Goldman et al. (2003) have surveyed expressed genes in the yeast phase of *P. brasiliensis*. These authors obtained the partial sequences of 4,692 expressed genes that were functionally classified by similarity to known genes. They have identified several *Candida albicans* virulence and pathogenicity homologues in *P. brasiliensis*. Furthermore, they have analysed the expression of some of these genes during the dimorphic yeast-mycelium-yeast transition by real-time RTPCR. The higher expression of *OLE1* (desaturase) the heat shock proteins (*HSP70, -82*, and *-104*) and ubiquitin (*UBI1*) supports the hypothesis that changes in membrane order and ability to monitor protein folding and degradation are important factors in determining the dynamics of entry into the yeast phase, since these genes display the same expression profile during the transition mycelium-yeast. Felipe et al (2003) extended this analysis by sequencing ESTs from both phases. Computer subtraction analysis of these ESTs revealed several genes possibly expressed in stage-specific forms of *P. brasiliensis*.

Marques et al. (2004) have undertaken supression subtraction hybridization (SSH). and macroarray analyses with the aim of identifying genes that are preferentially expressed in the yeast phase of *P. brasiliensis*. Genes identified by both procedures as being more highly expressed in the yeast phase are involved in basic metabolism, signal transduction, growth and morphogenesis, and sulfur metabolism. In order to test whether the observed changes in gene expression reflect the differences between the growth conditions rather than differences intrinsic to the cell types, these authors performed real-time RTPCR experiments using RNAs derived from both yeast cells and mycelia that had been cultured at 37 and 26 °C in either complete or minimum media. Twenty genes, including *AGS1* (alpha-1,3-glucan synthase) and *TSA1* (thiol-specific antioxidant) were shown to be more highly expressed in the yeast cells than in the hyphae. Although their levels of expression could be different in rich and minimal media, there was a general tendency for these genes to be more highly expressed in the yeast cells.

As above mentioned, the most striking change during the mycelium-yeast transition in all three fungi is the modification of the β-1,3-glucan to α-1,3-glucan. The main reason to this modifications in terms of the effects on virulence/pathogenicity is unknown but it is possible that α-(1,3)-glucan functions in a protective manner, or its effect on strain virulence might have more to do with structural alterations of the cell wall (e.g. allowing proper surface presentation of proteins). Brandhorst et al. (2002) considered α-(1,3)-glucan to be a virulence-associated trait rather than a virulence factor *per se*. Recently, Rappleye et al. (2004) used RNAi targeting *AGS1* (encoding α-(1,3)-glucan synthase) to deplete levels of α-(1,3)-glucan, a cell wall polysaccharide.

Table 1: List of some putative virulence/pathogenicity and phase-specific factors of dimorphic fungal pathogens.

Fungus	Gene	Virulence/pathogenicity; Expression phase	Function	References
P. brasiliensis	*GP43*	Yeast/Mycelium	Putative adhesin	Cisalpino et al. (1996).
	M32, M51, and M73	Mycelium	Unknown function	Venancio et al. (2002)
	M-Y1 and M-Y2	Upregulated during M to Y transition	Unknown function	Venancio et al. (2002).
H. capsulatum	*CBP1*	Yeast	Calcium-binding protein	Sebghati et al. (2000).
	yps-3	Yeast	Unknown	Keath and Abidi (1994).
	yps21:E9	Yeast	Unknown	Abidi et al. (1998).
	AGS1	Yeast	Synthesis of the α-1,3 glucan in the cell wall	Rappleye et al. (2004).
B. dermatitidis	*BAD1*	Yeast	Adhesion, phagocytosis, modulation of host response	Brandhorst et al. (1999).
	Bys1	Yeast	Host response unknown	Burg and Smith (1994).

Loss of α-(1,3) -glucan by RNAi yielded phenotypes indistinguishable from an *AGS1* deletion: attenuation of the ability to kill macrophages and colonize murine lungs. This demonstrates for the first time that α-(1,3) -glucan is an important contributor to virulence in dimorphic fungi, more specifically to *H. capsulatum*.

4. DATA MINING IN *P. BRASILIENSIS* AND *H. CAPSULATUM*

Although most of the *H. capsulatum* and *B. dermatitidis* isolates used in research are haploid, there is still a controversial issue about the ploidy of *P. brasiliensis*. The comparison of the genome sizes estimated by pulse-field gel electrophoresis of several *P. brasiliensis* isolates with those calculated by microfluorometry indicated the possible existence of haploid and diploid (or aneuploid) isolates of this fungus (Feitosa et al. 2003). The isolate Pb18 was considered by these authors as a diploid isolate. Recently, we described a collection of about 4,500 Expressed Sequence Tags (ESTs) for *P. brasiliensis* isolate Pb18 (Goldman et al. 2003). If the isolate Pb18 is really diploid and assuming that both alleles are transcribed, we thought that it could be a good opportunity to investigate the levels of polymorphism in this isolate. Of course, this will not provide a clear answer to the ploidy problem since the yeast cells are multinucleate and each nucleus could accumulate different point mutations in different loci. However, this could be a first approach to the identification of polymorphism in this fungus, what could have at least some utility for some typing procedures, such as multilocus sequence typing (MLST).

As a first step, we selected from a collection of 1,430 contigs those that had 4 or more reads. By using the program Polybayes (Marth et al. 1999) we chose the longest read (from a CAP3 clustering). and considered this read our query, aligning the other reads as subjects. Sixty-two percent of these polymorphisms are present in the central region of these alignments. In these 1,430 contigs, after filtering for paralogs, only 475 fulfill our criteria of having at least four reads. We were able to identify 322 contigs (68 %) without any SNPs (single nucleotide point mutations) and 153 contigs (32 %; 54 have one SNP and 99 have two or more SNPs). These figures are much lower than those observed for the diploid *C. albicans* that has an overall average frequency of one polymorphism in 237 bases (in *P. brasiliensis*, we have observed one polymorphism in 583 bases) considerably higher than those observed in human or *Anopheles* sequence (Jones et al. 2004). What is the nature of the SNPs observed ? Are they due to mutations in different nuclei or due to the diploid/aneuploid state ? The results are too preliminary to draw a conclusion but because of the low frequency of SNPs, we favour the hypothesis that the isolate Pb18 is rather aneuploid or the SNPs are due to the accumulation of point mutations in the different nuclei for a specific locus.

Another aspect that drove our attention was the number of clusters in our *P. brasiliensis* EST collection (Goldman et al. 2003) 1,808 (about 38 % of 4,692 clusters) that have not displayed any identity with other genes in the NCBI non-redundant databank. We hypothesized that among these clusters could be present a number of genes unique to *P. brasiliensis* and/or a number of genes that are present only in these three dimorphic fungi. In this case, their function could be related to specific features of this group, such as phase transition. To investigate this latter possibility, we made comparisons of this subset of clusters (here named "non hits") against *H. capsulatum* genome (http://genome.wustl.edu/projects/hcapsulatum/) by using BLASTn and tBLASTx algorithms (Altschul et al. 1997). As it can be seen in Figure 1, we were able to identify several *H. capsulatum* putative genes with high identity to some of the *P. brasiliensis* clusters, suggesting they are possibly homologs. What is the biological function of these genes ? As a first step to identify their biological function, we assessed their expression in the mycelium and yeast phases from both species. As can be seen in Table 2, the genes show an enriched mRNA expression in the corresponding phase in each species. These preliminary data demonstrate the great potential of comparative genomics for the comprehension of the biology of these organisms.

5. COMPARISON BETWEEN PROMOTER ACTIVITY AND GENES SPECIFICALLY PRESENT IN DIMORPHIC FUNGI

As above mentioned, several genes whose expression is specific to the yeast phase have been identified in these three dimorphic fungi (see Table 1). The expression of *B.*

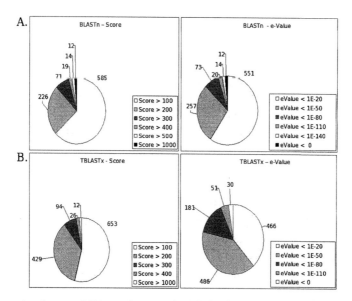

Fig. 1: Comparison between 1,808 putative genes from *P. brasiliensis* that do not show any similarity with other genes in public databanks (named "non hits") against the *H. capsulatum* genome (http://genome.wustl.edu/projects/hcapsulatum/).

Table 2 : Real-time RTPCR of the putative homologs of *P. brasiliensis* and *H. capsulatum*

Pb Cluster ID	X times more expressed*	Hc cluster ID	X times more expressed**
PbH1 (PbCS009-A12)	8.3 X Yeast	HcP1	Non determined
PbH2 (PbCS013-A05)	12.8 X Yeast	HcP2	2.7 x Yeast
PbH7 (PbCS018-F05)	7.5 X Mycelium	HcP7	Non determined
PbH8 (PbCS004-A02)	2.7 X Yeast	HcP8	19.0 X Yeast
PbH9 (PbCS011-E02)	1.7 X Yeast	HcP9	187.4 X Yeast

P. brasiliensis Pb18 was grown in minimal medium (Restrepo and Jimenez, 1981) for 10 days at 26 and 37 °C, respectively. The mRNA was extracted and cDNA synthesized. All the real-time PCR and RT-PCR reactions were performed using an ABI Prism 7700 Sequence Detection System (Perkin-Elmer Applied Biosystem, USA). The reactions and calculations were performed according to Semighini et al. (2002).
**H. capsulatum* was grown in Sabouraud medium and agar blood Muller-Hinton medium for 15-30 days at 26 °C and 37 °C, respectively. After mRNA extraction and cDNA synthesis, the real-time RTPCR was carried out as above mentioned.

dermatitidis BAD1 is restricted to the yeast phase and controlled transcriptionally by a mechanism that appears to be shared with at least one other systemic dimorphic fungus, *H. capsulatum*, suggesting that elements of this pathogenically crucial process may be conserved (Rooney et al. 2001). The *BAD1* upstream region showed two regions of high identity with the promoter of the *H. capsulatum* yeast phase-specific gene *yps-3* (Keath and Abidi 1994). The first is located from positions -175 to -237 upstream of the transcriptional start site (Hogan et al. 1995) of *BAD1*. This area, dubbed boxA, shares 84 % identity with a similarly spaced region in *yps-3* promoter. A second area, dubbed box B, is located from -325 to -360 upstream of the *BAD1* transcritional start site and shares 88 % identity with the corresponding region in the *yps-3* promoter (Rooney et al. 2001). The overall percentage similarity between 1 kb upstream of the *BAD1* transcriptional start site and that of *yps-3* in *H. capsulatum* is about 50 %, depending on the strain. In *H. capsulatum BAD1* transformants, yeast phase-specific expression of *BAD1* was conserved, and no transcript was detected in mycelia. *BAD1* β-galactosidase (*lacZ*) reporter fusions analysed in *B. dermatitidis* and *H. capsulatum* confirmed that *BAD1* is transcriptionally regulated in both fungi (Rooney et al. 2001). Serial truncations of the *BAD1* upstream region were fused to the *lacZ* reporter to define functional areas in the promoter (Rooney and Klein 2004). The 63-nucleotide box A region conserved in the *yps-3* upstream region was shown to be an essential component of minimal *BAD1* promoter. A fusion of the *yps-3* promoter to the *lacZ* indicated that this same region was needed for minimal *yps-3* promoter activity in *B. dermatitidis* tranformants. Reporter activity in *H. capsulatum* transformants similarly showed a requirement for box A in the minimal *BAD1* promoter (Rooney and Klein 2004).

Additional evidence for the biological role played by the box A came from experiments where two predicted sites within box A – a cAMP responsive element and a Myb binding site – were replaced in the *BAD1* promoter (Rooney and Klein 2004). This severely reduced transcriptional activity, indicating that these regions are critical for the yeast phase-specific expression of this gene. The conserved features of *BAD1* promoter activity and *BAD1* expression in *H. capsulatum* transformants lend further support to the theory that *BAD1* and *yps-3* are transcriptionally regulated by a similar mechanism (Rooney et al. 2001; Rooney and Klein 2004).

6. CONCLUSIONS

There are several evidences showing a high degree of conservation among *B. dermatitidis*, *H. capsulatum*, and *P. brasiliensis*. We are very close to finish the first genome of these three species, *H. capsulatum*. There is a large collection of ESTs already available for *P. brasiliensis*. Soon, *B. dermatitidis* will have a genome project with three-fold coverage started (Klein BS and Goldman WE, personal communication). This will open completely new avenues for research on these human pathogenic fungi. The occurrence of *P. brasiliensis* genes that do not show any similarity with the current databanks, but high similarity with putative *H. capsulatum* genes, and their concomitant expression at the same phase, demonstrate the power of comparative genomics in these species to understand their biology. Furthermore, promoters from different virulence and yeast phase-specific genes from *B. dermatitidis* and *H. capsulatum* demonstrate reciprocal functionality. Accordingly, the promoter region of *BAD1*, encoding a virulence factor from *B. dermatitidis*, is functional in *H. capsulatum*, and another promoter, from *H. capsulatum yps-3* gene, encoding a yeast phase-specific gene, is also functional in *B. dermatitidis*. This indicates the existence of common regulatory networks for virulence/pathogenicity and phase specificity among these three fungi. Since the number of scientists working on the molecular biology of these fungi is small, an integrated and comparative genomics approach would be extremely advantageous to the comprehension of their biology.

Acknowledgements: We would like to thank the Fundação de Amparo a Pesquisa do Estado de São Paulo (FAPESP) and Conselho Nacional de Desenvolvimento Científico e Tecnológico (CNPq), both from Brazil, for financial support for *P. brasiliensis* research.

REFERENCES

Abidi FE, Roh H and Keath EJ (1998). Identification and characterization of a phase-specific, nuclear DNA binding protein from the dimorphic pathogenic fungus *Histoplasma capsulatum*. Infect Immun 66: 3867-3873

Altschul SF, Madden TL, Schaffer AA, Zhang J, Zhang Z, Miller W and Lipman DJ (1997). Gapped blast and Psi-Blast: a new generation of protein database search programs. Nucleic Acids Res 25: 3389-3402

Bialek R, Ibricevic A, Fothergill A and Begerow D. Small subunit ribosomal DNA sequence shows *Paracoccidioides brasiliensis* closely related to *Blastomyces dermatitidis* (2000). J Clin Microbiol, 38: 3190-3193

Borges-Walmsley MI, Chen D, Shu X and Walmsley AR (2002). The pathobiology of *Paracoccidioides brasiliensis*. Trends Microbiol 10: 80-87

Brandhorst TT, Wuthrich M, Warner T and Klein B (1999). Targeted gene disruption reveals an adhesin indispensable for pathogenicity of *Blastomyces dermatitidis*. J Exp Med, 189: 1207-1216

Brandhorst TT, Rooney PJ, Sullivan TD and Klein BS (2002). Using new genetic tools to study the pathogenesis of *Blastomyces dermatitidis*. Trends in Microbiol 10: 25-30

Burg EFIII and Smith LH (1994). Cloning and characterization of *bys1*, a temperature-dependent cDNA specific to the yeast phase of the pathogenic dimorphic fungus *Blastomyces dermatitidis*. Infect Immun 62: 2521-2528

Choi, W, Yoo YJ, Kim M, Shin D, Jeon HB and Choi W (2003). identification of proteins highly expressed in the hyphae of *Candida albicans* by two-dimensional electrophoresis. Yeast 20: 1053-1060

Cisalpino PS, Puccia R, Yamauchi LM, Cano MIN, da Silveira JF and Travassos LR (1996). Cloning, characterization, and epitope expression of the major diagnostic antigen of *Paracoccidioides brasiliensis*. J Biol Chem 271: 4553-4560

Davidson JF and Schiestl RH (2001). Cytotoxic and genotoxic consequences of heat stress are dependent on the presence of oxygen in *Saccharomyces cerevisiae*. J Bacteriol 183: 4580-4587

De Backer MD, Magee PT and Pla J (2000). Recent developments in molecular genetics of *Candida albicans*. Annu Rev Microbiol 54: 463-498

Feitosa LS, Cisalpino PS, dos Santos MR, Mortara RA, Barros TF, Morais FV, Puccia R, da Silveira JF and de Camargo ZP. (2003). Chromosomal polymorphism, syntenic relationships, and ploidy in the pathogenic fungus *Paracoccidioides brasiliensis*. Fungal Genet Biol 39: 60-69

Felipe MS, Andrade RV, Petrofeza SS, Maranhão AQ, Torres FA, Albuquerque P, Arraes FB, Arruda M, Azevedo MO, Baptista AJ, Bataus LA, Borges CL, Campos EG, Cruz MR, Daher BS, Dantas A, Ferreira MA, Ghil GV, Jesuino RS, Kyaw CM, Leitao L, Martins CR, Moraes LM, Neves EO, Nicola AM, Alves ES, Parente JA, Pereira M, Pocas-Fonseca MJ, Resende R, Ribeiro BM, Saldanha RR, Santos SC, Silva-

Pereira I, Silva MA, Silveira E, Simões IC, Soares RB, Souza DP, De-Souza MT, Andrade EV, Xavier MA, Veiga HP, Venancio EJ, Carvalho MJ, Oliveira AG, Inoue MK, Almeida NF, Walter ME, Soares CM, and Brigido MM. (2003). Transcriptome characterization of the dimorphic and pathogenic fungus *Paracoccidioides brasiliensis* by EST analysis. Yeast 20: 263-271

Goldman GH, Marques EM, Ribeiro DCD, Bernardes LAS, Quiapin AC, Vitorelli PM, Savoldi M, Semighini CP, de Oliveira RC, Nunes LR, Travassos LR, Puccia R, Batista WL, Ferreira LE, Moreira JC, Bogossian AP, Tekaia F, Nobrega MP, Nobrega FG and Goldman MHS. (2003). EST analysis of the human pathogen *Paracoccidioides brasiliensis* yeast phase: identification of putative homologues of *Candida albicans* virulence/pathogenicity genes. Eukaryotic Cell 2: 34-48

Gueho E, Leclerc MC, de Hoog GS and Dupont B. (1997). Molecular taxonomy and epidemiology of *Blastomyces* and *Histoplasma* species. Mycosis 40: 69-81

Hogan, L.H., Josvai, S., and Klein, B.S. (1995). Genomic cloning, characterization, and functional analysis of the major surface adhesion WI-1 on *Blastomyces dermatitidis*. J Biol Chem 270: 39725-30732

Hwang L, Hocking-Murray D, Bahrami AK, Andersson M, Rine J, Sil A (2003). Identifying phase-specific genes in the fungal pathogen *Histoplasma capsulatum* by using a genomic shotgun microarray. Mol Biol Cell 14: 2314-2326

Jones T, Federspiel NA, Chibana H, Dungan J, Kalman S, Magee BB, Newport G, Thorstenson YR, Agabian N, Magee PT, Davis RW and Scherer S. (2004). The diploid genome sequence of *Candida albicans*. Proc Natl Acad Sci USA 101: 7329-7334

Keath EJ and Abidi FE (1994). Molecular cloning and sequence analysis of yps-3, a yeast-phase-specific gene in the dimorphic fungal pathogen *Histoplasma capsulatum*. Microbiology 140: 759-767

Kosugi A, Koizumi Y, Yanagida F and Udaka S (2001). MUP1, high affinity methionine permease, is involved in cysteine uptake by *Saccharomyces cerevisiae*. Biosci Biotechnol Biochem 65: 728-731

Levery SB, Toledo MS, Straus AH and Takahashi AK (1998). Structure elucidation of sphingolipids from the mycopathogen *Paracoccidioides brasiliensis*: an immunodominant β-galacto-furanose residue is carried by a novel glycosylinositol phosphorylceramide antigen. Biochemistry 37: 8764-8775

Magee PT, Gale C, Berman J and Davis D (2003). Molecular genetic and genomic approaches to the study of medically important fungi. Infect Immun 71: 2299-2309

Maresca B, Medoff J, Schlessinger D, Kobayashi GS and Medoff G (1977). Regulation of dimorphism in the pathogenic fungus *Histoplasma capsulatum*. Nature 266: 447-448

Maresca B, Johnson E, Medoff G and Kobayashi GS. 1978. Cystine reductase in the dimorphic fungus *Histoplasma capsulatum*. J Bacteriol 135: 987-992

Maresca B and Kobayashi GS (2000). Dimorphism in *Histoplasma capsulatum* and *Blastomyces dermatitidis*, p. 201-216. *In* J.F. Ernst, and A. Schimdyt (ed.), Dimorphism in Human Pathogenic and Apathogenic Yeasts. Contrib. Microbiol., Basel, Karger, vol.5

Magrini V and Goldman WE (2001). Molecular mycology: a genetic toolbox for *Histoplasma capsulatum*. Trends Microbiol 9: 541-546

Magrini V, Warren WC, Wallis J, Goldman WE, Xu J, Mardis ER and McPherson JD (2004). Fosmid-based physical mapping of the *Histoplasma capsulatum* genome. Genome Res 14: 1603-1609

Marques ER, Ferreira MES, Drummond RD, Felix JM, Menossi M, Savoldi M, Travassos LR, Puccia R, Batista WL, Carvalho KC, Goldman MHS and Goldman GH (2004). Identification of genes preferentially expressed in the pathogenic yeast phase of *Paracoccidioides brasiliensis* using suppression subtraction hybridization and macroarray differential analysis. Mol Gen Genom 271: 667-677

Marth GT, Korf I, Yandell MD, Yeh RT, Gu Z, Zakeri H, Stitziel NO, Hillier L, Kwok PY and Gish WR (1999). A general approach to single-nucleotide polymorphism discovery. Nat Genet 23: 452-456

Medoff G, Sacco M, Maresca B, Schlessinger D, Painter A, Kobayashi GS and Carratu L (1986). Irreversible block of the mycelial-to-yeast phase transition of *Histoplasma capsulatum*. Science 231: 476-479

Medoff, G, Kobayashi GS, Painter A and Travis S (1987). Morphogenesis and pathogenicity of *Histoplasma capsulatum*. Infect Immun 55: 1355-1358

Odds FC, Brown AJP and Gow NAR (2004). *Candida albicans* genome sequence: a platform for genomics in the absence of genetics. Genome Biology 5: 230

Paris S, Duran-Gonzalez S and Mariat F (1985). Nutritional studies on *Paracoccidioides brasiliensis*: the role of organic sulfur in dimorphism. J Med Vet Mycol 23: 85-9

Rappleye CA, Engle JT and Goldman WE (2004). RNA interference in *Histoplasma capsulatum* demonstrates a role for α-(1,3)-glucan in virulence. Mol Microbiol 53: 153-165

Restrepo A and Jimenez BE (1981). Growth of *Paracoccidioides brasiliensis* yeast phase in a chemically defined culture medium. J Clin Microbiol 12: 279-281

Rooney PJ, Sullivan TD and Klein BS (2001). Selective expression of the virulence factor BAD1 upon morphogenesis to the pathogenic yeast form of *Blastomyces dermatitidis*: evidence for transcriptional regulation by a conserved mechanism. Mol Microbiol 39: 875-889

Rooney PJ and Klein BS (2004). Sequence elements necessary for transcriptional activation of *BAD1* in the yeast phase of *Blastomyces dermatitidis*. Eukaryotic Cell 3: 785-794

San-Blas G (1982). The cell wall of fungal human pathogens: its possible role in host-parasite relationship. A review. Mycopathologia 79: 159-184

San-Blas G and Niño-Veja G (2001). *Paracoccidioides brasiliensis*: virulence and host response. *In* R.L. Cihlar and R.A. Calderone (eds.), Fungal pathogenesis: principles and clinical applications, Marcel Dekker, Inc. New York, N.Y.

Sebghati TS, Engle JT, Goldman WE (2000). Intracellular parasitism by *Histoplasma capsulatum*: fungal virulence and calcium dependence. Science 290: 1368-1372

Semighini CP, Marins M, Goldman MHS and Goldman GH (2002). Quantitative analysis of the relative transcript levels of ABC transporter *Atr* genes in *Aspergillus nidulans* by Real-Time Reverse Transcripition-PCR assay. Appl Environ Microbiol 68: 1351-1357.

Sudbery P, Gow N and Berman J (2004). The distinct morphogenic states of *Candida albicans*. Trends Microbiol 12: 317-324

Toledo MS, Levery SB, Straus AH, Suzuki E, Momany M, Glushka J, Moulton JM and Takahashi HK (1999). Characterization of sphingolipids from mycopathogens: factors correlating with expression of 2-hydroxy fatty acyl (*E*)-Δ[3]-unsaturation in cerebrosides of *Paracoccidioides brasiliensis* and *Aspergillus fumigatus*. Biochemistry 38: 7294-7306

Van Burik JA and Magee PT (2001). Aspects of fungal pathogenesis in humans. Ann Rev Microbiol 55: 743-772

Vanlerbeghe GC and McIntosh L (1997). Alternative oxidase: from gene to function. Annu Rev Plant Physiol Plant Mol Biol 48: 703-734

Venancio EJ, Kyaw CM, Mello CV, Silva SP, Soares CM, Felipe MS, Silva-Pereira L (2002). Identification of differentially expressed transcripts in the human pathogenic fungus *Paracoccidioides brasiliensis* by diferential display. Med Mycol, 40: 45-51

Vigh L, Maresca B and Horwood JL (1998). Does the membrane physical state control the expression of heat shock and other genes ? Trends Biochem Sci 23: 369-374

Woods J (2001). Pathogenesis of *Histoplasma capsulatum*. Seminars Respirat Infect 16: 91-101

Woods J (2002). *Histoplasma capsulatum* molecular genetics, pathogenesis, and responsiveness to its environment. Fungal Genetics Biol 35: 81-97

Zhao XM, Pujol C, Soll DR and Hoyer LL (2003). Allelic variation in the contiguous loci encoding *Candida albicans ALS5, ALS1* abd *ALS9*. Microbiology 149: 2947-2960

ELSEVIER

**Applied Mycology
and Biotechnology**
An International Series
Volume 5. Genes and Genomics
Published by Elsevier B.V.

14

Phanerochaete chrysosporium Genomics

Luis F. Larrondo[a], Rafael Vicuña[a] and Dan Cullen[b]
[a] Departamento de Genética Molecular y Microbiología, Facultad de Ciencias Biológicas,
Pontificia Universidad Católica de Chile, Santiago, Chile and Instituto Milenio de Biología
Fundamental y Aplicada, Santiago, Chile; [b]USDA Forest Products Laboratory, Madison,
Wisconsin 53705, USA (dcullen@facstaff.wisc.edu).

A high quality draft genome sequence has been generated for the lignocellulose-degrading
basidiomycete *Phanerochaete chrysosporium* (Martinez et al. 2004). Analysis of the genome in the
context of previously established genetics and physiology is presented. Transposable elements
and their potential relationship to genes involved in lignin degradation are systematically
outlined. Our current understanding of extracellular oxidative and hydrolytic systems is
described. Areas of uncertainty are highlighted and future prospects discussed in light of the
newly available genome data.

1. INTRODUCTION

Numerous microorganisms participate in the global conversion of organic carbon to
CO_2 with the concomitant reduction of molecular oxygen. The most abundant source of
carbon is plant biomass, composed primarily of cellulose, hemicellulose, and lignin.
Many fungi and bacteria are capable of degrading and utilizing cellulose and
hemicellulose as carbon and energy sources, but a much smaller group has evolved
with the ability to breakdown lignin, the most recalcitrant component of plant cell
walls. Collectively referred to as white rot fungi, these filamentous basidiomycetes
possess the unique ability to degrade lignin completely to CO_2 in order to gain access to
the carbohydrate polymers of plant cell walls for use as carbon and energy sources.
Such wood-decay fungi are common inhabitants of forest litter and fallen trees.

The enzymes from white rot fungi that catalyze degradation of lignin are
extracellular and unusually nonspecific. A constellation of oxidases, peroxidases, and
hydrogen peroxide are responsible for generating highly reactive and nonspecific free
radicals that can affect depolymerization and degradation of lignin. The nonspecific
nature and extraordinary oxidation potential of the enzymes from white rot fungi have
attracted considerable interest for industrial applications such as biological pulping of

Corresponding author: Dan Cullen

316

wood, fiber bleaching, and remediation of soils and effluents contaminated with a wide range of organopollutants.

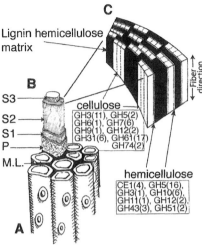

Fig. 1. Schematic illustration of wood tissue showing A. tracheid bundle; B. cell wall layers; and C. arrangement of carbohydrates and lignin within the S2 layer of the secondary wall (based on (Goring, 1977)). The bulk of the wall is in the S2 layer. In the model shown, lignin and hemicellulose form a matrix encrusting cellulose. The cellulose makes up approximately 45% of the weight of wood and is arranged in microfibril spirals along the long axis of the cell. Lignin is distributed throughout the S2 and middle lamella (M.L.). On the basis of their catalytic domains, enzymes hydrolysing the glycosidic bonds of cellulose and hemicellulose were assigned to glycosyl hydrolase (GH) families (http://afmb.cnrs-mrs.fr/CAZY/index.). Within families, the number of sequences detected by Martinez et al (2004) is indicated parenthetically. The precise function of genes within families is often unclear. Certain families, e.g. GH3, GH5, GH31 are quite diverse with respect to the biological function of members. Lignin depolymerization is believed to involve free radicals generated through the combined action of peroxidases, possibly small molecular weight mediators, and peroxide-generating oxidases (see Figure 2). P, primary wall; S1-S3, secondary wall layers.

The most intensively studied white rot organism is *P. chrysosporium*. Using a pure whole genome shotgun strategy, a high quality draft genome sequence has been assembled (www.jgi.doe.gov/whiterot). Initial analysis (Martinez et al. 2004) of the *P. chrysosporium* genome revealed features of importance to our understanding of lower eukaryotic gene structure and organization, identified hundreds of genes involved in lignocellulose degradation, and provided a framework for achieving a deeper understanding of degradative processes. In the following pages, we briefly summarize the microbiology and physiology of lignocellulose degradation. For more detail, readers are referred to previous reviews (Blanchette, 1991; Eriksson et al. 1990; Kirk and Farrell, 1987; Kirk and Cullen, 1998). Emphasis will be on the molecular genetics of *P. chrysosporium*. Other fungi are mentioned only as points of reference.

2. MICROBIOLOGY AND PHYSIOLOGY OF WOOD DECAY

The hyphae of white rot fungi rapidly invade wood cells and from within the lumen (Figure 1) secrete an array of enzymes and metabolites that depolymerize hemicelluloses, cellulose, and lignin. Constituting approximately 40% of the weight of wood, cellulose is a linear polymer of cellobiose units linked by β-1,4-glycosidic bonds. Individual cellulose molecules are arrayed in bundles known as microfibrils. The bulk of the cell wall is within the secondary wall (Figure 1; S1, S2 and S3 layers), and within these layers microfibrils have different parallel orientations with respect to the axis of the cell. Cellulose appears highly crystalline in diffraction measurements. Like cellulose, hemicelluloses are linear β-1,4-linked monosaccharide polymers. However hemicelluloses have mono-, di- or trisaccharides branches that may include sugars, sugar acids, acetylated sugars and sugar acid esters. Hemicellulose makes up 25 to 30% of the weight of wood and is covalently bound through infrequent linkages to lignin.

In contrast to the glycosidic linkages within cellulose and hemicellulose, lignin is comprised of carbon-carbon and ether bonds between phenylpropanoid residues (Higuchi, 1990; Lewis and Sarkanen, 1998). Consequently, lignin degradation involves oxidative mechanisms, as opposed to hydrolytic mechanisms. The polymer is stereoirregular, and the ligninolytic agents are generally assumed to be less specific relative to cellulases and hemicellulases. Extracellular peroxidases and oxidases are thought to play an important role in the initial depolymerization of lignin, and small molecular weight fragments are subsequently metabolized intracellularly, ultimately to water and carbon dioxide. It is generally believed that lignin depolymerization is necessary to gain access to cellulose and hemicellulose. No microbe, including any white rot species, is known to utilize lignin as a sole carbon source.

Only white rot basidiomycetes have been convincingly shown to efficiently mineralize lignin, although gross patterns of decay can differ substantially among species and strains (for review see Blanchette, 1991; Daniel, 1994; Eriksson et al. 1990). Microscopic analysis show *P. chrysosporium* strains simultaneously degrade cellulose, hemicellulose and lignin, whereas others such as *Ceriporiopsis subvermispora* tend to remove lignin in advance of cellulose and hemicellulose. How such selective degradation occurs is puzzling because enzymes are too large to penetrate sound, intact wood (Blanchette et al. 1997; Cowling, 1961; Flournoy et al. 1993; Srebotnik et al. 1988b; Srebotnik and Messner, 1991) Blanchette et al. (1997) have shown that during decay of pine by *C. subvermispora*, the walls gradually become permeable to insulin (5.7 kDa), and then to myoglobin (17.6 kDa), but not to ovalbumin (44.3 kDa), even in relatively advanced stages of decay. Because lignin-depolymerizing enzymes and many cellulases are in the same size range as ovalbumin, it has been proposed that small molecular weight oxidants penetrate from the lumens into the walls. Various diffusible oxidative species have been proposed.

Brown rot fungi, another category among homobasidiomycete wood decay fungi, do not degrade lignin but may be relevant to lignocellulose degradation by *P. chrysosporium*. These fungi rapidly depolymerize cellulose but only slowly modify lignin. Brown rot fungi are a major component of forest soils and litter, and they are responsible for most of the destructive decay of wood 'in service' (for review see Gilbertson, 1981; Worral et al. 1997). Recent molecular phylogeny suggests they have been repeatedly derived from white rot fungi (Hibbett and Donoghue, 2001). Depolymerization of crystalline cellulose proceeds long before wood porosity would admit cellulases, suggesting the participation of small molecular weight oxidants.

Hydroxyl radical, generated via the Fenton reaction (H_2O_2 + Fe^{2+} + H^+ → H_2O + Fe^{3+} + $\cdot OH$), has been strongly implicated as a diffusible oxidant in brown rot (Cohen et al. 2002; Cohen et al. 2004; Xu and Goodell, 2001), and to a lesser extent, in white rot wood degradation. The possibility of such a reactive oxygen species was long ago suggested in *P. chrysosporium* (Bes et al. 1983; Evans et al. 1984; Forney et al. 1982; Kirk and Nakatsubo, 1983; Kutsuki and Gold, 1982), but subsequent studies showed that Fenton reactions with lignin model compounds yielded products unlike those produced in ligninolytic cultures or by isolated peroxidases (Kirk et al. 1985). Still, some evidence supports Fenton system involvement in lignocellulose depolymerization by *P. chrysosporium* (Henriksson et al. 1995; Tanaka et al. 1999; Wood, 1994), particularly via cellobiose dehydrogenase (see below; for review see Henriksson et al. 2000a; Henriksson et al. 2000b). Current models for hydroxyl radical participation have been reviewed (Goodell, 2003) and typically involve generation of the highly reactive oxidant at or near the substrate. This might include small molecular weight chelators transferring iron along extracellular pH gradients (Xu and Goodell, 2001) or through cellulose binding as in the case of cellobiose dehydrogenase (Henriksson et al. 2000a; Henriksson et al. 2000b).

3. EXPERIMENTAL SYSTEMS
3.1. Experimental Tools

Advances on the molecular genetics of white rot fungi have been made possible by an array of experimental tools. For *P. chrysosporium*, methodology has been established for auxotroph production (Gold et al. 1982), recombination analysis (Alic and Gold, 1985; Gaskell et al. 1994; Krejci and Homolka, 1991; Raeder et al. 1989a), rapid DNA and RNA purification (Haylock et al. 1985; Raeder and Broda, 1985), differential display (Birch, 1998; Kurihara et al. 2002), pulsed field electrophoretic karyotyping (D'Souza et al. 1993; Gaskell et al. 1991; Orth et al. 1994), and genetic transformation by auxotroph complementation (Akileswaran et al. 1993; Alic et al. 1989; Alic et al. 1990; Alic et al. 1991; Alic, 1990; Randall et al. 1991; Zapanta et al. 1998) and by drug resistance markers (Gessner and Raeder, 1994; Randall et al. 1989; Randall et al. 1991; Randall and Reddy, 1992). Transformation efficiencies are relatively low and gene disruptions are difficult (Alic et al. 1993), but reporters for studying gene expression have been described (Birch et al. 1998; Gettemy et al. 1997; Ma et al. 2001). One of the most promising experimental approaches currently being adapted to *P. chrysosporium* is the use of two dimensional gel electrophoresis followed by mass spectrometry-based protein identification (Abbas et al. 2004; Shimizu et al. 2004).

As is common for basidiomycetes, the vegetative mycelium of *P. chrysosporium* is dikaryotic. However, clamp connections are absent and the cells are coenocytic

(Burdsall and Eslyn, 1974; Stalpers, 1984). The most widely studied strain, BKM-F-1767, produces abundant asexual spores, all of which are multinucleate and dikaryotic. Difficulties differentiating allelic variants from closely related genes and the lack of an accepted standardized nomenclature has complicated studies of gene families (Gaskell et al. 1994). Pulsed field gel electrophoresis identified 7-9 chromosomes with a haploid genome size of approximately 30 Mbp. Most allelic chromosomes differ in length (Gaskell et al. 1991; Gaskell et al. 1994; Kersten et al. 1995; Orth et al. 1994; Stewart et al. 1992). The underlying structure of such chromosome length polymorphisms (CLPs) is unknown although they are a common feature of fungal genomes (Zolan, 1995).

Single basidiospores of *P. chrysosporium* are fully viable and generally homokaryotic. Analyses of single basidiospore cultures have been used to differentiate alleles, and to create genetic and physical maps of *P. chrysosporium* (Covert et al. 1992a; Gaskell et al. 1992; Gaskell et al. 1994; Kersten et al. 1995; Li et al. 1997; Li and Renganathan, 1998; Raeder et al. 1989b; Schalch et al. 1989; Stewart et al. 1992; Stewart and Cullen, 1999). However, single basidiospore strains typically exhibit reduced sporulation, growth rate, and enzyme yields relative to the parental strain (Raeder et al. 1989b; Wyatt and Broda, 1995). Further, CLPs and other aspects of genome organization are not maintained through meiotic recombination, limiting the experimental value of basidiospores (Covert et al. 1992a; Gaskell et al. 1991; Kersten et al. 1995; Stewart et al. 1992; Zolan, 1995). A homokaryon of non-meiotic origin, RP-78, circumvents the disadvantages incurred by recombination and has greatly simplified assembly of genome sequence (Martinez et al. 2004; Stewart et al. 2000).

Beyond *P. chrysosporium*, *Pleurotus ostreatus* is probably the next best white rot experimental system offering transformation protocols (Honda et al. 2000; Irie et al. 2001; Sunagawa and Magae, 2002; Yanai et al. 1996) and methodology for physical (Larraya et al. 1999) and genetic mapping (Eichlerova and Homolka, 1999; Eichlerova-Volakova and Homolka, 1997; Larraya et al. 2000; Larraya et al. 2002). *Trametes versicolor* has also been transformed with drug resistance vectors (Bartholomew et al. 2001; Kim et al. 2002), and gene disruptions have been demonstrated (Dumonceaux et al. 2001). Aspects of the molecular biology of *P. chrysosporium* have been reviewed (Alic and Gold, 1991; Cullen and Kersten, 1996; Cullen, 1997; Gold and Alic, 1993; Pease and Tien, 1991).

3.2. Genome Sequencing

In a major research advance, the U.S. Department of Energy's Joint Genome Institute (JGI) has completed whole genome shotgun sequencing of *P. chrysosporium* strain RP-78 to 10.5X coverage. The draft assembly and interactive annotated browser are freely available at www.jgi.doe.gov/whiterot. The 30 Mb genome is distributed on 383 scaffolds greater than 2 kb, and the largest 165 scaffolds contain 90% of the assembled sequence. Contiguity of the largest scaffolds has been validated by genetic segregation analysis of terminal markers (Gaskell et al. 1994). Further support for the long-range structure of the assembly came from end sequencing cosmid clones. Of 1390 unique

ESTs derived from colonized wood, 98% were identified in the assembly and their positions noted on the browser.

Gene modeling predicted over 11,222 genes of which 8486 gave significant Smith-Waterman scores to known GenBank proteins. The taxonomic distribution and identification of conserved InterPro (Mulder et al. 2003) domains were reported and compared to the other published fungal genomes, *Saccharomyces cerevisiae* (Goffeau et al. 1996), *Schizosaccharomyces pombe* (Wood et al. 2002), and *Neurospora crassa* (Galagan et al. 2003; Mannhaupt et al. 2003; Schulte et al. 2002). Among other features, this analysis revealed a major expansion of the cytochrome P450s (below). Models not showing significant similarity to known proteins correspond to highly divergent, previously unrecognized genes perhaps unique to filamentous fungi, basidiomycetes, or white rot fungi, or spurious gene predictions.

One of the more distinguishing features of *P. chrysosporium* genome is the occurrence of large and complex families of structurally related genes. As described more fully below, families include cytochrome P450s, peroxidases, cellulases, copper radical oxidases and multicopper oxidases. In some cases, but not all, clustering is observed. The role of gene families in lignocellulose degradation remains uncertain. Structurally related genes may encode proteins with subtle differences in function, and such diversity may be needed to meet the challenges of changing environmental conditions (pH, temperature, ionic strength), substrate composition and accessibility, and wood species. Alternatively, some or all of the genetic multiplicity may simply reflect redundancy. Evidence against the latter view, albeit indirect, is that certain closely related genes are differentially regulated in response to substrate composition. It should also be mentioned that the genetic multiplicity of *P. chrysosporium* stands in stark contrast to *N. crassa*, where a repeat-induced point mutation system is believed to have greatly restricted the number and size of gene families. Providing further insight into these issues, analyses of the basidiomycetes *Filobasidiella neoformans* (= *Cryptococcus neoformans*) (http://www-sequence.stanford.edu/group/C.neoformans/index.html), *U. maydis* (http://www-genome.wi.mit.edu/seq/fgi/candidates.html), and *C. cinereus* (http://www-genome.wi.mit.edu/seq/fgi/candidates.html) will soon be published.

To be as current as possible, this review describes gene models recently 'mined' from the current database. However, we emphasize that these are generally automated predictions many of which are partially incorrect. This is a common problem in eukaryotic genomes, particularly among genes with multiple introns and short exons. Accordingly, proteins predicted from genomic sequence should be considered tentative until verified by cDNA analysis. Another qualification concerning the genome relates to inclusion of repeats. In short, whole genome shotgun assemblies such as *P. chrysosporium* typically exclude telomeres, rRNA clusters, and many repeats.

3.3. Repeats

Repetitive elements of *P. chrysosporium* have been associated with several genes encoding extracellular enzymes. The most thoroughly studied element had been Pce1, a repeat inserted within LiP allele *lipI2* (Gaskell et al. 1995). The 1747 nt sequence transcriptionally inactivates *lipl2* and three copies are distributed on the same chromosome. Sequence flanking these copies showed no evidence of recombination (Stewart et al. 2000). In addition to Pce1-like elements, a broad array of non-coding repetitive sequences and putative mobile elements have been identified by systematic examination of the genome database. Short repeats (<3kb) not clearly associated with transposons vary in copy number from >40 (GenBank accession number Z31724) to 4 (AF134289-AF134291)(Table 1).

Table 1. Simple non-coding repeats identified in *P. chrysosporium*

Location[1]	Type[2]	Accession	Probable copies	Comment
85:13061-14631	Pce1	L40593		Element inserted within lignin peroxidase gene *lipl*. Co-segregates with Pce2, Pce3, and Pce4.
223:16314-17591	Pce2	AF134289		Truncated at REND of scaffold.
57:144178-142237	Pce3	AF134290		
10:259013-257364	Pce4	AF134291		
91;115464-117260		Z31724	Portions distributed on >40 scaffolds. Most at termini, e.g. LEND s384, s28, s141; REND s292, s260, s136, s72.	Often associated with retroelements as direct or inverted repeats. First 350 nt of the Z31724 not located by blast.

[1]Location is defined by genome scaffold number : nucleotide coordinates on current assembly of the Joint Genome Institute's interactive browser (http://genome.jgi-psf.org/whiterot1/whiterot1.home.html) ;[2]The sequence, transcriptional impact, and genetic linkage of Pce elements have been reported (Gaskell et al. 1995; Stewart et al. 2000).

Several putative Class II elements, or DNA transposons, were identified in the *P. chrysosporium* genome (Table 2). Similar ascomycetous elements include *Aspergillus niger* Ant, *Cochiobolus carbonum* Fot1, *Nectria* "Restless", *Fusarium oxysporum* Tfo1, and *Cryphonectria parasitica* Crypt1 (Kempken and Kuck, 1998). Atypical of fungi but common in higher plant genomes, EN/Spm- and TNP-like elements were also found (Martinez et al. 2004). Interestingly, the *P. chrysosporium* DNA transposons are present in low copy numbers (1-4 copies) relative to Ascomycetes, where class II elements often exceed 50-100 copies.

A substantial number of multi-copy retrotransposons were identified in the database, some of which seem likely to impact expression of genes related to lignin degradation (Table 3). Typical of these elements, they often appear truncated and/or rearranged, and the long terminal repeats, characteristic of retroelements, often lie apart as "solo LTRs" (Goodwin and Poulter, 2000; Kim et al. 1998). Several non-LTR retrotransposons, similar to other fungal LINE-like retroelements, were also found by blast searches. Unusual for a fungal genome (Daboussi and Capy, 2003), *copia*-like retroelements are particularly abundant. In one case, the *copia* element gx.24.14.1 interrupts a cytochrome

P450 gene (models pc.24.16.1 + pc.24.17.1) within its seventh exon (gene model gx.24.14.1 (Martinez et al. 2004). A multicopper oxidase gene, *mco3* also has an inserted element (unpublished), and *gypsy* element pc.91.4.1 lies immediately adjacent to a gene encoding an extracellular peroxidase (gx.91.10.1).

Table 2. Putative class II elements identified[1] in *P. chrysosporium*

Gene model/location[2]	Type	Related[3] to:	Comments[4]
pc.197.6.1	Ac/hAT	*F. oxysporum* Folyt1 (AF057141)	Poor model. Possible ITRs.
pc.88.61.1	Ac/hAT	*F. oxysporum* Tfo1 (T00208)	Poor model. (210aa). Copy:pc.14.47.1 + pc.14.45.1
gw.65.44.1	Ac/hAT	*F. oxysporum* Tfo1 (T00208)	Poor model. (210aa).
pc.247.3.1+ pc.247.4.1	Ac/hAT	*C. parasitica* Crypt1 (AF283502)	Disjoint models. (76aa + 115aa).
pc.112.17.1	Tc1/Mariner	*A. niger* Ant1 (AF283502)	Partial model (320aa). Possible imperfect 65nt ITRs.
s213.13484-17985 (pc.213.8.1)	Tc1/Mariner	*Aspergillus niger* Ant1 (AF283502)	213 nt ITRs. Model overlaps dehydrogenase.
s128.64902-68226 (pc.128.43.1)	Tc1/Mariner	*Aspergillus niger* Ant1 (AF283502)	80 nt ITRs. Copy:pc.234.7.1 (REND; no ITRs)
pc.270.1.1(N-terminal) + gx.224.2.1(COOH region)	Fot1/Pogo	*Cochliobolus carbonum* Fot1-like. (JC5096)	N-terminus (239aa) & COOH (149aa) terminus on LEND & REND, respectively, of 2 scaffolds.
pc.25.43.1+gx.25.15.1	TNP	Putative *A. thaliana* TNP2 (AC005897). No fungal examples.	Poor models. Similar to higher plant TNP- & En/Spm-like elements. Copies:pc.15.120.1; pc.125.4.1
pc.90.8.1	TNP	Hypothetical carrot Tdc1(AB001569). No fungal examples.	Similar to higher plant TNP- and En/Spm-like elements.Copies:pc.125.4.1; pc.249.91.1

[1]Searches of Joint Genome Institute's interactive browser (http://genome.jgi-psf.org/whiterot1/whiterot1.home.html) by keywords (e.g. transposase, transpos, retroelement, transposon, retrotransposon) and by blast with known fungal class I and II elements; [2]Computer generated gene model designations (pc., gx., gw.) or, in cases of elements with terminal repeats, scaffold coordinates (s.nucleotide positions). Models are often truncated at either termini; [3]Following initial screening, trimmed sequences were identified by blast searches of NCBI database. Short sequences at scaffold termini and those located on short scaffolds were generally ignored; [4]Abbreviations: REND, right end of scaffold; LEND, left end of scaffold; ITRs, inverted terminal repeats.

Viewed together with the Pce1 mutant of *lipI*, it seems *P. chrysosporium* transposons have a proclivity for insertions within gene families. Recombination among these repetitive elements may be involved in the evolution of these families as well as in chromosome length polymorphisms (for review see Zolan, 1995). A novel class of eukaryotic DNA transposons, Helitrons, were recently identified in *P. chrysosporium*

(Poulter et al. 2003). None of the three *P. chrysosporium* Helitrons appear to be within or adjacent to functional genes. Tyrosine recombinase-encoding retrotransposons appear more abundant in the *P. chrysosporium* genome (Goodwin and Poulter, 2004), but again, the elements and their remnants are not inserted within recognizable structural genes.

Table 3. Identification[1] of class I elements in the *P. chrysosporium* genome.

Location or gene model[2]	Related[3] to:	LTRs / copies (c)/ comments (cm)
Copia-like retrotransposons (LTRs):		
s84.4772-9223 (gx.84.1.1)	*T. tabacum* retrotransposon TNT1 (P10978)	LTR: 333 nt c: pc.142.16.1; pc.209.1.1; pc.214.3.1; LTR at REND s23 cm: Few fungal examples, e.g. *Candida* (AF065434).
s24.47199-52675 (gx.24.14.1)	*T. tabacum* retrotransposon TNT1 (P10978)	LTR:420 nt c: LTR at REND s302 and s277; LTR at LEND s114 cm: Few fungal examples e.g. *S. cerevisiae* Ty2.
s235.5494-10983 (pc.235.3.1)	Copia type pol polypetptide rice (AC092553) and tobacco (T02206)	LTR: 127 nt c: gx.224.1.1 (w/LTR); gx.253.1.1 (w/LTR); gx.247.1.1 (w/LTR); pc.265.3.1 (w/LTR); gx.290.1.1(w/1LTR); gx.268.1.1; gx.84.1.1; gx.217.1.1(w/LTR); gx.293.1.1 cm: Few fungal examples e.g. *S. cerevisiae* Ty1 protein B (T29093).
gx.173.6.1	Gag protein of insects and higher plants. *Anopheles gambiae* (AF387862)	LTR: ~200 nt c: gw.14.1.1 cm: No pol polypeptide detected.
s24.202590-205821	*A. thaliana* pol polypeptide (AC006841)	LTR: 190 nt c: pc.324.2.1 + pc. 324.1.1 (w/LTR); cm: LTR at REND s145
Gypsy-like retrotransposons (LTRs):		
s166.34077-42356 (pc.166.16.1 + pc.166.17.1)	*Yarrowia lipolytica* retrotransposon Ylt1 (AJ310725)	LTR: 327 nt cm: Also similar to other fungal elements (e.g. MAGGY, CfT-1)
pc.233.1.1	*Yarrowia lipolytica* retrotransposon Ylt1 (AB310725)	c: s183.13150-27198 (w/LTR); pc.84.26.1; pc.141.13.1; s250. REND; gw.269.1.1; s152.LEND; gx.264.1.1; pc.65.56.1 + pc.65.57.1
s91.3017-11324 pc.91.4.1	*Magnaporthe grisea* MAGGY (L35053) and *Tricholoma matsuke* marY1 (AB028236)	LTR: 1023 nt c: gx.311.1.1; gx.241.2.1;pc.241.2.1; gx.219.2.1; s242.3501-8500 cm: Polymerase and integrase (91.4.1) similar to MAGGY and *Tricholoma* marY1. Gag polypeptide (pc.91.5.1) closer to *Glomeralla* and *Aspergillus*.

s21.40796-50402 (gx.21.9.1)	*Tricholoma matsuke* marY1 (AB028236)	LTR: 209 nt c:pc.14.4.1 cm: LTR flanking tandem duplication.
s91.69200-78439 (pc.91.37.1 + pc.91.36.1)	Most closely related to higher plant elements. Best fungal hit = *Glomerella* Cgret (AF264028).	LTR: 327 nt c: gw.195.3.1; gx.248.2.1; gx.204.1.1 + pc.204.3.1; gx.308.1.1; pc.98.1.1 cm: More distantly related to *S. pombe* Tf2.

Possible Gypsy-like Retrotransposon (LTRs not detected):

gx.305.1.1	*Tricholoma matsuke* marY1 (AB028236)	LTR: probably lost c: pc.291.1.1; gx.329.1.1; gx2.266.1.1; gx.170.1.1; gx.328.1.1 cm: No gag detected.

Retroposons (LINE-like):

gx.214.3.1 + gw.214.5.1	gx.214.3.1 most like *Candida albicans* RT AF139376.	cm: gw.214.5.1 encodes RnaseH with no fungal hits. Z31724 simple repeat located near COOH terminus.
gx.419.1.1	*Tricholoma matsuke* marY2 (AB047280)	cm: Similar to other fungal LINEs. Truncated at COOH terminus (LEND scaffold) & missing polyA.
pc.36.36.1(RnaseH) + pc.36.37.1(pol2)+ pc.36.38.1(pol1)+ pc.36.39.1(gag)	*Tricholoma matsuke* marY2 (AB047280)	c: pc.47.81.1 (pol2) + pc.47.80.1 (pol1); pc.44.90.1 (pol2); pc.206.5.1 (pol1) + pc.206.6.1 (pol2) + pc.206.7.1 (pol2) cm: Similar to other fungal LINEs, especially *N. crassa* Tad1. Appears complete (gag, pol, polyA)

[1]Searches of Joint Genome Institute's interactive browser (http://genome.jgi-psf.org/whiterot1/whiterot1.home.html) by keywords (e.g. transposase, transpos, retroelement, transposon, retrotransposon) and by blast with known fungal class I elements (e.g. MAGGY, Tad1); [2]Computer generated gene model designations (pc., gw., gx. prefixes) or, in cases of elements with extended terminal repeats, scaffold (s) coordinates. Models are often truncated at either termini; [3]Following initial screening, trimmed sequences were blasted against NCBI databases; [4]Following initial screening, trimmed sequences were used to blastn and tblastn search the *P. chrysosporium* database. Short sequences at scaffold termini and those located on short scaffolds were generally ignored.

4. EXTRACELLULAR OXIDATIVE SYSTEMS

As mentioned above, the polymers constituting plant cell walls are too large to be taken up by fungal hyphae and extracellular depolymerization must occur. Accordingly, the vast majority of research on white rot fungi has focused on the secreted enzymes.

4.1. Lignin Peroxidases and Manganese-dependent Peroxidases

Since their discovery (Glenn et al. 1983; Gold et al. 1984; Paszczynski et al. 1985; Tien and Kirk, 1983, 1984), lignin peroxidase (LiP) and manganese peroxidase (MnP) have been the most intensively studied extracellular enzymes of *P. chrysosporium*. Review articles summarize the biochemistry (Higuchi, 1990; Kirk and Farrell, 1987; Kirk, 1988; Schoemaker and Leisola, 1990) and genetics (Alic and Gold, 1991; Cullen and Kersten,

1996; Cullen, 1997, 2002; Gold and Alic, 1993) of these enzymes. Both are protoporphyrin IX peroxidases. Isozymic forms are encoded by families of structurally related genes and further modified posttranslationally.

LiP catalyzed reactions include C_α-C_β cleavage of the propyl side chains of lignin and lignin models, hydroxylation of benzylic methylene groups, oxidation of benzyl alcohols to the corresponding aldehydes or ketones, phenol oxidation, and even aromatic cleavage of nonphenolic lignin model compounds (Hammel et al. 1985; Leisola et al. 1985; Renganathan et al. 1985; Renganathan and Gold, 1986; Tien and Kirk, 1984; Umezawa et al. 1986)(Figure 2). The importance of lignin peroxidase in the depolymerization of lignin *in vivo* was convincingly shown by Leisola et al. (Leisola et al. 1988). Partial depolymerization of lignin *in vitro* has been demonstrated for LiP (Hammel and Moen, 1991) and MnP (Wariishi et al. 1991).

MnP oxidizes Mn^{2+} to Mn^{3+}, using H_2O_2 as oxidant (Gold et al. 1984; Paszczynski et al. 1985). Activity of the enzyme is stimulated by simple organic acids which stabilize the Mn^{3+}, thus producing diffusible oxidizing chelates (Glenn and Gold, 1985; Glenn et al. 1986). Kinetic studies with Mn^{2+} chelates support a role for oxalate in reduction of MnP Compound II by Mn^{2+}, and physiological levels of oxalate in *P. chrysosporium* cultures stimulate manganese peroxidase activity (Kishi et al. 1994; Kuan et al. 1993). In addition to the oxidases (reviewed below), extracellular H_2O_2 may also be generated by the oxidation of organic acids secreted by white-rot fungi. Specifically, Mn(II)-dependent oxidation of glyoxylate and oxalate generates H_2O_2 (Kuan et al. 1993; Urzua et al. 1998).

The crystal structures of LiP (Edwards et al. 1993; Piontek et al. 1993; Piontek et al. 2001) and MnP (Sundaramoorthy et al. 1997) show similarities; the active site has a proximal His ligand H-bonded to Asp, and a distal side peroxide-binding pocket consisting of a catalytic His and Arg. Kinetic studies of MnP variants indicate that manganese-binding involves Asp-179, Glu-35, Glu-39 (Kusters-van Someren et al. 1995; Sollewijn Gelpke et al. 1999; Whitwam et al. 1997; Youngs et al. 2001) and a heme propionate, consistent with x-ray crystallographic analysis (Sundaramoorthy et al. 1997).

Recent genome analyses (Martinez et al. 2004) located the 10 known LiP genes previously designated *lipA* through *lipJ* (Gaskell et al. 1994). Eight of the LiP genes were found clustered within 3% recombination (Gaskell et al. 1994; Stewart and Cullen, 1999), which corresponds to 96 kb (Martinez et al. 2004). cDNAs were previously reported for genes *mnp1*, *mnp2*, and *mnp3* (Alic et al. 1997; Orth et al. 1994; Pease et al. 1989; Pribnow et al. 1989), and two new MnP genes were revealed by Blast searches of the genome (Martinez et al. 2004). One of these new genes has been designated *mnp4* (gene model, 15.18.1) (Martinez et al. 2004). Unexpectedly, *mnp4* was found to lie approximately 5 kb upstream from *mnp1* (model 15.23.1), and a cytochrome P450 gene is located in the *mnp4*-*mnp1* intergenic region. Recent data shows that *mnp4* is actively transcribed when *P. chrysosporium* is grown on wood-containing soil samples (Stuardo et al. 2004). Gene model 9.126.1, *mnp5*, (Martinez et al. 2004) corresponds to the N-terminal amino acid sequence of an MnP purified from *P. chrysosporium*-colonized wood pulp (Datta et al.

1991). Future work requires a more detailed analysis of the expression profile of these newly identified *mnp* genes. The five *mnp* sequences are remarkably conserved (Figure 3). The number and positions of introns are also conserved, particularly *mnp1* and *mnp4*, and this suggests a recent duplication (Figure 4) (Stuardo et al. 2004).

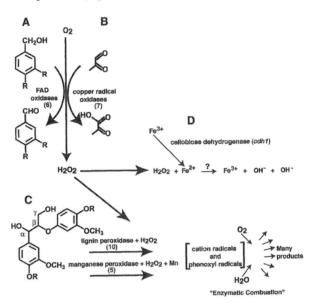

Fig. 2. Schematic representation of major extracellular oxidative enzymes produced by lignin degrading fungi. Generation of H_2O_2 is physiologically coupled to peroxidases. Benzyl alcohol derivatives (A) are substrates for FAD-dependent oxidases such as aryl alcohol oxidase (R= H or OCH_3). Methyl glyoxal (B) is a substrate for glyoxal oxidase and possibly for related copper radical oxidases. Peroxidase substrate C is a lignin model featuring the major β-O-4 linkage (R=H or ether linkage to additional monomeric units). Peroxidases abstract one electron from aromatic substrates which then undergo spontaneous degradation reactions or "enzymatic combustion" (Kirk and Farrell, 1987). The resulting small molecular weight fragments are further metabolized intracellularly to CO_2 and H_2O. Peroxide might also be a reactant in the spontaneous (non-enzymatic) generation of hydroxyl radical via Fenton's chemistry (D). Reduction of Fe^{3+} has been demonstrated for cellobiose dehydrogenase. Parenthetical numbers indicate structurally related sequences identified to date.

In addition to *mnp4* and *mnp5*, a partial *mnp*-like sequence encoding just the COOH-termimus was discovered. This partial 274 nt *mnp* sequence (gw.9.92.1), named *mnp6*, is located 85 kb downstream from *mnp5* (scaffold 9), and as observed for the latter, it is encoded by the minus strand. PCR amplification and sequencing verified *mnp6*, and excluded the possibility of assembly error (unpublished data). In addition, we confirmed that this partial sequence is present in both nuclei, and also in *P. chrysosporium* strain ME446 (unpublished data). When manually translated, this sequence shows the highest homology (65%) to aminoacids 299 to 370 of MnP5. An intron splits the codon for aa 355 in *mnp5*, and *mnp6* also possesses an 58 nt intervening sequence at that position. However, splicing of the intron adheres to the GT-AG rule only in its 5'; the 3' splice seems to be less well defined.

No evidence for the transcription of *mnp6* could be obtained. This, in addition to the presence of a stop codon in the middle of its sequence, and the low conservation of its intronic region suggest that *mnp6* is inactive and the possible product of an aberrant recombination or duplication event.

LiP and MnP genes have been identified in other species, and recent cladistic analysis by Martinez (Martinez, 2002) shows >50 invariant residues among approximately 30 known peroxidases. In general, the MnP and LiP genes fall within clearly defined clades and can be discriminated by certain key residues. As expected by its role in catalysis by long range electron transfer, Trp171 is common to LiPs, whereas Mn-binding residues (Glu35, Glu39, Asp179 in *mnp1*) are found in MnP sequences.

Certain peroxidases cannot be easily classified. Unusual *Pleurotus eryngii* sequences encode "versatile peroxidases," which have both LiP-like activities (oxidation of veratryl alcohol and an array of phenols) and MnP-like activities (Mn^{+2} oxidation) (Camarero et al. 2000; Ruiz-Duenas et al. 1999; Ruiz-Duenas et al. 2001). The *P. eryngii* genes have both Trp171 and the residues involved in Mn-binding. Searches of *P. chrysosporium* genome revealed a putative extracellular peroxidase related to the *Pleurotus* hybrid peroxidase (Ruiz-Duenas et al. 1999), but catalytic and Mn-binding residues are not conserved (Martinez et al. 2004). Designated *nop*, the *P. chrysosporium* gene (GenBank accession AY727765), shows novel structural features, which distinguish it from any of the aforementioned peroxidases (unpublished results).

It is now well established that the peroxidase genes of *P. chrysoporium* are differentially regulated by culture conditions. Holzbaur and Tien (Holzbaur and Tien, 1988) showed that steady state levels of *lipD* transcripts were far more abundant than those of *lipA* under carbon starvation. The situation was reversed under nitrogen starvation, i.e *lipA* transcripts dominated. Beyond these early Northern blot results, competitive RT-PCR and nuclease protection assays have been used for quantitative differentiation of the closely related transcripts (Reiser et al. 1993; Stewart et al. 1992) in defined media (Reiser et al. 1993; Stewart et al. 1992; Stewart and Cullen, 1999), in organopollutant contaminated soils (Bogan et al. 1996c), and in colonized wood (Janse et al. 1998). These studies have shown that differential regulation can exceed five orders of magnitude and that transcript profiles in defined media poorly predict profiles in complex substrates. Importantly, the observed patterns of expression show no clear relationship with genome organization. Post translational regulation by heme processing has been suggested (Johnston and Aust, 1994) but contradicted by the results of Li et al. (1994).

In addition to nutrient conditions, MnP production in *P. chrysosporium* is dependent upon Mn concentration (Bonnarme and Jeffries, 1990; Brown et al. 1990). The *P. chrysosporium* MnP genes *mnp1*, *mnp2*, and *mnp3* generally show coordinate regulation in colonized soil and wood (Bogan et al. 1996a; Janse et al. 1998). Putative metal

328

response elements (MREs) have been identified upstream of *mnp1* and *mnp2* and their transcript levels respond to Mn^{2+} supplements in low nitrogen media (Pease and Tien 1992a; Gettemy et al. 1998). In contrast, *mnp3*, lacks paired MREs and its transcript levels are not influenced by addition of Mn^{2+} (Gettemy et al. 1998; Pease and Tien, 1992). Taken together, these observation support a possible role for MREs in transcriptional regulation of *P. chrysosporium* MnP genes (Alic et al. 1997; Gettemy et al. 1998). However, MnP regulation appears not to involve MREs in *T. versicolor* (Johansson and Nyman, 1993; Johansson et al. 2002) or in *C.subvermispora* (Manubens et al. 2003). Another interesting *T. versicolor* gene, *npr*, appears repressed by Mn even though putative Mn-binding residues are present in the sequence (Collins et al. 1999).

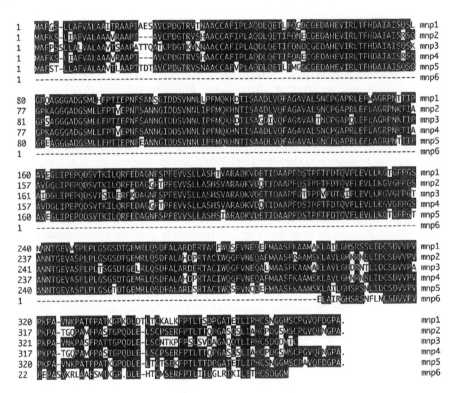

Fig. 3 Clustal W alignment of predicted manganese peroxidase proteins of *P. chrysosporium*. Genes *mnp1* through *mnp5* are supported by cDNA evidence. cDNAs corresponding to *mnp6* sequence have not been detected.

The role of MREs in Mn-regulated transcription of the *mnp* genes was recently examined by Ma et al (2004). Using a green fluorescent protein reporter system, a 48-bp sequence containing at least one Mn^{2+}-responsive *cis* element was identified. Further characterization suggests that a 33-nt portion is responsible for the observed regulation. None of the 6 putative MREs present in *mnp1* is contained in the aforementioned region, and functional evaluation of 4 of these MREs, show no significant effect in the Mn^{2+} response (Ma et al. 2004). Similar regulatory sequences were identified upstream of *mnp2* and *mnp3* (Ma et al. 2004). The new data suggest that in the absence of Mn, a negative control is exerted at that sequence, which is released in the presence of the metal. The existence of additional Mn-responsive *cis* acting sequences in the *mnp1* promoter is supported by the residual responsiveness to Mn when this *bona fide* sequence is absent (Ma et al. 2004).

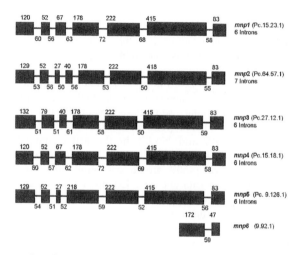

Fig. 4. Schematic representation of intron-exon composition of *P. chrysosporium* manganese peroxidase genes.

4.2. Copper Radical Oxidases

An important component of the ligninolytic system of *P. chrysosporium* is the H_2O_2 that is required as oxidant in the peroxidative reactions. A number of oxidases have been proposed to play a role in this regard. However, only one appears to be secreted in ligninolytic cultures; glyoxal oxidase (Kirk and Farrell, 1987). The temporal correlation of glyoxal oxidase (GLOX), peroxidase, and oxidase substrate appearances in cultures suggests a close physiological connection between these components (Kersten and Kirk, 1987; Kersten, 1990). Glyoxal oxidase is a glycoprotein of 68 kDa with two isozymic forms (pI 4.7 and 4.9). A number of simple aldehyde-, α-hydroxycarbonyl-, and α-dicarbonyl compounds are oxidized by GLOX. Lignin itself is a likely source of GLOX substrates. Oxidation of a β-O-4 model compound (representing the major substructure of lignin) by lignin peroxidase releases glycolaldehyde (Hammel et al. 1994). Glycolaldehyde is a substrate for GLOX and sequential oxidations yield oxalate and multiple equivalents of H_2O_2. Oxalate may be a source of chelate required for the manganese peroxidase reactions described above. Biochemical and spectroscopic investigations show structural similarities between GLOX and glactose oxidase and the correponding catalytic residues have been clearly identifed (Kersten et al. 1985; Kurek and Kersten, 1995; Whittaker, 2002; Whittaker et al. 1999).

The reversible inactivation of GLOX is of considerable physiological significance (Kersten, 1990; Kurek and Kersten, 1995). During enzyme turnover, GLOX becomes inactive in the absence of a coupled peroxidase system. The oxidase is reactivated, however, by lignin peroxidase and non-phenolic peroxidase substrates. Conversely, phenolics prevent the activation by lignin peroxidase. These observation show that glyoxal oxidase has a regulatory mechanism responsive to peroxidase, peroxidase substrates, and peroxidase products (e.g., phenolics resulting from ligninolysis). Also, lignin will also activate glyoxal oxidase in the coupled reaction with LiP (Kersten, 1990; Kurek and Kersten, 1995).

Glyoxal oxidase of *P. chrysosporium* is encoded by a single gene with two alleles (Kersten and Cullen, 1993; Kersten et al. 1995). On the basis of the catalytic similarities with *Dactylium dendroides* galactose oxidase, potential copper ligands were tentatively identified at Tyr377 and His378 (Kersten and Cullen, 1993). Subsequent studies also implicated Tyr135, Tyr70, and His471 in the active site (Whittaker et al. 1999). Surprisingly, Blast analysis of the genome has revealed 6 sequences with low overall sequence homology to *glx* (<50% amino acid similarity) but with highly conserved residues surrounding the catalytic site. Extended N-terminal domains of unknown function are present in new copper radical oxidase genes *cro3*, *cro4*, and *cro5*. On the basis of similarities to galactose oxidase, *cro1*, *cro2*, and possibly *cro6* may have propeptides which play a role in self-catalytic processing (Firbank et al. 2001; Rogers and Dooley, 2003; Xie and van der Donk, 2001). Our recent blast searches have found structurally related genes in a diverse array of fungi including, the ascomycete *Magnaporthe grisea* (www.broad.mit.edu/annotation/fungi/ ustilago_maydis/) and in the Badisiomycete, *Coprinus cinereus*. (www.broad.mit.edu/annotation/fungi/coprinus_cinereus/). Recently, one of these *M. grisea glx*-like sequences, *glo1*, was shown to be required for filamentous growth and pathogenicity (Leuthner et al. 2004). Elucidating the biological function of these copper radical oxidases remains a major challenge for future research.

One of the unexpected findings in the *P. chrysosporium* genome project was the identification of the aforemnentioned family of copper radical oxidases. In particular, copper radical oxidase genes *cro3*, *cro4*, and *cro5* were located within the cluster of LiP genes (Cullen and Kersten, 2004). The clustering of *lip* and *cro* genes seems consistent with a physiological connection between peroxidases and peroxide-generating oxidases. Of the seven copper radical oxidases of *P. chrysosporium*, only *glx* has been the focus of transcript analysis. Again consistent with a role in lignin degradation, *glx* transcripts are coincident with *lip* and *mnp* in defined media (Kersten and Cullen, 1993;

Stewart et al. 1992), soil (Bogan et al. 1996b), and wood (Janse et al. 1998).

4.3. FAD-dependent oxidases

Cellobiose dehydrogenase (CDH) is widely distributed among white-rot fungi. Its precise function is uncertain, but it may play a role in carbohydrate metabolism and in lignin degradation. The enzyme has two domains containing FAD and heme prosthetic groups, respectively. The two domains can be cleaved by *P. chrysosporium* proteases. CDH binds cellulose via a binding module in the flavin domain and oxidizes cellodextrins, mannodextrins, and lactose. Suitable electron acceptors include quinones, phenoxy radicals, molecular oxygen and Fe^{3+}. Several studies have emphasized the FeIII reductase activity of the heme domain and its implications in generating hydroxyl radicals via a Fenton reaction (Kremer and Wood, 1992a; Kremer and Wood, 1992b; Mason et al. 2003). One possible role for CDH may be enhancement of cellulases by relieving product inhibition (Cameron and Aust, 2001; Igarashi et al. 1998). Another possibility is that CDH generates hydroxyl radicals via Fenton type reactions thus oxidizing wood components including lignin. Potential CDH roles have been reviewed (Cameron and Aust, 2001; Henriksson et al. 2000a).

Genes encoding CDH have been cloned from several fungi including the white-rot fungi *P. chrysosporium* (Li et al. 1996; Raices et al. 1995), *Trametes versicolor* (Dumonceaux et al. 1998), and *Pycnosporus cinnabarinus* (Moukha et al. 1999). Sequences are highly conserved. All share a common architecture with separate FAD, heme, and cellulose binding domains (CBD), although the latter domain has no obvious similarity to functionally analogous bacterial or fungal CBDs. The heme ligands of *P. chrysosporium* CDH have been confirmed by site specific mutagenesis (Rotsaert et al. 2001). As mentioned above, the role of CDH in lignin degradation remains unsettled, but CDH gene disruptions in *T. versicolor* do not affected the ability to degrade synthetic lignin (Dumonceaux et al. 2001). A single CDH gene is present in *P. chrysosporium* as well as related white rot fungi, and full length sequences with non-trivial Smith-Waterman scores ($<e^{-20}$) are obvious in *A. nidulans*, *N. crassa*, and *M. grisea* genomes (http://www-genome.wi.mit.edu/annotation/fungi/).

Two glucose oxidases have been identified in *P. chrysosporium* cultures; glucose 1-oxidase from *P. chrysosporium* ME-446 (Kelley and Reddy, 1986) and glucose 2-oxidase or pyranose 2-oxidase from *P. chrysosporium* K3 (Eriksson et al. 1986). The peroxide-generating enzyme pyranose-2-oxidase is predominantly intracellular in liquid cultures of *P. chrysosporium*, but evidence supports an important role in wood decay (Daniel et al. 1994). The oxidase is preferentially localized in the hyphal periplasmic space and the associated membraneous materials. Pyranose 2-oxidase sequences have been reported from *C. versicolor* (Nishimura et al. 1996), related *Trametes* strains (Acc. Nos. P59097 and AAP40332) and most recently *P. chrysosporium* (de Koker et al. 2004). Transcript patterns for the *P. chrysosporium* pyranose 2-oxidase are similar to lignin peroxidases and glyoxal oxidase supporting a role in lignocellulose degradation (de Koker et al. 2004). A *P. chrysosporium* sequence highly similar (Smith-Waterman score = 480) to *A. niger* glucose-1-oxidase has been identified in the genome (Martinez et al. 2004).

Another strategy for peroxide generation may involve aryl alcohol oxidase (AAO) which is secreted by *Bjerkandera sp.* (de Jong et al. 1994). Chlorinated anisyl alcohols, synthesized *de novo* from glucose, are the preferred substrates. The oxidation products are reduced and recycled by the fungal mycelia. LiP does not oxidize the chlorinated anisyl alcohols. Various *Pleurotus* species may support a redox cycle supplying extracellular peroxide using AAO (or veratryl alcohol oxidase) coupled to intracellular aryl alcohol dehydrogenase (AAD) (Guillen and Evans, 1994; Marzullo et al. 1995; Varela et al. 2000b). In *Pleurotus ostreatus*, veratryl alcohol oxidase may participate in lignin degradation by supplying peroxide and by reducing quinones and phenoxy radical and thereby inhibiting the repolymerization of lignin degradation products (Marzullo et al. 1995). Genes encoding aryl alcohol oxidases have been characterized from *Pleurotus* spp. (Marzullo et al. 1995; Varela et al. 1999; Varela et al. 2000a; Varela et al. 2000b), and multiple AAO-like sequences ($\geq$4) have been identified in the *P. chrysosporium* genome (Martinez et al. 2004).

4.4. Multicopper Oxidases

Analysis of the genome has shown that, unlike other white rot fungi, *P. chrsysosporium* does not have any sequence encoding conventional laccases. Instead, it produces a multicopper oxidase possessing strong ferroxidase activity with catalytic parameters similar to those of yeast Fet3p (Larrondo et al. 2003). The physiological function of this protein (MCO1) is uncertain. The gene (*mco1*) is part of a cluster of 4 structurally related sequences located within 25 kb region. All 4 are transcribed, but only *mco1* has a clear secretion signal (Larrondo et al. 2004).

Multiple alignments of a large collection of fungal multicopper oxidase sequences, as well as structural comparison of MCO1, show that these MCOs are closer to Fet3 proteins, than to conventional laccases (Larrondo et al. 2003; Larrondo et al. 2004). Together with iron permease Ftr1 (Stearman et al. 1996), Fet3 ferroxidase (Askwith et al. 1994) plays a key role in iron homeostasis. Our recent clustal analysis of multicopper oxidases sequences in *P. chrysosporium*, *N. crassa* and *M. grisea*, databases (unpublished results) supports a new branch of the multicopper oxidase family in which the *P. chrysosporium mco* encoded sequences are in close association with two *M. grisea* sequences (MG00551.1, MG07771.1, www-genome.wi.mit.edu/annotation/fungi/ magnaporthe/) and with *C. neoformans* 'laccase". This branch is in close proximity to another harboring all known Fet3 proteins (Figure 5). Like *mco1*, the *C. neoformans* 'laccase' exhibits Fe^{2+} oxidation activity (Liu et al. 1999; Williamson, 1994).

Because of the intriguing similarity between *mco1* and *Fet3*, the *P. chrysosporium* genome database was searched for related sequences. A single gene highly homologous to *S. cerevisiae fet3* was identified at a separate locus less than 1 Kb from a *Ftr1*-like iron permease gene model. Transcripts of both are detected by RT-PCR. The *P. chrysosporium fet3* encodes a protein of 628 amino acids, 69 residues larger than *mco1* (unpublished results). Typical of ferroxidases, but unlike *mco1*, the COOH terminus has a predicted transmembrane domain.

Structural determinants that confer multicopper oxidases with ferroxidase activity have been identified (Askwith and Kaplan, 1998; Bonaccorsi di Patti et al. 2000; Bonaccorsi di Patti et al. 2001). Glu-185 and Tyr-354 are essential for the oxidation of Fe^{2+} by Fet3 from *S. cerevisiae*. These two residues are conserved in all known Fet3 proteins, whereas they are absent in ascorbate oxidases and laccases. The equivalent Glu residue is conserved in all but one of the MCO-like ferroxidases (Figure 6). However, the Tyr-354 is absent in most of these sequences, suggesting that Glu-185, but not Tyr, is essential for Fe^{2+} oxidation.

Our analysis supports the importance of the Glu-185 residue. The *C. neoformans* 'laccase' lacks this residue, which may explain its weak ferroxidase activity relative to MCO1 (Liu et al. 1999; Williamson, 1994). Supporting a new subfamily of multicopper oxidases, the abovementioned *M. grisea* sequences have the essential ferroxidase residues, as well as putative secretion signals. In addition, all these MCO-like sequences lack a COOH terminal transmembrane domain, which are common to Fet3 proteins. Consistent with separate subfamilies, blast analysis of the *M. grisea* genome reveals a *fet3* othologue (MG02156.1) at a locus separate from MCO1-like sequences. Thus, like *P. chrysosporium*, *M. grisea* seems to have extracellular ferroxidases, distinct from Fet3. Possibly, the similarities reflect the common requirements for attacking plant cell walls. In this context, it is interesting to note that both *P. chrysosporium* and *M. grisea* genome features an impressive number of glycosyl hydrolases. Our recent analysis of plant pathogens *Ustilago maydis* and *Fusarium graminearum* identified Fet3s, laccases, as well as MCO-like sequences. The saprophytes *Aspergillus nidulans* and *Coprinus cinereus* possess the first types of multicopper oxidases but lack MCO-like sequences in their genomes (unpublished results). A recently deposited multicopper oxidase sequence (GenBank AAR82933) from the basidiomicete *Auricularia auricula* falls directly into the MCO-like clade, posseses the discussed Glu residue, and does not have a COOH-terminal anchor.

The role of *mco1* remains unclear, although we have hypothesized that it might be involved in the control of Fenton-based chemical reactions in the extracellular medium (Larrondo et al. 2003). Recently, a similar function has been attributed to yeast Fet3p (Shi et al. 2003; Stoj and Kosman, 2003).

It should be noted that the absence of conventional laccases in *P. chrysosporium* does not exclude a role in lignin degradation in related fungi. Laccase genes, often occurring as multigene families, are widely distributed among lignin-degrading fungi (Cullen, 1997; Mayer and Staples, 2002; Thurston, 1994; Youn et al. 1995). Laccases oxidize the phenolic units in lignin to phenoxy radicals, which can lead to aryl-$C\alpha$ cleavage (Kawai et al. 1988). In the presence of certain mediators, the enzyme can depolymerize synthetic lignin (Kawai et al. 1999) and delignify wood pulps (Bourbonnais et al. 1997; Call and Muncke, 1997) suggesting a role in lignin biodegradation. White rot fungi such as *Pycnoporus cinnabarinus* efficiently degrade lignin, and in contrast to *P. chrysosporium*, secrete laccases but not peroxidases. Two laccase genes, closely related to sequences derived from other white rot fungi, have been characterized from *P. cinnabarinus* (Eggert et al. 1998; Temp et al. 1999). Also consistent with an important role for laccase

334

in *P. cinnabarinus*, "lac⁻" mutants are impaired in their ability to degrade ¹⁴C-labeled DHP (Eggert et al. 1997).

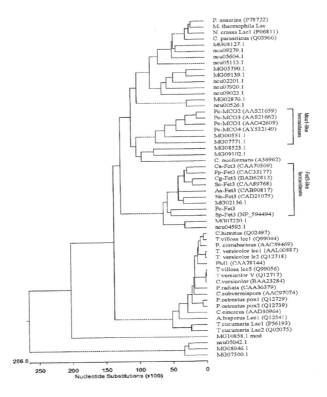

Fig. 5. Clustal W analysis of multicopper oxidases. GenBank accessions, where available, are given parenthetically. All other sequences are derived from publicly accessible databases for *N. crassa* (ncu), *M. grisea* (MG), *P. chrysosporium* (Pc).

5. EXTRACELLULAR CARBOHYDRATE ACTIVE ENZYMES

Relative to ligninolysis, the degradation of cellulose, hemicellulose and pectin by *P. chrysosporium* has received less attention. Enzyme activities imply a degradative strategy similar, but not identical, to other microbes, especially the intensively studied Ascomycete *Trichoderma reesei* (*Hypocrea jecorina*). Components of the cellulolytic system include multiple exocellobiohydrolase I (CBHI) isozymes, as well as an exocellobiohydrolase II (CBHII), and a β-glucosidase (reviewed by Kirk and Cullen, 1998). From a genetic point of view, the degradation of cellulose, hemicellulose and pectin is rather complicated in *P. chrysosporium*. More than 240 sequences encode putative carbohydrate active enzymes, and this includes a minimum of 50 cellulases

(Martinez et al. 2004). As in the case of peroxidases, many of these sequences are distributed within complex gene families.

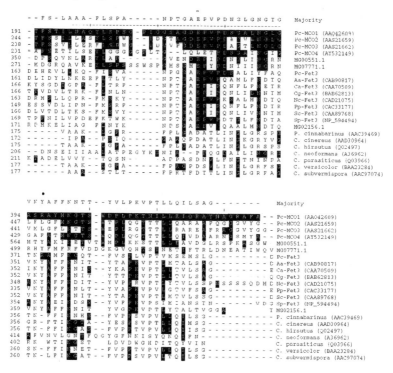

Fig. 6. Clustal W alignment of predicted multicopper oxidases: Selected region of the multiple alignment shows the location of Sc-Fet3 Glu185 (*) and Tyr354 (•), which are essential for ferroxidase activity in *S. cerevisiae* Fet3. Genebank accession numbers are provided. *M. grisea* (MG) sequences were obtained from NCBI. Numbering of the MG sequences might not correspond to the actual length of the proteins. Shaded residues match Pc-MCO1.

Eriksson and coworkers (Eriksson and Pettersson, 1975a, 1975b) characterized multiple endoglucanase (EG) and exocellobiohydrolase (CBH) isozymes in *P. chrysosporium* cultures containing cellulose as sole carbon source. Two CBHI isozymes, designated CBH62 and CBH58, and a single CBHII, designated CBH50, were later purified (Uzcategui et al. 1991c). Genetic analysis identified the corresponding genes, and following the glycosyl hydrolase nomenclature of Henrissat and coworkers ((Henrissat, 1991); http://afmb.cnrs-mrs.fr/CAZY/index.html), these have been named *cel7C*, *cel7D*, and *cel6A*. Four additional CBH1 cDNAs have been sequenced, and all six have been the subject of molecular modeling (Munoz et al. 2001). Transcripts of the *cel7* genes are present in cellulose-containing media (Covert et al. 1992b; Vanden Wymelenberg et al. 1993) and in colonized wood (Vallim et al. 1998), but protein identification to date has been limited to CEL7C and CEL7D (Uzcategui et al. 1991b; Uzcategui et al. 1991c). Transcript patterns among *cel7* genes are dramatically altered by substrate composition, and there is no apparent relationship between transcriptional regulation and genome organization. Extensive analysis of the genome has not revealed additional exocellobiohydrolase genes.

Five endoglucanase isozymes were partially resolved by Eriksson and colleagues (Eriksson and Pettersson, 1975a). Later, experimentally determined peptide sequence were reported for two glycosyl hydrolase family 5 (GH5) isozymes (Uzcategui et al. 1991a) and a single GH28 isozyme (Henriksson et al. 1999). Prior to completion of the genome, a single endoglucanase gene, *cel61A*, was known (Vanden Wymelenberg et al. 2002). *In silico* analysis of the genome has revealed more than 40 putative endoglucanases unevenly distributed in at least 5 glycosyl hydrolase families (Figure 1). Among these, glycosyl hydrolase family 61 encompasses 17 sequences, of which at least 5 contain highly conserved cellulose binding domains at their carboxy terminus. Alignment of their predicted proteins is shown in Figure 7.

Following the synergistic activities of CBHs and EGs, cellobiose and related oligosaccharides are converted to glucose by β-glucosidases. Several isozymes have been purified from *P. chrysosporium* cultures (Deshpande et al. 1978; Igarashi et al. 2003; Smith and Gold, 1979) and a single gene, *bgl1*, identified (Li and Renganathan, 1998). Again following the glycosyl hydrolase nomenclature of Henrissat and coworkers ((Henrissat, 1991); http://afmb.cnrs-mrs.fr/CAZY/index.html), *bgl1* is now designated *cel3A*. *cel3A* was shown to be expressed under cellulose induction (Li and Renganathan, 1998) at relatively low levels (Vanden Wymelenberg et al. 2002). Genome analysis (Martinez et al. 2004) indicates a minimum of 12 GH3-like sequences and recent investigations show that purified CEL3A has substantial glucan 1,3- β -glucosidase activity (Igarashi et al. 2003). Uncertainty regarding CEL3A substrate preference highlights the difficulties assigning function based solely on structure and family membership.

Relatively little is known about the enzymes involved in hemicellulose and pectin degradation in *P. chrysosporium* (Castanares et al. 1995; Copa-Patino et al. 1993; Kirk and Cullen, 1998). The complete conversion of major hemicelluloses of wood, glucuronoxylans and galactoglucomannans, requires the combined activities of numerous enzymes including endoxylanase, acetylxylan esterase, α-glucuronidase, β-xylosidase, α-arabinosidase, endomannanase, α-galactosidase, acetylglucomannan esterase, β -mannosidase, and β-glucosidase (Kirk and Cullen, 1998). Very few of these enzymes have been characterized from *P. chrysosporium* (Castanares et al. 1995; Copa-Patino et al. 1993). Two cDNAs encoding endoxylanases, *xyn10A* and *xyn11A*, were sequenced (GenBank accessions AAG44993, AAG44995) and recently expressed in a heterologous system (Decelle et al. 2004). An α-galactosidase and corresponding gene (*aga27A*) have been characterized (Brumer et al. 1999; Hart et al. 2000). Analysis of the genome reveals many putative genes involved in hemicellulose and pectin degradation (Martinez et al. 2004). Among probable xylanases, no additional members of glycoside family 10 were recognized, but a total of 6 family 11 sequences were identified. Sequences with significant Smith-Waterman scores to NCBI genes of known function include acetylxlan esterases (3), pectin methylesterase (1), β mannosidases (2), β-xylosidase (1), β-mannanase (3), xylanases (7), β-glucosidase (5), α-galactosidase (3), polygalacturonases (4), rhamnogalacturonase (1), β-xylosidase (1), and a-L-arabinofuranosidase (1). Because of problems associated with accurate gene model predictions, the relatively small number of NCBI representatives of certain enzyme classes, and sequence divergence, our assignments no doubt grossly underestimate the total number of genes.

Fig. 7. Clustal W alignment of predicted Cel61 proteins of *P. chrysosporium*. Only *cel61A* is supported by cDNA sequence (Vanden Wymelenberg et al. 2002). Others correspond to gene models available on the Joint Genome Institutes web browser (http://genome.jgi-psf.org/whiterot1/whiterot1.home.html). Models were manually adjusted to provide full length.

While our chapter focuses on genes involved in wood decay, it should be noted that numerous other sequences encoding carbohydrate active enzymes have been identified in the *P. chrysosporium* genome database. Among these are at least 25 putative enzymes involved in the degradation of β-1,3-glucan and mixed-linkage β-1,3-1,4-glucans. These polysaccharides are common constituents of cell walls of certain cereal grains, grasses and related plants as well as the cell walls of yeasts and fungi. Also present in the genome are hydrolase-encoding sequences probably involved in the degradation of starch and glycogen (amylase, glucoamylase, β-glucosidase), mutan (β-1, 3-glucanase), and chitin (chitinase). Chitin is an important structural component of fungal cell walls, and the 10 putative chitinases (GH family 18) may be involved in cell wall morphogenesis. In addition, a minimum of 57 putative glycosyltransferase encoding genes were identified in the genome of *P. chrysosporium*. Their precise function is unkown, but some are likely candidates for the biosynthesis of chitin, β-1,3-glucan, glycogen, cell wall mannan and N-glycans.

6. OTHER EXTRACELLULAR ENZYMES

Posttranslational processes regulate extracellular enzyme activity and contribute to isozyme multiplicity, but to date little progress has been made at the genetic level. Proteolytic processing of LiP has been shown in *P. chrysosporium* (Dass et al. 1995; Datta, 1992; Dosoretz et al. 1990a; Dosoretz et al. 1990b; Eriksson and Pettersson, 1982; Feijoo et al. 1995) and *T. versicolor* (Staszczak et al. 2000) cultures, and extracellular dephosphorylation of certain *P. chrysosporium* LiP isozymes is well established (Kuan and Tien, 1989; Rothschild et al. 1997; Rothschild et al. 1999). Proteases have also been implicated in regulating cellulase (Eriksson and Pettersson, 1982) and CDH (Eggert et al. 1996) activity. Dozens of putative extracellular protease genes are predicted from genome analysis (Martinez et al. 2004) including one that corresponds to the published N-terminal sequence of a pulp-derived protease (Datta, 1992). Recently, an unusual cluster of glutamic proteases have been identified in the *P. chrysosporium* genome database (Sims et al. 2004).

7. INTRACELLULAR ENZYMES RELATED TO LIGNOCELLULOSE DEGRADATION

The complete degradation of lignin requires many intracellular enzymes both for the complete mineralization of monomers to CO_2 and H_2O and for the generation of secondary metabolites (e.g. veratryl alcohol) supporting extracellular metabolism. Examples of enzymes that have been characterized from *P. chrysosporium* include methanol oxidase (Asada et al. 1995), 1,4-benzoquinone reductase (Brock and Gold, 1996), methyltransferases (Jeffers et al. 1997), cytochrome P450s (Kullman and Matsumura, 1997; Van Hamme et al. 2003; Yadav and Loper, 2000; Yadav et al. 2003), L-phenylalanine ammonia-lyase (Hattori et al. 1999), 1,2,4-trihydroxybenzene 1,2-dioxygenase (Rieble et al. 1994), glutathione transferases (Dowd et al. 1997), superoxide dismutase (Ozturk et al. 1999), catalase (Kwon and Anderson, 2001) and aryl alcohol dehydrogenase (Reiser et al. 1994).

Genome mining identified a large number of cytochrome P450s, 14 of which are located on a single scaffold (number 24). The startling enumeration of >148 partial or complete cytochrome P450 gene models (Martinez et al. 2004) represents a promising, if not daunting, framework for future investigation. The genetic complexity is reflected in recent studies demonstrating an impressive array of potential substrates and transformation products (Matsuzaki and Wariishi, 2004; Miura et al. 2004; Teramoto et al. 2004). In another recent study, Yadav and coworkers (Doddapaneni and Yadav, 2004) demonstrated differential regulation of two cytochrome P450 genes in response to various xenobiotics.

8. CONCLUSION

Analysis of the *P. chrysosporium* genome presents challenges and opportunities for future research. The biological role of impressive gene multiplicity among glycosyl hydrolases, cytochrome P450s, peroxidases and oxidases remains one of the most pressing issues. Are these closely related genes merely redundant or do they encode enzymes with subtle but important functional differences? Other significant questions revolve around the abundant transposable elements identified. Do these elements impact gene expression and/or the emergence of gene families?

The genome database provides a framework for addressing these and other questions. Transcript profiling using microarrays may provide indirect evidence for the role(s) of many genes. Tandem mass spectrometry and MALDI analysis, already begun in several laboratories, will also help identify key genes and enzymes. A major concern for such high throughput approaches centers on difficulties inherent in gene prediction, a problem especially common in fungal genomes where introns are present in most genes. As gene models are corrected, databases will need continual updating. A powerful approach to determining gene function would be gene knockouts, but unlike well established experimental systems, e.g. *S. cerevisiae*, gene disruptions are curently very difficult in *P. chrysosporium*. Hopefully, more efficient methodology will be forthcoming. A productive route for establishing the role of individual genes in lignocellulose degradation will continue to involve biochemical characterization of pure recombinant enzymes. Toward this end, more efficient heterologous expression systems are needed. Finally, whole genome comparisons, particularly among filamentous Ascomycetes and Basidiomycetes, may provide valuable information concerning gene function.

Acknowledgements: This research was supported by U.S. Department of Energy grant DE-FG02-87ER13712, by the Millenium Institute for Fundamental and Applied Biology, and by grant 1030495 from FONDECYT.

REFERENCES

Abbas A, Koc H, Liu F and Tien M (2004). Fungal degradation of wood: initial proteomic analyses of extracellular proteins of *Phanerochaete chrysosporium* grown on oak substrate. Curr Genet: In press.
Akileswaran L, Alic M, Clark E, Hornick J, and Gold MH (1993). Isolation and transformation of uracil auxotrophs of the lignin-degrading basidiomycete *Phanerochaete chrysosporium*. Curr. Genet. 23: 351-356.
Alic M, and Gold MH (1985). Genetic recombination in the lignin-degrading basidiomycete *Phanerochaete chrysosporium*. Appl. Env. Microbiol. 50: 27-30.

Alic M, Kornegay JR, Pribnow D, and Gold MH (1989). Transformation by complementation of an adenine auxotroph of the lignin-degrading basidiomycete *Phanerochaete chrysosporium*. Appl. Environ. Microbiol. 55: 406-411.

Alic M, Clark EK, Kornegay JR, and Gold MH (1990). Transformation of *Phanerochaete chrysosporium* and *Neurospora crassa* with adenine biosynthetic genes from *Schizophyllum commune*. Curr. Genet. 17: 305-311.

Alic M, and Gold MH (1991). Genetics and molecular biology of the lignin-degrading Basidiomycete *Phanerochaete chrysosporium*. In More Gene Manipulations in Fungi. Bennett, J. and Lasure, L. (eds).. New York: Academic Press, pp. 319-341.

Alic M, Mayfield MB, Akileswaran L, and Gold MH (1991). Homologous transformation of the lignin-degrading basidiomycete *Phanerochaete chrysosporium*. Curr. Genet. 19: 491-494.

Alic M, Akileswaran L, and Gold MH (1993). Gene replacement in the lignin-degrading basidiomycete *Phanerochaete chrysosporium*. Gene 136: 307-311.

Alic M, Akileswaran L, and Gold MH (1997). Characterization of the gene encoding manganese peroxidase isozyme 3 from *Phanerochaete chrysosporium*. Biochim. Biophys. Acta. 1338: 1-7.

Alic MM (1990). Mating system and DNA transformation of the lignin-degrading basidiomycete *Phanerochaete chrysosporium*. Diss. Abstr. Int. B 51: 3681.

Asada Y, Watanabe A, Ohtsu Y, and Kuwahara M (1995). Purification and characterization of an aryl-alcohol oxidase from the lignin-degrading basidiomycete *Phanerochaete chrysosporium*. Biosci. Biotech. Biochem. 59: 1339-1341.

Askwith C, Eide D, Van Ho A, Bernard PS, Li L, Davis-Kaplan S, Sipe DM, and Kaplan J (1994). The FET3 gene of *S. cerevisiae* encodes a multicopper oxidase required for ferrous iron uptake. Cell 76: 403-410.

Askwith CC, and Kaplan J (1998). Site-directed mutagenesis of the yeast multicopper oxidase Fet3p. J Biol Chem 273: 22415-22419.

Bartholomew K, Dos Santos G, Dumonceaux T, Charles T, and Archibald F (2001). Genetic transformation of *Trametes versicolor* to phleomycin resistance with the dominant selectable marker shble. Appl. Microbiol. Biotechnol. 56: 201-204.

Bes B, Ranjera R, and Bondet AM (1983). Evidence for the involvement of activated oxygen in fungal degradation of lignocellulose. Biochimie 65: 283-289.

Birch PR (1998). Targeted differential display of abundantly expressed sequences from the basidiomycete *Phanerochaete chrysosporium* which contain regions coding for fungal cellulose-binding domains. Curr. Genet. 33: 70-76.

Birch PR, Sims PF, and Broda P (1998). A reporter system for analysis of regulatable promoter functions in the basidiomycete fungus *Phanerochaete chrysosporium*. J. Appl. Microbiol. 85: 417-424.

Blanchette R (1991). Delignification by wood-decay fungi. Ann. Rev. Phytopath. 29: 381-398.

Blanchette R, Krueger E, Haight J, Akhtar M, and Akin D (1997). Cell wall alterations in loblolly pine wood decayed by the white-rot fungus, *Ceriporiopsis subvermispora*. J. Biotechnol. 53: 203-213.

Bogan B, Schoenike B, Lamar R, and Cullen D (1996a). Manganese peroxidase mRNA and enzyme activity levels during bioremediation of polycyclic aromatic hydrocarbon-contaminated soil with *Phanerochaete chrysosporium*. Appl. Environ. Microbiol. 62: 2381-2386.

Bogan B, Schoenike B, Lamar R, and Cullen D (1996b). Expression of *lip* genes during growth in soil and oxidation of anthracene by *Phanerochaete chrysosporium*. Appl. Environ. Microbiol. 62: 3697-3703.

Bogan BW, Schoenike B, Lamar RT, and Cullen D (1996c). Expression of lip genes during growth in soil and oxidation of anthracene by *Phanerochaete chrysosporium*. Appl. Environ. Microbiol. 62: 3697-3703.

Bonaccorsi di Patti MC, Felice MR, Camuti AP, Lania A, and Musci G (2000). The essential role of Glu-185 and Tyr-354 residues in the ferroxidase activity of *Saccharomyces cerevisiae* Fet3. FEBS Lett 472: 283-286.

Bonaccorsi di Patti MC, Paronetto MP, Dolci V, Felice MR, Lania A and Musci G (2001). Mutational analysis of the iron binding site of *Saccharomyces cerevisiae* ferroxidase Fet3. An *in vivo* study. FEBS Lett 508: 475-478.

Bonnarme P, and Jeffries T (1990). Mn(II). regulation of lignin peroxidases and manganese-dependent peroxidase from lignin-degrading white-rot fungi. Appl. Environ. Microbiol. 56: 210-217.

Bourbonnais R, Paice MG, Freiermuth B, Bodie E, and Borneman S (1997). Reactivities of various mediators and laccases with kraft pulp and lignin model compounds. Appl. Environ. Microbiol. 63: 4627-4632.

Brock BJ, and Gold MH (1996). 1,4-Benzoquinone reductase from basidiomycete Phanerochaete chrysosporium: spectral and kinetic analysis. Arch. Biochem. Biophys. 331: 31-40.

Brown JA Glenn JK, and Gold MH (1990). Manganese regulates expression of manganese peroxidase by Phanerochaete chrysosporium. J. Bacteriol. 172: 3125-3130.

Brumer H, 3rd, Sims PF, and Sinnott ML (1999). Lignocellulose degradation by Phanerochaete chrysosporium: purification and characterization of the main alpha-galactosidase. Biochem. J. 339 (Pt 1).: 43-53.

Burdsall HH, and Eslyn WE (1974). A new Phanerochaete with a chrysosporium imperfect state. Mycotaxon 1: 123-133.

Call HP, and Muncke I (1997). History, overeiw and applications of mediated lignolytic systems, especially laccase-mediator systems (lignozyme(R).-process).. J. Biotechnol. 53: 163-202.

Camarero S, Ruiz-Duenas FJ, Sarkar S, Martinez MJ, and Martinez AT (2000). The cloning of a new peroxidase found in lignocellulose cultures of Pleurotus eryngii and sequence comparison with other fungal peroxidases. FEMS Microbiol. Lett. 191: 37-43.

Cameron MD, and Aust SD (2001). Cellobiose dehydrogenase-an extracellular fungal flavocytochrome. Enzyme Microb Technol 28: 129-138.

Castanares A, Hay AJ, Gordon AH, McCrae SI, and Wood TM (1995). D-xylan-degrading enzyme system from the fungus Phanerochaete chrysosporium: isolation and partial characterisation of an alpha-(4-O-methyl).-D-glucuronidase. J. Biotechnol. 43: 183-194.

Cohen R, Jensen KA, Houtman CJ, and Hammel KE (2002). Significant levels of extracellular reactive oxygen species produced by brown rot basidiomycetes on cellulose. FEBS Lett 531: 483-488.

Cohen R, Suzuki MR, and Hammel KE (2004). Differential stress-induced regulation of two quinone reductases in the brown rot basidiomycete Gloeophyllum trabeum. Appl Environ Microbiol 70: 324-331.

Collins PJ, O'Brien MM, and Dobson AD (1999). Cloning and characterization of a cDNA encoding a novel extracellular peroxidase from Trametes versicolor. Appl. Environ. Microbiol. 65: 1343-1347.

Copa-Patino J, Kim YG, and Broda P (1993). Production and initial characterisation of the xylan-degrading enzyme system of Phanerochaete chrysosporium. Appl. Microbiol. Biotechnol. 40: 69-76.

Covert S, Bolduc J, and Cullen D (1992a). Genomic organization of a cellulase gene family in Phanerochaete chrysosporium. Curr. Genet. 22: 407-413.

Covert S, Vanden Wymelenberg A, and Cullen D (1992b). Structure, organization and transcription of a cellobiohydrolase gene cluster from Phanerochaete chrysosporium. Appl. Environ. Microbiol. 58: 2168-2175.

Cowling EB (1961). Comparative biochemistry of the decay of sweetgum sapwood by white-rot and brown-rot fungi. In Technical bulletin No, 1258 Washington. D.C.: U.S. Department of Agriculture.

Cullen D, and Kersten PJ (1996). Enzymology and molecular biology of lignin degradation. In The Mycota III. Bramble, R. and Marzluf, G. (eds).. Berlin: Springer-Verlag, pp. 297-314.

Cullen D (1997). Recent advances on the molecular genetics of ligninolytic fungi. J. Biotechnol. 53: 273-289.

Cullen D (2002). Molecular genetics of lignin-degrading fungi and their application in organopollutant degradation. In The Mycota. Vol. XI. Kempken, F. (ed).. Berlin: Springer-Verlag, pp. 71-90.

Cullen D, and Kersten PJ (2004). Enzymology and Molecular Biology of Lignin Degradation. In The Mycota III Biochemistry and Molecular Biology. Brambl, R. and Marzulf, G.A. (eds).. Berlin: Springer-Verlag, pp. In press.

D'Souza TM, Dass SB, Rasooly A, and Reddy CA (1993). Electrophoretic karyotyping of the lignin-degrading basidiomycete Phanerochaete chrysosporium. Mol. Microbiol. 8: 803-807.

Daboussi MJ, and Capy P (2003). Transposable elements in filamentous fungi. Annu Rev Microbiol 57: 275-299.

342

Daniel G (1994). Use of electron microscopy for aiding our understanding of wood biodegradation. FEMS Microbiol. Rev. 13: 199-233.

Daniel G, Volc J, and Kubatova E (1994). Pyranose oxidase, a major source of H_2O_2 during wood degradation by *Phanerochaete chrysosporium, Trametes versicolor*, and *Oudemansiella mucida*. Appl. Environ. Microbiol. 60: 2524-2532.

Dass SB, Dosoretz CG, Reddy CA, and Grethlein HE (1995). Extracellular proteases produced by the wood-degrading fungus *Phanerochaete chrysosporium* under ligninolytic and non-ligninolytic conditions. Arch. Microbiol. 163: 254-258.

Datta A, Bettermann A, and Kirk TK (1991). Identification of a specific manganese peroxidase among ligninolytic enzymes secreted by *Phanerochaete chrysosporium* during wood decay. Appl. Environ. Microbiol. 57: 1453-1460.

Datta A (1992). Purification and characterization of a novel protease from solid substrate cultures of *Phanerochaete chrysosporium*. J. Biol. Chem. 267: 728-732.

de Jong E, Cazemier A, Field J, and de Bont J (1994). Physiological role of chlorinated aryl alcohols biosynthesized de novo by the white rot fungus *Bjerkandera sp.* srain BOS55. Appl. Environ. Microbiol. 60: 271-277.

de Koker TH, Mozuch MD, Cullen D, Gaskell J, and Kersten PJ (2004). Pyranoxe 2-oxidase from *Phanerochaete chrysosporium*: isolation from solid substrate, protein purification, and characterization of gene structure and regulation. Appl Environ Microbiol 70: 5794-5800.

Decelle B, Tsang A, and Storms RK (2004). Cloning, functional expression and characterization of three *Phanerochaete chrysosporium* endo-1,4-beta-xylanases. Curr Genet 46: 166-175.

Deshpande V, Eriksson K-E, and Pettersson B (1978). Production, purification and partial characterization of 1,4-b-glucosidase enzymes from *Sporotrichum pulverulentum*. Eur. J. Biochem. 90: 191-198.

Doddapaneni H, and Yadav JS (2004). Differential regulation and xenobiotic induction of tandem P450 monooxygenase genes pc-1 (CYP63A1). and pc-2 (CYP63A2). in the white-rot fungus *Phanerochaete chrysosporium*. Appl Microbiol Biotechnol 65: 559-565.

Dosoretz C, Dass B, Reddy CA, and Grethlein H (1990a). Protease-mediated degradation of lignin peroxidase in liquid cultures of *Phanerochaete chrysosporium*. Appl. Environ. Microbiol. 56: 3429-3434.

Dosoretz CD, Chen H-C, and Grethlein HE (1990b). Effect of environmental conditions on extracellular protease activity in lignolytic cultures of *Phanerochaete chrysosporium*. App. Environ. Microbiol. 56: 395-400.

Dowd CA, Buckley CM, and Sheehan D (1997). Glutathione S-transferases from the white-rot fungus, *Phanerochaete chrysosporium*. Biochem. J. 324 (Pt 1).: 243-248.

Dumonceaux T, Bartholomew K, Valeanu L, Charles T, and Archibald F (2001). Cellobiose dehydrogenase is essential for wood invasion and nonessential for kraft pulp delignification by *Trametes versicolor*. Enzyme Microb Technol 29: 478-489.

Dumonceaux TJ, Bartholomew KA, Charles TC, Moukha SM, and Archibald FS (1998). Cloning and sequencing of a gene encoding cellobiose dehydrogenase from *Trametes versicolor*. Gene 210: 211-219.

Edwards S, Raag R, Wariishi H, Gold M, and Poulos T (1993). Crystal structure of lignin peroxidase. Proc. Nat. Acad. Sci. 90: 750-754.

Eggert C, Habu N, Temp U, and Eriksson K-EL (1996). Cleavage of *Phanerochaete chrysosporium* cellobiose dehydrogenase (CDH). by three endogenous proteases. In Biotechnology in the Pulp and Paper Industry. Srebotnik, E. and Messner, K. (eds).. Vienna: Fakultas-Universitatsverlag, pp. 551-554.

Eggert C, Temp U, and Eriksson K (1997). Laccase is essential for lignin degradation by the white-rot fungus *Pycnoporus cinnabarinus*. FEBS Lett. 407: 89-92.

Eggert C, LaFayette PR, Temp U, Eriksson KE, and Dean JF (1998). Molecular analysis of a laccase gene from the white rot fungus *Pycnoporus cinnabarinus*. Appl. Environ. Microbiol. 64: 1766-1772.

Eichlerova I, and Homolka L (1999). Preparation and crossing of basidiospore-derived monokaryons--a useful tool for obtaining laccase and other ligninolytic enzyme higher- producing dikaryotic strains of *Pleurotus ostreatus*. Antonie Van Leeuwenhoek 75: 321-327.

Eichlerova-Volakova I, and Homolka L (1997). Variability of ligninolytic enzyme activities in basidiospore isolates of the fungus *Pleurotus ostreatus* in comparison with that of protoplast-derived isolates. Folia Microbiol. 42: 583-588.

Eriksson K-E, and Pettersson B (1975a). Extracellular enzyme system utilized by the fungus *Sporotrichum pulverulentum (Chrysosporium lignorum)* for the breakdown of cellulose. 1. Separation, purification, and physio-chemical characterization of five endo-1,4-h-glucanases. Eur. J. Biochem. 51: 193-206.

Eriksson K-E, and Pettersson B (1975b). Extracellular enzyme system used by the fungus *Sporotrichum pulverulentum (Chrysosporium lignorum)* for the breakdown of cellulose. 3. Purification, physiochemical characterization of an exo-1,4-b-glucanases. Eur. J. Biochem. 51: 213-218.

Eriksson K-E, and Pettersson B (1982). Purification and partial characterization of two acidic proteases from the white rot fungus *Sporotrichium pulverulentum*. Eur. J. Biochem. 124: 635-642.

Eriksson K-EL, Blanchette RA, and Ander P (1990). Microbial and enzymatic degradation of wood and wood components. Berlin: Springer-Verlag.

Eriksson KE, Pettersson B, Volc J, and Musilek V (1986). Formation and partial characterization of glucose-2-oxidase, a hydrogen peroxide producing enzyme in *Phanerochaete chrysosporium*. Appl. Microbiol. Biotechnol. 23: 257-262.

Evans C, Farmer JY, and Palmer JM (1984). An extracellular haem protein from *Coriolus versicolor*. Phytochemistry 23: 1247-1250.

Feijoo G, Rothschild N, Dosoretz C, and Lema J (1995). Effects of addition of extracellular culture fluid on ligninolytic enzyme formation by *Phanerochaete chrysosporium*. J. Biotechnol. 40: 21-29.

Firbank SJ, Rogers MS, Wilmot CM, Dooley DM, Halcrow MA, Knowles PF, McPherson MJ, and Phillips SE (2001). Crystal structure of the precursor of galactose oxidase: an unusual self-processing enzyme. Proc Natl Acad Sci U S A 98: 12932-12937.

Flournoy D, Paul J, Kirk TK, and Highley T (1993). Changes in the size and volume of pore in sweet gum wood during simultaneous rot by *Phanerochaete chrysosporium*. Holzforschung 47: 297-301.

Forney LJ, Reddy CA, Tien M, and Aust SD (1982). The involvemment of hydroxyl radical derived from hydrogen peroxide in lignin degradation by the white rot fungus *Phanerochaete chrysosporium*. J. Biol. Chem. 257: 11455-11462.

Galagan JE, Calvo SE, Borkovich KA, Selker EU, Read ND, Jaffe D, FitzHugh W, Ma LJ, Smirnov S, Purcell S, Rehman B, Elkins T, Engels R, Wang S, Nielsen CB, Butler J, Endrizzi M, Qui D, Ianakiev P, Bell-Pedersen D, Nelson MA, Werner-Washburne M, Selitrennikoff CP, Kinsey JA, Braun EL, Zelter A, Schulte U, Kothe GO, Jedd G, Mewes W, Staben C, Marcotte E, Greenberg D, Roy A, Foley K, Naylor J, Stange-Thomann N, Barrett R, Gnerre S, Kamal M, Kamvysselis M, Mauceli E, Bielke C, Rudd S, Frishman D, Krystofova S, Rasmussen C, Metzenberg RL, Perkins DD, Kroken S, Cogoni C, Macino G, Catcheside D, Li W, Pratt RJ, Osmani SA, DeSouza CP, Glass L, Orbach MJ, Berglund JA, Voelker R, Yarden O, Plamann M, Seiler S, Dunlap J, Radford A, Aramayo R, Natvig DO, Alex LA, Mannhaupt G, Ebbole DJ, Freitag M, Paulsen I, Sachs MS, Lander ES, Nusbaum C, and Birren B (2003). The genome sequence of the filamentous fungus *Neurospora crassa*. Nature 422: 859-868.

Gaskell J, Dieperink E, and Cullen D (1991). Genomic organization of lignin peroxidase genes of *Phanerochaete chrysosporium*. Nucleic Acids Res. 19: 599-603.

Gaskell J, Vanden Wymelenberg A, Stewart P, and Cullen D (1992). Method for identifying specific alleles of a *Phanerochaete chrysosporium* encoding a lignin peroxidase. Appl. Environ. Microbiol. 58: 1379-1381.

Gaskell J, Stewart P, Kersten P, Covert S, Reiser J, and Cullen D (1994). Establishment of genetic linkage by allele-specific polymerase chain reaction: application to the lignin peroxidase gene family of *Phanerochaete chrysosporium*. Bio/Technology 12: 1372-1375.

Gaskell J, Vanden Wymelenberg A, and Cullen D (1995). Structure, inheritance, and transcriptional effects of Pce1, an insertional element within *Phanerochaete chrysosporium* lignin peroxidase gene *lipI*. Proc. Natl. Acad. Sci. USA 92: 7465-7469.

Gessner M, and Raeder U (1994). A histone promoter for expression of a phleomycin-resistance gene in *Phanerochaete chrysosporium*. Gene 142: 237-241.

Gettemy JM, Li D, Alic M, and Gold MH (1997). Truncated-gene reporter system for studying the regulation of manganese peroxidase expression. Curr. Genet. 31: 519-524.

Gettemy JM, Ma B, Alic M, and Gold MH (1998). Reverse transcription-PCR analysis of the regulation of the manganese peroxidase gene family. Appl. Environ. Microbiol. 64: 569-574.

Gilbertson RL (1981). North American wood-rotting fungi that cause brown rots. Mycotaoxon 12: 372-416.

Glenn JK, Morgan MA, Mayfield MB, Kuwahara M, and Gold MH (1983). An extracellular H_2O_2-requiring enzyme preparation involved in lignin biodegradation by the white-rot basidiomycete *Phanerochaete chrysosporium*. Biochem. Biophys. Res. Comm. 114: 1077-1083.

Glenn JK, and Gold MH (1985). Purification and characterization of an extracellular Mn(II).-dependent peroxidase from the lignin-degrading basidiomycete Phanerochaete chrysosporium. Arch. Biochem. Biophys. 242: 329-341.

Glenn JK, Akileswaran L, and Gold MH (1986). Mn(II). oxidation is the principal function of the extracellular Mn-peroxidase from *Phanerochaete chrysosporium*. Arch. Biochem. Biophys. 251: 688-696.

Goffeau A, Barrell BG, Bussey H, Davis RW, Dujon B, Feldmann H, Galibert F, Hoheisel JD, Jacq C, Johnston M, Louis EJ, Mewes HW, Murakami Y, Philippsen P, Tettelin H, and Oliver SG (1996). Life with 6000 genes. Science 274: 546, 563-547.

Gold M, and Alic M (1993). Molecular biology of the lignin-degrading basidiomycete *Phanerochaete chrysosporium*. Microbiol. Rev. 57: 605-622.

Gold MH, Cheng TM, and Mayfield MB (1982). Isolation and complementation studies of auxotrophic mutants of the lignin-degrading basidiomycete *Phanerochaete chrysosporium*. Appl. Environ. Microbiol. 44: 996-1000.

Gold MH, Kuwahara M, Chiu AA, and Glenn JK (1984). Purification and characterization of an extracellular H_2O_2-requiring diarylpropane oxygenase from the white rot basidiomycete, *Phanerochaete chrysosporium*. Archives. Biochem. Biophys. 234: 353-362.

Goodell B (2003). Brown rot fungal degradation of wood: our evolving view. In Wood deterioration and preservation. Goodell, B., Nicholas, D. and Schultz, T. (eds).. Washington, DC.: American Society of Microbiology, pp. 97-118.

Goodwin TJ, and Poulter RT (2000). Multiple LTR-retrotransposon families in the asexual yeast *Candida albicans*. Genome Res 10: 174-191.

Goodwin TJ, and Poulter RT (2004). A new group of tyrosine recombinase-encoding retrotransposons. Mol Biol Evol 21: 746-759.

Goring DA (1977). A speculative picture of the delignification process. In Cellulose Chemistry and Technology. Arthur, J.C. (ed).. Washington: American Chemical Society, pp. 273-277.

Guillen F, and Evans C (1994). Anisaldehyde and veratraldehyde acting as redox cycling agents for peroxide prodcution by *Pleurotus eryngii*. Appl. Environ. Microbiol. 60: 2811-2817.

Hammel KE, Tien M, Kalyanaraman B, and Kirk TK (1985). Mechanism of oxidative Ca-Cb cleavage of a lignin model dimer by *Phanerochaete chrysosporium* ligninase: Stochiometry and involvement of free radicals. J. Biol. Chem. 260: 8348-8353.

Hammel KE, and Moen MA (1991). Depolymerization of a synthetic lignin *in vitro* by lignin peroxidase. Enzyme Microb. Technol. 13: 15-18.

Hammel KE, Jensen KA, and Kersten PK (1994). H_2O_2 recycling during oxidation of the arylglycerol b-aryl ether lignin structure by lignin peroxidase and glyoxal oxidase. Biochem. 33: 13349-13354.

Hart DO, He S, Chany CJ, 2nd, Withers SG, Sims PF, Sinnott ML, and Brumer H, 3rd (2000). Identification of Asp-130 as the catalytic nucleophile in the main alpha-galactosidase from *Phanerochaete chrysosporium*, a family 27 glycosyl hydrolase. Biochem. 39: 9826-9836.

Hattori T, Nishiyama A, and Shimada M (1999). Induction of L-phenylalanine ammonia-lyase and suppression of veratryl alcohol biosynthesis by exogenously added L-phenylalanine in a white-rot fungus *Phanerochaete chrysosporium*. FEMS Microbiol. Lett. 179: 305-309.

Haylock R, Liwicki R, and Broda P (1985). The isolation of mRNA from the basidiomycete fungi *Phanerochaete chrysosporium* and *Coprinus cinereus* and its *in vitro* translation. Appl. Environ. Microbiol. 46: 260-263.

Henriksson G, Ander P, Pettersson B, and Pettersson G (1995). Cellobiose dehydrogenase (cellobiose oxidase). from *Phanerochaete chrysosporium* as a wood-degrading enzyme. Studies on cellulose, xylan, and synthetic lignin. Appl. Biochme. Biotechnol. 42.

Henriksson G, Nutt A, Henriksson H, Pettersson B, Stahlberg J, Johansson G, and Pettersson G (1999). Endoglucanase 28 (Cel12A)., a new *Phanerochaete chrysosporium* cellulase. Eur. J. Biochem. 259: 88-95.

Henriksson G, Johansson G, and Pettersson G (2000a). A critical review of cellobiose dehydrogenases. J. Biotechnol. 78: 93-113.

Henriksson G, Zhang L, Li J, Ljungquist P, Reitberger T, Pettersson G, and Johansson G (2000b). Is cellobiose dehydrogenase from *Phanerochaete chrysosporium* a lignin degrading enzyme? Biochim. Biophys. Acta. 1480: 83-91.

Henrissat B (1991). A classification of glycosyl hydrolases based on amino acid sequence similarities. Biochem. J. 280 (Pt 2).: 309-316.

Hibbett DS, and Donoghue MJ (2001). Analysis of character correlations among wood decay mechanisms, mating systems, and substrate ranges in homobasidiomycetes. Syst. Biol. 50: 215-242.

Higuchi T (1990). Lignin biochemistry: biosynthesis and biodegradation. Wood Sci. Technol. 24: 23-63

Holzbaur E, and Tien M (1988). Structure and regulation of a lignin peroxidase gene from *Phanerochaete chrysosporium*. Biochem. Biophys. Res. Commun. 155: 626-633.

Honda Y, Matsuyama T, Irie T, Watanabe T, and Kuwahara M (2000). Carboxin resistance transformation of the homobasidiomycete fungus *Pleurotus ostreatus*. Curr. Genet. 37: 209-212.

Igarashi K, Samejima M, and Eriksson KE (1998). Cellobiose dehydrogenase enhances *Phanerochaete chrysosporium* cellobiohydrolase I activity by relieving product inhibition. Eur. J. Biochem. 253: 101-106.

Igarashi K, Tani T, Kawai R, and Samejima M (2003). Family 3 b-glucosidase from celulose-degrading cultures of the white-rot fungus *Phanerochaete chrysosporium* is a glucan 1,3-b-glucosidase. J. Biosci. Bioeng. 95: 572-576.

Irie T, Honda Y, Watanabe T, and Kuwahara M (2001). Efficient transformation of filamentous fungus *Pleurotus ostreatus* using single-strand carrier DNA. Appl. Microbiol. Biotechnol. 55: 563-565.

Janse BJH, Gaskell J, Akhtar M, and Cullen D (1998). Expression of *Phanerochaete chrysosporium* genes encoding lignin peroxidases, manganese peroxidases, and glyoxal oxidase in wood. Appl. Environ. Microbiol. 64: 3536-3538.

Jeffers MR, McRoberts WC, and Harper DB (1997). Identification of a phenolic 3-O-methyltransferase in the lignin-degrading fungus *Phanerochaete chrysosporium*. Microbiol. 143 (Pt 6).: 1975-1981.

Johansson T, and Nyman PO (1993). Isozymes of lignin peroxidase and manganese(II). peroxidase from the white-rot basidiomycete *Trametes versicolor*. I. Isolation of enzyme forms and characterization of physical and catalytic properties. Arch. Biochem. Biophys. 300: 49-56.

Johansson T, Nyman PO, and Cullen D (2002). Differential regulation of mnp2, a new manganese peroxidase-encoding gene from the ligninolytic fungus *Trametes versicolor* PRL 572. Appl. Environ. Microbiol. 68: 2077-2080.

Johnston C, and Aust S (1994). Transcription of ligninase H8 by *Phanerochaete chrysosporium* under nutrient nitrogen sufficient conditions. Biochem. Biophys. Res. Commun. 200: 108-112.

Kawai S, Umezawa T, and Higuchi T (1988). Degradation mechanisms of phenolic b-1 lignin substructure and model compounds by laccase of *Coriolus versicolor*. Arch. Biochem. Biophys. 262: 99-110.

Kawai S, Asukai M, Ohya N, Okita K, Ito T, and Ohashi H (1999). Degradation of non-phenolic b-O-4 susbstructure and of polymeric lignin model compounds by laccase of *Coriolus versicolor* in the presence of 1-hydroxybenzotriazole. FEMS Microbiol. Lett. 170: 51-57.

Kelley RL, and Reddy CA (1986). Purification and characterization of glucose oxidase from ligninolytic cultures of *Phanerochaete chrysosporium*. J. Bacteriol. 166: 269-274.

Kempken F, and Kuck U (1998). Transposons in filamentous fungi-facts and perspectives. BioEssays 20: 652-659.

Kersten P, and Cullen D (1993). Cloning and characterization of a cDNA encoding glyoxal oxidase, a peroxide-producing enzyme from the lignin-degrading basidiomycete *Phanerochaete chrysosporium*. Proc. Nat. Acad. Sci. USA 90: 7411-7413.

Kersten PJ, Tien M, Kalyanaraman B, and Kirk TK (1985). The ligninase of *Phanerochaete chrysosporium* generates cation radicals from methoxybenzenes. J. Biol. Chem. 260: 2609-2612.

Kersten PJ, and Kirk TK (1987). Involvement of a new enzyme, glyoxal oxidase, in extracellular H_2O_2 production by *Phanerochaete chrysosporium*. J. Bacteriol. 169: 2195-2201.

Kersten PJ (1990). Glyoxal oxidase of *Phanerochaete chrysosporium*; Its characterization and activation by lignin peroxidase. Proc. Natl. Acad. Sci. USA 87: 2936-2940.

Kersten PJ, Witek C, Vanden Wymelenberg A, and Cullen D (1995). *Phanerochaete chrysosporium* glyoxal oxidase is encoded by two allelic variants: structure, genomic organization and heterologous expression of *glx1* and *glx2*. J. Bact. 177: 6106-6110.

Kim JM, Vanguri S, Boeke JD, Gabriel A, and Voytas DF (1998). Transposable elements and genome organization: a comprehensive survey of retrotransposons revealed by the complete *Saccharomyces cerevisiae* genome sequence. Genome Res 8: 464-478.

Kim K, Leem Y, and Choi HT (2002). Transformation of the medicinal basidiomycete *Trametes versicolor* to hygromycin B resistance by restriction enzyme mediated integration. FEMS Microbiol. Lett. 209: 273-276.

Kirk TK, and Nakatsubo F (1983). Chemical mechanism of an important cleavage reaction in the fungal degradation of lignin. Biochem. Biophys. Acta. 756: 376-384.

Kirk TK, Mozuch MD, and Tien M (1985). Free hydroxyl radical is not involved in an important reaction of lignin degradation by *Phanerochaete chrysosporium* Burds. Biochem J 226: 455-460.

Kirk TK, and Farrell RL (1987). Enzymatic "combustion": the microbial degradation of lignin. Ann. Rev. Microbiol. 41: 465-505.

Kirk TK (1988). Lignin degradation by *Phanerochaete chrysosporium*. ISI Atlas of Science: Biochem. 1: 71-76.

Kirk TK, and Cullen D (1998). Enzymology and molecular genetics of wood degradation by white-rot fungi. In Environmentally Friendly Technologies for the Pulp and Paper Industry. Young, R.A. and Akhtar, M. (eds).. New York: John Wiley and Sons, pp. 273-308.

Kishi K, Wariishi H, Marquez L, Dunford HB, and Gold MH (1994). Mechanism of manganese peroxidase compound II reduction. Effect of organic chelators and pH. Biochem. 33: 8694-8701.

Krejci R, and Homolka L (1991). Genetic mapping in the lignin-degrading basidiomycete *Phanerochaete chrysosporium*. Appl. Environ. Microbiol. 57: 151-156.

Kremer P, and Wood PM (1992a). Evidence that cellobiose oxidase from *Phanerochaete chrysosporium* is primarily an Fe(III). reductase. Eur. J. Biochem. 205: 133-138.

Kremer SM, and Wood PM (1992b). Production of Fenton's reagent by cellobiose oxidase from cellulolytic cultures of *Phanerochaete chrysosporium*. Eur. J. Biochem. 208: 807-814.

Kuan I-C, Johnson KA, and Tien M (1993). Kinetic analysis of manganese peroxidase: the reaction with manganese complexes. J. Biol. Chem. 268: 20064-20070.

Kuan IC, and Tien M (1989). Phosphorylation of lignin peroxidase from *Phanerochaete chrysosporium*. J. Biol. Chem. 264: 20350-20355.

Kullman SW, and Matsumura F (1997). Identification of a novel cytochrome P-450 gene from the white rot fungus *Phanerochaete chrysosporium*. Appl. Environ. Microbiol. 63: 2741-2746.

Kurek B, and Kersten P (1995). Physiological regulation of glyoxal oxidase from *Phanerochaete chrysosporium* by peroxidase systems. Enz. Microb. Technol. 17: 751-756.

Kurihara H, Wariishi H, and Tanaka H (2002). Chemical stress-responsive genes from the lignin-degrading fungus *Phanerochaete chrysosporium* exposed to dibenzo-p-dioxin. FEMS Microbiol. Lett. 212: 217-220.

Kusters-van Someren M, Kishi K, Ludell T, and Gold M (1995). The manganese binding site of manganese peroxidase: characterization of an Asp179Asn site-directed mutant protein. Biochem. 34: 10620-10627.

Kutsuki H, and Gold MH (1982). Generation of hydroxyl radical and its involvement in lignin degradation by *Phanerochaete chrysosporium*. Biochem. Biophys. Res. Commun. 109: 320-327.

Kwon SI, and Anderson AJ (2001). Catalase activities of *Phanerochaete chrysosporium* are not coordinately produced with ligninolytic metabolism: catalases from a white-rot fungus. Curr. Microbiol. 42: 8-11.

Larraya LM, Perez G, Penas MM, Baars JJ, Mikosch TS, Pisabarro AG, and Ramirez L (1999). Molecular karyotype of the white rot fungus *Pleurotus ostreatus*. Appl. Environ. Microbiol. 65: 3413-3417.

Larraya LM, Perez G, Ritter E, Pisabarro AG, and Ramirez L (2000). Genetic linkage map of the edible basidiomycete *Pleurotus ostreatus*. Appl. Environ. Microbiol. 66: 5290-5300.

Larraya LM, Idareta E, Arana D, Ritter E, Pisabarro AG, and Ramirez L (2002). Quantitative trait loci controlling vegetative growth rate in the edible basidiomycete *Pleurotus ostreatus*. Appl. Environ. Microbiol. 68: 1109-1114.

Larrondo L, Salas L, Melo F, Vicuna R, and Cullen D (2003). A novel extracellular multicopper oxidase from *Phanerochaete chrysosporium* with ferroxidase activity. Appl. Environ. Microbiol. 69: 6257-6263.

Larrondo L, Gonzalez B, Cullen D, and Vicuna R (2004). Characterization of a multicopper oxidase gene cluster in *Phanerochaete chrysosporium* and evidence for altered splicing of the *mco* transcript. Microbiology 150: 2775-2783.

Leisola M, Schmidt B, Thanei-Wyss T, and Fiechter A (1985). Aromatic ring cleavage of veratryl alcohol by *Phanerochaete chrysosporium*. FEBS 189: 267-270.

Leisola MSA, Haemmerli SD, Waldner R, Schoemaker HE, Schmidt HWH, and Fiechter A (1988). Metabolism of a lignin model compound, 3,4-dimethoxybenzyl alcohol by *Phanerochaete chrysosporium*. Cellulose Chem. Technol. 22.

Leuthner B, Aichinger C, Oehmen E, Koopmann E, Muller O, Muller P, Kahmann R, Bolker M, and Schreier PH (2004). A peroxide producing glyoxal oxidase is required for filamentous growth and pathogenicity in *Ustilago maydis*. Mol. Gen. Genomics: In press.

Lewis NG, and Sarkanen S, (eds). (1998). Lignin and lignan biosynthesis. Washington, DC: ACS Symposium Series 697.

Li B, Nagalla SR, and Renganathan V (1996). Cloning of a cDNA encoding cellobiose dehydrogenase, a hemoflavoenzyme from *Phanerochaete chrysosporium*. Appl. Environ. Microbiol. 62: 1329-1335.

Li B, Nagalla SR, and Renganathan V (1997). Cellobiose dehydrogenase from *Phanerochaete chrysosporium* is encoded by two allelic variants. Appl. Environ. Microbiol. 63: 796-799.

Li B, and Renganathan V (1998). Gene cloning and characterization of a novel cellulose-binding beta-glucosidase from *Phanerochaete chrysosporium*. Appl. Environ. Microbiol. 64: 2748-2754.

Li D, Alic M, and Gold M (1994). Nitrogen regulation of lignin peroxidase gene transcription. Appl. Environ. Microbiol. 60: 3447-3449.

Liu L, Tewari RP, and Williamson PR (1999). Laccase protects *Cryptococcus neoformans* from antifungal activity of alveolar macrophages. Infect. Immun. 67: 6034-6039.

Ma B, Mayfield MB, and Gold MH (2001). The green fluorescent protein gene functions as a reporter of gene expression in *Phanerochaete chrysosporium*. Appl. Environ. Microbiol. 67: 948-955.

Ma B, Mayfield MB, Godfrey BJ, and Gold MH (2004). Novel promoter sequence required for manganese regulation of manganese peroxidase isozyme 1 gene expression in *Phanerochaete chrysosporium*. Eukaryot Cell 3: 579-588.

Mannhaupt G, Montrone C, Haase D, Mewes HW, Aign V, Hoheisel JD, Fartmann B, Nyakatura G, Kempken F, Maier J, and Schulte U (2003). What's in the genome of a filamentous fungus? Analysis of the *Neurospora genome* sequence. Nucleic Acids Res 31: 1944-1954.

Manubens A, Avila M, Canessa P, and Vicuna R (2003). Differential regulation of genes encoding manganese peroxidase (MnP). in the basidiomycete *Ceriporiopsis subvermispora*. Curr Genet 43: 433-438.

Martinez AT (2002). Molecular biology and structure-function of lignin-degrading heme peroxidases. Enzyme Microb. Technol. 30: 425-444.

Martinez D, Larrondo LF, Putnam N, Sollewijn Gelpke MD, Huang K, Chapman J, Helfenbein KG, Ramaiya P, Detter JC, Larimer F, Coutinho PM, Henrissat B, Berka R, Cullen D, and Rokhsar D (2004). Genome sequence of the lignocellulose degrading fungus *Phanerochaete chrysosporium* strain RP78. Nature Biotechnology 22: 695-700.

Marzullo L, Cannio R, Giardina P, Santini MT, and Sannia G (1995). Veratryl alcohol oxidase from *Pleurotus ostreatus* participates in lignin biodegradation and prevents polymerization of laccase-oxidized substrates. J. Biol. Chem. 270: 3823-3827.

Mason MG, Nicholls P, Divne C, Hallberg BM, Henriksson G, and Wilson MT (2003). The heme domain of cellobiose oxidoreductase: a one-electron reducing system. Biochim Biophys Acta 1604: 47-54.

Matsuzaki F, and Wariishi H (2004). Functional diversity of cytochrome P450s of the white-rot fungus *Phanerochaete chrysosporium*. Biochem Biophys Res Commun 324: 387-393.

Mayer AM, and Staples RC (2002). Laccase: new functions for an old enzyme. Phytochemistry 60: 551-565.

Miura D, Tanaka H, and Wariishi H (2004). Metabolomic differential display analysis of the white-rot basidiomycete *Phanerochaete chrysosporium* grown under air and 100% oxygen. FEMS Microbiol Lett 234: 111-116.

Moukha SM, Dumonceaux TJ, Record E, and Archibald FS (1999). Cloning and analysis of *Pycnoporus cinnabarinus* cellobiose dehydrogenase. Gene 234: 23-33.

Mulder NJ, Apweiler R, Attwood TK, Bairoch A, Barrell D, Bateman A, Binns D, Biswas M, Bradley P, Bork P, Bucher P, Copley RR, Courcelle E, Das U, Durbin R, Falquet L, Fleischmann W, Griffiths-Jones S, Haft D, Harte N, Hulo N, Kahn D, Kanapin A, Krestyaninova M, Lopez R, Letunic I, Lonsdale D, Silventoinen V, Orchard SE, Pagni M, Peyruc D, Ponting CP, Selengut JD, Servant F, Sigrist CJ, Vaughan R, and Zdobnov EM (2003). The InterPro Database, 2003 brings increased coverage and new features. Nucleic Acids Res 31: 315-318.

Munoz IG, Ubhayasekera W, Henriksson H, Szabo I, Pettersson G, Johansson G, Mowbray SL, and Stahlberg J (2001). Family 7 cellobiohydrolases from *Phanerochaete chrysosporium*: crystal structure of the catalytic module of Cel7D (CBH58). at 1.32 A resolution and homology models of the isozymes. J Mol Biol 314: 1097-1111.

Nishimura I, Okada K, and Koyama Y (1996). Cloning and expression of pyranose oxidase cDNA from *Coriolus versicolor* in *E. coli*. J. Biotechnol. 52: 11-20.

Orth A, Rzhetskaya M, Cullen D, and Tien M (1994). Characterization of a cDNA encoding a manganese peroxidase from *Phanerochaete chrysosporium*: genomic organization of lignin and manganese peroxidase genes. Gene 148: 161-165.

Ozturk R, Bozhaya I, Atav E, Saglam N, and Tarhan L (1999). Purification and characterization of superoxide dismutase from *Phanerochaete chrysosporium*. Enzyme Microb Technol 25: 392-399.

Paszczynski A, Huynh V-B, and Crawford RL (1985). Enzymatic activities of an extracellular, manganese-dependent peroxidase from *Phanerochaete chrysosporium*. FEMS Microbiol. Lett. 29: 37-41.

Pease E, and Tien M (1992). Heterogeneity and regulation of manganese peroxidases from *Phanerochaete chrysosporium*. J. Bact. 174: 3532-3540.

Pease EA, Andrawis A, and Tien M (1989). Manganese-dependent peroxidase from *Phanerochaete chrysosporium*. Primary structure deduced from complementary DNA sequence. J. Biol. Chem. 264: 13531-13535.

Pease EA, and Tien M (1991). Lignin-degrading enzymes from the filamentous fungus *Phanerochaete chrysosporium*. In Biocatalysts for Industry. Dordick, J.S. (ed).. New York: Plenum Press, pp. 115-135.

Piontek K, Glumoff T, and Winterhalter K (1993). Low pH crystal structure of glycosylated lignin peroxidase from *Phanerochaete chrysosporium* at 2.5 A resolution. FEBS Lett 315: 119-124.

Piontek K, Smith AT, and Blodig W (2001). Lignin peroxidase structure and function. Biochem Soc Trans 29: 111-116.

Poulter RT, Goodwin TJ, and Butler MI (2003). Vertebrate helentrons and other novel Helitrons. Gene 313: 201-212.

Pribnow D, Mayfield MB, Nipper VJ, Brown JA, and Gold MH (1989). Characterization of a cDNA encoding a manganese peroxidase, from the lignin-degrading basidiomycete *Phanerochaete chrysosporium*. J. Biol. Chem 264: 5036-5040.

Raeder U, and Broda P (1985). Rapid preparation of DNA from filamentous fungi. Lett. Appl. Microbiol. 1: 17-20.

Raeder U, Thompson W, and Broda P (1989a). RFLP-based genetic map of *Phanerochaete chrysosporium* ME446: lignin peroxidase genes occur in clusters. Mol. Microbiol. 3: 911-918.

Raeder U, Thompson W, and Broda P (1989b). Genetic factors influencing lignin peroxidase activity in *Phanerochaete chrysosporium* ME446. Mol. Microbiol. 3: 919-924.

Raices M, Paifer E, Cremata J, Montesino R, Stahlberg J, Divne C, Szabo IJ, Henriksson G, Johansson G, and Pettersson G (1995). Cloning and characterization of a cDNA encoding a cellobiose dehydrogenase from the white rot fungus *Phanerochaete chrysosporium*. FEBS Lett. 369: 233-238.

Randall T, Rao TR, and Reddy CA (1989). Use of a shuttle vector for the transformation of the white-rot basidiomycete, *Phanerochaete chrysosporium*. Biochem. Biophys. Res. Commun. 161: 720-725.

Randall T, Reddy CA, and Boominathan, K (1991). A novel extrachromosomally maintained transformation vector for the lignin-degrading basidiomycete *Phanerochaete chrysosporium*. J. Bacteriol. 173: 776-782.

Randall TA, and Reddy CA (1992). The nature of extra-chromosomal maintenance of transforming plasmids in the filamentous basidiomycete *Phanerochaete chrysosporium*. Curr. Genet. 21: 255-260.

Reiser J, Walther I, Fraefel C, and Fiechter A (1993). Methods to investigate the expression of lignin peroxidase genes by the white-rot fungus *Phanerochaete chrysosporium*. Appl. Environ. Microbiol. 59: 2897-2903.

Reiser J, Muheim A, Hardegger M, Frank G, and Feichter A (1994). Aryl-alcohol dehydrogenase from the white-rot fungus *Phanerochaete chrysosporium* : gene cloning, sequence analysis, expression and purification of recombinant protein. J. Biol. Chem 269: 28152-28159.

Renganathan V, Miki K, and Gold MH (1985). Multiple molecular forms of diarylpropane oxygenase, an H_2O_2-requiring, lignin-degrading enzyme from *Phanerochaete chrysosporium*. Arch. Biochem. Biophys. 241: 304-314.

Renganathan V, and Gold MH (1986). Spectral characterization of the oxidized states of lignin peroxidase, an extracellular enzyme from the white-rot basidiomycete *Phanerochaete chrysosporium*. Biochem. 25: 1626-1631.

Rieble S, Joshi DK, and Gold MH (1994). Purification and characterization of a 1,2,4-trihydroxybenzene 1,2-dioxygenase from the basidiomycete *Phanerochaete chrysosporium*. J Bacteriol 176: 4838-4844.

Rogers MS, and Dooley DM (2003). Copper-tyrosyl radical enzymes. Curr Opin Chem Biol 7: 189-196.

Rothschild N, Hadar Y, and Dosoretz C (1997). Lignin peroxidase isozymes from *Phanerochaete chrysosporium* can be enzymatically dephosphorylated. Appl. Environ. Microbiol. 63: 857-861.

Rothschild N, Levkowitz A, Hadar Y, and Dosoretz C (1999). Extracellular mannose-6-phosphatase of *Phanerochaete chrysosporium*: a lignin peroxidase-modifying enzyme. Arch. Biochem. Biophys. 372: 107-111.

Rotsaert FA, Li B, Renganathan V, and Gold MH (2001). Site-directed mutagenesis of the heme axial ligands in the hemoflavoenzyme cellobiose dehydrogenase. Arch. Biochem. Biophys. 390: 206-214.

Ruiz-Duenas FJ, Martinez MJ, and Martinez AT (1999). Molecular characterization of a novel peroxidase isolated from the ligninolytic fungus *Pleurotus eryngii*. Mol. Microbiol. 31: 223-235.

Ruiz-Duenas FJ, Camarero S, Perez-Boada M, Martinez MJ, and Martinez AT (2001). A new versatile peroxidase from *Pleurotus*. Biochem Soc Trans 29: 116-122.

Schalch H, Gaskell J, Smith TL, and Cullen D (1989). Molecular cloning and sequences of lignin peroxidase genes of *Phanerochaete chrysosporium*. Mol. Cell. Biol. 9: 2743-2747.

Schoemaker HE, and Leisola MSA (1990). Degradation of lignin by *Phanerochaete chrysosporium*. J. Biotechnol. 13: 101-109.

Schulte U, Becker I, Mewes HW, and Mannhaupt G (2002). Large scale analysis of sequences from *Neurospora crassa*. J Biotechnol 94: 3-13.

Shi X, Stoj C, Romeo A, Kosman DJ, and Zhu Z (2003). Fre1p Cu2+ reduction and Fet3p Cu1+ oxidation modulate copper toxicity in *Saccharomyces cerevisiae*. J Biol Chem 278: 50309-50315.

Shimizu M, Yuda N, Nakamura T, Tanaka H, and Wariishi H (2004). Metabolic regulation at the TCA and glyoxylate cycles of the lignin-degrading basidiomycete *Phanerochaete chrysosporium* against exogenous addition of vanillin. Proteomics 4: In press.

Sims AH, Dunn-Coleman NS, Robson GD, and Oliver SG (2004). Glutamic protease distribution is limited to filamentous fungi. FEMS Microbiol Lett 239: 95-101.

Smith MH, and Gold MH (1979). *Phanerochaete chrysosporium* b-glucosidase: induction, cellular localization, and physical characterization. Appl. Environ. Microbiol. 37: 938-942.

Sollewijn Gelpke MD, Mayfield-Gambill M, Lin Cereghino GP, and Gold MH (1999). Homologous expression of recombinant lignin peroxidase in *Phanerochaete chrysosporium*. Appl. Environ. Microbiol. 65: 1670-1674.

Srebotnik E, Messner K, Foisner R, and Petterson B (1988b). Ultrastructural localization of ligninase of *Phanerochaete chrysosporium* by immunogold labeling. Curr. Microbiol. 16: 221-227.

Srebotnik E, and Messner KE (1991). Immunoelectron microscopical study of the porosity of brown rot wood. Holzforschung 45: 95-101.

Stalpers JA (1984). A revision of the genus *Sporotrichum*. Studies in Mycology 24: 1-105.

Staszczak M, Zdunek E, and Leonowicz A (2000). Studies on the role of proteases in the white-rot fungus *Trametes versicolor*: effect of PMSF and chloroquine on ligninolytic enzymes activity. J. Basic Microbiol. 40: 51-63.

Stearman R, Yuan DS, Yamaguchi-Iwai Y, Klausner RD, and Dancis A (1996). A permease-oxidase complex involved in high-affinity iron uptake in yeast. Science 271: 1552-1557.

Stewart P, Kersten P, Vanden Wymelenberg A, Gaskell J, and Cullen D (1992). The lignin peroxidase gene family of *Phanerochaete chrysosporium*: complex regulation by carbon and nitrogen limitation, and the identification of a second dimorphic chromosome. J. Bact. 174: 5036-5042.

Stewart P, and Cullen D (1999). Organization and differential regulation of a cluster of lignin peroxidase genes of *Phanerochaete chrysosporium*. J. Bact. 181: 3427-3432.

Stewart P, Gaskell J, and Cullen D (2000). A homokaryotic derivative of a *Phanerochaete chrysosporium* strain and its use in genomic analysis of repetitive elements. Appl. Environ. Microbiol. 66: 1629-1633.

Stoj C, and Kosman DJ (2003). Cuprous oxidase activity of yeast Fet3p and human ceruloplasmin: implication for function. FEBS Lett 554: 422-426.

Stuardo M, Larrondo L, Vasquez M, Vicuna R, and Gonzalez B (2004). Incomplete processing of peroxidase transcripts in the lignin degrading fungus *Phanerochaete chrysosporium*. FEMS Microbiol Lett: In press.

Sunagawa M, and Magae Y (2002). Transformation of the edible mushroom *Pleurotus ostreatus* by particle bombardment. FEMS Microbiol. Lett. 211: 143-146.

Sundaramoorthy M, Kishi K, Gold MH, and Poulos TL (1997). Crystal structures of substrate binding site mutants of manganese peroxidase. J Biol Chem 272: 17574-17580.

Tanaka H, Itakura S, and Enoki A (1999). Hydroxyl radical generation by an extracellular low-molecular weight substance and phenol oxidase activity during wood degradation by the white rot basidiomycete, *Phanerochaete chrysosporium*. Holzforschung 52: 21-28.

Temp U, Zierold U, and Eggert C (1999). Cloning and characterization of a second laccase gene from the lignin- degrading basidiomycete *Pycnoporus cinnabarinus*. Gene 236: 169-177.

Teramoto H, Tanaka H, and Wariishi H (2004). Degradation of 4-nitrophenol by the lignin-degrading basidiomycete *Phanerochaete chrysosporium*. Appl Microbiol Biotechnol.

Thurston CF (1994). The structure and function of fungal laccases. Microbiol. 140: 19-26.

Tien M, and Kirk TK (1983). Lignin-degrading enzyme from the Hymenomycete *Phanerochaete chrysosporium* Burds. Science (Wash. D. C.). 221: 661-663.

Tien M, and Kirk TK (1984). Lignin-degrading enzyme from *Phanerochaete chrysosporium*: Purification, characterization, and catalytic properties of a unique H_2O_2-requiring oxygenase. Proc. Natl. Acad. Sci. USA. 81: 2280-2284.

Umezawa T, Kawai S, Yokota S, and Higuchi T (1986). Aromatic ring cleavage of various b-O-4 lignin model dimers by *Phanerochaete chrysosporium*. Wood Research 73: 8-17.

Urzua U, Kersten PJ, and Vicuna R (1998). Kinetics of Mn3+-oxalate formation and decay in reactions catalyzed by manganese peroxidase of *Ceriporiopsis subvermispora*. Arch. Biochem. Biophys. 360: 215-222.

Uzcategui E, Johansson G, Ek, B, and Pettersson G (1991a). The 1,4-D-glucan glucanohydrolases from *Phanerochaete chrysosporium*. Re-assessment of their significance in cellulose degradation mechanisms. J. Biotechnol. 21: 143-160.

Uzcategui E, Raices M, Montesino R, Johansson G, Pettersson G, and Eriksson K-E (1991b). Pilot-scale production and purification of the cellulolytic enzyme system from the white-rot fungus *Phanerochaete chrysosporium*. Biotechnol. Appl. Biochem. 13: 323-334.

Uzcategui E, Ruiz A, Montesino R, Johansson G, and Pettersson G (1991c). The 1,4-b-D-glucan cellobiohydrolase from *Phanerochaete chrysosporium*. I. A system of synergistically acting enzymes homologous to *Trichoderma reesei*. J. Biotechnol. 19: 271-286.

Vallim MA, Janse BJ, Gaskell J, Pizzirani-Kleiner AA, and Cullen D (1998). *Phanerochaete chrysosporium* cellobiohydrolase and cellobiose dehydrogenase transcripts in wood. Appl. Environ. Microbiol. 64: 1924-1928.

Van Hamme JD, Wong ET, Dettman H, Gray MR, and Pickard MA (2003). Dibenzyl sulfide metabolism by white rot fungi. Appl Environ Microbiol 69: 1320-1324.

Vanden Wymelenberg A, Covert S, and Cullen D (1993). Identification of the gene encoding the major cellobiohydrolase of the white-rot fungus *Phanerochaete chrysosporium*. Appl. Environ. Microbiol. 59: 3492-3494.

Vanden Wymelenberg AV, Denman S, Dietrich D, Bassett J, Yu X, Atalla R, Predki P, Rudsander U, Teeri TT, and Cullen D (2002). Transcript analysis of genes encoding a family 61 endoglucanase and a putative membrane-anchored family 9 glycosyl hydrolase from *Phanerochaete chrysosporium*. Appl Environ Microbiol 68: 5765-5768.

Varela E, Martinez AT, and Martinez MJ (1999). Molecular cloning of aryl-alcohol oxidase from the fungus *Pleurotus eryngii*, an enzyme involved in lignin degradation. Biochem. J. 341: 113-117.

Varela E, Bockle B, Romero A, Martinez AT, and Martinez MJ (2000a). Biochemical characterization, cDNA cloning and protein crystallization of aryl-alcohol oxidase from *Pleurotus pulmonarius*. Biochim. Biophys. Acta. 1476: 129-138.

Varela E, Martinez JM, and Martinez AT (2000b). Aryl-alcohol oxidase protein sequence: a comparison with glucose oxidase and other FAD oxidoreductases. Biochim. Biophys. Acta. 1481: 202-208.

Wariishi H, Valli K, and Gold MH (1991). *In vitro* depolymerization of lignin by manganese peroxidase of *Phanerochaete chrysosporium*. Biochem. Biophys. Res. Comm. 176: 269-275.

Whittaker JW (2002). Galactose oxidase. Adv. Protein Chem. 60: 1-49.

Whittaker MM, Kersten PJ, Cullen D, and Whittaker JW (1999). Identification of catalytic residues in glyoxal oxidase by targeted mutagenesis. J. Biol. Chem. 274: 36226-36232.

Whitwam RE, Brown KR, Musick M, Natan MJ, and Tien M (1997). Mutagenesis of the Mn2+-binding site of manganese peroxidase affects oxidation of Mn2+ by both compound I and compound II. Biochemistry 36: 9766-9773.

Williamson PR (1994). Biochemical and molecular characterization of the diphenol oxidase of *Cryptococcus neoformans*: identification as a laccase. J. Bact. 176: 656-664.

Wood PM (1994). Pathways for the production of Fenton's reagen by wood-rotting fungi. FEMS Microbiol. Rev. 13: 313-320.

Wood V, Gwilliam R, Raj MA, Lyne M, Lyne R, Stewart A, Sgouros J, Peat N, Hayles J, Baker S, Basham D, Bowman S, Brooks K, Brown D, Brown S, Chillingworth T, Churcher C, Collins M, Connor R, Cronin A, Davis P, Feltwell T, Fraser A, Gentles S, Goble A, Hamlin N, Harris D, Hidalgo J, Hodgson G, Holroyd S, Hornsby T, Howarth S, Huckle, EJ, Hunt S, Jagels K, James K, Jones L, Jones M, Leather S, McDonald S, McLean J, Mooney P, Moule S, Mungall K, Murphy L, Niblett D, Odell C, Oliver K, O'Neil S, Pearson D, Quail MA, Rabbinowitsch, E, Rutherford, K, Rutter, S, Saunders, D, Seeger, K, Sharp, S, Skelton J, Simmonds M, Squares R, Squares S, Stevens K, Taylor K, Taylor RG, Tivey A, Walsh S, Warren T, Whitehead S, Woodward J, Volckaert G, Aert R, Robben J, Grymonprez B, Weltjens I, Vanstreels E, Rieger M, Schafer M, Muller-Auer S, Gabel C, Fuchs M, Dusterhoft A, Fritzc C, Holzer E, Moestl D, Hilbert H, Borzym K, Langer I, Beck A, Lehrach H, Reinhardt R, Pohl TM, Eger P, Zimmermann W, Wedler H, Wambutt

R, Purnelle B, Goffeau A, Cadieu E, Dreano S, Gloux S, Lelaure V, Mottier S, Galibert F, Aves SJ, Xiang Z, Hunt C, Moore K, Hurst SM, Lucas M, Rochet M, Gaillardin C, Tallada VA, Garzon A, Thode G, Daga RR, Cruzado L, Jimenez J, Sanchez M, del Rey F, Benito J, Dominguez A, Revuelta JL, Moreno S, Armstrong J, Forsburg SL, Cerutti L, Lowe T, McCombie, WR, Paulsen I, Potashkin J, Shpakovski GV, Ussery D, Barrell BG, Nurse P, and Cerrutti L (2002). The genome sequence of *Schizosaccharomyces pombe*. Nature 415: 871-880.

Worral JJ, Anagnost SE, and Zabel RA (1997). Comparison of wood decay among diverse lignicolous fungi. Mycologia 89: 199-219.

Wyatt AM, and Broda P (1995). Informed strain improvement for lignin degradation by *Phanerochaete chrysosporium*. Microbiol. 141 (Pt 11): 2811-2822.

Xie L, and van der Donk WA (2001). Homemade cofactors: self-processing in galactose oxidase. Proc Natl Acad Sci U S A 98: 12863-12865.

Xu G, and Goodell B (2001). Mechanisms of wood degradation by brown-rot fungi: chelator-mediated cellulose degradation binding of iron by cellulose. J Biotechnol 87: 43-57.

Yadav JS, and Loper JC (2000). Cytochrome P450 oxidoreductase gene its differentially terminated cDNAs from the white rot fungus *Phanerochaete chrysosporium*. Curr. Genet. 37: 65-73.

Yadav JS, Soellner MB, Loper JC, Mishra PK (2003). T em cytochrome P450 monooxygenase genes splice variants in the white rot fungus *Phanerochaete chrysosporium*: cloning, sequence analysis, regulation of differential expression. Fungal Genet Biol 38: 10-21.

Yanai K, Yonekura K, Usami H, Hirayama M, Kajiwara S, Yamazaki T, Shishido K, Adachi T (1996). The integrative transformation of *Pleurotus ostreatus* using bialaphos resistance as a dominant selectable marker. Biosci. Biotechnol. Biochem. 60: 472-475.

Youn H, Hah Y, Kang S (1995). Role of laccase in lignin degradation by white-rot fungi. FEMS Microbiol. Lett. 132: 183-188.

Youngs HL, Sollewijn Gelpke MD, Li D, Sundaramoorthy M, Gold MH (2001). The role of Glu39 in MnII binding oxidation by manganese peroxidase from *Phanerochaete chrysosporium*. Biochem. 40: 2243-2250.

Zapanta LS, Hattori T, Rzetskaya M, Tien M (1998). Cloning of *Phanerochaete chrysosporium leu2* by complementation of bacterial auxotrophs transformation of fungal auxotrophs. Appl. Environ. Microbiol. 64: 2624-2629.

Zolan M (1995). Chromosome-length polymorphisms in fungi. Microbiol. Rev. 59: 686-698.

**Applied Mycology
and Biotechnology**
An International Series
Volume 5. Genes and Genomics
© 2005 Elsevier B.V. All rights reserved

15

ELSEVIER

Genetics of Morphogenesis in Basidiomycetes

**J. Stephen Horton◊, Guus Bakkeren§, Steven J. Klosterman*, Maria Garcia-Pedrajas*,
Scott E. Gold***
◊Department of Biological Sciences, Science and Engineering Center, Union College,
Schenectady, NY 12308-2311, USA; §Agriculture & Agri-Food Canada, Pacific Agri-Food
Research Centre, 4200 Highway 97, Summerland, B.C., Canada V0H 1Z0. *Department of Plant
Pahtology, University of Georgia, Athens, GA 30602-7274, USA. Corresponding author: Scott E.
Gold (sgold@uga.edu).

In this chapter, our aim is to discuss the current knowledge of the genetics of morphogenesis in
basidiomycetes. We begin by outlining some features that are shared among fungi in general
and those that are unique to basidiomycetes. With this background of basic fungal morphology
and physiology, we focus our discussion on the genetics of morphogenesis and the fascinating
biology of three broad groups of basidiomycetes: the smuts, the rusts and the mushrooms.
Where the smuts and rusts are considered, there is ample discussion on pathogenesis, as these
two broad groupings of fungi are particularly destructive to plants worldwide. Of course,
mushrooms represent a valuable commodity in their own right. Each of these broad sections
encapsulates a current assessment of the environmental cues and the genes regulating
morphological change during development. Each of these three taxon-focused sections also
concludes with a section discussing current trends and/or future directions.

1. INTRODUCTION

The kingdom fungi lies on the opisthokont lineage of the Eukaryotes and thus is a
closer relative to the animals than to the plants. The fossil record of the fungi is
estimated to date back as far as 900 million years with all major classes represented
about 300 million years ago (Taylor et al. 1994). The basidiomycetes are one of the two
ultimate branches of the true fungi, the other being the ascomycetes. About 30,000
basidiomycete species have been described. Among the basidiomycetes are the well
known mushrooms of the Hymenomycetes, the rusts of the Urediniomycetes known
through recorded history as devastating pathogens of staple grain crops, and the smuts
of the Ustilaginomycetes that cause persistent crop losses. In this chapter, we present an
overview of some of the central processes particular to the basidiomycetes. We then

Corresponding author: Scott E. Gold

delve more deeply into exemplars of each of the three above major groups and provide a current snap shot of what is known regarding the genetics of their developmental processes.

2. SPECIALIZED FEATURES

In the following sections, we present overviews of some of the developmental features common among the basidiomycetes. Some of these such as dimorphic growth and sporulation are found in other fungi while others like clamp connections, peculiarities of their mating systems and the surreal complexity of some of their host interactions, are specific to the basidiomycetes.

2.1 Budding vs. Filamentous Growth

Basidiomycetes display wide morphological diversity, having both multicellular filamentous forms as well as budding unicellular yeast forms. Dimorphic basidiomycetes exhibit a switch between yeast-like growth and filamentous growth (Alexopolous et al. 1996). Alternating phases of budding and filamentous growth are commonly observed in the dimorphic tremelloid and smut fungi. On the other hand, in most basidiomycetes, such as those that are found in the rusts and gasteromycetes, a yeast-like budding phase is not present. Because the dimorphic smut fungus *Ustilago maydis* will be discussed in great detail below in the exemplary model systems section 3.1, interesting examples of other fungi exhibiting dimorphic growth will be presented here.

One group of dimorphic fungi is found in the order Tremellales (Alexopolous et al. 1996). The well-studied fungus *Cryptococcus neoformans*, a microbe found on trees and a facultative pathogen of animals including humans, is a tremelloid fungus that infects the central nervous system, causing meningoencephalitis. *C. neoformans* grows as a budding yeast with a conspicuous polysaccharide capsule (Alexopolous et al. 1996; Hull and Heitman 2002). But in response to nutrient-limiting conditions or mating pheromone, yeast cells of opposite mating type produce conjugation tubes and fuse, forming a filamentous heterokaryon (Hull and Heitman 2002). Under the appropriate conditions, yeast cells of the alpha mating type can undergo haploid filamentation, producing monokaryotic filaments with unfused clamp connections (discussed below). Like the ascomycete fungal pathogens of animals, *C. neoformans* exhibits thermal dimorphism (Sia et al. 2000). In *C. neoformans*, yeast cell fusion and maintenance at 37° C leads to the formation of stable diploid yeasts. A switch to 25° C leads to the formation of monokaryotic diploid hyphae with unfused clamp connections. In contrast, yeast cell fusion and maintenance at 25° C leads to the formation of dikaryotic filaments with fused clamp connections (Sia et al. 2000).

Another group of dimorphic basidiomycete fungi are the Sporidiales. These fungi, represented here by *Rhodosporidium sphaerocarpum*, are taxonomically more closely related to the rust fungi than either the Ustilaginales or Tremellales based on septa structure and rDNA analyses, and exhibit both a yeast phase and a filamentous phase in their lifecycles (Alexopolous et al. 1996). Compatible yeast cells of *R. sphaerocarpum* fuse, forming a filamentous dikaryon that eventually gives rise to intercalary and terminal teliospores. Teliospores germinate and meiosis occurs, giving rise to a four-celled septate promycelium from which basidiospores develop and bud in a manner strikingly similar to the developmental program characterized in the smuts. As a result, these have been mistakenly identified as such (Alexopolous et al. 1996).

Some ants in the genus *Cyphomyrmex* maintain basidiomycetes in the family lepiotaceae as unicellular masses of yeasts. These yeasts form a monophyletic clade within a larger filamentous clade (Mueller et al. 1998) suggesting that these yeasts were derived from mycelial ancestors and switched to yeast-like growth upon association with *Cyphomyrmex* yeast specialists (Mueller 2002). As Mueller (2002) pointed out, the ants may have evolved a feature to manipulate their cultivars, allowing the ants to keep their fungus in a yeast-like state. It would certainly be of interest to examine the growth of these fungi for filamentation under various culture conditions, away from the nests. In contrast, most other fungus growing ants propagate and maintain their cultivars in the filamentous form (Mueller et al. 1998; Mueller 2002). Although some filaments have been observed with clamp connections, these fungi traditionally have defied identification because they rarely produce a sexual form (Alexopolous et al. 1996). But DNA-based phylogenetic analysis has revealed a close relationship between many of these fungi and a group of poorly understood tropical mushrooms (Mueller et al. 1998).

2.2 Mating Types

Basidiomycetes generally have four functional mating specificities that combine to generate full sexual compatibility. Their cousins of the Ascomycota classically have 2 mating types (eg. a and α in *Saccharomyces* and A and a in *Neurospora*). Many excellent reviews are available on the topic of fungal mating types (Brown and Casselton 2001; Casselton and Olesnicky 1998; Casselton 2002; Fraser and Heitman 2004) and so the issue will not be belabored here, rather we will give a brief outline for the basidiomycetes as the mating type genes are crucial developmental factors for these fungi. The basidiomycetes have taken sexual promiscuity to new heights. For example it is estimated that *Coprinopsis cinerea* (*Coprinus cinereus*) and *Schizophyllum commune* have 12,000 and 20,000 different mating specificities all of which are inter-compatible but self-incompatible, respectively (Brown and Casselton 2001). This makes the probability of coming upon a compatible partner nearly 100%. Most mushrooms species have no inhibition to anastomosis in the homokaryotic stage and this makes biological sense since nearly all fusions will be sexually productive.

There are two functional classes of mating loci in the basidiomycetes. These are classically described as the *a* and *b* or *A* and *B* loci. Unfortunately in the smut model *Ustilago maydis* and related species the mating loci are named in reverse for function when compared to the Hymenomycetes including the mushrooms. In the mushrooms, the A loci encode homeodomain containing transcription factors while the B loci encode lipopeptide pheromones and transmembrane pheromone receptors. In the smuts the *a* locus encodes the pheromones and receptors and the *b* locus encodes the homeodomain containing proteins. In the smuts the *a* and *b* loci may be genetically linked or unlinked. This situation leads to two major groups the bipolar smuts in which there are only two mating specificities that segregate at meiosis and the tetrapolar smuts in which meiosis produces four mating specicities in the progeny. It has been shown that at least for some bipolar smuts the *a* and *b* loci are both present but are not separable by recombination (Bakkeren et al. 1992; Bakkeren and Kronstad 1994; Lee er al. 1999). The tetrapolar mating system of the smuts is described later in this chapter. The arrangement of the *A* and *B* genes in the mushrooms is quite complex often with linked sub-loci. In some species the *A* and *B* genes may be distributed between distinct subloci known as *A*α, *A*β, *B*α and *B*β (Fowler et al. 2004; Pardo et al. 1996). The paradigm in C. *cinerea* is that each *A* locus encodes three pairs on homeodomain containing proteins while each *B* locus encodes three cassettes each with a pheromone receptor gene and usually two pheromones (Brown and Casselton, 2001). The pairs are made up of canonical members called HD1 and HD2, which must form at least one nonself heterodimer with those of another monokaryon for completetion of successful *A* mating function. None of the HD1 proteins are able to form a productive heterodimer with the HD2 proteins encoded by the same monokaryon. Similarly at the *B* locus the pheromones produced by a monokaryon will not interact with any of the receptors produced by that strain. Thus the basidiomycetes are exquisitely designed to find compatible partners in nature and yet recognize self as incompatible.

2.3 Nuclear Condition

In the basidiomycetes, nuclear condition alternates between haploid forms and dikaryotic forms generated by mating proccesses. Karyogamy takes place in the basidium, generating the diploid stage of the cell cycle. Also in these structures, meiosis occurs, resulting in the production of haploid basidiospores. Interesting questions can be asked when comparing for example rusts and smuts and trying to correlate nuclear state, ploidy level and developmental program and/or infection of hosts. A simple pattern is found in the smut fungi. The haploid, mononucleate smut basidiospores are mostly saprobic and although able to live off their host, are unable to cause disease (i.e., sporulate). Mating and the formation of a binucleate dikaryon are necessary for pathogenicity. There are interesting variations among smut fungi regarding the putative role of the haploid sporidia in nature. Thus, in *Tilletia* species mating of compatible basidiospores takes place while they are still attached to the promycelium and upon discharge and germination the secondary sporidia develop a dikaryotic mycelium, which is capable of infecting plants. Therefore, the duration of the haploid phase in these species is reduced in time. It has been shown that very early in the infection of corn by *U. maydis*, the hyphae can have multiple (more than 2) nuclei per cellular compartment but that soon after a proper dikaryotic mycelium is produced (Snetselaar and Mims 1993). In the macrocyclic rusts, such as in *Puccinia* spp., both dikaryotic aeciospores and uridiniospores are pathogenic on the primary host. But in addition, haploid basidiospores, which are often binucleate after a subsequent mitotic division (Gold and Mendgen 1991; Anikster 1983; Heath et al. 1996), are also pathogenic on the alternate host or on the same host as for autoecious *Uromyces* spp. (see section 3.2.2 for a description of a rust life cycle and spore stages). Haploid basidiospores often produce binucleate infection hyphae early during infections but one nucleus disintegrates soon after during subsequent fungal growth *in planta* producing haploid, mono-nuclear cells, as has been convincingly shown for *U. vignae* (Heath et al. 1996). The fertilization of haploid, mono-nuclear pycniospores with ones of a different (opposite or "non-self") mating type in specialized structures, the pycnia, recreates the dikaryotic (n+n) mycelium which reprograms the fungus to produce aeciospores in analogy with the smuts (although these produce teliospores at that point). Dikaryotic urediniospores produce infection structures containing variable numbers of nuclei per "cell" compartment, depending on the species and/or differentiation stage, e.g., 2 nuclei in the germ tube, 4 in the appressorium, and various numbers in the substomatal vesicle (Heath et al. 1996). This can occur through various rounds of mitosis without septum formation, asynchronous nuclear divisions and/or septation and nuclear degradation. However, this is mostly seen during early stages of fungal development and established (n+n) mycelium appears to be uniformly binucleate/dikaryotic. Unusual situations exist in haustoria of *P. coronata* which have only a single nucleus whereas the multinucleate condition (up to 24 nuclei per cell) apparently persists in one studied isolate of *P. striiformis* (Chong et al. 1992). It seems that the nuclear state is not synchronized with the rapidly changing developmental programming necessary during the early infection. Karyogamy takes place within the plant producing diploid teliospores. Teliospores arise from dikaryotic hyphal compartments by different mechanisms. In some cases karyogamy occurs early during teliospores development so that even immature teliospores are uninecleate. In other species however, nuclear fusion happens at late stages of teliospore development. In smut and rust fungi, teliospores are considered the probasidium. Upon germination, the teliospore gives rise to a promycelium, into which the nucleus moves and undergoes meiosis, giving rise to the formation of haploid basidiospores directly on the surface of the promycelium (O'Donell and McLaughlin 1984).

2.4 Clamp Connections

In many basidiomycetous fungi, a stable dikaryotic state is established after a mating between two compatible homokaryons. The dikaryon is a specialized form of heterokaryon in which each cell has two haploid nuclei, one from each parent of the cross. These nuclei do not fuse, but are maintained as a pair in each cell in an "n + n" arrangement. This genetically balanced condition is characteristic of many, but not all basidiomycete fungi (Raper 1966; Raper 1983). The dikaryon is capable of indefinite propagation, and is found in such well-studied mushroom-producing fungi such as *Coprinopsis cinerea* (*Coprinus cinereus*) and *Schizophyllum commune*, but not in the white button mushroom, *Agaricus bisporus* (Raper et al. 1972) or rusts and smuts (which form stable dikaryons following mating and in host tissues). The cellular processes involved in maintaining the dikaryon are described below, and in Figure 1.

During the course of a compatible mating, the dikaryon is established by means of the reciprocal migration of nuclei from each mating partner into the hyphae of the other. This is followed by a precise pairing of the two different nuclei in each cell. In order to maintain a strict 1:1 ratio of the two nuclear types in each new cell of the dikaryon, each subsequent cell division is accompanied by the coordinated processes of hook cell formation and conjugate nuclear division. The hook cell, or clamp connection, starts as a lateral bulge that quickly grows in a manner curving away from the apex of the hyphal tip, forming a characteristic hook shape (Buller 1933; Niederpruem et al. 1971). The two parental nuclei in the apical cell divide synchronously, but the mitotic spindles are of different length. The nuclear division arising from the more apical nucleus has a shorter spindle, and generates a daughter nucleus that migrates into the hook cell. The companion daughter nucleus of this pair remains in the tip cell, and migrates towards its apex. A longer spindle is generated by the nuclear division of the more distal of the two parental nuclei, and is oriented in the same plane as the long axis of the dividing cell. This event generates one daughter nucleus that migrates towards the hyphal tip, during which time it moves past the nucleus of the other parental type. The other daughter nucleus moves backwards in a distal direction, in the soon to be subapical cell (Runeberg et al. 1986; Salo et al. 1989; Tanabe and Kamada 1994). There is a consequence of these precise patterns of nuclear division and movement. With each new conjugate division, there is an alternation in the order of which nuclear type takes the leading position in the apical cell (Iwasa et al. 1998). The next cellular event is that a septum is formed at the base of the hook cell, trapping the single nucleus within. A second septum is laid down in the hypha just distal to the hook cell, the result of which is a new apical cell with two haploid nuclei of each parental type (Buller 1933; Niederpruem et al. 1971). A lateral bulge or peg then forms in the subapical cell in close proximity to the septum (Buller 1933; Badalyan et al. 2004; Kues et al. 2002). This event is quickly followed by the fusion of the free end of the hook cell to the subapical cell at the location of the peg. The previously entrapped nucleus in the hook then moves into the subapical cell, restoring the dikaryotic state to this cell (Buller 1933; Tanabe and Kamada 1994). The presence of fused clamp connections is diagnostic of the dikaryotic state.

The processes involved in establishing and maintaining the dikaryotic state are genetically controlled at a fundamental level by the activity of the A and B mating-type genes (Raper 1966). This concept has been confirmed at the molecular level by a host of studies (Casselton and Olesnicky 1998; Kamada 2002; Kothe 1996; Kues 2000). Nuclear pairing, conjugate nuclear division, hook cell formation, and septation are regulated by the products of the A genes. Nuclear migration and hook cell fusion are controlled by the action of the B gene products. The formation of subapical pegs is attributed to the activity of both the A and B genes, with the latter likely to exert a greater influence on the process (Kues et al. 2002; Badalyan et al. 2004). The genes clp1 and pcc1 are downstream elements in the A-regulated pathway leading to clamp connection development (Kamada 2002). These two genes will be discussed later in the section on the genetic control of mushroom development.

2.5 Sporulation

In order to reproduce and to protect precious genetic material and safeguard it to allow for future generations, many eukaryotic organisms have evolved elaborate mechanisms. Fungi produce several kinds of spores, which can be divided roughly into asexual and sexual spores. The production of asexual spores, mainly meant for dispersal to increase the population under favorable conditions, seems to be more prevalent among ascomycetes although these play a major role in the rusts and the human pathogen *Cryptococcus neoformans* and can cause epidemics. Under certain

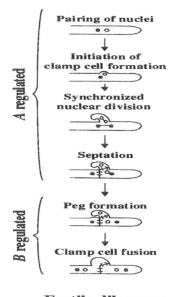

Fertile dikaryon

Fig. 1: Cellular processes involved in dikaryon formation (from Kues 2000). Involvement of A and B loci depicted for homobasidiomycete/mushroom growth.

(sudden, unfavorable) conditions, some fungi produce chlamydospores, thick-walled cells that develop from single (dikaryotic) hyphal compartments. In some higher mushroom species like *Coprinopsis cinerea*, monokaryotic (aerial) mycelium can produce abundant single-celled mitotic spores (oidia) on oidiophores (Polak et al. 1997) (See Figures 4 and 5). *C. cinerea* can also produce chlamydospores, which are submerged, thick-walled mitotic spores found in older cultures of dikaryons and certain monokaryons (Kues 2000). These structures arise in areas of compressed cytoplasm, and are not found in the other model mushroom-forming basidiomycete, *S. commune*. Sexual spores are mainly produced under unfavorable conditions as a "resting" stage, e.g., to prepare for draught or overwintering. A hall mark of the basidiomycetes is the production of a basidium in which karyogamy in the otherwise dikaryotic mycelial cells takes place, followed by meiosis and generally the production of four haploid basidiospores in which different (often two opposite) mating types also segregate (Alexopolous et al. 1996). Ironically for this chapter, the phytopathogenic rust and smut fungi are different in that they produce an additional spore type, the teliospore which is essentially a probasidium. The teliospore is a true resting structure with a thick, often pigmented or melanized (Butler and Day 1998) and ornamented wall in which karyogamy has taken place. This cell type represents therefore a true diploid stage. Teliospores can survive for many years and are also meant for dispersal such as for the rusts and smuts. There is a tremendous variation in morphology and ornamentation of teliospores and other spore types such as the urediniospores in rusts (e.g., see Fig. 20-19 in Alexopolous et al. 1996). For example, *Puccinia* species often produce two-celled teliospores whereas in *Uromyces* spp. mostly one-celled teliospores are found. In general, spore features are often used for taxonomic purposes. Teliospore germination can take place under favorable conditions either directly after they have been produced or after a required, prolonged dormancy period. Germination produces a promycelium (metabasidium) in which meiosis takes place. Four resulting haploid nuclei then move into four cells, but one or four mature basidiospores might eventually form on the basidium (Alexopolous et al. 1996; Goates and Hoffmann 1987; Mendgen 1984). Haploid basidiospores from many smuts reproduce by budding and are saprobic allowing them to be cultured and making them amenable to molecular techniques. In the obligate rusts, the probasidium from the germinating teliospore and the resulting basidiospores are the only stages not requiring a host plant. However, cultivating rusts *in vitro* has been unsuccessful or is difficult at best (Bose and Shaw 1974; Williams 1984; Boasson and Shaw 1988; Fasters et al. 1993).

2.6 Pathogenicity and Virulence Factors in Biotrophic Interactions of Smuts and Rusts

By definition, pathogens have the ability to cause disease on a host. Successfully completing their life cycle (i.e., producing viable progeny/survival structures) is the pathogen's only goal, selected for during evolution. A large arsenal of proteins and metabolites contribute to its armor to allow it to recognize certain organisms as hosts, be induced to germinate and risk its precious genetic material, and to wage chemical warfare in order to breach barriers, gain access, suppress recognition and defense responses and coax the invaded organism in supporting its development. The pathogen will achieve this at various levels of success resulting in the attacked host being classified as susceptible, partially resistant, resistant, or as a nonhost and the pathogen correspondingly as more or less virulent, as avirulent, or as an inappropriate pathogen. Traditionally, these interactions are described from a plant geneticist's perspective and often inadequately describe the wide range of interactions that can be observed microscopically in plant pathogenesis. Indeed, "modern" agriculture, selection and breeding (starting more than 10,000 years ago) have revealed a sizeable repertoire of such pathogenicity and virulence factors contributing to their respective classification (reviewed in Bakkeren and Kronstad 1994). In some (rare) instances, the host-pathogen interaction leads to complete incompatibility in an otherwise compatible pathogen species. Genetic analysis of both pathogen and host identified (often dominant) single genes superimposed on compatibility which have been termed "avirulence" and resistance genes, respectively. It is now generally believed that these "avirulence" genes are maintained in pathogen species and populations because they are among the general arsenal of virulence factors of the pathogen, but have been co-opted by the plant's innate surveillance system to be recognized as "non-self" and trigger a resistance (immune) response. Thus, even though harboring "avirulence" genes bestows an obvious disadvantage on its bearer with respect to host infection, deleting or mutating them also comes at a cost, sometimes subtly and only apparent on a larger population level (Tian et al. 2003; Brown 2003; Wichmann and Bergelson 2004). Surveying and monitoring of population dynamics of large numbers of races (harboring different combinations of avirulence genes in a pathogen species) uncovered through numerous cultivars (resulting from breeding programs and harboring an equal number of cognate resistance genes capable of triggering defense after recognition) form the basis of modern crop resistance breeding.

To distinguish between virulence factors *in sensu stricto*, that is, contributing to the disease process, and genes merely involved in fitness and/or general metabolism is extremely difficult in the biotrophs and might be no more than a matter of definition and semantics (reviewed in Bakkeren and Gold 2004). During the infection process, including the establishment of intercellular hyphae and haustoria, true biotrophs inflict minimal damage to host cells and tissue integrity is maintained. Moreover, detection of 'pathogen associated molecular patterns' (PAMPs) or 'non-self' molecules by the host is minimized by cloaking or biochemical changes when *in planta*, and by active suppression of host defenses. An excellent recent review discusses the attributes of 'true biotrophs' vs. hemibiotrophs and necrotrophs (Oliver and Ipcho 2004).

3. EXEMPLARY MODEL SYSTEMS
3.1 Smuts
3.1.1 Introduction

The basidiomycete smut fungi belong to the order Ustilaginales, an order that includes plant pathogenic fungi. Smuts are facultative obligate biotrophs that affect approximately 4000 species of angiosperms belonging to over 75 families (Alexopolous et al. 1996). While these fungi have a saprobic, budding yeast phase that can be easily maintained in culture, these fungi employ an obligate dikaryotic phase to grow filamentously within the plant host, sometimes producing spectacular symptoms. One of the most striking symptoms (and particularly pronounced in the *U. maydis*-maize interaction) is the induction of galls or tumors in the host (Figure 2).

At maturity, these galls are associated with black dusty masses of teliospores that resemble soot or smut, hence the name. In this chapter section, the smut fungus *Ustilago maydis* is discussed in detail as an exemplary model system of development and morphogenesis, a discussion propelled forward at the genetic level in the context with the fascinating biology of this fungus. *U. maydis* is among the most important models for fungal plant pathogenesis, morphogenesis, mating and signaling. It is currently the only well developed genetic model of plant pathogens among the basidiomycetes. *U. maydis* shares many features with the mushroom fungi and is an important comparative model for the important human disease caused by the basidiomycete *Cryptococcus neoformans*. In the following treatment we present the central aspects of *U. maydis* under current and continuing study.

3.1.2 Dimorphism

Fungal dimorphism is an interconversion of the yeast and mycelial morphologies, commonly termed the dimorphic switch. Although various genetic determinants and environmental stimuli have been characterized that govern this dramatic morphological alteration, one central theme in the control of dimorphism in fungi is the interplay between a pheromone-responsive mitogen-activated protein (MAP) kinase and a cyclic AMP (cAMP) pathways. Through signaling via the pheromone-responsive MAP kinase pathway, the yeast form of the smuts mate and form a filamentous dikaryon. In *U. maydis*, mate recognition is controlled by the master regulatory *a* locus, encoding both pheromone and pheromone receptors (Bolker et al. 1992). Cells of opposite mating types, differing at their *a* loci, secrete pheromone which induces the production of mating hyphae that grow toward the opposing mating partner and eventually fuse (Snetsalaar et al. 1996). Once cell fusion has occurred, production and maintenance of the filamentous dikaryon is dependent on master regulatory *b* genes from opposite mating types, e.g. *b1* and *b2* (Kamper et al. 1995) dependent on master regulatory *b* genes from opposite mating types, e.g. *b1* and *b2* (Kamper et al. 1995). Filamentous growth may facilitate penetration of plant cells, thereby aiding in the collection of nutrients from the host and providing an ideal environment for sexual maturation. Environmental signals also trigger filamentous growth or are involved in the maintenance of a yeast phase. As opposed to the filamentous dikaryon, the precise role of the haploid yeast phase remains enigmatic in *U. maydis*. A yeast phase, in which cell division occurs rapidly by budding, leading to geometric cell population growth, may potentially aid in the dispersal of these plant pathogens by rain and wind. The production of large numbers of budding cells increases the overall number of propagules available in the environment, ensuring that at least some compatible members of the population meet at the right place and time under conditions suitable for mating. Therefore, yeast-like growth, as opposed to filamentous growth, may be advantageous in the saprophytic phase of smuts.

As mentioned above, interplay between the pheromone-responsive MAP kinase pathway and the cAMP pathway controls dimorphism in *U. maydis*. In this section, we discuss dimorphic switching as it relates to environmental signals as well as alternative pathways regulating dimorphism in the exemplary model system of *U. maydis*. These signals impinge on various pathways or proposed regulatory components that will be discussed in great depth in upcoming sections of this chapter (see sections 3.1.3 through

3.1.5). The genetic determinants of dimorphism in the smuts induced by compatible mating partners will be described in the Mating section. Ultimately, the signaling pathways that control dimorphism must converge on the regulation of the cytoskeleton, the genetic control of which will be discussed in detail under Cell Cycle and Cytoskeletal Regulation.

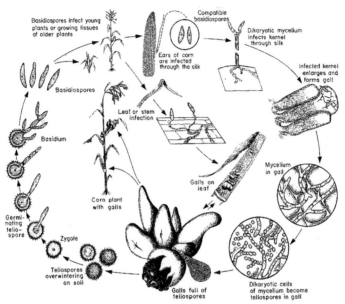

Fig. 2. Disease cycle of corn smut, caused by Ustilago maydis (from Agrios 1997)

Control of the dimorphic switch has been an area of intensive study in *U. maydis* and other pathogenic fungi (Sanchez-Martinez and Perez-Martin 2001). In *U. maydis*, cAMP is required for the maintenance of the budding form (Gold et al. 1994). Disruption of the *U. maydis* adenylate cyclase (*uac1*) gene led to the formation of constitutive filamentous strains from budding haploids (Barrett et al. 1993). A number of mutations, termed *ubc* for *Ustilago* bypass of cyclase, suppress the filamentous phenotype of the *uac1* mutant (Gold et al. 1994; Mayorga and Gold 1998). Complementation of one of these *ubc* suppressor mutations led to the restoration of the filamentous *uac1⁻* phenotype and enabled the isolation of the *ubc1* gene encoding the regulatory subunit of cAMP-dependent protein kinase A (PKA) (Gold et al. 1994). In addition to *ubc1*, four additional *ubc* genes (*ubc2, ubc3, ubc4* and *ubc5*) were cloned by mutant complementation (Mayorga and Gold 1998; Andrews et al. 2000). The *ubc5/fuz7* gene was identified in an independent screen (Banuett and Herskowitz, 1994). The *ubc3* and *ubc4* genes were also idenified in screens independently as *kpp2* and *kpp4*, respectively (Muller et al. 1999; Muller et al. 2003 b). The *ubc3/kpp2, ubc4/kpp4*, and *fuz7/ubc5* genes encode a MAP kinase, MAPKK kinase and a MAPK kinase respectively and all are members of the pheromone-responsive MAP kinase cascade involved in mating, morphogenesis, and pathogenic development (Mayorga and Gold 1998; Mayorga and Gold 1999; Banuett and Herskowitz 1994; Andrews et al. 2000; Muller et al. 2003 b). The cAMP and MAP kinase pathways impinge on the transcription factor Prf1, which is differentially phosporylated by PKA and the MAP kinase Kpp2/Ubc3 to activate pheromone and receptor gene expression as well as genes regulating filamentous growth (Kaffarnik et al. 2003).

The number and types of stimuli that can flip the dimorphic switch in fungi appears to defy a unifying theme. In U. maydis, in addition to pheromone response-induced and b gene-regulated filamentous growth, pH changes (Garrido and Perez-Martin, 2003; Martinez-Espinoza et al. 2004) exposure to air (Gold et al. 1994) and nutrient deprivation (Smith et al. 2003) and the presence of lipids (Klose et al. 2004) can trigger morphogenic changes. In U. maydis, the cAMP and MAPK pathways play a key role in pH-regulated dimorphism (Martinez-Espinoza et al. 2004). On solid acid medium (pH3), haploid wild-type strains of U. maydis exhibit a mycelial phenotype. However, ubc2 and ubc5 MAPK pathway mutants grow as yeasts at pH 3. In contrast, prf1 mutants remained in the mycelial phase at pH 3. These results strongly implicate the pheromone-responsive MAPK cascade in pH-regulated dimorphism and that this signaling is independent of the transcription factor Prf1, suggesting a branching of this signaling pathway upstream of Prf1. Further analyses revealed that ubc1 mutants, which display a mycelial morphology at neutral pH characteristic of cAMP pathway mutants, grew as yeasts at pH 3. Also, upon addition of cAMP to the medium, the ubc2 and ubc5 mutant strains grew as yeasts. Taken together, these results provide a strong indication of the role of the cAMP pathway in signaling pH-regulated dimorphism (Martinez-Espinoza et al. 2004). Interestingly, in the smut fungus Ustilago hordei, wild-type strains behave oppositely in response to pH changes. In contrast with U. maydis, U. hordei primarily maintains a yeast-like morphology at acid pH but more filamentous growth is observed at neutral and basic pH (Lichter and Mills 1998).

Like pH changes, nutrient sensing plays a critical role in modulating morphogenic changes in fungi. In U. maydis, a gene encoding a protein with high similarity to the high-affinity ammonium permease, Mep2p, from S. cerevesiae was identified as highly expressed in budding yeast cells as opposed to filamentous cells (Smith et al. 2003). mep2 is involved in cAMP signaling and is required for pseudohyphal growth in response to low ammonium in S. cerevisiae. Interestingly, ump2 was able to complement the pseudohyphal defect characteristic of the yeast mep2 mutant. Furthermore, deletion of this gene in U. maydis eliminated the filamentous phenotype of haploid cells on low ammonium. Additionally, the importance of a PKA phosphorylation site in Ump2p was revealed by site-directed mutagenesis and complementation studies (Smith et al. 2003). Overall this work indicates a potential role of the cAMP-PKA pathway in signaling morphogenic changes through nutrient stress and specifically the important role of ammonium permeases in response to low nitrogen conditions. An additional example of nutrient sensing affecting morphogenesis in U. maydis is the presence of lipids, which promote filamentous growth dependent on the cAMP and MAPK signaling pathways (Klose et al. 2004).

In addition to the cAMP and MAPK pathways and their relatively defined roles in regulating dimorphism in *U. maydis*, other pathways may operate in conjunction with or independently in signaling the dimorphic switch. Guevara-Olvera et al. (1997) revealed a role of polyamine biosynthesis as a determinant of the dimorphic switch. Mutants of the ornithine decarboxylase (*odc*) gene, which encodes a product that catalyzes the first step in polyamine biosysnthesis, behave as polyamine auxotrophs. The dimorphic switch was inhibited in medium containing the minimum concentration of polyamines to support growth. Supplementation of the medium with additional polyamines led to the dimorphic transition (Guevara-Olvera et al. 1997). In contrast with the central role of polyamine biosynthesis, analyses of single and double chitin synthase (*chs1-6*) mutants revealed only slight alterations in morphology (reviewed in Martinez-Espinoza et al. 2002). Finally, and likely an additional contributor of the dimorophic switch in *U. maydis*, what is known of the role of calcium homeostasis in the morphogenesis will be discussed in detail in the Cell Cycle and Cytoskeletal Regulation section.

3.1.3 Mating

Genetic analysis of the sexual system has been a major focus of work on *Ustilago maydis* and this species has become a paradigm for the higher basidiomycetes (Casselton and Olesnicky, 1998; Casselton 2002). The heterothallism of *U. maydis* was recognized as early as 1927 (Stakman and Christensen 1927). Employing tetrad analysis, Hana (1929) showed conclusively that there were at least two pairs of segregating sex factors because in some cases four distinct mating types were derived in a single tetrad. He designated the diploid nucleus as possessing an AaBb genotype and the progeny as AB, Ab, aB and ab. Similar notations for the mating-type genes are still used. No distinction was made for the function of A or B in Hana's work (Hana 1929). Rowell and DeVay (1954), designated these two mating factors as "a" and "b" and determined that 2 specificities of *a* and multiple specificities of *b* existed. Different specificities were necessary at both the *a* and *b* loci to generate productive maize infection in which teliospores (chlamydospores) were generated, and the currently used designations for these mating-type loci was established. In this same study it was determined that amphisexual progeny were occasionally generated that had both *a* specificities such that they could productively be paired with any strain possessing a different *b* allele. However, these amphisexual strains were not solopathogenic while diploids heterozygous at both *a* and *b* were solopathogens. This indicated to the authors that the *a* compatibility factor was clearly not a primary pathogenicity characteristic, a fact further corroborated by Banuett and Herskowitz (1989). Rowell clearly demonstrated that the roles of heterozygosity at *a* and *b* were in fusion and dikaryon vigor/stability, respectively (Rowell 1955). He also noted that alleles of both factors had to be different in the mating partners to generate the virulent pathogen.

Trueheart and Herskowitz (1992) generated a cytoduction assay in which cell fusion was strictly controlled by possession of unequal alleles at the *a* locus while the *b* locus played no role. Later, Laity et al. (1995) complemented this work showing that heterozygosity at the *b* locus within a strain inhibits further mating (Laity et al. 1995). In summary then, *a1* strains will fuse with *a2* strains regardless of the condition at the *b* locus except that once *b* becomes heterozygous the cell will fuse no further with any other strain. Additionally, $a1a2b_n$ stains fuse promiscuously with any normal haploid mating type. By addition of charcoal to solid medium, the development of the functional dikaryon can easily be monitored in culture (Day and Anagnostakis 1971; Holliday 1974). White filamentous growth is observed on this medium only when both *a* and *b* differ in cospotted compatible strain pairs.

The *b* locus is multiallelic (Rowell and DeVay, 1954). Puhalla (1970) found only 2 *a* but 18 different *b* mating-type alleles in 62 lines amongst 33 different isolations from the U.S. and Canada. He predicted that there should be no more than 25 distinct alleles of *b* in the population (Puhalla 1970).

The master control genes of mating and pathogenicity, the *a* and *b* mating-type genes, have now been cloned and characterized (Kronstad and Leong 1989; Kronstad and Leong 1990; Schulz et al. 1990; Froeliger and Leong 1991; Bolker 1992). Holliday had noted that the *pan1* gene was tightly linked at about 2.5 map units from the *a* mating-type locus (Holliday 1961; Holliday 1974). Using this information, Froeliger and Leong (1991), cloned the *a2* mating-type determinant by complementation of a *pan1* mutant *a1* strain with a cosmid from a prototrophic *a2* strain. By mating *a1b1/a2* transformant strains with an *a1b2* strain to generate a filamentous and pathogenic dikaryon, the presence of the *a2* allele was confirmed. The initial cloning of the *a* locus indicated that the *a1* and *a2* allelic sequences were idiomorphs, (i.e. they lacked sequence homology). Homologous flanks were then employed to isolate the *a2* mating-type idiomorph and similar methods used to confirm function (Froeliger and Leong 1991). The *a* mating-type genes were then sequenced and Bolker et al. (1992). demonstrated that the mating-type specificity in each idiomorph was determined by two genes. One gene encodes a lipopeptide mating factor, and the other a pheromone receptor. Thus the *a* locus encompasses *mfa* and *pra*, two tightly linked genes that encode secreted pheromone and membrane spanning pheromone receptors, respectively (Bolker 1992). The pheromone encoded by the *mfa* gene is thought to interact directly with the pheromone receptor product encoded by the *pra* gene of the opposite *a* mating specificity (Spellig et al. 1994). Synthetic pheromone causes cell cycle arrest in the G2 phase (Garcia-Muse et al. 2003). This is in contrast with the situation described in ascomycete yeasts such as *S. cerevisiae* and *Schizosaccharomyces pombe* where pheromone induces cell cycle arrest at G1. The function of the genes at the *a* locus helps explain the fact noted above that a≠ (possession of two different allelic specificities of *a*) is required for a diploid heterozygous at *b* to become filamentous on charcoal mating media (Banuett and Herskowitz 1989). In addition to its function as a mating attraction system, dikaryon heterozygosity at the *a* locus (in addition to heterozygosity at *b*) also contributes to the *in vitro* production of the post mating dikaryotic filamentous form through an autocrine response in which the pheromones and receptors of opposite allelic specificity are present within the same cell and therefore may continually interact (Banuett and Herskowitz 1989; Spellig et al. 1994). Smuts having a bipolar mating system, such as the barley covered smut *U. hordei*, possess both *a* and *b* gene complexes homologous in sequence and function. When the gene complexes were tested across the species *U. maydis* and *U. hordei*, the *a* gene complexes were shown to determine interspecies compatibility as well and the *b* gene complexes to function properly in triggering filamentous growth thereby proving the existence of conserved pathways (Bakkeren and Kronstad 1993; Bakkeren and Kronstad 1994; Bakkeren and Kronstad 1996). The *a* and *b* gene complexes are separated by approximately 500 kb in *MAT-1* and 430 kb in *MAT-2* strains and the region in between these complexes displays some inversions and deletions compared to the other mating type possibly giving an explanation for the lack of recombination over this part of the chromosome and alleviating the need for multiple mating types; apart from the two opposite mating specificities for *a* similar to *U. maydis*, only two specificities seem to exist for *b* as well in Nature (Bakkeren and Kronstad 1994; Lee et al. 1999). The *U. hordei MAT-1* region has been sequenced and analyzed and harbors apart from several genes, a large number of (partial) transposable element and repeat sequences which could have been involved in the evolution of this part of the chromosome (Jiang, Bakkeren and Kronstad, manuscript in preparation). The *a* and *b* gene complexes separated in the tetrapolar *U. maydis*, could have become fixed in bipolar smuts during the genesis of primitive sex chromosomes as has been suggested happened in several *Cryptococcus* species (Fraser and Heitman 2004).

Downstream events generating the final response to pheromone appear to involve components similar to those encountered in *S. cerevisiae*. In this budding yeast, signal transduction from the pheromone-receptor interaction to the final cellular responses involves a trimeric G protein and a MAP kinase cascade with the final phosphorylation and activation of two critical proteins. These proteins are the Ste12p transcription factor, which when activated regulates transcription of target genes, and Far1p which causes cell cycle arrest by inhibition of the kinase activity of the G1 cyclin complex Cdc28-Cln (Banuett 1998; Valdivieso et al. 1993). In *U. maydis* none of the four cloned Gα subunits of the trimeric G proteins appear to be directly involved in transmission of the pheromone signal (Regenfelder et al. 1997; Kruger et al. 1998). As is the case in *Schizosaccharomyces pombe* (Sipiczki 1988), a *ras* gene (*ras2*) functions to stimulate filamentous growth through the pheromone responsive MAP kinase cascade (Lee and Kronstad 2002). Additional work suggested that the cdc25 homolog Sql1 may function as an activator of Ras2 (Muller et al. 2003 a). An additional finding in this work was that activated Ras1, the product of a second *ras* gene, increased pheromone gene expression. The three members of the pheromone responsive MAP kinase cascade have been identified. These are *ubc4/kpp4* encoding the ste11p MAPKK kinase homolog (Mayorga and Gold 1998; Andrews et al. 2000; Muller et al. 2003 b) *fuz7/ubc5* encoding the ste7p MAPK kinase homolog (Banuett and Herskowitz 1994; Mayorga and Gold 1998; Andrews et al. 2000), and the fus3p and kss1p MAP kinase homolog *ubc3/kpp2*

(Mayorga and Gold 1998; Mayorga and Gold 1999; Muller et al. 1999). A putative adaptor protein Ubc2 may link the MAP kinase cascade with the upstream components of signaling through Ras proteins (Mayorga and Gold 2001). Preliminary two-hybrid screens with Ubc2 as bait identified a strong interaction of Ubc2 and Ubc4 MAPKKK SAM domains (Klosterman et al. unpublished) confirming a previously observed genetic interaction (Mayorga and Gold 1998). This is analogous to the interaction of the SAM domains of Ste11p and Ste50p in *S. cerevisiae* (Jansen et al. 2001). A gene designated *prf1* encodes an HMG family transcription factor that links the pheromone response pathway to the expression of the *b* locus and thus to pathogenicity (Hartmann et al. 1996). The *prf1* protein has potential phosphorylation sites for both a MAP kinase (presumably *ubc3/kpp2*, see below) and for the cyclic AMP dependent protein kinase (Kahmann et al. 1999; Muller et al. 1999; Kaffarnik et al. 2003). The putative MAP kinase phosphorylation sites appear important for the biological function of the protein (Muller et al. 1999). The *prf1* gene is required for pathogenicity due to its essential function in the regulation of the *b* mating-type genes. Constitutive expression of the *b* genes restores pathogenicity in *prf1* mutants (Hartmann et al. 1996). Additional transcription factor(s) are likely involved in transmitting the pheromone responsive MAP kinase and/or cAMP pathway signals besides Prf1. As noted by Lee and Kronstad (2002) epistasis experiments indicated that Ras2 may regulate filamentation via the pheromone responsive MAP kinase cascade including Ubc3, but not through the activation of Prf1. Recently it has also been shown that although the pheromone response pathway is required for conjugation tube formation in the mating reaction, Prf1 is not required for this process to occur (Muller et al. 2003 b). Additionally, work from our laboratory indicates that the MAP kinase cascade is required for acid-induced filamentation while prf1 is not (Martinez-Espinoza et al. 2004).

The *b* locus controls events after cell fusion necessary for establishment of the infectious filamentous dikaryon. *lga2*, a gene of unknown function (Urban 1996 b) located within the *a2* idiomorph, is directly and positively regulated by the *b*-heterodimer (Romeis et al. 2000). Employing inducible promoters to replace those native to *b*, both positively and negatively *b* transcriptionally regulated genes have been identified (Brachmann et al. 2001). However, deletion of a number of these genes did not produce any discernible effect on morphology or pathogenicity, indicating that the ones characterized so far do not individually play a major role in pathogenesis and development. Additionally, genes, that when mutated, induced expression of the *b* genes in haploid cells as well as other dikaryon specific genes, have been identified using another reporter system (Quadbeck-Seeger et al. 2000; Reichmann et al. 2002). Mutants deleted for these genes (*rum1* and *hda1*) are able to colonize plants apparently normally but are fully defective in teliosporogenesis a process discussed in more detail below.

3.1.4 Signaling

Signaling pathways in plant pathogenic fungi clearly play central roles in environmental sensing, mating processes, morphogenesis and communication with the host. The pathways most well studied and most significant in the fungi, as judged by the frequency with which they are encountered in mutants affected in these processes, are the cAMP activated protein kinase A (PKA) and MAP kinase pathways. The importance of signaling involving calcium as a messenger molecule for these processes is clearly a common phenomenon. These pathways tend not to stand alone but rather frequently crosstalk and can be viewed as a web of interconnected pathways, often several of which impact a single phenotypic outcome. A common problem with interpreting these studies is that perturbing one pathway is like cutting a single strand of a spider's web, consequently deforming many unintended interconnected strands. Several excellent reviews have recently been published that deal with these topics in great detail. cAMP signaling and its interaction with MAP kinase pathways in phytopathogenic fungi was very recently well reviewed (Lee et al. 2003). Thus here we provide a few well-characterized illustrative examples.

In general, mutations inactivating PKA lead to debilitation of a fungus. For example, in *U. maydis*, mutation in adenylate cyclase, Uac1 (Barrett et al. 1993; Gold et al. 1994), the catalytic subunit of PKA, Adr1 (Durrenberger et al. 1998) or in a G-protein alpha subunit Gpa3, required for activation of adenylate cyclase (Regenfelder et al. 1997) cause similar consequences, freezing this dimorphic fungus in the filamentous phase and eliminating the ability to colonize maize. Nonetheless, upon close comparison of these mutants, they do differ in several subtle phenotypes. For example, the adenylate cyclase mutant is much more invasive on agar than is the PKA catalytic subunit mutant (Gold, unpublished). This indicates that cAMP likely plays a role beyond the function of *adr1*.

Mutations that activate PKA such as mutations in the regulatory subunit of PKA (a PKA inhibitor protein) or activated alleles of specific G-alpha proteins tend to cause phenotypes of opposite character to those of inactivating mutations. For example, in *U. maydis*, *ubc1* encoding the regulatory subunit of PKA, is an epistatic suppressor to the *uac1* mutation (Gold et al. 1994). Likewise an activated allele of *gpa3* generates several phenotypes similar to a *ubc1* mutant but is hypostatic to *uac1* (Regenfelder et al. 1997; Kruger et al. 1998). However, mutations activating the PKA pathway still tend to be detrimental to virulence. Null *ubc1* mutants are able to colonize maize leaves but are unable to induce gall formation (Gold et al. 1997). Mutants with intermediate activation of PKA caused further progression toward the wild type infection (Kruger et al. 2000). These results indicate that a delicate balance of PKA activation must be maintained for progression through the various phases of the infection cycle. Perturbation in one or another direction, activation or inactivation, leads to detrimental effects on the fungus.

Mitogen activated protein kinases have often been encountered in phytopathogenic fungi as required for morphogenesis and for full virulence. This topic has been reviewed in some detail (Xu 2000). *S. cerevisiae* has five functional (and partially overlapping) MAP kinase cascades. These are involved in mating, filamentation, cell integrity, high osmotic growth stress response, and ascospore formation. The environmental triggers activating these pathways are commensurate with their function, eg. pheromone for mating. In the genomes of plant pathogenic fungi (e.g., in

M. grisea), and filamentous fungi in general for that matter (e.g., in *N. crassa*), often only three MAP kinase enzymes appear to be present. These fall into three families relative to yeast: pheromone responsive (Fus3/Kss1), osmoregulation (Hog1) and cell integrity (Slt2). Mutations in these MAP kinases tend to cause differential defects in pathogenicity. For example, in *M. grisea* PMK1 (pheromone response) and MPS1 (cell integrity) are both required for disease on unwounded leaves but *mps1* mutants can colonize wounded leaves while *pmk1* mutants cannot (Xu and Hamer, 1996; Xu et al. 1998). In this fungus the third MAP kinase, OSM1, is not required for virulence under laboratory conditions. Pheromone responsive MAP kinase genes have been repeatedly found to be critical for full virulence in the fungi (Xu 2000). MAP kinase genes from other families have also been shown to be important for virulence in plant pathogenic fungi but tend to be more variable in their effect than the Fus3 family. The pheromone responsive MAP kinase in *U. maydis ubc3/kpp2* (Mayorga and Gold 1999; Muller et al. 1999) was initially defined as of relatively minor importance in virulence. However, it now appears that this was primarily due to partial functional redundancy because when only the kinase activity is mutated but the protein still synthesized a much more dramatic effect on virulence is observed (Muller et al. 2003b). This situation is reminiscent of the complementation of *fus3* deletion by *kss1*, which is unable to complement the *fus3* kinase mutant (Madhani et al. 1997). In *U. maydis*, formation of swollen appressorium-like structures and their production of invading hyphae which penetrate epidermal cells, appear to be distinct steps in the infection process. A mutant strain defective in Kpp6 activity, a *b* mating-type gene-regulated MAP kinase, has recently been constructed, which is able to produce appressoria but is unable to penetrate plant cells (Brachmann et al. 2003). Microscopic observations of plant surfaces after inoculation with compatible strains both carrying an inactivated mutant allele, *kpp6*[T355A,Y357F], showed appressorium formation. However, from the majority of those appressoria only short filaments that failed to penetrate plant cells emerged.

Cross talk between the cAMP and MAP kinase cascade signaling pathways has been well documented (D'Souza and Heitman 2001; Lee et al. 2003). In *U. maydis* the interaction of the cAMP and MAP kinase pathways was made evident by the finding that a number of suppressors of the filamentous phenotype of an adenylate cyclase (*uac1*) mutant were members of the pheromone responsive MAP kinase cascade (Mayorga and Gold 1998; Mayorga and Gold 1999; Andrews et al. 2000; Mayorga and Gold 2001). Also in *U. maydis*, it has recently been demonstrated that the differential phosphorylation of Prf1 by the pheromone responsive MAPK and the cAMP dependent protein kinase regulates the activity of this transcription factor toward its various promoter targets including the *a* and *b* mating-type genes (Kaffarnik et al. 2003). Induction of the *a* mating-type genes requires the PKA phosphorylation sites while induction of the *b* genes requires both the PKA and MAPK phosphorylation sites. Another integrator of cAMP and MAP kinase signaling is apparently the *crk1* encoded kinase (Garrido and Perez-Martin 2003). The trascription of this kinase is oppositely affected by the function of the MAPK and cAMP pathways.

3.1.5 Cell cycle and cytoskeletal regulation

Detailed analyses of the cell cycle in *U. maydis* have been conducted in relation to budding cell morphology. Cells of *U. maydis* produce one polar bud per cell cycle (Jacobs et al. 1994). When the bud nears maturity, the nucleus divides in the bud of the daughter cell and following retrograde movement of one nucleus to the mother cell, cytokinesis occurs (Banuett and Hirskowitz 2002; O'Donell and McLaughlin 1984). Analyses of nuclear density in asynchronously growing populations revealed that *U. maydis* cells exhibit both prereplicative DNA content (1C) typical of G1 phase of the cell cycle and postreplicative DNA content (2C) of G2 phase (Snetselaar and McCann 1997). When haploid *U. maydis* cells are exposed to mating pheromone, they undergo G2 arrest and produce conjugation tubes (Garcia-Muse et al. 2003) in contrast with the budding yeast *S. cereviseae*, which generally lacks a G2 phase (O'Farrell 2001).

The genetic determinants of cell cycle control are beginning to be uncovered in *U. maydis*. In eukaryotes, cyclin dependent kinases (CDKs) and their cognate regulatory subunits known as cyclins control cell cycle events such as the onset of mitosis and S phase. In *S. pombe* and *S. cereviseae*, one CDK in each of these fungi and different cyclins control these events of the cell cycle (O'Farrell 2001). In *U. maydis*, using a PCR-based approach, the CDK, Cdk1, was identified along with two B-type cyclins, Clb1 and Clb2 (Garcia-Muse et al. 2004). The *cdk1* gene product was isolated from cell lysates together with the Clb1 and Clb2 proteins, which together form the mitotic CDK. Levels of Clb1 and Clb2 were determined to fluctuate during the cell cycle, falling in G1 and rising at entry to S/G2/M. Conditional mutants revealed Clb1 depletion resulted in cell cycle arrest at two points, one at which cells exhibited a prereplicative 1C DNA content and one in which the cells exhibited a postreplicative 2C DNA content. Clb2 depletion on the other hand arrested the cells when they exhibited a 2C DNA content. This suggested that Clb1 is required for the G1 to S and G2 to M transitions while Clb2 is required for entry into mitosis, a finding supported by the observation that overexpression of *clb2* caused early entry into mitosis. Additionally, cells overexpressing *clb2* or those with a single dosage of *clb2*, divided by septation and grew filamentuosly but did not form buds, nor could they produce tumors in the plant. However, both of these mutants were capable of filamentous growth within the plant (Garcia-Muse et al. 2004). Interestingly, filamentous growth within the plant but lack of tumor formation is also observed in *ubc1* and *ukb1* mutants that exhibit defects in bud site selection (Gold 1997; Abramovitch 2002).

Further searches for Cdk genes in *U. maydis* revealed the presence of a novel gene, *crk1* (cdk related kinase 1), that bears limited identity with mitotic cyclin-dependent kinases and the putative encoded protein most closely matches Ime2, a kinase involved in morphogenic control in *S. cerevisiae* (Garrido and Perez-Martin 2003). Disruption of the *crk1* gene in *U. maydis* revealed morphogological defects such as the production of shorter and rounder cells. At low pH, *crk1* defective cells were unable to form the multicellular chains like wild type cells. However, complementation with wild-type *crk1* was capable of inducing filamentation. The activity of CRK1 was also correlated with both the cAMP and MAP kinase pathways since inactivation of *crk1* suppresses filamentous growth of cAMP pathway defective mutants. Strains deficient in *gpa3* produce highly elevated levels of the *crk1* message, suggesting that the cAMP pathway negatively regulates *crk1* expression. Further experiments revealed that the MAPK pathway counteracts the negative regulation of *crk1* by the cAMP pathway (Garrido and Perez-Martin 2003).

Coupled with the findings of the Cdk and Cdk-related genes in *U. maydis*, two genes involved in cell separation also have been identified. Weinzierl et al. (2002) identified *don1* and *don3* (donuts) by complementation of a mutant colony morphology that resembled a donut. Sequence analysis revealed that *don1* encodes a protein containing a nucleotide exchange domain (GEF), a PH domain and a FYVE zinc finger domain while *don3* encodes a highly conserved Ser/Thr-protein kinase domain characteristic of Ste20-like kinases, targets of small GTP-binding proteins of the Rho/Rac family, including Cdc42. Two-hybrid assays revealed Don1 and Don3 both interact with *U. maydis* Cdc42. Because *don1/don3* double mutants and individual mutants exhibited the same phenotypes, it was suggested that Don1 and Don3 both interact with Cdc42 in the same signaling pathway. Further analyses revealed that both wild-type and *don1/don3* mutant budding cells form a primary septum at the bud neck between mother and daughter cells following bud formation and mitosis. *don1* and *don3* mutants were unable to form a secondary septum and consequently did not form a fragmentation zone. In conclusion, *don1* and *don3* appear necessary for secondary septum formation and cell separation. The targets of Don3 phosphorylation are presently unknown (Weinzierl 2002).

In *U. maydis*, the cell cycle has been studied in relation to changes in cytoskeletal organization (Steinberg et al. 2001; Banuett and Herskowitz 2002). These studies have revealed that transitions in the cell cycle require rearrangement of the microtubule (MT)- and F-actin-based cytoskeleton. These changes in MTs are highly dynamic and determine cell polarity (Steinberg et al. 2001; Bannuett and Hirskowitz 2002). The appearance of lateral budding and buds at opposing poles of the cell following benomyl (known to affect MTs and frequently used as a systemic fungicide) treatment or in a *U. maydis* α tubulin conditional mutant strain indicates MTs play a key role in determining cell polarity (Steinberg et al. 2001). The assembly of the assymetric dimers of α and β tubulin subunits confer MT polarization, having a plus end and a minus end (reviwed in Steinberg and Fuchs 2004). Motors such as dynein and kinesin use ATP for movement along microtubules (MTs) (Straube et al. 2001; Wedlich-Soldner et al. 2002 a; Wedlich-Soldner 2002 b) in a directional manner.

Morphological changes in *U. maydis* and other fungi are ultimately linked to the molecular motors that drive cytoskeletal rearrangement and move cargo to the growing tip to support polar growth of the cell. Searches of the *U. maydis* genome sequence revealed the presence of at least 14 such motors (Basse and Steinberg 2004). A PCR-based approach was used to identify the *dyn1* and *dyn2* genes (Straube et al. 2001) encoding the two components of the dynein heavy chain, a minus-end-directed MT motor (reviewed in Xiang and Plamann 2003). The *dyn1* gene encodes the predicted site of ATP hydrolysis while *dyn2* encodes the putative MT binding site. Interaction between Dyn1 and Dyn2 was confirmed by coprecipitation from growing budding cells and colocalization was confirmed by fluorescence microscopic analyses of Dyn1 and Dyn2 proteins. Following a shift of conditional mutants to restrictive conditions whereby *dyn1* and *dyn2* expression was repressed, cells with two or more nuclei began to appear. Additionally, under these conditions, the mutants exhibited more and longer MTs, suggesting a function of dynein in MT dynamics. These studies indicated Dyn1 and Dyn2 control nuclear migration to the neck region of budding cells where the minus ends of MTs are localized in polar MT organizing centers (Straube et al. 2001).

Other motors essential for appropriate control of morphogenesis have been identified in *U. maydis*. Lehmler et al. (1997) identified the *kin2* gene encoding the heavy chain of kinesin, a plus-end-directed MT motor important for pathogenicity (Lehmler et al. 1997; Steinberg et al. 1998) and discussed later. *kin3* was identified by a PCR-based approach and encodes a member of the UNC-104/KIF1 family, consisting of an N-terminal kinesin motor domain, a forkhead-associated domain, and a C-terminal pleckstrin homology (PH) domain (Wedlich-Soldner et al. 2002 b). The *kin3* mutant cells form tree-like aggregates due to a cell separation defect and a defect in the bipolar budding pattern. These morphological defects most likely arise from the disruption of bidirectional transport of early endosomes, evident from the finding that unlike wild-type cells, *kin3* mutants do not exhibit endosomal clustering at the septa and distal cell pole. Kin3-GFP was localized to endosomes, organelles likely involved in the delivery of components for cell growth, and its movement was associated with MTs in a plus-end-directed manner. Further analysis of a *dyn2* temperature-sensitive mutant in a *kin3* mutant background revealed a role of dynein in minus-end-directed movement of endosomes. Thus, a balance of between Kin3 and dynein activity organizes polar endosomes in opposing directions during budding (Wedlich-Soldner et al. 2002 b). This is in agreement with previous studies of the *U. maydis yup1* gene, encoding a target soluble N-ethylmaleimide-sensitive fusion protein attachment protein receptor (t-SNARE) (Wedlich-Soldner et al. 2000). In this study, a Yup1-GFP fusion protein was localized at vesicles, revealing rapid bidirectional motion. At nonpermissive temperature, *yup1ts* mutants do not accumulate Yup1 carrying vesicles at cell poles like that observed in wild type, resulting in an abnormal distribution of wall components and striking morphological defects characterized by elongated multicellular structures with multiple growth sites. Taken together, these studies indicate a recycling of membanes via Yup1-mediated endo- and exocytosis and that this action, driven by dynein and Kin3, contributes to polar growth (Wedlich-Soldner et al. 2000; Wedlich-Soldner et al. 2002 b).

To examine the overall pattern of MT organization during the cell cycle of *U. maydis*, Straube et al. (2003) used fluorescent protein variants fused to MT plus-end-binding (Peb1) and α-tubulin (Tub1) proteins and minus-end-SPB-specific γ-tubulin (Tub2) from *U. maydis*. Fluorescence microscopic analysis of these fusion proteins at various stages of the cell cycle revealed that MT bundles containing antipolar-oriented MTs span the length of the unbudding cells of G1 and S phase and those with small buds in G2. When the daughter cell reaches about 15 percent of the length of the mother cell, MTs are nucleated and anchored at the bud neck. This nucleation occurs at a polar microtubule organizing center (MTOC) so that polarization occurs and most MT plus ends grow away from the bud neck in both mother and daughter cells in G2 (Straube et al. 2003). Given this arrangement of MTs in G2, Garcia-Muse et al. (2003) suggested that G2 phase may be especially suited to serve as a decision point for a polar morphogenic response because the polar MT arrangement of G2 could support conjugation tube formation. Disruption of MTs with benomyl and following their regrowth revealed there are multiple dispersed MT nucleating centers in unbudded cells of G1 and S, indicating the spindle pole body (SPB) is inactive during interphase. In contrast, multiple MT nucleation sites were localized near the neck region of budding cells. At the onset of mitosis, the cytoplasmic network of MTs disassembles and SPBs nucleate MTs, forming the spindle and asters (Steinberg et al. 2001; Straube et al. 2003; Bannuett and Hirskowitz 2002). Following mitosis and septum formation, the MTs are again elongated toward both poles of each cell. This work also clearly illustrated that MTOCs organize MTs during budding and thus MTs do not reach the growth region by stochastic chance, evident from the finding that lateral budding mutants contain MTs that bend out of the growing bud and emanate from paired tubulin structures. Overall, this work indicates that MT nucleation in *U. maydis* occurs away from the nucleus at an MTOC near the bud neck (Straube et al. 2003). In many fungi, including *S. cereviseae*, MTs are nucleated at the SPB (reviewed in Xiang and Plamann 2003; Steinberg and Fuchs 2004). In contrast, in the fission yeast *S. pombe* and in *U. maydis*, the SPB is inactive during interphase and MT nucleation begins at the onset of mitosis where MTs are important for chromosomal segregation. In support of this idea, overexpression of the *U. maydis* B-cyclin Clb1 resulted in hypersensivity to benomyl and FACS analysis revealed that these cells had unusual DNA contents less than 1C or greater than 2C, suggesting that interference with microtubule assembly may be responsible for alterations in chromosomal segregation (Garcia-Muse et al. 2004).

Calcium homeostasis plays a role in modulating morphogenesis in fungi and other eukaryotes. *U. maydis* is no exception as indicated by the analysis of *ucn1* and *upa2* mutants, which encode the catalytic subunit (protein phosphatase 2B) of the Ca^{2+}-dependent holoenzyme encoding and the calcineurin catalytic subunit (protein phosphatase 2A), respectivley (Egan and Gold, upublished). The *U. maydis ucn1* gene encodes a calcineurin phosphatase. Mutation of *ucn1* leads to a dramatic phenotype characterized by large clusters of cells that exhibit a multiple budding pattern (Egan and Gold unpublished) similar to the *eca1* mutants (Adamikova et al. 2004), originally identified by the complementation of a temperature sensitive mutant. The gene *eca1* encodes a protein sharing highest identity with endoplasmic reticulum (ER)-resident Ca^{2+} ATPases (SERCAs). Analysis of *eca1* deletion mutants revealed a temperature-dependent morphological defect. At 30° C, *eca1* mutant cells exhibited growth at both poles and multiple septa were formed while a switch to 22° C restored wild type growth. Low external concentrations of Ca^{2+} suppressed the *eca1* mutant phenotype at 30° C while high Ca^{2+} levels increased morhological defects at 22° C. As an underlying basis for this morphological defect, *eca1* mutants displayed longer and highly disordered MTs at 30° C as opposed to 22° C. Further studies revealed that the calcineurin phosphatase activity was not inhibited in *eca1* mutant cells. But the altered Ca^{2+} homeostasis in *eca1* mutants caused increases in Ca^{2+}/calmodulin dependent kinase (CamK) activity. This increased activity was proposed to deregulate MT dynamics, which underly defects in morphology in *eca1* mutants. Consistent with this notion, suppression of CamK activity had a dramatic effect, restoring wild type budding morphology and MT organization. Similarity was observed in the phenotypes of *eca1* and dynein mutants. Moreover, *eca1*/dynein double mutants do not exhibit an increased mutant phenotype, suggesting both Eca1 and dynein participate in the same pathway. Dynein was previously shown to modulate motility of the peripheral tubular ER network in *U. maydis* (Wedlich-Soldner 2002 a) and *eca1* mutants are defective in this organellar motility (Adamikova et al. 2004).

In addition to MTs, studies have indicated a role of F-actin in polar growth of fungi (reviewed in Xiang and Plamann 2003; Steinberg and Fuchs 2004). In *U. maydis*, studies of Bannuett and Hirskowitz (2002) revealed the presence of actin patches that are concentrated at sites of bud emergence, the small bud, and the bud tip in large buds. Overall, this resembles the localization pattern of actin patches observed in yeasts at sites of polarized growth and secretion (reviewed in Xiang and Plamann 2003). Weber et al. (2003) also noted actin patches at sites of apical growth in *U. maydis*. The *U. maydis* actin cytoskeleton also consists of fine and thick cables (Banuett and Hirskowitz 2002). Some of these cables may cooperate with a newly identified *U. maydis* class-V myosin (Myo5) motor involved in actin transport of vesicles and organelles to actively growing regions of the buds and filaments (Weber et al. 2003). Deletion of *myo5* results in a phenotype characterized by thicker cells that fail to separate, forming cellular aggregates that are divided by septa and retain growth polarity. Localization studies revealed that a GFP-Myo5 fusion protein accumulates in a polar manner near the bud tip. This localization was disrupted by addition of the F-actin inhibitor lantrunculin A, indicating a dependency on F-actin. Further studies indicated that temperature sensitive *myo5* mutants exhibit reduced dikaryon formation likely attributable to impaired pheromone perception and conjugation tube formation. As a result, *myo5* mutants were also reduced in mating and pathogenicity (Weber et al. 2003). While MT and actin-based transport must function cooperatively in morphological transitions in *U. maydis*, the level of this cooperativity is presently unknown (Basse and Steinberg 2004).

3.1.6 Differential gene expression

Due to its central role in regulating morphogenesis and pathogenicity and its postulated function as a transcriptional regulator of gene expression, considerable effort has been made to identify genes whose expression is regulated by the bW/bE heterodimer. This follows from the hypothesis that these downstream genes could play important roles in morphogenesis and pathogenicity. Several studies identified a number of genes either up- or downregulated upon dikaryon formation; *egl1* encoding an endoglucanase (Schauwecker et al. 1995) and *rep1* and *hum1*, coding for a repellent and a hydrophobin, respectively (Wosten et al. 1996) were all identified as upregulated in dikaryotic cells. *lga2*, a gene of unknown function present at the *a2* locus, is also strongly upregulated in the presence of an active bE/bW heterodimer (Urban et al. 1996 a). However, deletion of these genes did not affect dikaryon formation/stability and/or pathogenicity. Among the genes known to be repressed in the presence of an active bW/bE heterodimer are the pheromone and pheromone receptor genes (Urban et al. 1996 a). Recently, higher throughput differential screening methods have been applied to identify genes whose expression is controlled by the b heterodimer. A general limitation to identify genes differentially expressed in budding versus dikaryotic cells is the lack of synchronization of the mating process and the fact that only a small fraction of cells fuse when two compatible strain are crossed *in vitro* (Urban et al. 1996 a). To overcome this problem, engineered haploid strains were constructed in which a functional bW/bE heterodimer can be induced in the appropriate carbon or nitrogen sources (Brachmann et al. 2001). Using these strains and an RNA fingerprint method, Brachmann et al. (2001) identified 10 new b regulated genes; five b-induced and five b-repressed. In a follow up study using the same approach, Brachmann et al. (2003). identified the *kpp6* gene, which encodes a MAP kinase required for efficient plant penetration. Among all the genes reported as b-regulated so far, *kpp6* is the only one shown to be crucial for pathogenicity.

To characterize binding sequences for the b heterodimer, binding *in vitro* of a synthetic b heterodimer to the putative promoter region of *lga2* was investigated and a b-protein binding sequence, termed *bbs1*, was identified (Romeis et al. 2000). This work furnished the first direct evidence that the bW/bE heterodimer can function as a transcriptional activator. A similar sequence (identical for 16 out of 23 nucleotides), present in the putative promoter region of the b-induced gene *frb52*, is bound by an active bW/bE heterodimer *in vitro* (Brachmann et al. 2001). However, these or similar sequences are absent in some of the b-regulated genes identified. It may be possible that sequences very divergent from the ones identified so far are also targets for the b heterodimer. However, a more plausible explanation is that some of the genes whose expression is highly induced upon dikaryon formation are only indirectly regulated by the bW/bE heterodimer through the action of other transcription factors in the b-regulatory cascade. Some indirect regulation is likely to go through the transcription factor Prf1. For instance, *kpp6* contains no potential binding sites for bW/bE; however, two putative binding sites for Prf1 (PREs) are found upstream of the start codon.

Several studies have identified downstream components of the b-dependent regulatory cascade. As explained above, *egl1* is specifically expressed in the dikaryon, identification of mutants that bypass the requirement of an active b heterodimer for *egl1* expression in haploid cells and complementation of such mutants led to the identification of two genes encoding putative repressors of b-regulated genes, *rum1* and *hda1* (Quadbeck-Seeger et al. 2000; Reichmann et al. 2002). Deletion of either gene resulted in expression in haploid cells of several genes known to be b-regulated as well as induction of the bE and bW genes themselves. These mutations also had an effect on disease development leading to arrest of teliospore formation after karyogamy and therefore producing galls lacking mature teliospores. It has been hypothesized that

Hda1 functions in a complex with Rum1. However, detailed microscopic observations show that the block in teliospore development appears to occur earlier in *hda1* mutants than in *rum1* mutants. This, together with the fact that repression of plant induced genes such as *mig1* and *ssp1* in haploid cells is relieved in a *hda1* but not in a *rum1* mutant background (Huber et al. 2002; Torreblanca et al. 2003) suggest a Rum1-independent function for Hda1. These studies show that temporal or spatial misexpression of a set of genes prevents the completion of the disease cycle, illustrating how tight regulation of gene expression is critical for disease development.

Another major regulator of morphogenesis and pathogenicity in *U. maydis* is the cAMP signaling pathway. Two recent papers from our laboratory have specifically focused on the identification of genes either up- or downregulated in filamentation induced by disruption of cAMP production (Andrews et al. 2004; Garcia-Pedrajas and Gold 2004). For these screenings, subtractive cDNA libraries were constructed in which the two growth conditions compared were budding haploid wild type cells and the constitutively filamentous *uac1* mutant. These studies led to the identification of 26 genes upregulated in filamentous growth and 37 downregulated in filamentous growth, the vast majority of which have not previously been reported in *U. maydis*. Interestingly, *rep1*, encoding a repellent protein and identified as highly induced in b-dependent filamentous growth was repeatedly encountered in our screening for genes upregulated in the constitutively filamentous *uac1* mutant (Andrews et al. 2004). Similarly, in screening for genes downregulated in the filamentous *uac1* mutant strain *frb124*, previously identified as b-repressed gene (Brachmann et al. 2001) was found. This suggests that these genes are specific for particular growth forms, budding or filamentous, and that there is some overlap in gene expression in filamentation induced by either a low level of cAMP or an active b-heterodimer. Deletion of two genes upregulated in filamentous growth; *uor1* encoding a putative member of the aldo-keto reductase family and *ufu1* with no similarity to known genes, produced no detectable mutant phenotype for morphology, mating or pathogenicity (Andrews et al. 2004). Among the genes downregulated in the filamentous *uac1* mutant an interesting case is *ump2* with similarity to ammonium transporters. Although highly expressed in budding cells and repressed in filaments its deletion impairs filamentous growth in response to low nitrogen (Smith et al. 2003). Interestingly, several genes with putative roles in extracellular matrix formation and adhesion were identified as dowregulated in the *uac1* filamentous mutant (Garcia-Pedrajas and Gold, 2004). These genes could play a role on adhesion to plant surfaces prior mating. Deletion of one of these genes, which exhibit high similarity to UDP-glucose dehydrogenase, in a wild type haploid background, resulted in loss of adhesion properties *in vitro* (Garcia-Pedrajas and Gold, unpublished).

A topic that has received considerable attention due to its implication in development during parasitic growth is the study of *U. maydis* genes whose expression is induced during growth within plant tissue. The identification and characterization of these genes will be discussed in section 3.1.7. Although not directly aimed to screen for differentially expressed genes, production of EST libraries from various developmental stages is also useful in the identification of genes differentially expressed in various developmental conditions. Two such libraries have been produced and analyzed recently; one containing ESTs from a diploid strain (Nugent et al. 2004) and another containing ESTs from genes expressed in germinating teliospores (Sacadura and Saville 2003). The identification and deletion of genes whose transcription appears to be regulated by pathways playing critical roles in morphogenesis and pathogenicity has made it obvious that not all differentially expressed genes play detectable roles in these processes. Indeed, to date, for most identified differentially expressed genes, deletions have not produced any discernable mutant phenotype. This could be the result of functional redundancy; alternatively, these genes may play roles dispensable for pathogenicity or morphological transitions.

3.1.7 Making a home for one's self

U. *maydis* is fully dependent on the plant to complete its life cycle. After mating the dikaryon can only develop parasitically. An approximate time frame of plant colonization in laboratory inoculations has been established. The early stage of colonization after penetration is characterized by rapid growth of the fungal tip leaving behind compartments devoid of cytoplasm that are sealed off and collapse. It is interesting to note that disease symptoms such as chlorosis and anthocyanin production are not uncommon at this stage, sometimes observed well in advance of the colonizing hyphae (Callow and Ling 1973) suggesting release of toxins and/or degradative enzymes by the fungus. Three or 4 days post infection dikaryotic hyphae start branching and are filled with cytoplasm. This change in growth mode coincides with the beginning of tumor development, which is induced approximately 5 days after inoculation. Branch primordia that resemble the clamp connections of other basidiomycetes are observed. However these structures do not appear to play the role of true clamp connections in maintaining the dikaryotic stage; nuclear migration into these structures has not been observed and they do not fuse with adjacent cells. As the fungus proliferates within the plant tumor it branches profusely with the formation of increasingly shorter branches. These changes appear to signal the switch from vegetative to sporogeneous hyphae. About nine days after inoculation hyphae are embedded in a mucilaginous material, presumably derived from hyphal walls, and tend to stick together. The tip of the hyphae became lobed followed by hyphal fragmentation into segments of one to several cells. Karyogamy probably takes placed at this stage, followed by rounding of individual cells and deposition of ornamented secondary cell walls (Snetselaar and Mims 1992; Snetselaar and Mims 1993; Snetselaar and Mims 1994; Banuett and Herskowitz 1994). Little is known about how U. *maydis* acquires nutrients during this biotrophic development inside the host. Intracellular structures somewhat resembling haustoria described in rust fungi have been observed (Luttrell 1987; Snetselaar and Mims 1994). However, these structures do not show a clear demarcation of an interface analogous to the one present in true haustoria; they are not consistently observed and when present are very irregularly branched. Conceivably, these irregular structures may correspond to the multilobed sporogenous hyphae that give rise to spores (Banuett and Herskowitz 1996) rather than to structures formed to obtain nutrients. Although the identification and characterization of the genetic components governing the switch from saprophytic to pathogenic development have been major research topics in *U. maydis*, characterization of the genetic programs acting during fungal growth within plant tissue is still at an early stage. Questions such as how the fungus perforates cell walls to travel within plant tissue, what signals triggers the switch between the various morphological stages observed during colonization, or how the fungus acquires nutrients during the growth *in planta*, remain unanswered.

Characterization of mutant strains affected in cAMP signaling has made it obvious that this regulatory pathway play a critical role during penetration, induction of galls and teliosporogenesis. An active cAMP pathway is required for formation of infection structures and penetration. Thus, mutants with low PKA activity as a result of inactivation of genes in this pathway do not produce any symptoms in inoculated plants. Mutant strains lacking adenylate cyclase activity (*uac1*⁻), the catalytic subunit of PKA (*adr1*⁻), or the α subunit of the G-protein Gpa3 (*gpa3*⁻) are all non-pathogenic (Barret et al. 1993; Gold et al. 1994; Regenfelder et al. 1997; Durrenberger et al. 2001). On the other hand, mutations leading to situations mimicking cAMP level above normal do not affect the early stages of infection but have a profound effect on gall formation and teliosporogenesis. In this sense, mutant strains with constitutive PKA activity achieved by disruption of *ubc1*, encoding the regulatory subunit of PKA, are able to infect and colonize plant tissue but they do not induce gall formation (Gold et al. 1997). Even lower levels of activation of cAMP pathway obtained in mutants with a constitutively active *gpa3* allele result in alteration of gall morphology. Tumors induced by these mutant strains show very reduced fungal proliferation and lack teliospores (Kruger et al. 2000). Taken together these results suggest that the necessary level of cAMP and PKA activity for initial colonization of plant tissue is relatively high, while a decreased cAMP concentration and PKA activity is necessary for induction and normal gall morphology. Few targets for PKA phosphorylation have been identified in *U. maydis*. Interestingly the *hgl1* gene encoding one such putative target was found to be important for teliospore maturation (Durrenberger et al. 2001). Inoculation of plants with *hgl1* mutant strains induced galls but they lacked darkly pigmented teliospores.

Identification of a gene encoding the heavy chain of conventional kinesin, *kin2*, and analysis of *kin2* mutant strains provided valuable information about the putative mechanism behind the mode of growth of the dikaryon in the initial steps of colonization. In both, dikaryons at early stage of plant colonization and in those formed *in vitro* a rapid growth with all the cytoplasm migrating to the hyphal tip leaving behind empty collapsed cells is observed. In contrast with dykaryons formed by wild type compatible strains, Lehmler et al. (1997) observed that in dikaryons formed by mutant *kin2* strains, hyphal structures remained short and filled with cytoplasm. Further investigation of dikaryotic hyphae formed by *kin2* mutants showed that they lack the large basal vacuole present in wild-type dikaryons and that instead they contain more 200-400 nm vesicles scattered within the hyphae (Steinberg et al. 1998).

These results strongly suggest that Kin2 is involved in vacuole formation and that the accumulations of these vacuoles at the basal end of the tip play a critical role in supporting cytoplasmic migration. Kin2 mutant strains are severely reduced in pathogenicity indicating that vacuolization plays an important role during normal dikaryon development *in planta*.

Another line of research likely to generate useful data to understand parasitic growth in *U. maydis* is the identification of plant upregulated genes. Among the genes identified as highly induced during fungal growth within plant tissue are *mig1* and the *mig2* gene cluster, all coding for secreted proteins (Basse et al. 2000; Basse et al. 2002). The *mig* genes have a number of features reminiscent of avirulence genes such as secretion, plant-inducible expression and an even number of cysteines presumably indicating the likely importance of desulfide bonds. However, they do not appear to play a critical role during pathogenic development since their deletion does not produce a discernible mutant phenotype. This makes it difficult to discern the roles of *mig* genes during pathogenic development. Recently, an approach that combines REMI (restriction enzyme mediated integration) mutagenesis with enhancer trapping by using green fluorescent protein as a reporter for *in planta* detection identified a new set of *in planta*-induced genes, the *pig* genes (Aichinger et al. 2003). As previously found for the *mig* genes, deletion of *pig* genes did not have an effect on virulence. The same is true for *ssp1*, a gene with similarity to dioxygenases identified as highly induced in mature teliospores (Huber et al. 2002). Again it has been difficult to determine the role of these genes in the biotrophic growth of *U. madyis* because deletions of these did not have an effect on pathogenicity. The search for genes upregulated in the plant continues, preliminary microarray analysis data indicate that more than 500 genes are plant-regulated (Kahmann and Kamper, 2004).

3.1.8 Lessons from the genome

The *Ustilago maydis* genome sequence was made publicly accessible in June, 2003 (http://www.broad.mit.edu/annotation/fungi/ustilago_maydis/). The genomic data is of very high quality and although the genome is not "finished" it is an outstanding tool for researchers of this fungus as well as the mycological research community at large. The genome data is primarily from a 10X shotgun assembly produced at the Broad Institute (previously the Whitehead). There is also data contributed from two private sector sequencing efforts, one from Bayer CropScience and the other from Excelixis, Inc. 15,389 ESTs have been generated primarily through the Bayer sequencing project and the efforts of Dr. Barry Saville at the University of Toronto. These ESTs are critical for the validation of the gene annotation. The genome is currently predicted to contain 6,522 coding genes. Most genes have no introns and where present introns are predicted to be relatively shortl with the bulk being less that 200 bp. Codon bias is quite apparent with nearly 6-fold differences in wobble base usage for several amino acids. A manual annotation project is currently underway at the Munich Information Center for Protein Sequences (MIPS) available at (http://mips.gsf.de/genre/proj/ustilago/). This effort was funded primarily through the efforts of Drs. R. Kahmann and J. Kamper. As of November, 2004, 3.8 of the 20 Mb genome has been manually annotated at mips. The MIPS site offers a number of analytical tools for navigating the genomic data.

Functional genomics efforts are proposed and in some cases underway. Affymetrix microarrays have been generated based on the earlier Bayer CropScience sequencing efforte and are being employed to observe gene transcription under a number of specific conditions (Kahmann and Kamper 2004). Unfortunately, arrays are not currently available to the general research community. Construction of a publicly deposited gene deletion set has been discussed by the *U. maydis* research community but the logistics of this process and necessary funding have not yet been solidified. The International *Ustilago maydis* Research Conferences organized by Drs. F. Banuett and R. Kahmann are greatly helping in the planning and coordination of utilization of the genomic information.

Overall the availability of the genome is a permanent resource available to the scientific community at large and is of outstanding value to those of us that primarily work on this extraordinary fungus.

3.2 Rusts
3.2.1 Introduction

Rust fungi belong to the order *Uredinales*, estimated to include from 4,000 to 6,000 species, belonging to 140-150 genera (Alexopolous et al. 1996; Hahn, 2000). Several well-known genera are *Puccinia*, *Uromyces*, *Gymnosporangium* and *Cronartium*. The divergence of the rust lineage from related basidiomycetes is estimated to have occurred 310 M years ago (Berbee and Taylor 1993). Savile and Baum (Savile 1976; Baum and Savile 1985; Savile 1990) hypothesized that rusts were very early parasites on early vascular plants and therefore have had a long time to co-evolve with their hosts resulting in a very intimate life style. The various rusts attack a wide variety of unrelated mono- and dicot plants. Many are economically important world-wide, such as the cereal, bean, pine, coffee, carnation and peanut rusts (Agrios 1997; Staples 2000). In particular *Puccinia* spp., infecting mainly monocots (as do most smuts), have been known since Biblical times because of the devastation they cause due to expanding cultivation of grains. Several rusts have been studied extensively over the last 100 years, such as cereal rusts *Puccinia triticina* (formerly, *P. recondita*) causing leaf or brown rust on wheat and rye, *P. hordei* on barley and *P. sorghi* on corn; stem rusts *P. graminis* f. sp *tritici*, *avenae*, or *secalis* on their respective hosts; *P. striiformis* causing stripe or yellow rusts of wheat, barley and rye; and *P. coronata* f. sp. *avenae* causing oat crown rust (Johnson et al. 1967; Bushnell and Roelfs 1984; Roelfs and Bushnell 1985; Kolmer 1996) flax rust, *Melampsora lini* (Agrios 1997; Ellis et al. 1997) bean rusts, *Uromyces phaseoli/appendiculatus* (Stavely 1984; Stavely et al. 1989); cowpea rust, *U. vignae* (Heath and Heath 1971); soybean rust, *Phakopsora pachyrhizi* (Bonde et al. 1976; Kuchler et al. 1984); cedar-apple rusts, *Gymnosporangium* spp. (Mims and Richardson 1989) and pine rusts, (*Endo*)*cronartium* spp (Hirt 1964: Hiratsuka et al. 1991).

Rusts generally cannot be cultured or with difficulty (Bose 1974; Williams 1984; Boasson and Shaw 1988; Fasters et al. 1993) although transient transformation of some rusts has been achieved by biolistic methods (Bhairi and Staples 1992; Li et al. 1993; Schillberg et al. 2000) or microinjection (Barja et al. 1998). Attempts using *Agrobacterium*, commonly used for the genetic transformation of many fungi, have not yet been reported. General lack of such tools has been a major bottleneck in advancing the field, making common molecular genetic techniques to create targeted gene deletions or mutations or to clone genes by complementation currently impractical. In addition, even though some natural, easy-to-score color mutations have been described, no characterized rust mutants are available at this time. Within populations, changes in virulence revealed through extensive testing against large collections of host differential cultivars, have been noted many times in many rust species and it is assumed that these occur via mutation of specific ("avirulence") genes. This is thought to occur more readily in sexual populations (Kolmer 1992) but somatic mutations are likely frequent in asexual populations such as from wheat leaf rust (Samborski 1985). Although tedious, genetic analysis by (back-) crossing has been achieved in certain rust species. In this manner, the presence of (dominant) avirulence genes and their interactions with cognate host resistance genes ("gene-for-gene" interaction (Flor 1971; Samborski 1985) and species-specific elicitors (Chen and Heath 1993) has been validated. Crosses have also been used to construct genetic maps (Zambino et al. 2000; Dodds et al. 2004). It should be possible to mutagenize rust populations by other means but characterizing (morphological or metabolic) mutants in these obligate biotrophs will be very difficult. Despite the challenges in research on the rusts, there is hope and recent developments exploiting more random genomics approaches, seem to herald a breakthrough and are reviewed here. There are several excellent recent overviews on rust research (Staples 2000; Heath 2002; Hahn 2000).

3.2.2 Life cycles and spore stages

The rusts have the most complicated life cycles in the fungal kingdom and can include up to five stages and five different spore types produced on two unrelated plant hosts for the heteroecious, macrocyclic forms (Figure 3). According to Savile (1976) this was the result of different adaptations throughout evolution caused by environmental stresses; the aecial spore stage was the final spore state to develop and arose as a result of the adaptation of heteroecism (host-alternation), which itself was stimulated by climatic stress. Increased or diminished selection pressures caused by changing environmental climatic conditions also gave rise to less complex cyclic forms from which various spore stages and/or one of the hosts have been lost. This led to so-called demi- and micro-cyclic and autoecious rusts. It is thought that the sexual cycle is ancestral because it is more adaptable (Savile 1976) but some species are very successful by mainly relying on their asexual urediniospores (epidemic cycle), such as some cereal rusts that seem to survive in Mexico and travel each year north to Canada on the so-called 'Puccinia path', following the cereal growing season (Nagarajan and Singh 1990). It is easy to see how some species can occupy a niche successfully without the need for a sexual cycle, ultimately losing that dependency altogether or because it plays such a minor role that it has not been discovered. The cereal yellow rust, *P.*

striiforms, has no (known) sexual cycle and asexual populations have been described for *M. lini* (Burdon and Roberts 1995) and *P. triticina* (Liu and Kolmer 1998).

Very extensive cell biological, light- and ultrastructural microscopic work has been done on the many morphogenic stages, in particular of several cereal rusts and the bean

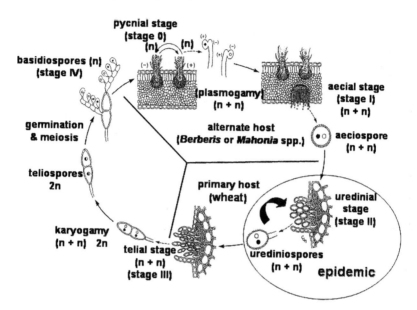

Fig. 3. Life cycle of *Puccinia graminis*, wheat stem rust. Example of a macrocyclic (full-cycled), heteroecious rust. See text for description of stages. (Adapted from Alexopoulos et al. 1996)

and cowpea rust systems (Littlefield and Heath 1979; Gold and Littlefield 1979; Gold and Mendgen 1984; Mendgen 1984; Harder 1984; Harder and Chong 1991; Hu and Rijkenberg 1998 a; Hu and Rijkenberg 1998 b). Their life cycles have been described in great detail. However, when comparing these rusts to the other model organisms described in this chapter, no obvious sudden cell morphological switch from budding to filamentous growth, such as described for the smuts when changing from monokaryon to dikaryon after mating, or other morphologic mutants are apparent.

3.2.2 .1 Stage 0 & I: pycniospores, sex and aeciospores

Interactions on and with hosts that produce pycnio- and aeciospores (in many systems the alternate host) have been studied to a lesser degree even though these often include the sexual stage (Harder 1984). Several studies have attempted to shed light on the mating-type system in the rusts. Conclusions and speculations vary from rusts having a simple bipolar system in several *Puccinia* and *Uromyces* species (Anikster et al. 1999) to a more complicated tetra-polar system with multiple allelic specificities in *M.*

lini (Lawrence 1980) and P. coronata (Narisawa 1994) similar to that in *S. commune* (see section 2.3). The mating system in *Cronartium* species seems to promote outcrossing (Yamazaki and Katsuya 1988; Gitzendanner et al. 1996). In the smuts, both bi- and tetrapolar systems exist, depending on the genetic arrangement and linkage of the two *a* and *b* gene complexes (see section 3.1.3). Rust pycnia consist of haploid, monokaryotic cells and produce haploid pycniospores of only one mating type embedded in a sugar-containing solution, called nectar. When transferred (by insects) to a pycnium of a different (opposite) mating type, among the very early mating events is the induction after 10 minutes of a dark-staining, proteinaceous polar cap on these pycniospores by a factor of a (glyco)-protaneceous nature found in the nectar of several *Puccinia* and *Uromyces* species. Among the many pairings of pycniospores, spore-free nectar and pycnia performed, somewhat less than 50% produced caps and aecia suggesting this factor was mating type-specific and eluded to the existence of a simple bipolar mating system (Anikster et al. 1999). However, hyphal confrontation-fusion assays of pure basidiospore cultures and microscopic observation of the production of dikaryons, suggested the existence of a tetra-polar system with multiple alleles for a similar rust, *P. coronata* (Narisawa 1994). It is conceivable that different rusts have different systems. More research is necessary and it will be very interesting to unravel the molecular basis of these mating systems to see whether they also harbor similar gene complexes in various arrangements as in the smuts.

After fertilization, that is, the fusion of one pycniospore to a receptive hypha in the pycnium of a different mating type and nuclear transfer, the newly formed dikaryon undergoes developmental reprogramming. Mycelium traverses the leaf and often forms aecia on the underside (Harder 1984). Aeciospores are dispersal structures and for heteroecious rusts need to land on the primary host.

3.2.2.2 Stage II: urediniospores and the "increase or epidemic" cycle

Rust pustules, ruptured through host surfaces, produce enormous numbers of urediniospores and generate self-inhibitors preventing their premature germination. In *M. lini*, the production of such self-inhibitors seems to be under the control of a recessively inherited, single gene trait (Ayliffe et al. 1997). Volatile chemicals have been described that can stimulate the germination of uredinio- and teliospores (Macko 1981; French 1992; French et al. 1993). Hydrophobic interactions contribute to the attachment of spores and sporelings to host surfaces (Clement et al. 1994) and glycoproteins and β-1,3-glucans are apparently also involved in adhesion (Epstein et al. 1987; Chaubal et al. 1991). Upon hydration, spores of *U. viciae-fabae* seem to exude from pre-made surface components and actively secrete an adhesive substance; it is unclear whether recognition of the correct host, that is, chemical sensing of host components (possibly produced as a reaction to these fungal products) is involved. Cellulolytic enzymes, e.g., (endo-)cellulose, have been found on dormant spores (Heiler et al. 1993). Spores produce serine esterases, one of which is a cutinase (Deising et al. 1992; Clement et al. 1997). In *Uromyces* germlings, recognition and the mediation of extracellular signals occurs via transmembrane glycoproteins known as integrins, found in extracellular matrix components and often exhibiting specific affinities to the tripeptide sequence Arg-Gly-Asp (RGD); competing synthetic RGD peptides inhibit the thigmostimulated cell differentiation (Correa et al. 1996).

Urediniospores of most rust fungi germinate on the host, form infection hyphae, which grow over the surface until they encounter a stomatal lip, triggering the formation of an appressorium. Extensive research has described the requirement of leaf topographical features and involvement of K^+ and Ca^{2+} signaling, pH and host compounds such as sucrose to initiate this process (reviewed in Staples 2000, and see below). Presumably, perceived changes in or active reorientation of the microtubules and actin filaments making up the cytoskeleton, induce differentiation and dictate changes in cell morphology. However, appressoria and subsequently differentiated morphological structures during the infection process, such as the substomatal vesicle, infection hyphae and even haustorial mother cells can also be induced *in vitro* on artificial substrates, and by heat shock and/or chemicals; this might not represent the exact same differentiation process as that *in planta* because different signal transduction pathways and genes seem involved (Macko et al. 1978; Wanner et al. 1985; Hoch et al. 1986; Heath and Perumalla 1988; Bhairi et al. 1990; Wietholter et al. 2003).

Once an appressorium has formed over a stoma, an infection peg forms basaly, which gains entry into the underlying cavity where a subsequent substomatal vesicle is produced (reviewed in Mendgen et al. 1996). Several fungal-produced enzymes have been implicated in the infection process such as superoxide dismutase, a catalase, a peroxidase and other, cell wall degrading enzyme classes such as extracellular metallo-protease, cellulase, xylanase, pectin methylesterase and polygalacturonate lyase. Some of these are linked to the fungal developmental program, expressed after appressorium formation and their expression gradually increases upon penetration of host cells by the haustorial mother cell (hmc); for reviews on this, see (Deising et al. 1995; Heiler et al. 1993; Lamboy et al. 1995; Rauscher et al. 1995; Xu and Mendgen 1997).

Once the vesicle is produced, new infection hyphae are protruding which when encountering mesophyll cells, differentiate into the haustorial mother cells. Here the first real contact is made with the host upon actual penetration of a mesophyl cell to produce an intracellular (D-, for dikaryotic) haustorium. Importantly, this leaves the host cell membrane intact and produces a seal, the neckband and a highly specific extra-haustorial membrane (Heath 1972; Harder and Chong 1991; Mendgen et al. 1996; Mendgen et al. 2000). It is often at this stage when incompatible interactions are visible and the Hypersensitive Response (HR) is activated, either due to avirulence gene products or nonhost combinations. However, recent studies indicate that host and nonhost responses are distinct and can occur as soon as spores land on leaf surfaces. Overall, differences between compatible and incompatible host and nonhost interactions have been described extensively and they involve host nuclear behavior, protoplast/vacuolar streaming, cell wall modifications and appositions, fungal encasement etc. (Niks and Dekens 1991; Heath and Skalamera 1997; Heath 1997; Munch-Garthoff et al. 1997; Hu and Rijkenberg 1998 a; Hu and Rijkenberg 1998 b; Mould and Heath 1999; Heath 2002; Christopher Kozjan and Heath 2003; Neu et al.

2003). These studies typically involve cytological descriptions sometimes using highly specific antibodies and including some pharmacological data but molecular data are mostly limited to host response genes. The study of the role of fungal genes during these interactions is in its infancy (see molecular interactions section 3.2.3).

Not all rust urediniospores or aeciospores for that matter gain entry through stomatal openings; for example, those from the soybean pathogen *Phakopsora pachyrhizi* penetrate host epidermis cells directly (Bonde et al. 1976). In the latter case, mesophyl cells are invaded and haustoria formed, similarly to the process described above (Littlefield and Heath1979)

3.2.2 .3 Stage III: telial stage

Teliospores are produced in the uredium, often but not always under adverse conditions (e.g. upon draught or host senescence). Upon production of teliospores the uredium is by definition converted into a telial sorus and can be open or covered by the host epidermis. Teliospore ontogeny is not very different from that of the urediniospores but they subsequently differentiate into more condensed cells with thicker walls and in which vacuoles are absent, lipid droplets and glycogen-like material is present and the two nuclei pair if not fuse. Rust teliospores may be one- to five-celled and the germination of teliospores to produce the basidium may require dormancy periods of various lengths, depending on the species (Harder 1984; Mendgen 1984).

3.2.2 .4 Stage IV: basidiospores

When formed on the basidium, mature basidiospores typically contain two haploid, homokaryotic nuclei (Gold and Mendgen 1991) and can be forcefully ejected and dispersed by wind. They can survive for a moderate period of time (days) and will readily germinate under humid conditions to form a short delicate germ tube. For many rust species such as for those in the genera *Puccinia* and *Uromyces*, these germ tubes penetrate the host cuticle and epidermal cells directly after forming an appressorium-like structure whereas others such as *Cronartium* spp., enter through stomata (Flor 1971; Mendgen 1984; Gold and Mendgen 1984; Hoch et al. 1987 b; Hoch and Staples 1987). Following penetration, an (M-, for monokaryotic) haustorium is produced which is merely an intracellular extension of intercellular hyphae without the significant morphological specialization as seen for the D-haustorium (Gold and Mendgen 1991).

Upon establishing a feeding relationship and presumably suppression of (alternate) host defense responses, the specialized pycnia are produced which generate the pycniospores from stage 0. See section 2.3 for a discussion of this topic.

3.2.3 Infection process: molecular aspects of interactions

Most genes described to date have been revealed through protocols using differential screening of cDNA libraries focusing on stage II: urediniospore infection and resulting biotrophic phase in the primary host. Urediniospores probably draw most of their energy from stored lipids when germinating (Staples and Wynn 1965). Indeed, ultrastructural observations suggest that large lipid bodies in the cytoplasm of several spore types seem to degrade during germination (reviewed in Mendgen 1984). In the bean rust, *U. appendiculatus*, several genes were identified by expression during appressorium formation *in vitro* such as *Inf56* and *Inf24* (Bhairi et al. 1989; Xuei et al. 1992). They have been characterized to some degree but their role in the differentiation process is unknown. Microinjection of an antisense fragment to the ORF of *Inf24* strongly inhibited appressorium formation but it did not inhibit continued development of subsequent infection structures (a penetration peg and a substomatal vesicle) into already formed appressoria (Barja et al. 1998). Incidentally, this study also demonstrated that a gene knock-down approach could be successful in this rust. Several rif- ("rust infection-specific") genes of unknown function have been revealed by differential hybridization in *U. fabae* (Deising et al. 1995). *P. graminis* harbors a small gene family, *usp*, coding for small hydrophobin-like proteins that most likely function extracellularly (Liu et al. 1993). Two members are highly expressed during urediniospore germination but one seems to be a sporulation-specific gene.

3.2.4 Signaling

Cyclic AMP signaling plays an important role during differentiation in many fungi including in *U. maydis* (as detailed in section 3.1 of this chapter). Cyclic AMP and cGMP can induce differentiation in *U. appendiculatus* such as mitosis and septum formation and generally regulate appressorium development in urediniospore germlings (Epstein et al. 1989; Hoch and Staples 1984; Hoch et al. 1987 a). Ca^{2+} signaling is very important in fungi (Zelter, 2004) and it was shown early on that K^+ and Ca^{2+} signaling was involved in the morphological changes during the infection process (Hoch et al. 1987 a). In an early attempt to study the importance of Ca^{2+} signaling, calmodulin was isolated from *U. appendiculatus* and shown to stimulate Ca^{2+}-dependent cyclic nucleotide phosphodiesterase (Laccetti et al. 1987).

3.2.5 Establishment of the biotrophic phase

In an effort to reveal biotrophic phase-specific genes, essential for "acclimatization" in a hostile environment such as establishing a feeding relationship and host defense-suppression strategies, several protocols to isolate haustoria from host leaves infected with *Puccinia* and *Uromyces* species have been developed (Hahn and Mendgen 1992; Tiburzy et al. 1992; Cantrill and Deverall 1993). Differential screening of a constructed haustorium-specific *U. fabae* cDNA library yielded many Plant-Induced Genes (PIGs) the analysis of which has produced some insight into the way the bean rust establishes a biotrophic state (Hahn and Mendgen 1997). Some genes might play a role in adapting to the hostile host environment such as overcoming oxidative and osmotic stress and detoxification of host compounds. Several genes involved in nutrient uptake and metabolism appear to be differentially regulated (Wirsel et al. 2001). Amino acid transporters have been identified (Hahn et al. 1997; Struck et al. 2002; Struck et al. 2004 a). of which *AAT1* and *AAT3* encode general amino acid permeases. Each prefers uptake of a different subset of scarce amino acids and are expressed in all infection structures but upregulated in haustoria. In contrast, the amino acid transporter, AAT2p (formerly PIG2p), is exclusively expressed in haustoria and localizes to their plasma membranes (Mendgen et al. 2000). A gene involved in sugar uptake (HeXose Transporter 1, *HXT1*) is exclusively expressed in haustoria and HXT1p also localizes to their plasma membranes (Voegele et al. 2001). Interestingly, a *Saccharomyces cerevisiae* glucose uptake mutant and *Xenopus laevis* oocytes were used to functionally test this rust protein which was characterized as a proton-motive force driven monosaccharide (glucose, fructose, mannose) transport system. A fungal invertase was also found residing in the same subcellular location and suggested to convert sucrose to the monosaccharide substrates for HXT1p (Voegele and Mendgen 2003). Another proton-dependent transporter, AAT1p for ammonium, was discovered and cloned (Struck et al. 2002). This transporter is expressed during all stages of infection and thought to scavenge limited N-metabolites in the host. The presence of a H(+)-ATPase in the haustorial membrane has been suspected but the cloning and analysis of its gene (Struck et al. 1996; Struck et al. 1998) suggests that the fungus exerts control over the nutrient flux. The *BGL1* gene was revealed in a search for genes involved in sugar mobilization. It is expressed in all stages of growth, including haustoria. BGL1p is a fungal α-glucosidase, probably secreted in the extracellular space where it is found in the periphery of intercellular hyphae and haustoria. It may be involved in cellulose/cellobiose degradation and/or possibly defense (Haerter and Voegele 2004). Two of the PIGs (*THI1* and *THI2*) are involved in thiamine (vitamin B1) biosynthesis as revealed by complementation of *Schizosaccharomyces pombe* thiamine auxotrophic mutants. It was suggested that these essential co-factors for central carbon metabolism indicate the need for a very active metabolism through *de novo* biosynthesis (Sohn et al. 2000). Although many relevant genes have and will be revealed, such purified haustoria unless fixed immediately at the beginning of the extraction procedure, have undoubtedly changed their expression patterns during the isolation procedure which will hamper the construction of representative cDNA libraries or skew mRNA populations for transcript profile analyses.

Using suppression subtractive hybridization, Thara et al. (2003) constructed a fungal, *in planta*-specific cDNA library from *P. triticina*-infected wheat leaves 4 days after infection of compatible wheat and subtracted with mock-inoculated host sequences (Thara et al. 2003). They obtained 104 unique random sequences of which 69 were likely fungal. Among them, 25 represented ribosomal proteins and the remaining 44 encoded non-ribosomal fungal proteins. Some of these were novel and some represented previously found PIGs or virulence genes from other fungi. A targeted cDNA-AFLP technique allowed the selection of genes with interesting expression patterns during the infection process of *P. triticina* on wheat leaves from urediniospore inoculation up to sporulation (Zhang et al. 2003). Both up- and down regulated genes from both fungus and host were revealed, 50% of which showed no homology to known genes. Another technique, differential display, was used to reveal 9 genes expressed in galls during the infection of southern pine by *Cronartium quercuum* f. sp. *fusiforme*; genes involved in metabolism and stress were identified (Warren and Covert

2004). The above-mentioned molecular studies confirm that rusts undergo a major reprogramming of their metabolism and transcriptome once inside the host. Illustrative is the apparent induction of a large number of ribosomal protein genes upon establishment in the host (Thara et al. 2003; Hu and Bakkeren unpublished). Moreover, these recent molecular studies have substantiated an old hypothesis that the haustoria divert nutrients and play a major role in the source-sink relationship with the host (Hahn 2000; Staples 2001; Szabo and Bushnell 2001; Wirsel et al. 2001; Struck et al. 2002; Voegele and Mendgen 2003).

3.2.6 Virulence factors

Many of the genes described in this section can undoubtedly be classified as pathogenicity or virulence genes, although their function as such has not been directly verified in a (targeted) mutational analysis as is common in other, more tractable pathogens. Factors eliciting the host defense as measured by the induction of (cell wall) autofluorescence and/or host PR genes, up to a full-blown HR including necrosis and DNA laddering, have been described for some rusts but their primary role in virulence is unknown (reviewed in Hahn 2000). The unfortunate elicitor functionality can reside within a true virulence factor produced to facilitate the infection process or be caused by "by-products" of the infection process (reviewed in Bakkeren and Gold 2004). Such elicitors triggering defense responses can be general, produced by a whole genus or by an inappropriate pathogen during a nonhost interaction, or more specific to a species or a certain race or isolate within a species. The mentioned avirulence factors elicit defenses resulting in complete restriction of the fungal life cycle on the host. For example, general chitin oligosaccharides, as well as a more specific glycopeptide elicitor have been described for *P. graminis*, which induce hypersensitive-like responses in several but not all wheat genotypes and stimulate lipoxygenase (LOX) activity (Sutherland et al. 1989; Beissmann et al. 1992; Bohland et al. 1997; Tada et al. 2001). In the latter category, race-specific elicitors have been isolated from *U. vignae* (Chen and Heath 1990; D'Silva and Heath 1997) and *P. triticina* (Saverimuttu and Deverall 1998).

In the field of plant-microbe interactions, avirulence genes have long attracted attention because they represent often single dominant genes and trigger defense reactions. 'Race-cultivar specialization' has been described for many biotrophic fungi including cereal rusts (Kolmer 1996; Chong et al. 2000; Chen and Line 2003; Brown and Casselton 2001; Long and Kolmer 1989), flax rust (Dodds et al. 2004) cowpea rust (Chen and Heath 1993) and the white pine-blister rust (Kinloch and Dupper 2002). The large number of races revealed for some rusts suggest the presence of many fungal factors that the respective host cultivars can detect. Efforts to construct genetic maps are underway for several rusts, e.g. *P. graminis* (Zambino et al. 2000) *M. lini* (Dodds et al. 2004) and *P. triticina* (McCallum and Mulock unpublished) and this has recently resulted in the isolation of the first basidiomycete avirulence gene, *AvrL567*, from *M. lini* (Dodds et al. 2004). The *AvrL567* gene is expressed in rust haustoria and encodes a 127 amino acid preprotein of unknown function. The N-terminal 23-amino acid signal sequence presumably is cleaved off and the mature protein is secreted in the plant cell where it induces a hypersensitive response-like necrosis that is dependent on co-expression of the L5, L6, or L7 resistance gene. Elucidating the intended (virulence) function of such "avirulence" gene products is a biological challenge and of utmost importance.

Double-stranded RNAs have been found in many rust species. The number and size of these dsRNA molecules seem to be highly species-specific, but contrary to the effect they have on many other phytopathogenic fungi such as causing hypovirulence (Dawe and Nuss 2001), they don't seem to be involved in pathogenicity, virulence, fitness or toxin production in the rusts (Pryor et al. 1990; Zhang et al. 1994).

3.2.7 Countering host defense

The existence of suppression of host defense response during rust infections is descriptive (microscopy; Harder and Chong 1991) or indirect through a phenomenon called 'induced susceptibility' (Niks 1989; Arz and Grambow 1995; Skalamera et al. 1997) and reviewed in Voegele and Mendgen (2003). Molecular studies are only just beginning for the rusts, but fungal factors suppressing host defense and increasing host susceptibility will by definition be encoded by virulence genes. In analogy with bacterial "effector" molecules that are transferred to the host cytoplasm/nucleus, a possible candidate is the U. fabae Rust Transferred Protein, RTP1, which can be found in the nucleus of infected bean cells. Its function, however, is unknown and there are no known homologs (Struck et al. 2002). Cro rI from C. ribicola is a small secreted protein that is preferentially expressed by the fungus growing in susceptible host seedlings and could have a role in moderating the interaction (Yu et al. 2002). Another, probably major strategy of biotrophs is avoidance of recognition of pathogen-associated molecular patterns or PAMPs and/or of elicitation of host defense responses. In U. appendiculatus it was shown that the walls of germ tubes and appressoria contain chitin, the amount of which is reduced or masked in infection hyphae and haustoria. This may occur through deacetylation of chitin converting it to chitosan and it was suggested that this would prevent recognition by the host (Heath 1989; Freytag and Mendgen 1991; Deising et al. 1995; El Gueddari et al. 2002). The U. fabae β-glucosidase, BGL1p, could be involved in countering general host defenses by detoxifying phytoalexins such as saponins (Haerter and Voegele 2004).

3.2.8 Comparative biology, genomics and future direction

Whether genes shown to be involved in certain developmental stages on the primary host such as, sporulation, infection, metabolite diversion and the establishment of the source-sink relationship during biotrophism, or suppression of host defense, play roles in equivalent stages on the alternate host and the production of the other spore types is currently unknown. "Genomics", the large-scale analysis of the genome, its genes and their expression, is in its infancy for the rusts but is expected to provide breakthroughs for these hard to manipulate obligate parasites. Large scale EST projects are underway for P. triticina (Hu and Bakkeren unpublished; Zhang et al. 2003). A large P. triticina EST database generated from cDNA libraries representing five distinct morphological phases in Stage II, has revealed homologs of some of the genes discussed here and others that will expand our insights. For example, we identified sugar transporters (from resting and germinating urediniospores), a high-affinity leucine-specific transport protein, a manganese ABC transporter and a β-glucosidase, a homolog of *U. fabae BGL1* (from germinating urediniospores). With respect to signaling, we identified a putative calcium transporting ATPase in resting spores and a calcium binding signal transducing protein, a putative calcium transporting ATPase, calmodulin, a calmodulin-dependent protein kinase, a calcium/proton exchanger and a potassium channel subunit, all in resting and germinating urediniospores. The discovery of several enzymes involved in lipid metabolism, such as lipases, acyl CoA dehydrogenase and fatty acid hydroxylase, in cDNA libraries from germinating urediniospores, might substantiate hypotheses regarding the physiology of germinating spores. However, no functional or transcript analyses have been done (Hu and Bakkeren manuscript in preparation). The comparison of global gene expression profiles using microarrays harboring large sets of genes representing several life cycle stages of *P. triticina* (Bakkeren lab in progress) might shed light on such questions. In additon to EST projects, total genome sequencing is underway for *P. graminis* f. sp. *tritici* (Dr. L. Szabo et al., USDA and the Broad Institute) and two soybean rust species, *Phakopsora pachyrhizi* and *P. meibomiae* (Dr. R. D. Frederick et al. USDA).

Ultimately, apart from the academic interest, we need to understand delicate interactions in order to develop "smart", very specific (bio)-fungicides to be used as sprays, or expressed in antagonists or in host plants to be used as crop protectants. In addition, if we find critical factors perturbed in the host during establishment of the feeding relationship, we can counteract these in the plant and possibly attain durable resistance. Understanding the molecular basis of nonhost resistance, widely thought to underpin more durable resistance, should open new avenues for crop protection.

3.3 Mushrooms
3.3.1 Introduction

The differentiation of fruiting bodies (mushrooms) in Basidiomycetous fungi represents what is arguably the most important developmental event in the life cycle of the species capable of forming such structures. Despite its' importance to these fungi, the process of mushroom development, also called fruiting, is poorly understood at the genetic level. It is not the purpose of this brief overview to systematically list and detail all of the pertinent research in the field: a number of excellent and thorough reviews have been recently published to which the reader is referred (Fischer and Kues 2003; Kamada 2002; Kues 2000; Kues and Liu 2000; Kues et al. 2004; Walser et al. 2003). This section briefly outlines mushroom development as it occurs in the two best-studied model systems, *Coprinopsis cinerea* (*Coprinus cinereus*) and *Schizophyllum commune*, and then describes what is currently known about the genetic control of this process. In an effort to make connections between genetic controls and signaling networks, special reference will be made to two well-studied non-mushroom forming Basidiomycetes, *U. maydis*, and *C. neoformans*.

3.3.2 A double threat for use as models in the molecular study of mushroom development: *Coprinopsis cinerea* and *Schizophyllum commune*

For the most part, commercially valuable mushroom-producing fungi have been difficult subjects for both genetic and molecular biological studies of fruiting body development. However, it can be argued that this potential cloud has resulted in a silver lining, because not one, but two model systems have been studied to fill the gap. The concurrent utilization of two distinct model organisms for the study of mushroom development allows for the distinguishing of elements common to the process from those that are species-specific. The ink cap mushroom *Coprinopsis cinerea* and the common split-gill mushroom *Schizophyllum commune* both have a rich history of genetic study dating back to the early twentieth century (Buller 1933; Essig 1922). *C. cinerea* forms what can be termed a typical mushroom with a long stipe and well-developed cap, while the fruiting body of *S. commune* appears as a fan-shaped cup close to the surface of the substrate. Both of these fungi have life cycles (Figures 4 and 5) that can be completed on artificial media in the laboratory within two weeks. Each has independently propagatible haploid and dikaryotic (n + n) phases, facilitating genetic analyses. The molecular study of gene function has been enabled by DNA-mediated transformation (Binninger et al. 1987; Munoz-Rivas et al. 1986), and by the recent genomic sequencing of *C. cinerea* (www.genome.wi.mit.edu/cgi-bin/annotation/fungi/Coprinopsis_cinerea). Targeted gene disruption by homologous integration has been achieved in both systems (Aime and Casselton 2002; Binninger et al. 1991; Horton et al. 1999; Lengeler and Kothe 1999) although the frequency of these events are considerably lower than in other Basidiomycetes such as the plant pathogen *U. maydis* (Kahmann and Kamper 2004), or the human pathogen *C. neoformans* (Hull and Heitman 2002). Overexpression of genes has been demonstrated in *S. commune* through the use of recombinant constructs using strong promotors endogenous to this fungus (Yamagishi et al. 2002; Yamagishi et al. 2004). Current technical limitations include the lack of episomal plasmids for use in non-integrative transformation, and a well-developed GFP-tagging system for use as a reporter in cytological studies.

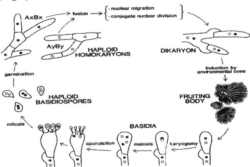

Fig. 4: Life cycle of the mushroom-forming Basidiomycte *Schizophyllum commune* (Stankis et al. 1990)

3.3.3 Other mushroom-producing fungi of agricultural value

Mushroom species of agricultural importance include the button mushroom *Agaricus bisporus*, the oyster mushroom *Pleurotus ostreatus*, and the Shiitake mushroom *Lentinus edodes*. The worldwide market for these and other mushrooms for use as food, as dietary supplements and nutriceuticals, and in medicine was estimated to be over US $13 billion per year in 1996 (Chang 1996). The value of this market has spurred interest in the application of molecular genetic techniques for use in improving the characteristics of these fungi. Classical and molecular genetic studies have been hampered by the slow growth and the lack of fruiting of these fungi on synthetic media. An additional complication for genetic analyses is that in *A. bisporus*, basidiospores are binucleate and produce heterokaryotic mycelia upon germination. Recent advances in transformation technologies using *Agrobacterium tumefaciens* (de Groot et al. 1998; Chen et al. 2000) have opened up possibilities for molecular-based strain improvement, although these methods are still inefficient as compared to the model organisms. To date, there are no genomic sequencing projects for these fungi.

3.3.4 Overview of mushroom development

From a macroscopic point of view, the development of mushrooms represents the single most dramatic event in the life cycle of a fungus capable of producing such a structure. Under normal circumstances, only the dikaryotic phase is competent to form mushrooms: however, haploid fruiting strains do exist (Esser et al. 1979; Leslie and Leonard 1979; Murata et al. 1998; Uno and Ishikawa 1971; Verrinder-Gibbins and Lu 1984; Yli-Mattila et al. 1989). Mushrooms are also produced in haploids with

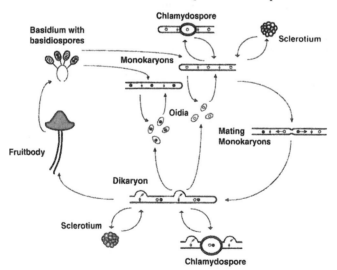

Fig. 5: Life cycle of of the mushroom-forming Basidiomycte *Coprinopsis cinerea* (Kues 2000)

constitutive mutations in the *A* and *B* mating-type loci in both *C. cinerea* and *S. commune* (Raper et al. 1965; Swamy and Ishikawa 1984). One interesting difference between these two fungi is that the activation of the *A* pathway alone (along with light) is sufficient to induce fruiting in *C. cinerea* (Kues et al. 1994; Kues et al. 1998) but not in *S. commune* (C. Raper and T. Fowler, personal communication).

The process of mushroom development can be divided into three main cellular events: aggregation, differentiation, and expansion. Since the cytological processes are best described in *C. cinerea*, these are outlined here. For a more comprehensive description, see the excellent review by Kues (2000). Differences between what occurs in *C. cinerea* and *S. commune* will be noted. In both species, proper environmental conditions are required for normal fruiting to occur. These include nutrient limitation, periods of light and dark, and low CO_2 levels (reviewed in Kues 2000; Wessels 1992).

Mushroom development begins with the formation of what is called a primary hyphal knot. This undifferentiated structure is characterized by intense localized formation of short hyphal branches and restricted tip growth (Matthews and Niederprum 1972). The resulting cellular compartments are short, and in *C. cinerea* the hyphal knot is also characterized by higher order side branches. A blue light signal results in hyphal aggregation and the formation of a compact ball called a secondary hyphal knot (Matthews and Niederprum 1973), which is the first fruiting-body specific structure (Lu 1974; Uno et al. 1974). In *S. commune*, hyphal aggregation results in a stalk-like structure because of parallel growth of the adhering hyphae (Raudaskoski and Vauras 1982; Raudaskoski and Viitanen 1982; van der Valk and Marchant 1978).

The development of fruiting body primordia from secondary hyphal knots in *C. cinerea* requires additional light signals, and it is within these structures that cellular differentiation occurs. There is an observable polarity of cells in the primordia, with roughly the upper third forming the cap, the middle third the stipe, and the lower third the basal plechtenchema. Cells in the latter region are randomly oriented, and do not get incorporated into the stipe, where cells have a varied spatial arrangement, depending upon their function (Hammad et al. 1993; Kues et al. 2004). From a genetic point of view, the key differentiation event is the development of the spore-bearing basidia within the gills of the cap, in a layer called the hymenium. It is within the basidia that karyogamy and meiosis occur, which in *C. cinerea* is a synchronous event induced by a light signal (Lu 1974). At the same time, the mushroom grows by cell expansion and elongation (Kamada and Takemaru 1977). In *C. cinerea*, the fruiting body is a short-lived structure because of autolysis of the cap tissue, which releases the basidiospores into the deliquescent milieu. The process of fruiting body formation in this fungus is quite rapid: from the first visible appearance of primordia to the autolysis of the mushroom cap takes about 4 to 5 days (Moore et al. 1979). In contrast, fruiting bodies of *S. commune* are essentially perennial, and undergo karyogamy, meiosis, and sporulation in an asynchronous manner. The fan-shaped fruiting bodies enlarge primarily by cellular proliferation on the periphery, and not by cellular inflation. Spores are borne on the upper surface, and are dispersed directly into the air, and the mushrooms do not undergo autolysis.

C. cinerea exhibits much more plasticity than does *S. commune* in the program of fruiting body development. At various points in the program, other multicellular structures such as sclerotia and capless etiolated stipes can result in *C. cinerea* if light signals are not received at the appropriate times (Kamada et al. 1978; Kues et al. 1998; Lu 1974; Moore 1981). While light is necessary for the development of mushrooms in *S. commune* (Perkins 1969; Perkins and Gordon 1969) these other developmental "options" have not been observed.

3.3.5 Genetic control of mushroom development

The master regulatory genes responsible for the initiation of fruiting in the dikaryon reside at the mating-type loci *A* and *B*. In both *C. cinerea* and *S. commune*, the *A* mating-type genes encode homeodomain (HD) transcription factors, while the *B* genes encode lipopeptide pheromones and their receptors. The molecular genetic analysis of these genes in these fungi has been recently reviewed extensively elsewhere (Casselton and Olesnicky 1998; Kamada 2002; Kothe 1996; Kues 2000). The focus here will be on what is known about the genetic regulation of mushroom development downstream of the mating-type genes. Given the eclectic nature of the "fruiting" genes so far analyzed, it is informative to make reference to the genetics of sexual development in the better-studied non-mushroom forming Basidiomycetes. It is hoped that the concepts illustrated in these other species may inform us about the many gaps in our understanding with regards to the genetic control of mushroom development.

3.3.6 Experimental approaches for gene isolation

The genes characterized to date that are implicated in the genetic control of mushroom development are a heterogeneous lot: in many cases these genes do not have an obvious connection to one another. This may be a reflection of the fact that relatively few laboratories are currently studying this problem, and also the diverse manner of the approaches by which the relevant genes have been isolated. These approaches include: (1) complementation of recessive mutants defective or altered in the fruiting process, (2) differential mRNA expression under fruiting and non-fruiting conditions, (3) degenerate PCR, and (4) two-hybrid protein interaction screens. A high priority in the future will be to greatly expand our knowledge of the "players" found to be regulating mushroom development, and to integrate these genes into a set of genetic pathways that is much more definitive than is known at present.

The complementation of mutants defective in fruiting body development has been an effective genetic tool in efforts to isolate genes regulating this developmental process. Various mutagenesis procedures have been employed, including UV irradiation, mutagenic chemicals, and REMI or restriction enzyme mediated integration (reviewed in Kues 2000). Homokaryotic strains of *C. cinerea* and *S. commune* with constitutive mutations in the *A* and *B* mating-type loci reproduce the phenotype of the dikaryon, including the formation of clamp connections and fruiting bodies (Raper et al. 1965; Swamy and Ishikawa 1984). These self-compatible strains are called *Amut Bmut* and *Acon Bcon*; in *C. cinerea* and *S. commune*, respectively. There are considerable advantages of using such strains over dikaryons in mutagenesis screens for fruiting-defective mutants. Both recessive and dominant mutations can be detected in the haploid genetic background, unlike in dikaryons. Without the complication of two different nuclei present in the same cytoplasm, the homokaryotic strain is by definition isogenic for all loci, including those genes that are necessary for fruiting. In *C. cinerea* (but not *S. commune*), uninucleate asexual spores called oidia can be used for mutagenesis (Pukkila 1993). Large scale UV or REMI mutagenesis of oidia from homokaryotic fruiting strains of *C. cinerea* results in a very high proportion (30-40%) of surviving clones with defects in fruiting (Granado et al. 1997; Muraguchi et al. 1999). These findings suggest that in addition to fruiting-specific genes, many other genes involved in cellular metabolism, hyphal growth, and general housekeeping play important roles in mushroom development (Kues 2000).

3.3.7 Early-acting genes in the control of fruiting

The master controlling elements of sexual development in mushroom-producing fungi are the products of the *A* and *B* mating-type genes. In the corn pathogen *Ustilago maydis*, the mating-type genes encode the same types of protein products (reviewed in Kahmann and Kamper 2004) as they do in *C. cinerea* and *S. commune*, but the nomenclature of *a* and *b* are "reversed" in *U. maydis*. DNA microarray analyses have set the number of *b*-regulated genes in *U. maydis* at 246 (Scherer and Kamper unpublished, cited in Kahmann and Kamper 2004). Only three genes to date have been shown to be direct targets of the bE/bW heterodimer, suggesting that the majority of the remaining genes are likely to be indirectly regulated. These genes have a *b* binding sequence (bbs) in their promoter regions, and have been designated class 1 *b*-regulated genes. It has been proposed that the bE/bW heterodimer triggers a regulatory cascade, with some of the class 1 genes encoding regulatory proteins, as well as being targets themselves (Kahmann and Kamper 2004). Unfortunately, none of the three class 1 genes characterized to date encode a likely regulatory protein, and none are essential for pathogenicity (Brachmann et al. 2001; Romeis et al. 2000; Weinzierl and Kamper, unpublished, cited in Kahmann and Kamper 2004).

In *C. cinerea*, one gene likely to be an immediate downstream target of the HD heterodimer encoded by the *A* mating-type loci is *clp1* (Inada et al. 2001). The *clp1-1* recessive mutant allele results in a truncated protein product, and has a "clampless" phenotype in a genetic background where the *A* mating-type pathway is activated (Inada et al. 2001). The promoter region of *clp1* contains a sequence (GATGX$_{11}$ACA) that is similar to the conserved sequence (GATGX$_9$ACA) found in the *b*-binding sequence or bbs of *U. maydis* (Romeis et al. 2000) and also to the hsg motif that is the target of the MATα2/MATa1 heterodimer in the budding yeast *S. cerevisiae* (Goutte and Johnson 1988). While it is a good candidate for a class 1 gene, *clp1* is unlikely to encode a transcription factor, as it has no obvious structural motifs, and does not show any extensive similarity to known DNA-binding proteins when used as a query sequence in BLAST (Altschul et al. 1997) searches. The most significant hit (E value = 3 X 10^{-9}) is to a hypothetical protein of unknown function found in the *U. maydis* genome. As is the case in *U. maydis*, other class 1 genes from *C. cinerea* need to be identified as potential regulatory genes in the HD-regulated pathway.

The gene *pcc1* was first identified in a homokaryotic strain of *C. cinerea* that was observed to produce fruiting bodies after prolonged culture (Murata et al. 1998). This spontaneous recessive mutation induced the formation of unfused clamp connections, also called pseudoclamps. This cell type is typical of *A*-activated heterokaryons, as is fruiting body development, although the occurrence of the latter is also dependent upon the genetic background of the individual (Kues et al. 1994; Kues et al. 1998; Kues et al. 2002). *pcc1* is predicted to encode a likely DNA-binding protein, and has an HMG domain with significant similarity to those found in ascomycete mating-type gene products (Murata et al. 1998). This region of the Pcc1 protein is also 43% and 29% identical to the HMG domain of the Rop1 and Prf1 proteins of *U. maydis*, respectively. Prf1 encodes a transcription factor that is a central regulator of both cell fusion and pathogenicity in this fungus (Hartmann et al. 1999). These effects are mediated by Prf1 binding to pheromone response elements, or PREs. PREs are present in the promoter of *prf1* itself, and also in the regulatory regions of the *a* and *b* mating-type genes (Hartmann et al. 1996; Urban et al. 1996 a). The former finding is highly suggestive of *prf1* autoregulation, while the latter result indicates that expression of the mating-type genes is regulated by Prf1. Interestingly, the promoter of *pcc1* also has a PRE-like sequence (Murata et al. 1998). Rop1 has been shown to regulate *prf1* gene transcription by directly binding the *prf1* promoter *in vitro* (Brefort et al. 2004). Rop1 is also essential for mating-type gene expression (Brefort et al. 2003). Given all of this information, it is possible that Pcc1 could represent the *C. cinerea* equivalent of either Rop1 or Prf1. If either of these is the case, it suggests that pheromone response as regulated by Pcc1 may be influenced by both cAMP and MAP kinase signaling, as is the case in *U. maydis* (Kahmann and Kamper 2004).

Pcc1 has been hypothesized to act as a repressor of *A*-regulated development in the homokaryon, and is also likely to be downstream of *clp1* (Kamada 2002). The latter idea is supported by the fact that *clp1* is not transcribed in the *pcc1* mutant, despite the observation that clamps are observed (Murata et al. 1998). It is important to note that forced expression of *clp1* with a β1-tubulin promoter induces clamp development in the absence of *A*-activation, thus releasing the *A* pathway from Pcc1 repression (Inada et al. 2001). This de-repression is likely to be post-transcriptional, as *pcc1* transcript levels are actually higher in *A*-on than in *A*-off mycelia (Murata et al. 1998). There exists the possibility that an as yet unidentified protein(s) mediates the regulation of Pcc1 by Clp1 (Kamada 2002). In addition to *pcc1*, the genes *hmg1* and *hmg2* from *C. cinerea* also encode proteins with HMG domains that have strong homology to that of Prf1 (Milner 1999). Gene disruption of *hmg1* indicates that the Hmg1 protein is necessary for nuclear migration, but not for clamp cell fusion (Aime and Casselton 2002). The *B* pathway of pheromone signaling induces both of these cellular events. Taken together, these findings suggest that HMG domain proteins might occupy central regulatory roles in both *A*-regulated (Pcc1) and *B*-regulated (Hmg1 and/or Hmg2) development in *C. cinerea*.

A UV-induced mutant blocked after hyphal knot formation in the *Amut Bmut* homokaryotic fruiting strain of *C. cinerea* was complemented by the gene *cfs1*. This gene is predicted to encode a eukaryotic counterpart to bacterial cyclopropane fatty acid synthases, which are a specific sub-family of S-adenosylmethionine-dependent methyltransferases (Liu et al. 2001; Liu 2001). This category of enzymes convert membrane-localized unsaturated phospholipids into cyclopropane fatty acids (Grogan and Cronan 1997), which are believed to be part of stress defense in bacteria. In mushroom-producing fungi, membrane alteration has been postulated to be a stress signal that helps tell the fungus to make the switch from vegetative to reproductive growth (Magae 1999; Oita and Yanagi 1993).

An unnamed gene has been identified that is essential for primary hyphal knot formation, the first step in fruiting body development (Clergeot et al. 2003). It appears to encode a protein related to *het-e* gene product of the Ascomycete Podospora anserine, and encodes a protein with a GTP-binding site and a WD-40 repeat domain similar to that seen in Gβ subunits (Saupe et al. 1995). In addition to being defective in hyphal knot formation, another phenotype of this mutant is a fast dark-brown staining of the agar around the growing cultures, which could indicate a de-regulation of laccase(s) (Clergeot and Ruprich-Robert unpublished results, cited in Kues et al. (2004). There are eight members of the laccase gene family in *C. cinerea*, the largest number identified so far in a single haploid fungal genome (Hoegger et al. 2004). Although the literature is certainly not conclusive on the matter, it has been proposed that laccases could function in mushroom development by mediating hyphal aggregation, acting to form chemical crosslinks by oxidative polymerization of phenolic cell wall components (Broxholme et al. 1991; Zhao and Kwan 1999).

3.3.8 Signaling and fruiting body development

Signaling cascades involving the cAMP and MAP kinase pathways play definitive roles in mating, morphogenesis, and pathogenicity in the plant pathogen *U. maydis* (Kahmann and Kamper 2004), and in the human pathogen *C. neoformans* (Hull and Heitman 2002). It is likely that this will also be the case for the processes of mating and sexual development in the mushroom-forming Basidiomycetes as well. In *U. maydis*, the cAMP signaling pathway is controlled through the Gα subunit of a heterotrimeric G-protein, Gpa3 (Kruger et al. 1998; Kruger et al. 2000; Regenfelder et al. 1997). Neither the signal nor the receptor responsible for activating Gpa3 have been identified. The receptor that activates the Gα protein Gpa1 in a cAMP signaling pathway relevant to mating and virulence in *C. neoformans* has also eluded identification (Hull and Heitman 2002). There are likely to be four distinct Gα genes in *S. commune*; SCGP-1 (GenBank accession AF157495, A. Pardo, M. Gorfer and M. Raudaskoski), ScGP-A, ScGP-B, and ScGP-C (Yamagishi et al. 2002). Dominant activating mutations were made in three of these Gα genes, and were introduced into homokaryotic strains under the control of a strong promoter. Over half of the recipients transformed with mutated ScGP-A or ScGP-C exhibited suppressed aerial hyphae formation (Yamagishi et al. 2002). Mutated ScGP-B transformants did not show any differences with wild-type strains. Contrary to expectations, none of the constitutive mutant Gα constructs stimulated the pheromone response (B-regulated) pathway in A-on heterokaryotic matings, as evidenced by the absence of observable clamp connections in these mated strains. Both the mutated ScGP-A and ScGP-C genes also markedly suppressed fruit-body formation in A-on, B-on dikaryons (Yamagishi et al. 2002). The suppression of both fruiting and of aerial hyphae is reminiscent of the phenotype observed for strains carrying the transposon-mediated loss-of-function mutation in the thn1 gene of S. commune (Fowler and Mitton 2000). The Thn1 protein is a likely member of the RGS domain family that regulate the activity of G-protein α-subunits, and it is possible that it may act upon the ScGP-A and/or the ScGP-C gene products. It is unclear exactly how the mutated G-proteins act to repress fruiting body formation, but it could be because of a disruption in cAMP signaling. Measured intracellular cAMP levels were elevated by 160-200% in monokaryons and dikaryons containing mutated ScGP-A and ScGP-C genes (Yamagishi et al. 2004). Taken together, these results suggest that both of these two Gα subunits may play a role in cAMP signaling, and that this pathway helps to regulate mushroom development in S. commune. Two cDNAs encoding Gα subunits have been isolated from the mushroom-producing fungus Coprinellus congregatus, but it has not been determined if either of these genes participate in cAMP signaling (Kozak et al. 1995).

In C. cinerea, it has been established that cAMP levels rise at the time of light-induced secondary knot and primordia formation, and then decline during fruiting body maturation until karyogamy and meiosis. Mature fruiting bodies (after maturation of basidiospores) are low in cAMP content (Uno et al. 1974; Uno and Ishikawa 1974; Kues et al. 2004). Indeed, addition of exogenous cAMP has been shown to induce fruiting in receptive strains of this fungus (Uno and Ishikawa 1971; Uno and Ishikawa 1973). Changes in cAMP levels have also been observed during the process of fruiting body formation in S. commune (Kinoshita et al. 2002). Genes encoding adenylate cyclase, responsible for cAMP production, have been isolated from both C. cinerea (Bottoli et al. 1999) and S. commune (Horton et al. unpublished results). Transcripts for the C. cinerea gene cac1 are constitutively expressed in vegetative mycelia, regardless of light conditions (Bottoli 2001). This finding suggests that enzyme activity is likely regulated at the post-translational level. The gene encoding a S. commune adenylate cyclase was isolated by means of a two-hybrid screen as an expressed cDNA potentially interacting with a Frt1 "bait" (Horton et al. unpublished results). No experiments manipulating this adenylate cyclase have been reported to date. However, the gene frt1 has been implicated in the genetic pathway controlling fruiting, see below (Horton and Raper 1991).

Rising cellular cAMP levels that occur as a result of the activity of adenylate cyclase allows for the activation of cAMP–dependent protein kinases (PKAs). Binding of cAMP to the regulatory subunits of the PKA tetramer induces conformational changes that result in the release and subsequent activation of the catalytic subunits. In the U. maydis pheromone response pathway, the PKA catalytic subunit Adr1 phosphorylates Prf1, which in turn induces the expression of both the a and b mating-type genes (Kaffarnik et al. 2003). A cDNA encoding a catalytic subunit of PKA has been isolated from S.

commune, but to date there is no definitive evidence that links this gene with the fruiting pathway (Horton et al. unpublished results).

In *U. maydis*, a MAP kinase module is necessary for pheromone response, and is intimately associated with both mating and pathogenicity. Prf1 is a direct target of the MAPK enzyme Kpp2/Ubc3 (Muller 2003 b). No member of a MAPK signaling cascade has been isolated to date from a mushroom-producing Basidiomycete. The availability of the sequenced *C. cinerea* genome can clearly be of value in this effort. The use of two-hybrid technologies using Pcc1, Hmg1, or Hmg2 as bait could help to determine which MAPK is likely to interact with these putative transcription factors.

Ras proteins belong to a conserved family of small GTP-binding proteins, acting as molecular switches controlling a wide range of cellular processes (reviewed in Takai et al. 2001). Two Ras proteins are proposed to play roles in the MAPK and cAMP signaling pathways in *U. maydis*. Ras1 likely affects the activity of the adenylate cyclase protein Uac1, while Ras2 has been assigned to the MAPK cascade, upstream of the MAPK Fuz7/Ubc5 (Lee and Kronstad 2002; Muller 2003 b). In *C. neoformans*, the protein Ras1 is required for both pheromone production and pheromone response in *C. neoformans*, and has been implicated in the MAPK signaling pathway (Alspaugh et al. 2000; Waugh et al. 2002). A single *ras* gene has been isolated to date from *C. cinerea* (Ishibashi and Shishido 1993). A dominant active allele of *ras* has been constructed, and was observed to cause changes in both directional and invasive growth of homokaryotic hyphae (Bottoli et al. 2001). When this mutant allele of *ras* was expressed in the homokaryotic *Amut Bmut* mutant strain that mimics the dikaryon, the formation of primary hyphal knots was found to be repressed (Bottoli et al. 2001). This finding suggests that this Ras protein may participate in the genetic control of fruiting. Interestingly, expression of the dominant active *ras* allele also appears to cause defects in mating linked to the *B* mating-type pathway (Kues et al. 2004). A dominant activating allele of a *S. commune ras* gene had a similar repressing effect on fruiting in dikaryons, although cAMP levels were found not to be different from controls (Yamagishi et al. 2004). This suggests that this particular *ras* gene may not act through the cAMP signaling pathway, but perhaps through the MAPK cascade. A second *ras* gene is likely to exist in this fungus, as it appears that two different *ras* genes are in *C. cinerea* from BLAST searches of the translated genome.

As mentioned earlier, the gene *frt1* has been implicated as impinging upon the fruiting pathway in *S. commune*. This is because the gene was originally isolated by its' ability to induce fruiting in certain homokaryotic transformation recipients (Horton and Raper 1991). The transgene also enhanced the mainstream developmental process of dikaryotic fruiting. It was hypothesized that homokaryotic fruiting was induced due to a difference in the *frt1* allelic type between the transgene and the endogenous gene residing in the recipient. A rigorous test of this hypothesis would be to isolate the proposed alternate allele of *frt1*, and to demonstrate that it had the ability to induce fruiting in a homokaryotic strain containing the characterized allele (*i.e.* the reciprocal experiment to the original). Unfortunately, extensive efforts to isolate an alternate allele of *frt1* have proven to be unsuccessful to date (Horton et al. unpublished). The *frt1* translation product is predicted on the basis of motif analysis to be a small ATP-binding protein of 192 amino acids (Horton and Raper 1995). BLAST searches have shown extensive similarity to the C-terminal half of XPMC2, a *Xenopus laevis* protein that has been shown to rescue a mitotic catastrophe phenotype in the fission yeast *S. pombe* (Su and Maller 1995). A number of translated ORFs of unknown function derived from genomic sequencing projects (including *C. neoformans*, but interestingly not *U. maydis* or *C. cinerea*) also exhibit significant similarity to Frt1. All of these other polypeptides are considerably larger than Frt1: there are anywhere from 100 to over 200 additional amino acids upstream of the region of similarity. The overlap region between Frt1 and these other proteins contains both the putative nucleotide-binding region and a lengthy exonuclease domain found in proteins involved in a wide variety of cellular functions, such as transcription, DNA replication, repair and recombination, cell cycle progression, and RNA processing (Moser et al. 1997). However, the localization of Frt1 to the cell wall in hyphae of *S. commune* (Palmer and Horton in preparation), and the absence of several key conserved residues at active sites of the exonuclease domain make it unlikely that Frt1 has a functional domain of this type. In contrast, the other predicted proteins all have the conserved active site residues that Frt1 lacks, and the XPMC2 protein has been localized to the nucleus (Su and Maller 1995). To further define gene function, Δ*frt1* null strains were constructed by homologous recombination and gene replacement. It was found that dikaryons homoallellic for Δ*frt1* fruited normally, indicating that *FRT1* is not essential to the production of fruiting bodies (Horton et al. 1999). Homokaryotic Δ*frt1* null strains exhibited a profuse growth of aerial hyphae, yielding characteristic fluffy colony morphology. Steady-state transcript levels of normally dikaryotic-expressed genes such as *Sc1*, *Sc4*, and *Sc7* were greatly elevated in these Δ*frt1* null strains as compared to wild type (Horton et al. 1999). These findings suggested a rethinking of the role of *frt1*. It is likely that a major function of this gene in homokaryons is to act as a negative regulator of dikaryotic gene expression, an important event in the control of fruiting. Given this function for *frt1*, its' absence in dikaryons would not be expected to impair mushroom development. The peripheral cellular location of Frt1 and its' likely role as a negative regulator of fruiting suggest that there must be additional regulatory elements downstream of *frt1* responsible for the genetic control of mushroom development in this fungus.

Although the mechanism by which *frt1* affects the process of mushroom formation is unclear, the gene has proven useful in identifying some potential players in the genetic control of fruiting. *frt1* cDNA was used as bait in a yeast two-hybrid screen of a *S. commune* cDNA library in order to select clones encoding proteins that potentially interact with Frt1 (Horton et al. 2001). Approximately 7×10^6 clones of the library were screened using the CytoTrap *Ras*-rescue two-hybrid approach (Aronheim et al. 1997; Aronheim 2004). Amongst the putative interactors isolated was a cDNA encoding a likely membrane-spanning sugar transporter with strong similarities to those isolated from yeasts and filamentous fungi. While some members of the sugar transporter superfamily act as receptors for monosaccharides such as glucose, others act as sensors for a variety of nutritional signals (Lalonde et al. 1999; Moriya and Johnston 2004;

Ozcan and Johnston 1999; Versele et al. 2001). For example, the product of the *rco-3* gene of *Neurospora crassa* encodes a glucose sensor protein (Madi et al. 1997). mutants in which affect macroconidia formation, which can serve as either asexual spores or as male gametes. Transcript levels of a sugar transporter were found to markedly increase during mushroom growth in the commercially important species *Agaricus bisporus* (Molloy et al. 2001). What is clear from a variety of studies on yeast and filamentous fungi is that multiple signaling events (pheromones, nutritional signals) are coordinated to regulate developmental processes. The possible interaction between Frt1 and a sugar transporter/sensor protein may provide a molecular link with previous studies describing carbon starvation as a necessary trigger for mushroom development in Basidiomycetes (reviewed in Kues 2000; Kues and Liu 2000).

As stated earlier, light is a necessary environmental cue that regulates mushroom development in both *C. cinerea* and *S. commune*. Several "blind" mutants of *C. cinerea* have been identified (Yuki et al. 2003) which produce etiolated, non-fertile fruiting bodies in the presence of light. This phenotype is normally observed only if cultures are incubated in constant darkness after secondary hyphal knot formation (reviewed in Kues 2000). The *dst1* gene from *C. cinerea* has been identified as a result of the complementation of one of these mutants (Yuki et al. 2003) and appears to encode a protein of high similarity to WC-1, a blue light receptor in the Ascomycete *Neurospora crassa* (Ballario et al. 1996; Froehlich et al. 2002; He et al. 2002). It is hoped that further characterization of this gene will result in the elucidation of important molecular connections between light signaling and mushroom development.

3.3.9 Genes affecting mushroom morphogenesis

In addition to those genes that are instrumental in the initiation of mushroom development, a number of genes have been isolated that play a role in the later stages of mushroom morphogenesis. The *ich1* gene of *C. cinerea* was isolated from a spontaneous recessive mutant that when present in homozygous form in a dikaryon results in a cap-less fruiting body void of spore production. The gene encodes a 1353 amino acid protein predicted to have three nuclear targeting signals, and is expressed specifically in the mushroom cap in wild-type strains (Muraguchi and Kamada 1998). NCBI domain searches revealed a possible S-adenosylmethionine (SAM)-dependent methyltransferase domain in the Ich1 protein. It is unclear how the presence of both this domain and the putative nuclear localization signals in Ich1 can be reconciled. An enzyme from *Aspergillus flavus* sharing this methyltransferase domain is responsible for the conversion of sterigmatocystin to O-methylsterigmatocystin in the aflatoxin biosynthetic pathway (Keller et al. 1993). Aflatoxin is likely to play a role in *Aspergillus* development: when production is blocked, development of sclerotia and basidiospores are also inhibited (Guzman-de Pena and Ruiz-Herrera 1997; Trail et al. 1995). Kues (2000) speculates that Ich1 might help produce an analogous substance to aflatoxin, and that this substance may play a role in signaling within the fruiting body primordia. It is interesting to point out that the predicted protein product for the essential fruiting gene *cfs1* from *C. cinerea* also has a SAM-dependent methyltransferase domain, although the substrate for this enzyme is likely some sort of membrane phospholipid (Kues et al. 2004). BLAST searches revealed that the most significant similarity (E value = 5×10^{-34}) to Ich1 was to a hypothetical protein of unknown function translated from the *U. maydis* genome. Other significant hits were largely to plant cell surface proteins, due to the proline-rich C-terminal third of the Ich1 protein.

Two different mutants producing mushrooms with abnormally short stipes have been used to isolate the *eln2* and *eln3* genes in *C. cinerea*. *eln2* encodes a novel type of microsomal cytochrome P450, and is constitutively expressed in both vegetative mycelia and developing mushrooms (Muraguchi and Kamada, 2000). The substrate(s) and the metabolite(s) of this particular enzymatic activity are unknown at present, as is the mechanism of how this protein participates in the process of mushroom morphogenesis. A cytochrome P450 gene has been reported to be specifically expressed in fruiting bodies of the commercial mushroom *Agaricus bisporus* (de Groot et al. 1997). *eln3* encodes a predicted transmembrane protein with a general glycotransferase domain, and has high similarity to a number of hypothetical proteins from filamentous fungi (Arima et al. 2004). Kamada and co-workers have put forth the notion that the disorganization of stipe tissue in the loss of function mutant *eln3* is the result of a defect in the mechanism responsible for cell-to-cell connections in that tissue. Eln3 might be involved in adding a carbohydrate moiety onto extracellular matrix proteins, which might be necessary for proper cellular organization within the stipe of the mushroom (Arima et al. 2004).

3.3.10 Structural genes

As mentioned earlier, the aggregation of hyphae is a prerequisite for fruiting body formation, and most certainly involves cell wall and extracellular matrix proteins (reviewed by Walser et al. 2003). Carbohydrate-binding lectins have been strongly implicated in both hyphal aggregation and mushroom development (reviewed in Wang et al. 1998). The genes *cgl1* and *cgl2* encode fruiting-specific galectins regulated by the *A* mating-type pathway in *C. cinerea* (Cooper et al. 1997; Boulianne et al. 2000). Galectins are a class of lectin that specifically bind β-galactoside sugars in a calcium-independent manner, and are conserved in sequence within their carbohydrate-recognition domain (Barondes et al. 1994 a; Barondes et al. 1994 b). The two galectin genes are differentially expressed during fruiting body development. Cgl1 is specifically expressed in primordia and mature fruiting bodies, while Cgl2 is initiated earlier, at the time of hyphal knot formation, and maintained until fruiting body maturation (Boulianne et al. 2000). This differential expression between the two galectins may in part be explained by the presence of two cAMP response elements (CREs, Conkright et al. 2003) in the promoter of *cgl2*, and their absence in *cgl1* (Bertossa et al. submitted). *cgl2* expression occurs concomitantly with a rise in cAMP levels at the time of hyphal knot formation, whereas *cgl1* expression is only observed during the formation of fruiting body primordia, after cAMP levels peak (Bottoli 2001; Boulianne et al. 2000). Interestingly, each of the *C. cinerea* galectin genes have a promoter element ($GATGX_{11}CAA$) that is a potential binding site for the heterodimer of HD protein products of the *A* mating-type genes, and identical to that seen in the *C. cinerea* regulatory gene *clp1*. However, the replacement of this element by an unrelated linker sequence in the *cgl2* promoter did not influence transcription of a heterologous reporter gene (Bertossa et al. submitted). This result suggests that transcriptional regulation of *cgl2* (and perhaps by inference, *cgl1*) by A mating-type proteins is likely to be indirect. Both galectins have been localized primarily to the cell wall and extracellular matrix (ECM), but because they lack an obvious signal sequence, they are secreted by a non-classical pathway (Boulianne et al. 2000). Given the developmental regulation of their expression and their peripheral cellular localization, it would seem reasonable to propose that galectins may play a key role in the hyphal aggregation process necessary for fruiting body development. To date, neither *cgl1* nor *cgl2* have been disrupted, so no galectin-deficient strains have been generated to assess whether or not they are essential to mushroom development.

Another important class of structural proteins found in mushrooms, and indeed aerial structures of many fungi are the hydrophobins (reviewed in Wosten 2001; Wessels 1996). Hydrophobins were originally discovered in *S. commune* as products of genes that were highly expressed during the formation of fruiting bodies, and are small secreted proteins that contain eight cysteine residues in a conserved pattern (Mulder and Wessels 1986; Schuren and Wessels 1990). These proteins self-assemble at the outer surface of hyphae, covering them with an amphipathic membrane that lowers surface tension at the water/air interface, which in turn enables the hyphae to emerge into the air (Wosten et al. 1999). *S. commune* contains at least four hydrophobin genes, *sc1*, *sc6*, *sc3*, and *sc4*, with the latter two being the best characterized. The SC3 protein is expressed in both homokaryons and dikaryons, while SC4 is dikaryon-specific (Wessels et al. 1995). SC3 is found in hyphae covering the outer surface of the fruiting bodies (Asgeirsdottir et al. 1995) whereas SC4 lines the air channels within (Lugones et al. 1999). Knockouts of *sc3* and *sc4* have shown that neither gene is absolutely necessary for the formation of aerial hyphae and fruiting bodies. In Δ*sc3* and Δ*sc3*Δ*sc4* dikaryons fruiting was delayed, and the number of mushrooms were decreased by 50-80%, as compared to wild-type or Δ*sc4* strains (van Wetter et al. 2000). The greater effect of the Δ*sc3* knockout on fruiting is somewhat surprising, given that SC4 is localized exclusively in fruiting bodies, and SC3 is not.

3.3.11 Future prospects

While it is safe to say that the molecular genetic study of mushroom development is still in it's early stages, the next decade promises to be a time where great strides will be made in our understanding of this process so fundamental to sexual reproduction in many Basidiomycete species. Connections are starting to be made between known elements of signaling pathways and the molecular mechanisms controlling fruiting. The long documented link between cAMP metabolism and mushroom development is finally getting some molecular genetic underpinning. Of high importance will be defining the roles of the individual elements of the cAMP and MAPK pathways, and integrating these into a coherent conceptual framework as they relate to each other, and mushroom development. There are now some encouraging initial forays into exploring the molecular links between known environmental cues of fruiting, such as light and nutrition, and the initiation of the process itself. The comparison of the molecular genetic regulation of fruiting in two discrete model systems will be of particular use in applying these concepts to commercially important mushroom species.

Many of the tools used to date to explore the genetics of mushroom development will continue in their useful role. Mutants defective in the process will help elucidate still more of the key elements, especially in those cases concerning genes that might be unique to the mushroom-producing fungi, and therefore recalcitrant to comparative genomics approaches. The isolation of suppressor mutants has been under-utilized in the molecular study of fruiting. Existing experimental approaches will be accompanied by new resources such as the *Coprinopsis cinerea* genome, microarray analyses, the development of GFP-tagging for cytological studies, and RNAi for examining gene function, amongst others. There is every reason to be optimistic that our understanding of the genetic mechanisms controlling mushroom development will grow substantially in the near future.

4. CONCLUSIONS

In this chapter, we have presented an overview of the current knowledge of the genetic control of morphogenesis in several exemplary basidiomycete systems: smuts, rusts and mushrooms. The discussion presented herein reveals a staggering level of complexity and a multitude of questions that remain unanswered in this fascinating field. Therefore, it is highly encouraging and quite apparent that the future directions of research using the above-mentioned model systems will utilize the newly acquired and/or rapidly expanding genomic and EST database resources. Coupled with newly acquired tools such as microarray-based studies and general improved technology for transformation and gene specific mutagenesis, these resources will further enhance and rapidly accelerate these studies, providing a more unified model of basidiomycete development. As discussed in this chapter, such approaches have already begun to yield rapid advances of our knowledge in this area.

Acknowledgements: J.S.H. would like to acknowledge Gail Palmer, Bill Smith, Alan Hebert, Alaap Shah, Ben Wormer, Esther Ndungo, Josh White, and Pete Sage for their contributions to published and unpublished work on the genetics of mushroom development in *Schizophyllum*. Ursula Kues is acknowledged for making her unpublished work on *Coprinus* available. J.S.H. is supported by an NIH AREA grant 1 R15 GM63507-01. G.B. would like to acknowledge the actual experimentation and bioinformatics analysis on the preliminary data of the *P. triticina* ESTs from Guanggan Hu and Rob Linning, respectively, and helpful discussions with Drs. Les Szabo, Michele Heath and James Chong. Although the scope of this review is extensive, the authors apologize for unintended omissions of valuable and relevant works from our colleagues in the field.

REFERENCES

Abramovitch RB, Yang G and Kronstad JW (2002). The ukb1 gene encodes a putative protein kinase required for bud site selection and pathogenicity in Ustilago maydis. Fungal Genet Biol 37:98-108.

Adamikova L, Straube A, Schulz I and Steinberg, G (2004). Calcium signaling is involved in dynein-dependent microtubule organization. Mol Biol Cell 15:1969-1980.

Agrios GN (1997). Plant Pathology. 4th ed San Diego, CA, Academic Press.

Aichinger, C, Hansson, K, Eichhorn, H, Lessing, F, Mannhaupt, G, Mewes, W, and Kahmann R (2003). Identification of plant-regulated genes in Ustilago maydis by enhancer-trapping mutagenesis. Mol Genet Genomics 270:303-314.

Aime MC and Casselton L (2002). *hmg1* encodes an HMG domain protein implicated in nuclear migration in *Coprinus cinereus*. Paper presented at: 6th European Conference on Fungal Genetics (Pisa, Italy).

Alexopolous CJ, Mims CW, Blackwell M (1996). Introductory Mycology 4th ed. NY, John Wiley and Sons.

Alspaugh JA, Cavallo LM, Perfect JR and Heitman J (2000). RAS1 regulates filamentation, mating and growth at high temperature of *Cryptococcus neoformans*. Mol Microbiol 36:352-365.

Altschul SF, Madden TL, Schaffer AA, Zhang J, Zhang Z, Miller W, and Lipman DJ (1997). Gapped BLAST and PSI-BLAST: a new generation of protein database search programs. Nucleic Acids Res 25:3389-3402.

Andrews DL, Egan JD, Mayorga ME and Gold SE (2000). The Ustilago maydis ubc4 and ubc5 genes encode members of a MAP kinase cascade required for filamentous growth. Mol Plant Microbe Interact 13:781-786.

Andrews DL, Garcia-Pedrajas MD and Gold SE (2004). Fungal dimorphism regulated gene expression in Ustilago maydis: I. Filament up-regulated genes. Molecular Plant Pathology 5:281-293.

Anikster Y (1983). Binucleate basidiospores; a general rule in rust fungi. Trans Br Mycol Soc 81:624-626.

Anikster Y, Eilam T, Mittelman L, Szabo LJ and Bushnell WR (1999). Pycnial nectar of rust fungi induces cap formation on pycniospores of opposite mating type. Mycologia 91:858-870.

Arima T, Yamamoto M, Hirata A, Kawano S and Kamada T (2004). The *eln3* gene involved in fruiting body morphogenesis of *Coprinus cinereus* encodes a putative membrane protein with a general glycosyltransferase domain. Fungal Genet Biol 41:805-812.

Aronheim A (2004). Ras signaling pathway for analysis of protein-protein interactions in yeast and mammalian cells. Methods Mol Biol 250:251-262.

Aronheim A, Zandi E, Hennemann H, Elledge SJ and Karin M (1997). Isolation of an AP-1 repressor by a novel method for detecting protein-protein interactions. Mol Cell Biol 17:3094-3102.

Arz MC and Grambow HJ (1995). Elicitor and suppressor effects on phospholipase C in isolated plasma membranes correlate with alterations in phenylalanine ammonia-lyase activity of wheat leaves. J Plant Physiol 146:64-70.

Asgeirsdottir SA, van Wetter MA and Wessels JGH (1995). Differential expression of genes under control of the mating-type genes in the secondary mycelium of *Schizophyllum commune*. Microbiology 141:1281-1288.

Ayliffe MA, Lawrence GJ, Ellis JG, and Pryor AJ (1997). Production of a self-inhibitor of urediospore germination in Melampsora lini (flax rust) segregates as a recessive, single gene trait. Can J Bot 75:74-76.

Badalyan SM, Polak E, Hermann R, Aebi M and Kues U (2004). Role of peg formation in clamp cell fusion of homobasidiomycete fungi. J Basic Microbiol 44:167-177.

Bakkeren G, Gibbard B, Yee A, Froeliger E, Leong S and Kronstad J (1992). The a and b loci of Ustilago maydis hybridize with DNA sequences from other smut fungi. Mol Plant Microbe Interact 5:347-355.

Bakkeren G and Gold S (2004). The path in fungal plant pathogenicity: many opportunities to outwit the intruders? In: Genet Eng, J. K. Setlow, ed. New York: Kluwer Academic/Plenum Publishers, pp. 175-223.

Bakkeren G and Kronstad, JW (1993). Conservation of the b mating-type gene complex among bipolar and tetrapolar smut fungi. Plant Cell 5:123-136.

Bakkeren G and Kronstad JW (1994). Linkage of mating-type loci distinguishes bipolar from tetrapolar mating in basidiomycetous smut fungi. Proc Natl Acad Sci U S A 91:7085-7089.

Bakkeren G and Kronstad JW (1996). The pheromone cell signaling components of the Ustilago a mating-type loci determine intercompatibility between species. Genetics 143:1601-1613.

Ballario P, Vittorioso P, Magrelli A, Talora C, Cabibbo A and Macino G (1996). White collar-1, a central regulator of blue light responses in *Neurospora*, is a zinc finger protein. Embo J 15:1650-1657.

Banuett F (1998). Signalling in the yeasts: an informational cascade with links to the filamentous fungi. Microbiol Mol Biol Rev 62:249-274.

Banuett F and Herskowitz I (1989). Different a alleles of Ustilago maydis are necessary for maintenance of filamentous growth but not for meiosis. Proc Natl Acad Sci USA 86:5878-5882.

Banuett F and Herskowitz I (1994). Identification of fuz7, a Ustilago maydis MEK/MAPKK homolog required for a-locus-dependent and -independent steps in the fungal life cycle. Genes Dev 8:1367-1378.

Banuett F and Herskowitz I (1996). Discrete developmental stages during teliospore formation in the corn smut fungus, Ustilago maydis. Development 122:2965-2976.

Banuett F and Herskowitz I (2002). Bud morphogenesis and the actin and microtubule cytoskeletons during budding in the corn smut fungus, Ustilago maydis. Fungal Genet Biol 37:149-170.

Barja F, Correa A Jr, Staples RC and Hoch HC (1998). Microinjected antisense Inf24 oligonucleotides inhibit appressorium development in Uromyces. Mycological Res 102:1513-1518.

Barondes SH, Castronovo V, Cooper DN, Cummings RD, Drickamer K, Feizi T, Gitt MA, Hirabayashi J, Hughes C and Kasai K (1994 a). Galectins: a family of animal beta-galactoside-binding lectins. Cell 76:597-598.

Barondes SH, Cooper DN, Gitt MA and Leffler H (1994 b). Galectins. Structure and function of a large family of animal lectins. J Biol Chem 269:20807-20810.

Barrett KJ, Gold SE and Kronstad JW (1993). Identification and complementation of a mutation to constitutive filamentous growth in Ustilago maydis. Mol Plant Microbe Interact 6:274-283.

Basse CW, Kolb S and Kahmann R (2002). A maize-specifically expressed gene cluster in Ustilago maydis. Mol Microbiol 43:75-93.

Basse CW and Steinberg G (2004). Ustilago maydis, model system for analysis of the molecular basis of fungal pathogenicity. Mol Plant Pathol 5:83-92.

Basse CW, Stumpferl S and Kahmann R (2000). Characterization of a Ustilago maydis gene specifically induced during the biotrophic phase: Evidence for negative as well as positive regulation. Mol Cell Biol 20:329-339.

Baum BR and Savile DBO (1985). Rusts (Uredinales) of Triticeae: evolution and extent of coevolution, a cladistic analysis. Bot J Linn Soc 91:367-394.

Beissmann B, Engels W, Kogel K, Marticke KH and Reisener HJ (1992). Elicitor-active glycoproteins in apoplastic fluids of stem-rust-infected wheat leaves. Physiol Mol Plant Pathol 40:79-89.

Berbee ML and Taylor JW (1993). Dating the evolutionary radiations of the true fungi. Can J Bot 71:1114-1127.

Bertossa RC, Kues U, Aebi M and Kunzler M (submitted). Promoter analysis of cgl2, a galectin encoding gene transcribed during fruiting body formation in Coprinopsis cinerea (Coprinus cinereus).

Bhairi SM, Laccetti L and Staples RC (1990). Effect of heat shock on expression of thigmo-specific genes from a rust fungus. Exp Mycology 14:94-98.

Bhairi SM and Staples RC (1992). Transient expression of the beta-glucuronidase gene introduced into Uromyces appendiculatus uredospores by particle bombardment. Phytopathology 82:986-989.

Bhairi SM, Staples RC, Freve P and Yoder OC (1989). Characterization of an infection structure-specific gene from the rust fungus Uromyces appendiculatus. Gene 81:237-243.

Binninger DM, Le Chevanton L, Skrzynia C, Shubkin CD and Pukkila PJ (1991). Targeted transformation in Coprinus cinereus. Mol Gen Genet 227:245-251.

Binninger DM, Skrzynia C, Pukkila PJ and Casselton LA (1987). DNA-mediated transformation of the basidiomycete Coprinus cinereus. Embo J 6:835-840.

Boasson R and Shaw M (1988). The effects of glutamine citrate and PH on the growth and sporulation of Melampsora lini in axenic culture. Can J Bot 66:1230-1236.

Bohland C, Balkenhohl T, Loers G, Feussner I and Grambow HJ (1997). Differential induction of lipoxygenase isoforms in wheat upon treatment with rust fungus elicitor, chitin oligosaccharides, chitosan, and methyl jasmonate. Plant Physiol 114:679-685.

Bolker M (1992). The a mating type locus of U. maydis specifies cell signaling components. Cell 68:441-450.

Bonde MR, Melching JS and Bromfield KR (1976). Histology of the suscept pathogen relationship between Glycine max and Phakopsora pachyrhizi the cause of soybean rust. Phytopathology 66:1290-1294.

408

Bose A and Shaw M (1974). Growth of rust fungi of wheat and flax on chemically-defined media. Nature 251:646-648.

Bottoli AP, Kertesz-Chaloupkova K, Boulianne RP, Granado JD, Aebi M and Kues U (1999). Rapid isolation of genes from an indexed genomic library of C. cinereus in a novel pab1+ cosmid. J Microbiol Methods 35:129-141.

Bottoli APF (2001). Metabolic and environmental control of development in Coprinus cinereus. Ph.D. thesis, ETH Zurich, Zurich, Switzerland.

Bottoli APF, Boulianne RP, Aebi M and Kues U (2001). Characterization of ras mutant alleles from Coprinus cinereus homokaryon Amut Bmut. Paper presented at: 21st Fungal Biology Conference (Pacific Grove, CA, USA).

Boulianne RP, Liu Y, Aebi M, Lu BC and Kues U (2000). Fruiting body development in Coprinus cinereus: regulated expression of two galectins secreted by a non-classical pathway. Microbiology 146 (Pt 8):1841-1853.

Brachmann A, Schirawski J, Muller P and Kahmann R (2003). An unusual MAP kinase is required for efficient penetration of the plant surface by Ustilago maydis. Embo Journal 22:2199-2210.

Brachmann A, Weinzierl G, Kamper J and Kahmann R (2001). Identification of genes in the bW/bE regulatory cascade in Ustilago maydis. Mol Microbiol 42:1047-1063.

Brefort T, Muller P and Kahmann R (2003). The HMG-box protein Rop1 is essential for pheromone-responsive gene expression in Ustilago maydis. Paper presented at: 22nd Fungal Biology Conference (Pacific Grove, CA, USA).

Brefort T, Muller P and Kahmann R (2004). There is more than one HMG-box transcription factor regulating mating in Ustilago maydis. Paper presented at: 7th European Conference on Fungal Genetics (Copenhagen, Denmark).

Brown AJ and Casselton LA (2001). Mating in mushrooms: increasing the chances but prolonging the affair. Trends in Genetics 17:393-400.

Brown JKM (2003). A cost of disease resistance: Paradigm or peculiarity? Trends in Genetics 19:667-671.

Brown WM Jr, Hill JP and Velasco VR (2001). Barley yellow rust in North America. Annu Rev Phytopathol 39:367-384.

Broxholme SJ, Read ND and Bond DJ (1991). Developmental regulation of proteins during fruiting body morphogenesis of Sordaria brevicollis. Mycol Res 95:958-969.

Buller AHR (1933). Researches on Fungi. V. Hyphal Fusions and Protoplasmic Streaming in the Higher Fungi, Together with an Account of the Production and Liberation of Spores in Sporobolomyces, Tilletia, and Sphaerobolus. New York, NY: Hafner Publishing Co.

Burdon JJ and Roberts JK (1995). The population genetic structure of the rust fungus Melampsora lini as revealed by pathogenicity, isozyme and RFLP markers. Plant Pathology 44:270-278.

Bushnell WR and Roelfs AP eds. (1984). The cereral rusts. Vol. I. Origins, specificity, structure and physiology. Orlando, FL: Academic Press, Inc.

Butler MJ and Day AW (1998). Fungal melanins: A review. Can J Microbiol 44:1115-1136.

Callow JA and Ling IT (1973). Histology of neoplasms and chlorotic lesions in maize seedlings following injection of sporidia of Ustilago-maydis (Dc) Corda. Physiol Plant Pathol 3:489-&.

Cantrill LC and Deverall BJ (1993). Isolation of haustoria from wheat leaves infected by the leaf rust fungus. Physiol Mol Plant Pathol 42:337-341.

Casselton LA (2002). Mate recognition in fungi. Heredity 88:142-147.

Casselton LA and Olesnicky NS (1998). Molecular genetics of mating recognition in basidiomycete fungi. Microbiol Mol Biol Rev 62:55-70.

Chang ST (1996). Mushroom research and development - equality and mutual benefit. Paper presented at: 2nd International conference on mushroom biology and mushroom products. University Park, PA: Penn State University Press.

Chaubal R, Wilmot VA and Wynn WK (1991). Visualization adhesiveness and cytochemistry of the extracellular matrix produced by urediniospore germ tubes of Puccinia-sorghi. Can J Bot 69:2044-2054.

Chen CY and Heath MC (1990). Cultivar-specific induction of necrosis by exudates from basidiospore germlings of the cowpea rust fungus. Physiol Mol Plant Pathol 37:169-178.

Chen CY and Heath MC (1993). Inheritance of resistance to Uromyces vignae in cowpea and the correlation between resistance and sensitivity to a cultivar-specific elicitor of necrosis. Phytopathology 83:224-230.

Chen X and Line R-F (2003). Identification of genes for resistance to Puccinia striiformis f. sp. hordei in 18 barley genotypes. Euphytica 129:127-145.

Chen X, Stone M, Schlagnhaufer C and Romaine CP (2000). A fruiting body tissue method for efficient Agrobacterium-mediated transformation of Agaricus bisporus. Appl Environ Microbiol 66:4510-4513.

Chong J, Kang Z, Kim WK and Rohringer R (1992). Multinucleate condition of Puccinia-striiformis in colonies isolated from infected wheat leaves with macerating enzymes. Can J Bot 70:222-224.

Chong J, Leonard KJ and Salmeron JJ (2000). A North American system of nomenclature for Puccinia coronata f. sp. avenae. Plant Disease 84:580-585.

Christopher Kozjan R and Heath MC (2003). Cytological and pharmacological evidence that biotrophic fungi trigger different cell death execution processes in host and nonhost cells during the hypersensitive response. Physiol Mol Plant Pathol 62:265-275.

Clement JA, Porter R, Butt TM and Beckett A (1994). The role of hydrophobicity in attachment of urediniospores and sporelings of Uromyces viciae-fabae. Mycol Res 98:1217-1228.

Clement JA, Porter R, Butt TM and Beckett A (1997). Characteristics of adhesion pads formed during imbibition and germination of urediniospores of Uromyces viciae-fabae on host and synthetic surfaces. Mycological Res 101:1445-1458.

Clergeot P-H, Ruprich-Robert G, Liu Y, Loos S, Srivilai P, Velagapudi R, Goebel S, Kunzler M, Aebi M and Kues U (2003). Mutants in initiation of fruiting body development of the basidiomycete Coprinus cinereus. Paper presented at: 22nd Fungal Biology Conference (Pacific Grove, CA, USA).

Conkright MD, Guzman E, Flechner L, Su AI, Hogenesch JB and Montminy M (2003). Genome-wide analysis of CREB target genes reveals a core promoter requirement for cAMP responsiveness. Mol Cell 11:1101-1108.

Cooper DN, Boulianne RP, Charlton S, Farrell EM, Sucher A and Lu BC (1997). Fungal galectins, sequence and specificity of two isolectins from Coprinus cinereus. J Biol Chem 272:1514-1521.

Correa A Jr, Staples RC and Hoch HC (1996). Inhibition of thigmostimulated cell differentiation with RGD-peptides in Uromyces germlings. Protoplasma 194:91-102.

D'Silva I and Heath MC (1997). Purification and characterization of two novel hypersensitive response-inducing specific elicitors produced by the cowpea rust fungus. J Biol Chem 272:3924-3927.

D'Souza CA and Heitman J (2001). Conserved cAMP signaling cascades regulate fungal development and virulence. Fems Microbiology Reviews 25:349-364.

Dawe AL and Nuss DL (2001). Hypoviruses and chestnut blight: exploiting viruses to understand and modulate fungal pathogenesis. Annu Rev Genet 35:1-29.

Day PR and Anagnostakis SL (1971). Corn Smut Dikaryon in Culture. Nature-New Biology 231:19-20.

de Groot MJ, Bundock P, Hooykaas PJ and Beijersbergen AG (1998). Agrobacterium tumefaciens-mediated transformation of filamentous fungi. Nat Biotechnol 16:839-842.

de Groot PW, Schaap PJ, Van Griensven LJ and Visser J (1997). Isolation of developmentally regulated genes from the edible mushroom Agaricus bisporus. Microbiology 143 (Pt 6):1993-2001.

Deising H, Nicholson RL, Haug M, Howard RJ and Mendgen K (1992). Adhesion pad formation and the involvement of cutinase and esterases in the attachment of uredospores to the host cuticle. Plant Cell 4:1101-1111.

Deising H, Rauscher M, Haug M and Heiler S (1995). Differentiation and cell wall degrading enzymes in the obligately biotrophic rust fungus Uromyces viciae-fabae. Can J Bot 73:S624-S631.

Dodds PN, Lawrence GJ, Catanzariti AM, Ayliffe MA and Ellis JG (2004). The Melampsora lini AvrL567 avirulence genes are expressed in haustoria and their products are recognized inside plant cells. Plant Cell 16:755-768.

Durrenberger F, Wong K and Kronstad JW (1998). Identification of a cAMP-dependent protein kinase catalytic subunit required for virulence and morphogenesis in Ustilago maydis. Proc Natl Acad Sci U S A 95:5684-5689.

El Gueddari NE, Rauchhaus U, Moerschbacher BM and Deising HB (2002). Developmentally regulated conversion of surface-exposed chitin to chitosan in cell walls of plant pathogenic fungi. New Phytol 156:103-112.

Ellis J, Lawrence G, Ayliffe M, Anderson P, Collins N, Finnegan J, Frost D, Luck J and Pryor T (1997). Advances in the molecular genetic analysis of the flax-flax rust interaction. Ann Rev Phytopathol 35:271-291.

Epstein L, Laccetti LB, Staples RC and Hoch HC (1987). Cell-substratum adhesive protein involved in surface contact responses of the bean rust fungus. Physiol Mol Plant Pathol 30:373-388.

Epstein L, Staples RC and Hoch HC (1989). Cyclic AMP, Cyclic GMP and bean rust uredospore germlings. Exp Mycol 13:100-104.

Esser K, Saleh F and Meinhardt F (1979). Genetics of fruit body production in higher Basidiomycetes. II. Monokayotic and dikaryotic fruiting in Schizophyllum commune. Curr Genet 1:85-88.

Essig FM (1922). The morphology, development, and economic aspects of Schizophyllum commune Fries. University of California Publications in Botany 7:447-498.

Fasters MK, Daniels U and Moerschbacher BM (1993). A simple and reliable method for growing the wheat stem rust fungus, Puccinia graminis f. sp. tritici, in liquid culture. Physiological and Mol Plant Pathol 42:259-265.

Fischer R and Kues U (2003). Developmental Processes in Filamentous Fungi. In Genomics of Plants and Fungi, R. A. Prade, and H. J. Bohnert, eds. New York: Marcel Dekker, Inc. pp. 41-118.

Flor HH (1971). Current status of the gene-for-gene concept. Annu Rev Phytopathol 9:275-296.

Fowler TJ and Mitton MF (2000). Scooter, a new active transposon in Schizophyllum commune, has disrupted two genes regulating signal transduction. Genetics 156:1585-1594.

Fowler TJ, Mitton MF, Rees EI and Raper CA (2004). Crossing the boundary between the B alpha and B beta mating-type loci in Schizophyllum commune. 41:89-101.

Fraser JA and Heitman J (2004). Evolution of fungal sex chromosomes. 51:299-306.

French RC (1992). Volatile chemical germination stimulators of rust and other fungal spores. Mycologia 84:277-288.

French RC, Nester SE and Stavely JR (1993). Stimulation of germination of teliospores of Uromyces appendiculatus by volatile aroma compounds. J agric food chem 41:1743-1747.

Freytag S and Mendgen K (1991). Carbohydrates on the surface of urediniospore- and basidiospore-derived infection structures of heteroecious and autoecious rust fungi. New Phytol 119:527-534.

Froehlich AC, Liu Y, Loros JJ and Dunlap JC (2002). White Collar-1, a circadian blue light photoreceptor, binding to the frequency promoter. Science 297:815-819.

Froeliger EH and Leong SA (1991). The a mating-type alleles of Ustilago maydis are idiomorphs. Gene 100:113-122.

Garcia-Muse T, Steinberg G and Perez-Martin J (2003). Pheromone-induced G2 arrest in the phytopathogenic fungus Ustilago maydis. Eukaryot Cell 2:494-500.

Garcia-Muse T, Steinberg G and Perez-Martin J (2004). Characterization of B-type cyclins in the smut fungus Ustilago maydis: roles in morphogenesis and pathogenicity. J Cell Sci 117:487-506.

Garcia-Pedrajas MD and Gold SE (2004). Fungal dimorphism regulated gene expression in Ustilago maydis: II. Filament down-regulated genes. Mol Plant Pathol 5:295-307.

Garrido E and Perez-Martin J (2003). The crk1 gene encodes an Ime2-related protein that is required for morphogenesis in the plant pathogen Ustilago maydis. Mol Microbiol 47:729-743.

Gitzendanner MA, White EE, Foord BM, Dupper GE, Hodgskiss PD, and Kinloch BB Jr. (1996). Genetics of Cronartium ribicola. III. Mating system. Can J Bot 74: 1852-1859.

Goates BJ and Hoffmann JA (1987). Nuclear behavior during teliospore germination and sporidial development in Tilletia caries, Tilletia foetida, and Tilletia controversa. Can J Bot 65:512-517.

Gold RE and Littlefield LJ (1979). Light microscopy and scanning electron microscopy of the telial, pycnial and aecial stages of Melampsora-lini. Can J Bot 57:629-638.

Gold, RE, and Mendgen, K (1984). Cytology of basidiospore germination, penetration, and early colonization of Phaseolus vulgaris by Uromyces appendiculatus var. appendiculatus. Can J Bot 62:1989-2002.

Gold RE and Mendgen K (1991). Rust basidiospore germlings and disease initiation. The Fungal spore and disease initiation in plants and animals / edited by Garry T Cole and Harvey C Hoch New York: Plenum Press, c1991 p:67-99.

Gold S, Duncan G, Barrett K and Kronstad J (1994). cAMP regulates morphogenesis in the fungal pathogen Ustilago maydis. Genes Dev 8:2805-2816.

Gold SE, Brogdon SM, Mayorga ME and Kronstad JW (1997). The Ustilago maydis regulatory subunit of a cAMP-dependent protein kinase is required for gall formation in maize. Plant Cell 9:1585-1594.

Goutte C and Johnson AD (1988). a1 protein alters the DNA binding specificity of alpha 2 repressor. Cell 52:875-882.

Granado JD, Kertesz-Chaloupkova K, Aebi M and Kues U (1997). Restriction enzyme-mediated DNA integration in Coprinus cinereus. Mol Gen Genet 256:28-36.

Grogan DW and Cronan JE Jr (1997). Cyclopropane ring formation in membrane lipids of bacteria. Microbiol Mol Biol Rev 61:429-441.

Guevara-Olvera L, Xoconostle-Cazares B and Ruiz-Herrera J (1997). Cloning and disruption of the ornithine decarboxylase gene of Ustilago maydis: evidence for a role of polyamines in its dimorphic transition. Microbiol 143:2237-2245.

Guzman-de-Pena D and Ruiz-Herrera J (1997). Relationship between aflatoxin biosynthesis and sporulation in Aspergillus parasiticus. Fungal Genet Biol 21:198-205.

Hana WF (1929). Studies in the physioloogy and cytology of Ustilago zeae and Sorosporium reilianum. Phytopathology 19:415-442.

Haerter AC and Voegele RT (2004). A novel beta-glucosidase in Uromyces fabae: feast or fight? Curr Genet 45:96-103.

Hahn M (2000). The rust fungi. Cytology, physiology and molecular biology of infection. In: Fungal pathology., J. W. Kronstad, ed. Dordrecht, Kluwer: Academic Publishers, pp. 267-306.

Hahn M and Mendgen K (1992). Isolation of ConA binding of haustoria from different rust fungi and comparison of their surface qualities. Protoplasma 170:95-103.

Hahn M and Mendgen K (1997). Characterization of in planta-induced rust genes isolated from a haustorium -specific cDNA library. Mol Plant Microbe Interact 10:427-437.

Hahn M, Neef U, Struck C, Gottfert M and Mendgen K (1997). A putative amino acid transporter is specifically expressed in haustoria of the rust fungus Uromyces fabae. Mol Plant Microbe Interact 10:438-445.

Hammad F, Watling R and Moore D (1993). Cell population dynamics in Coprinus cinereus: narrow and inflated hyphae in the basidiome stipe. Mycological Res 97:275-282.

Harder, DE (1984). Developmental ultrasructure of Hyphae and spores. In: The cerreal rusts. Vol. I. Origins, specificity, structure and physiology., W. R. Bushnell, and A. P. Roelfs, eds. Orlando, FL: Academic Press, Inc. pp. 333-373.

Harder DE (1989). Rust fungal haustoria--past, present, future. Can J Plant Pathol 11:91-99.

Harder DE and Chong J (1991). Rust haustoria. In Electron microscopy of plant pathogens., K. Mendgen, and D. E. Lesemann, eds. Berlin: Springer-Verlag, pp. 235-250.

Hartmann HA, Kahmann R and Bolker M (1996). The pheromone response factor coordinates filamentous growth and pathogenicity in Ustilago maydis. Embo J 15:1632-1641.

Hartmann HA, Kruger J, Lottspeich F and Kahmann R (1999). Environmental signals controlling sexual development of the corn Smut fungus Ustilago maydis through the transcriptional regulator Prf1. Plant Cell 11:1293-1306.

He Q, Cheng P, Yang Y, Wang L, Gardner KH and Liu Y (2002). White collar-1, a DNA binding transcription factor and a light sensor. Science 297:840-843.

Heath MC (1972). Ultrastructure of host and nonhost reactions to cowpea rust. Phytopathology 62:27-38.

Heath MC (1989). In-itro formation of haustoria of the cowpea rust fungus Uromyces-Vignae in the absence of a living plant cell, I. Light Microscopy. Physiol Mol Plant Pathol 35:357-366.

Heath MC (1997). Signalling between pathogenic rust fungi and resistant or susceptible host plants. Ann bot 80:713 -720.

Heath MC (2002). Cellular interactions between biotrophic fungal pathogens and host or nonhost plants. Can J Plant Pathol 24:259-264.

Heath MC and Heath IB (1971). Ultrastructure of an immune and a susceptible reaction of cowpea leaves to rust infection. Physiol Plant Pathol 1:277-287.

Heath MC and Perumalla CJ (1988). Haustorial mother cell development by Uromyces vignae on collodion membranes. Can J Bot 66:736-741.

Heath MC and Skalamera D (1997). Cellular interactions between plants and biotrophic fungal parasites. Adv Bot Res 24:195-225.

Heath MC, Xu H and Eilam T (1996). Nuclear behavior of the cowpea rust fungus during the early stages of basidiospore- or urediospore-derived growth in resistant or susceptible cowpea cultivars. Phytopathology 86:1057-1065.

Heiler S, Mendgen K and Deising H (1993). Cellulolytic enzymes of the obligately biotrophic rust fungus Uromyces viciae fabae are regulated differentiation-specifically. Mycological Res 97:77-85.

Hiratsuka Y, Samoil JK, Blenis PV, Crane PE and Laishley BL, eds. (1991). Rusts of pine. Edmonton, Forestry Canada, Northern Forestry Centre.

Hirt RR (1964) Cronartium ribicola: Its growth and reproduction in the tissues of eastern white pine. Vol 86 STAT U Coll., Syracuse.

Hoch HC, Bourett TM and Staples RC (1986). Inhibition of cell differentiation in Uromyces appendiculatus with deuterium oxide and taxol. Eur J Cell Biol 41:290-297.

Hoch HC and Staples RC (1984). Evidence that cyclic AMP initiates nuclear division and infection structure formation in the bean rust fungus Uromyces-Phaseoli. Exp Mycology 8:37-46.

Hoch HC and Staples RC (1987). Structural and chemical changes among the rust fungi during appressorium development. Annu Rev Phytopathol 25:231-247.

Hoch HC Staples RC and Bourett T (1987 a). Chemically induced appressoria in Uromyces appendiculatus are formed aerially apart from the substrate. Mycologia 79:418-424.

Hoch HC, Staples RC, Whitehead B Comeau J and Wolf ED (1987 b). Signaling for growth orientation and cell differentiation by surface topography in Uromyces. Science 235:1659-1662.

Hoegger PJ, Navarro-Gonzalez M, Kilaru S, Hoffmann M, Westbrook ED and Kues U (2004). The laccase gene family in Coprinopsis cinerea (Coprinus cinereus). Curr Genet 45:9-18.

Holliday R (1961). Induced mitotic crossing-over in Ustilago maydis. Genet Res 2:231-248.

Holliday R (1974). Ustilago maydis. In: Handbook of genetics, R. C. King, ed. New York: Plenum pp. 575-595.

Horton JS, Palmer GE and Shah A (2001). Mushroom development and cAMP signalling in Schizophyllum commune. Paper presented at: The genetics and cellular biology of basidiomycetes V. Mississauga, ON, Canada.

Horton JS, Palmer GE and Smith WJ (1999). Regulation of dikaryon-expressed genes by FRT1 in the basidiomycete Schizophyllum commune. Fungal Genet Biol 26:33-47.

Horton JS and Raper CA (1991). A mushroom-inducing DNA sequence isolated from the basidiomycete, Schizophyllum commune. Genetics 129:707-716.

Horton JS and Raper CA (1995). The mushroom-inducing gene Frt1 of Schizophyllum commune encodes a putative nucleotide-binding protein. Mol Gen Genet 247:358-366.

Hu G and Rijkenberg FHJ (1998 b). Scanning electron microscopy of early infection structure formation by Puccinia recondita f. sp. tritici on and in susceptible and resistant wheat lines. Mycological Res 102:391-399.

Hu GG and Rijkenberg FHJ (1998 a). Development of early infection structures of Puccinia recondita f. sp. tritici in nonhost cereal species. J Phytopathol 146:1-10.

Huber SM, Lottspeich F and Kamper J (2002). A gene that encodes a product with similarity to dioxygenases is highly expressed in teliospores of Ustilago maydis. Mol Genet Genomics 267:757-771.

Hull CM and Heitman J (2002). Genetics of cryptococcus neoformans. Annu Rev Genet 36:557-615.

Inada K, Morimoto Y, Arima T, Murata Y and Kamada T (2001). The clp1 gene of the mushroom Coprinus cinereus is essential for A-regulated sexual development. Genetics 157:133-140.

Ishibashi O and Shishido K (1993). Nucleotide sequence of a *ras* gene from the basidiomycete *Coprinus cinereus*. Gene 125:233-234.

Iwasa M, Tanabe S and Kamada T (1998). The two nuclei in the dikaryon of the homobasidiomycete *Coprinus cinereus* change position after conjugate division. Fungal Genet Biol 23:110-116.

Jacobs C, Mattichak SJ and Knowles JF (1994). Budding patterns during the cell cycle of the maize smut pathogen Ustilago maydis. Can J Bot 72:1675-1680.

Jansen G, Buhring F, Hollenberg CP and Rad MR (2001). Mutations in the SAM domain of STE50 differentially influence the MAPH-mediated pathways for mating, filamentous growth and osmotolerance in Saccharomyces cerevisiae. Mol Genet Genom 265:102-117.

Johnson T, Green GJ and Samborski DJ (1967). The world situation of the cereal rusts. Annu Rev Phytopathol 5:183-200.

Kaffarnik F, Muller P, Leibundgut M, Kahmann R and Feldbrugge M (2003). PKA and MAPK phosphorylation of Prf1 allows promoter discrimination in Ustilago maydis. Embo J 22:5817-5826.

Kahmann R, Basse C and Feldbrugge M (1999). Fungal-plant signalling in the Ustilago maydis-maize pathosystem. Curr Opin Microbiol 2:647-650.

Kahmann R and Kamper J (2004). Ustilago maydis: how its biology relates to pathogenic development. New Phytol 164:31-42.

Kamada T (2002). Molecular genetics of sexual development in the mushroom Coprinus cinereus. Bioessays 24:449-459.

Kamada T, Kurita R and Takemaru T (1978). Effects of light on basidiocarp maturation in *Coprinus macrorhizus*. Plant Cell Physiol 19:263-275.

Kamada T and Takemaru T (1977). Stipe elongation during basidiocarp maturation in *Coprinus macrorhizus*: mechanical properties of the stipe cell wall. Plant Cell Physiol 18:831-840.

Kamper J, Reichmann M, Romeis T, Bolker M and Kahmann R (1995). Multiallelic recognition: nonself-dependent dimerization of the bE and bW homeodomain proteins in Ustilago maydis. Cell 81:73-83.

Keller NP, Dischinger HCJ, Bhatnagar D, Cleveland TE and Ullah AH (1993). Purification of a 40-kilodalton methyltransferase active in the aflatoxin biosynthetic pathway. Appl Environ Microbiol 59:479-484.

Kinloch BB, and Dupper GE (2002). Genetic specificity in the white pine-blister rust pathosystem. Phytopathol 92:278-280.

Kinoshita H, Sen K, Iwama H, Samadder PP, Kurosawa S and Shibai H (2002). Effects of indole and caffeine on cAMP in the *ind1* and *cfn1* mutant strains of *Schizophyllum commune* during sexual development. FEMS Microbiol Lett 206:247-251.

Klose J, de Sa MM and Kronstad JW (2004). Lipid-induced filamentous growth in Ustilago maydis. Mol Microbiol 52:823-835.

Kolmer JA (1992). Effect of sexual recombination in two populations of the wheat leaf rust fungus Puccinia-Recondita. Can J Bot 70:359-363.

Kolmer JA (1996). Genetics of resistance to wheat leaf rust. Annu Rev Phytopathol 34:435-455.

Kothe E (1996). Tetrapolar fungal mating types: sexes by the thousands. FEMS Microbiol Rev 18:65-87.

Kozak KR, Foster LM and Ross IK (1995). Cloning and characterization of a G protein alpha-subunit-encoding gene from the basidiomycete, *Coprinus congregatus*. Gene 163:133-137.

Kronstad JW and Leong SA (1989). Isolation of two alleles of the b locus of Ustilago maydis. Proc Natl Acad Sci U S A 86:978-982.

Kronstad JW and Leong SA (1990). The b mating-type locus of Ustilago maydis contains variable and constant regions. Genes Dev 4:1384-1395.

Kruger J, Loubradou G, Regenfelder E, Hartmann A and Kahmann R (1998). Crosstalk between cAMP and pheromone signalling pathways in *Ustilago maydis*. Mol Gen Genet 260:193-198.

Kruger J, Loubradou G, Wanner G, Regenfelder E, Feldbrugge M and Kahmann R (2000). Activation of the cAMP pathway in *Ustilago maydis* reduces fungal proliferation and teliospore formation in plant tumors. Mol Plant Microbe Interact 13:1034-1040.

Kuchler F, Duffy M, Shrum RD and Dowler WM (1984). Potential economic consequences of the entry of an exotic fungal pest the case of soybean rust. Phytopathology 74:916-920.

414

Kues U (2000). Life history and developmental processes in the basidiomycete *Coprinus cinereus*. Microbiol Mol Biol Rev 64:316-353.

Kues U, Gottgens B, Stratmann R, Richardson WV, O'Shea SF and Casselton LA (1994). A chimeric homeodomain protein causes self-compatibility and constitutive sexual development in the mushroom *Coprinus cinereus*. Embo J 13:4054-4059.

Kues U, Granado JD, Hermann R, Boulianne RP, Kertesz-Chaloupkova K and Aebi M (1998). The *A* mating type and blue light regulate all known differentiation processes in the basidiomycete *Coprinus cinereus*. Mol Gen Genet 260:81-91.

Kues U, Kunzler M, Bottoli APF, Walser PJ, Granado JD, Liu Y, Bertossa RC, Ciardo D, Clergeot P-H, Loos S and Ruprich-Robert G (2004). Mushroom development in higher basidiomycetes; implications for human and animal health. In: Fungi in Human and Animal Health, R. K. S. Kushwaha, ed. Jodhpur, India.

Kues U and Liu Y (2000). Fruiting body production in basidiomycetes. Appl Microbiol Biotechnol 54:141-152.

Kues U, Walser PJ, Klaus MJ and Aebi M (2002). Influence of activated *A* and *B* mating-type pathways on developmental processes in the basidiomycete *Coprinus cinereus*. Mol Genet Genomics 268:262-271.

Laccetti L, Staples RC and Hoch HC (1987). Purification of Calmodulin from Bean Rust Uredospores. Exp Mycology 11:231-235.

Laity C, Giasson L, Campbell R and Kronstad J (1995). Heterozygosity at the b mating-type locus attenuates fusion in Ustilago maydis. Curr Genet 27:451-459.

Lalonde S, Boles E, Hellmann H, Barker L, Patrick JW, Frommer WB and Ward JM (1999). The dual function of sugar carriers. Transport and sugar sensing. Plant Cell 11:707-726.

Lamboy JS, Staples RC and Hoch HC (1995). Superoxide dismutase: A differentiation protein expressed in Uromyces germlings during early appressorium development. Exp Mycology 19:284-296.

Lawrence GL (1980). Multiple mating-type specificities in the flax rust Melampsora lini. Science 209:501-503.

Lee N, Bakkeren G, Wong K, Sherwood JE and Kronstad JW (1999). The mating-type and pathogenicity locus of the fungus Ustilago hordei spans a 500-kb region. Proc Natl Acad Sci U S A 96:15026-15031.

Lee N, De'Souza CA and Kronstad JW (2003). Of Smuts, Blasts, Mildews, and blights: cAMP signaling in phytopathogenic fungi. Annu Rev Phytopathol 41:399-427.

Lee N and Kronstad JW (2002). *ras2* controls morphogenesis, pheromone response, and pathogenicity in the fungal pathogen *Ustilago maydis*. Eukaryot Cell 1:954-966.

Lehmler C, Steinberg G, Snetselaar KM, Schliwa M, Kahmann R and Bolker M (1997). Identification of a motor protein required for filamentous growth in Ustilago maydis. Embo J 16:3464-3473.

Lengeler KB and Kothe E (1999). Mated: a putative peptide transporter of *Schizophyllum commune* expressed in dikaryons. Curr Genet 36:159-164.

Leslie JF and Leonard TJ (1979). Monokaryotic fruiting in *Schizophyllum commune*: Genetic control of the response to mechanical injury. Mol Gen Genet 175:5-12.

Li A, Altosaar I, Heath MC and Horgen PA (1993). Transient expression of the beta-glucuronidase gene delivered into urediniospores of Uromyces appendiculatus by particle bombardment. Can J Plant Pathol Rev Can Phytopathol 15:1-6.

Lichter A and Mills D (1998). Control of pigmentation of Ustilago hordei: the effect of pH, thiamine, and involvement of the cAMP cascade. Fungal Genet Biol 25:63-74.

Littlefield LJ and Heath MC (1979). Ultrastructure of rust fungi (New York, Academic Press).

Liu JQ and Kolmer JA (1998). Molecular and virulence diversity and linkage disequilibria in asexual and sexual populations of the wheat leaf rust fungus, Puccini recondita. Genome 41:832-840.

Liu L (2001) Fruiting body initiation in the basidiomycete *Coprinus cinereus*., ETH Zurich, Zurich, Switzerland.

Liu Y, Loos S, Aebi M and Kues U (2001). An essential gene for fruiting body initiation in the basidiomycete *Coprinus cinereus* is homologous to bacterial cyclopropane fatty acid synthesis genes. Paper presented at: 21st Fungal Biology Conference (Pacific Grove, CA, USA).

Liu Z, Szabo LJ and Bushnell WR (1993). Molecular cloning and analysis of abundant and stage-specific mRNAs from Puccinia graminis. Mol Plant Microbe Interact 6:84-91.

Long DL and Kolmer JA (1989). A North American system of nomenclature for Puccinia recondita f. sp. tritici. Phytopathology 79:525-529.

Lu BC (1974). Meiosis in Coprinus. V. The role of light on basidiocarp initiation, meiosis, and hymenium differentiation in Coprinus lagopus. Can J Bot 52:299-305.

Lugones LG, Wosten HAB, Birkenkamp KU, Sjollema KA, Zager J and Wessels JGH (1999). Hydrophobins line air channels in fruiting bodies of Schizophyllum commune and Agaricus bisporus. Mycological Res 103:635-640.

Luttrell ES (1987). Relations of hyphae to host cells in smut galls caused by bpecies of Tilletia, Tolyposporium, and Ustilago. Can J Bot-Revue Canadienne De Botanique 65:2581-2591.

Macko V (1981). Inhibitors and stimulants of spore germination and infection structures in the wheat stem rust. In The fungal spore: Morphogenic controls., G. Turian, and H. R. Hohl, eds. (New York, Academic Press), pp. 565-584.

Macko V, Renwick JAA and Rissler JF (1978). Acrolein induces differentiation of infection structures in the wheat stem rust fungus. Science 199:442-443.

Madhani HD, Styles CA and Fink GR (1997). MAP kinases with distinct inhibitory functions impart signaling specificity during yeast differentiation. Cell 91:673-684.

Madi L, McBride SA, Bailey LA and Ebbole DJ (1997). rco-3, a gene involved in glucose transport and conidiation in Neurospora crassa. Genetics 146:499-508.

Magae Y (1999). Saponin stimulates fruiting of the edible basidiomycete Pleurotus ostreatus. Biosci Biotechnol Biochem 63:1840-1842.

Martinez-Espinoza AD, Garcia-Pedrajas MD and Gold SE (2002). The Ustilaginales as plant pests and model systems. Fungal Genet Biol 35:1-20.

Martinez-Espinoza AD, Ruiz-Herrera J, Leon-Ramirez CG and Gold SE (2004). MAP kinase and cAMP signaling pathways modulate the pH-induced yeast-to-mycelium dimorphic transition in the corn smut fungus Ustilago maydis. Curr Microbiol 49:274-281.

Matthews TR and Niederpruem DJ (1972). Differentiation in Coprinus lagopus. I. Control of fruiting and cytology in initial events. Arch Mikrobiol 87:257-268.

Matthews TR and Niederpruem DJ (1973). Differentiation in Coprinus lagopus. II. Histology and ultrastructural aspects of developing primordia. Arch Mikrobiol 88:169-180.

Mayorga ME and Gold SE (1998). Characterization and molecular genetic complementation of mutants affecting dimorphism in the fungus ustilago maydis. Fungal Genet Biol 24:364-376.

Mayorga ME and Gold SE (1999). A MAP kinase encoded by the ubc3 gene of Ustilago maydis is required for filamentous growth and full virulence. Mol Microbiol 34:485-497.

Mayorga ME and Gold SE (2001). The ubc2 gene of Ustilago maydis encodes a putative novel adaptor protein required for filamentous growth, pheromone response and virulence. Mol Microbiol 41:1365-1379.

Mendgen K (1984). Development and physiology of teliospores. In: The cereal rusts, W. R. Bushnell, and A. P. Roelfs, eds. New York: Academic Press, Inc. pp. 375-398.

Mendgen K, Hahn M and Deising H (1996). Morphogenesis and mechanisms of penetration by plant pathogenic fungi. Annu Rev Phytopathol 34:367-386.

Mendgen K, Struck C, Voegele RT and Hahn M (2000). Biotrophy and rust haustoria. Physiol Mol Plant Pathol 56:141-145.

Milner MJ (1999). HMG domain proteins involved in mating and sexual development of Coprinus cinereus. Paper presented at: 20th Fungal Biology Conference (Pacific Grove, CA, USA).

Mims CW and Richardson EA (1989). Ultrastructure of appressorium development by basidiospore germlings of the rust fungus Gymnosporangium juniperi virginianae. Protoplasma 148:111-119.

Molloy S, Burton KS and Sreenivasaprasad S (2001). Characterisation of a sugar transporter gene associated with Agaricus bisporus morphogenesis. Paper presented at: The genetics and cellular biology of basidiomycetes V (Mississauga, ON, Canada).

Moore D (1981). Developmental genetics of *Coprinus cinereus*: genetic evidence that carpophores and sclerotia share a common pathway of initiation. Curr Genet 3:145-150.

Moore D, Elhiti MMY and Butler RD (1979). Morphogenesis of the carpophore of *Coprinus cinereus*. New Phytol 66:377-382.

Moriya H and Johnston M (2004). Glucose sensing and signaling in *Saccharomyces cerevisiae* through the Rgt2 glucose sensor and casein kinase I. Proc Natl Acad Sci U S A 101:1572-1577.

Moser MJ, Holley WR, Chatterjee A and Mian IS (1997). The proofreading domain of *Escherichia coli* DNA polymerase I and other DNA and/or RNA exonuclease domains. Nucleic Acids Res 25:5110-5118.

Mould MJR and Heath MC (1999). Ultrastructural evidence of differential changes in transcription, translation, and cortical microtubules during in planta penetration of cells resistance or susceptible to rust infection. Physiol Mol Plant Pathol 55:225-236.

Mueller UG (2002). Ant versus fungus versus mutualism: ant-cultivar conflict and the deconstruction of the attine ant-fungus symbiosis. American naturalist 160:S67-S98.

Mueller UG, Rehner SA and Schultz TR (1998). The evolution of agriculture in ants. Science 281:2034-2038.

Mulder GH and Wessels JGH (1986). Molecular cloning of RNAs differentially expressed in monokaryons and dikaryons of *Schizophyllum commune*. Exp Mycol 10:214-227.

Muller P, Aichinger C, Feldbrugge M and Kahmann R (1999). The MAP kinase kpp2 regulates mating and pathogenic development in Ustilago maydis. Mol Microbiol 34:1007-1017.

Muller P, Katzenberger JD, Loubradou G and Kahmann R (2003 a). Guanyl nucleotide exchange factor Sql2 and Ras2 regulate filamentous growth in *Ustilago maydis*. Eukaryot Cell 2:609-617.

Muller P, Weinzierl G, Brachmann A, Feldbrugge M and Kahmann R (2003 b). Mating and pathogenic development of the Smut fungus *Ustilago maydis* are regulated by one mitogen-activated protein kinase cascade. Eukaryot Cell 2:1187-1199.

Munch-Garthoff S, Neuhaus JM, Boller T, Kemmerling B and Kogel KH (1997). Expression of beta-1,3-glucanase and chitinase in healthy, stem-rust-affected and elicitor-treated near-isogenic wheat lines showing Sr5-or Sr24-specified race-specific rust resistance. Planta 201:235-244.

Munoz-Rivas A, Specht CA, Drummond BJ, Froeliger E, Novotny CP and Ullrich RC (1986). Transformation of the basidiomycete, *Schizophyllum commune*. Mol Gen Genet 205:103-106.

Muraguchi H and Kamada T (1998). The *ich1* gene of the mushroom *Coprinus cinereus* is essential for pileus formation in fruiting. Development 125:3133-3141.

Muraguchi H and Kamada T (2000). A mutation in the *eln2* gene encoding a cytochrome P450 of *Coprinus cinereus* affects mushroom morphogenesis. Fungal Genet Biol 29:49-59.

Muraguchi H, Takemaru T and Kamada T (1999). Isolation and characteriztion of developmental mutants in fruiting using a homokaryotic fruiting strain of *Coprinus cinereus*. Mycoscience 40:227-235.

Murata Y, Fujii M, Zolan ME and Kamada T (1998). Molecular analysis of *pcc1*, a gene that leads to A-regulated sexual morphogenesis in *Coprinus cinereus*. Genetics 149:1753-1761.

Nagarajan S and Singh DV (1990). Long-distance dispersion of rust pathogens. Annu Rev Phytopathol 28:139-153.

Narisawa K, Yamaoka Y and Katsuya K (1994). Mating type of isolates derived from the spermogonial state of Puccinia coronata var. coronata. Mycoscience 35(2): 131-135.

Neu C, Keller B and Feuillet C (2003). Cytological and molecular analysis of the Hordeum vulgare-Puccinia triticina nonhost interaction. Mol Plant Microbe Interact 16:626-633.

Niederpruem DJ, Jersild RA and Lane PL (1971). Direct microscopic studies of clamp connection formation in growing hyphae of *Schizophyllum commune*. I. The dikaryon. Arch Mikrobiol 78:268-280.

Niks RE (1989). Induced accessibility and inaccessibility of barley cells in seedling leaves inoculated with two leaf rust species. J Phytopathol 124:296-308.

Niks RE and Dekens RG (1991). Prehaustorial and posthaustorial resistance to wheat leaf rust in diploid wheat seedlings. Phytopathology 81:847-851.

Nugent KG, Choffe K and Saville BJ (2004). Gene expression during Ustilago maydis diploid filamentous growth: EST library creation and analyses. Fungal Genet Biol 41:349-360.

417

O'Donell KL and McLaughlin DJ (1984). Postmeitotic mitosis, basidiospore development, and septation in Ustilago maydis. Mycologia 76:486-502.

O'Farrell PH (2001). Triggering the all-or-nothing switch into mitosis. Trends Cell Biol 11:512-519.

Oita S and Yanagi SO (1993). Stimulation of *Schizophyllum commune* fruit body formation by inhibitor of membrane function and cell wall synthesis. Biosci Biotechnol Biochem 57:1270-1274.

Oliver RP and Ipcho SVS (2004). Arabidopsis pathology breathes new life into the necrotroph-vs.-biotroph classification of fungal pathogens. Mol Plant Pathol 5:347-352.

Ozcan, S, and Johnston, M (1999). Function and regulation of yeast hexose transporters. Microbiol Mol Biol Rev 63:554-569.

Pardo EH, Oshea SF and Casselton LA (1996). Multiple versions of the a mating type locus of Coprinus cinereus are generated by three paralogous pairs of multiallelic homeobox genes. Genetics 144:87-94.

Perkins JH (1969). Morphogenesis in *Schizophyllum commune*. I. Effects of white light. Plant Physiol 44:1706-1711.

Perkins JH and Gordon SA (1969). Morphogenesis.in *Schizophyllum commune*. II. Effects of monochromatic light. Plant Physiol 44:1712-1716.

Polak E, Hermann R, Kues U and Aebi M (1997). Asexual sporulation in Coprinus cinereus: Structure and development of oidiophores and oidia in an Amut Bmut homokaryon. Fungal Genet Biol 22:112-126.

Pryor A, Boelen MG, Dickinson MJ and Lawrence GJ (1990). Widespread incidence of double-stranded RNA of unknown function in rust fungi. Can J Bot 68:669-676.

Puhalla JE (1970). Genetic studies on the *b* incompatibility locus of *Ustilago maydis*. Genet Res Camb 16:229-232.

Pukkila PJ (1993). Methods of genetic manipulation in *Coprinus cinereus*. In: Genetics and breeding of edible mushrooms. S.-T. Chang, J. A. Buswell, and P. G. Miles, eds. Y-Parc, Switzerland: Gordon and Breach Science Publishers. pp. 249-264.

Quadbeck-Seeger C, Wanner G, Huber S, Kahmann R and Kamper J (2000). A protein with similarity to the human retinoblastoma binding protein 2 acts specifically as a repressor for genes regulated by the b mating type locus in Ustilago maydis. Mol Microbiol 38:154-166.

Raper CA (1983). Controls for development and differentiation in the dikaryon in Basidiomycetes. In: Secondary Metabolism and Differentiation in Fungi., J. Bennett, and A. Ciegler, eds. New York, NY: Marcel Dekker, Inc. pp. 195-238.

Raper CA, Raper JR and Miller RE (1972). Genetic analysis of the life cycle of *Agaricus bisporus*. Mycologia 64:1088-1117.

Raper JR (1966). Genetics of Sexuality in Higher Fungi. (New York, NY, USA, The Ronald Press).

Raper JR, Boyd DH and Raper CA (1965). Primary and secondary mutations at the incompatibility loci in *Schizophyllum*. Proc Natl Acad Sci U S A 53:1324-1332.

Raudaskoski M and Vauras R (1982). Scanning electron microscope study of fruit body differentiation in *Schizophyllum commune*. Trans Br Mycol Soc 78:475-481.

Raudaskoski M and Viitanen H (1982). Effects of aeration and light on fruit body induction in *Schizophyllum commune*. Trans Br Mycol Soc 78:89-96.

Rauscher M, Mendgen K and Deising H (1995). Extracellular proteases of the rust fungus Uromyces viciae-fabae. Exp Mycol 19:26-34.

Regenfelder E, Spellig T, Hartmann A, Lauenstein S, Bolker M and Kahmann R (1997). G proteins in Ustilago maydis: transmission of multiple signals? Embo J 16:1934-1942.

Reichmann M, Jamnischek A, Weinzierl G, Ladendorf O, Huber S, Kahmann R and Kamper J (2002). The histone deacetylase Hda1 from Ustilago maydis is essential for teliospore development. Mol Microbiol 46:1169-1182.

Roelfs AP and Bushnell WR, eds. (1985). The Cereal Rusts Vol. II: Diseases, distribution, epidemiology and control. Orlando, FL: Academic Press.

Romeis T, Brachmann A, Kahmann R and Kamper J (2000). Identification of a target gene for the bE-bW homeodomain protein complex in Ustilago maydis. Mol Microbiol 37:54-66.

Rowell JB (1955). Functional role of compatibility factors and an in vitro test for sexual compatibility of haploid lines of Ustilago zea. Phytopathology 45:370-374.

Rowell JB and Devay JE (1954). Genetics of Ustilago-Zeae in Relation to Basic Problems of Its Pathogenicity. Phytopathology 44:356-362.

Runeberg P, Raudaskoski M and Virtanen I (1986). Cytoskeletal elements in the hyphae of the homobasidiomycete *Schizophyllum commune* visualized with indirect immunofluorescence and NBD-phallicidin. Eur J Cell Biol 41:25-32.

Sacadura NT and Saville BJ (2003). Gene expression and EST analyses of Ustilago maydis germinating teliospores. Fungal Genet Biol 40:47-64.

Salo V, Niini SS, Virtanen I and Raudaskoski M (1989). Comparative immunocytochemistry of the cytoskeleton in filamentous fungi with dikaryotic and multinucleate hyphae. J Cell Sci 94:11-24.

Samborski DJ (1985). Wheat leaf rust. In The Cereal Rusts Vol. II: Diseases, distribution, epidemiology and control., A. P. Roelfs, and W. R. Bushnell, eds. Orlando, Fl.: Academic Press. pp. 39-59.

Sanchez-Martinez C and Perez-Martin J (2001). Dimorphism in fungal pathogens: Candida albicans and Ustilago maydis--similar inputs, different outputs. Curr Opin Microbiol 4:214-221.

Saupe S, Turcq B and Begueret J (1995). A gene responsible for vegetative incompatibility in the fungus *Podospora anserina* encodes a protein with a GTP-binding motif and a Gb homologous domain. Gene 162:135-139.

Saverimuttu N and Deverall BJ (1998). A cytological assay reveals pathotype and resistance gene specific elicitors in leaf rust infections of wheat. Physiol Mol Plant Pathol 52:25-34.

Savile DBO (1976). Evolution of the rust fungi (Uredinales) as reflected by their ecological problems. In Evolutionary Biology, Vol. 9., M. K. Hecht, W. C. Steere, and B. Wallace, eds. New York, NY: Plenum. pp. 137-207.

Savile DBO (1990). Coevolution of Uredinales and Ustilaginales with Vascular Plants. Reports of the Tottori Mycological Institute 28:15-24.

Schauwecker F, Wanner G and Kahmann R (1995). Filament-specific expression of a cellulase gene in the dimorphic fungus Ustilago maydis. Biol Chem Hoppe Seyler 376:617-625.

Schillberg S, Tiburzy R and Fischer R (2000). Transient transformation of the rust fungus Puccinia graminis f. sp. tritici. Mol Gen Genet 262:911-915.

Schulz B, Banuett F, Dahl M, Schlesinger R, Schafer W, Martin T, Herskowitz I, and Kahmann R (1990). The b alleles of U. maydis, whose combinations program pathogenic development, code for polypeptides containing a homeodomain-related motif [published erratum appears in Cell 1990 Feb 9;60(3):following 520]. Cell 60:295-306.

Schuren FH and Wessels JG (1990). Two genes specifically expressed in fruiting dikaryons of *Schizophyllum commune*: homologies with a gene not regulated by mating-type genes. Gene 90:199-205.

Sia RA, Lengeler KB and Heitman J (2000). Diploid strains of the pathogenic basidiomycete Cryptococcus neoformans are thermally dimorphic. Fungal Genet Biol 29:153-163.

Sipiczki M (1988). The role of sterility genes (ste and aff) in the initiation of sexual development in Schizosaccharomyces pombe. Mol Gen Genet 213:529-534.

Skalamera D, Jibodh S and Heath MC (1997). Callose deposition during the interaction between cowpea (Vigna unguiculata) and the monokaryotic stage of the cowpea rust fungus (Uromyces vignae). New Phytologist 136:511-524.

Smith DG, Garcia-Pedrajas MD, Gold SE and Perlin MH (2003). Isolation and characterization from pathogenic fungi of genes encoding ammonium permeases and their roles in dimorphism. Mol Microbiol 50:259-275.

Snetselaar KM (1993). Microscopic observation of Ustilago-maydis mating interactions. Exp Mycology 17:345-355.

Snetselaar KM, Bolker M and Kahmann R (1996). Ustilago maydis mating hyphae orient their growth toward pheromone sources. Fungal Genet Biol 20:299-312.

Snetselaar KM and McCann MP (1997). Using microdensitometry to correlate cell morphology with the nuclear cycle in Ustilago maydis. Mycologia 89:689-697.

Snetselaar KM and Mims CW (1992). Sporidial fusion and infection of maize seedlings by the smut fungus Ustilago-Maydis. Mycologia 84:193-203.

Snetselaar KM and Mims CW (1993). Infection of maize stigmas by Ustilago maydis: Light and electron microscopy. Phytopathology 83:843-850.

Snetselaar KM and Mims CW (1994). Light and electron-microscopy of Ustilago-maydis hyphae in maize. Mycological Res 98:347-355.

Sohn J, Voegele RT, Mendgen K and Hahn M (2000). High level activation of vitamin B1 biosynthesis genes in haustoria of the rust fungus Uromyces fabae. Mol Plant Microbe Interact 13:629 636.

Spellig T, Bolker M, Lottspeich F, Frank RW and Kahmann R (1994). Pheromones trigger filamentous growth in Ustilago maydis. Embo J 13:1620-1627.

Stakman EC and Christensen JJ (1927). Heterothallism in Ustilago zeae. Phytopathology 17:827-834.

Stankis MM, Specht CA and Giasson L (1990). Sexual incompatibility in *Schizophyllum commune*: from classical genetics to a molecular view. In: Seminars in Developmental Biology, C. A. Raper, and D. I. Johnson, eds. Philadelphia, PA: Saunders Scientific Publishers. pp. 195-206.

Staples, RC (2000). Research on the Rust Fungi During the Twentieth Century. Annu Rev Phytopathol 38:49-69.

Staples RC (2001). Nutrients for a rust fungus: the role of haustoria. Trends Plant Sci 6:496-498.

Staples RC and Wynn WK (1965). The physiology of uredospores of the rust fungi. Bot Rev 31:537-564.

Stavely JR (1984). Pathogenic specialization in Uromyces-phaseoli in the USA and rust resistance in beans. Plant Disease 68:95-99.

Stavely JR, Steadman JR and McMillan RT (1989). New pathogenic variability in Uromyces-appendiculatus in North America. Plant Disease 73:428-432.

Steinberg G and Fuchs U (2004). The role of microtubules in cellular organization and endocytosis in the plant pathogen Ustilago maydis. J Microsc 214:114-123.

Steinberg G, Schliwa M, Lehmler C, Bolker M, Kahmann R and McIntosh JR (1998). Kinesin from the plant pathogenic fungus Ustilago maydis is involved in vacuole formation and cytoplasmic migration. J Cell Sci 111 (Pt 15):2235-2246.

Steinberg G, Wedlich-Soldner R, Brill M and Schulz I (2001). Microtubules in the fungal pathogen Ustilago maydis are highly dynamic and determine cell polarity. J Cell Sci 114:609-622.

Straube A, Brill M, Oakley BR, Horio T and Steinberg G (2003). Microtubule organization requires cell cycle-dependent nucleation at dispersed cytoplasmic sites: polar and perinuclear microtubule organizing centers in the plant pathogen Ustilago maydis. Mol Biol Cell 14:642-657.

Straube A, Enard W, Berner A, Wedlich-Soldner R, Kahmann R and Steinberg G (2001). A split motor domain in a cytoplasmic dynein. Embo J 20:5091-5100.

Struck C, Ernst M and Hahn M (2002). Characterization of a developmentally regulated amino acid transporter (AAT1p) of the rust fungus Uromyces fabae. Mol Plant Pathol 3:23-30.

Struck C, Hahn M and Mendgen K (1996). Plasma membrane H+-ATPase activity in spores, germ tubes, and haustoria of the rust fungus Uromyces viciae-fabae. Fungal Genet Biol 20:30-35.

Struck C, Mueller E, Martin H and Lohaus G (2004 a). The Uromyces fabae UfAAT3 gene encodes a general amino acid permease that prefers uptake of in planta scarce amino acids. Mol Plant Pathol 5:183-189.

Struck C, Siebels C, Rommel O, Wernitz M and Hahn M (1998). The plasma membrane H(+)-ATPase from the biotrophic rust fungus Uromyces fabae: molecular characterization of the gene (PMA1) and functional expression of the enzyme in yeast. Mol Plant Microbe Interact 11:458-465.

Struck C, Voegele RT, Hahn M and Mendgen K (2004 b). Rust haustoria as sinks in plant tissues or - how to survive in leaves. In Biology of Plant-Microbe Interactions - Proceedings of the 11th International congress on molecular plant-microbe interactions., I. Tikhonovich, B. Lugtenberg, and N. Provorov, eds. (St. Paul, MN, Int. Soc. for Mol. Plant-Microbe Int.), pp. 177-179.

Stumpf MA, Leinhos GME, Staples RC and Hoch HC (1991). The effect of pH and potassium on appressorium formation by Uromyces appendiculatus urediospore germlings. Exp Mycol 15:356-360.

Su JY and Maller JL (1995). Cloning and expression of a *Xenopus* gene that prevents mitotic catastrophe in fission yeast. Mol Gen Genet 246:387-396.

Sutherland MW, Deverall BJ, Moerschbacher BM and Reisener HJ (1989). Wheat cultivar and chromosomal selectivity of two types of eliciting preparations from rust pathogens. Physiol Mol Plant Pathol 35:535-542.

Swamy S, Uno I and Ishikawa Y (1984). Morphogenetic effects of mutations at the A and B incompatibility factors of Coprinus cinereus. J Gen Microbiol 130:3219-3224.

Szabo LJ and Bushnell WR (2001). Hidden robbers: the role of fungal haustoria in parasitism of plants. Proc Natl Acad Sci USA 98:7654-7655.

Tada Y, Hata S, Takata Y, Nakayashiki H, Tosa Y and Mayama S (2001). Induction and signaling of an apoptotic response typified by DNA laddering in the defense response of oats to infection and elicitors. Mol Plant Microbe Interact 14:477-486.

Takai Y, Sasaki T and Matozaki T (2001). Small GTP-binding proteins. Physiol Rev 81:153-208.

Tanabe S and Kamada T (1994). The role of astral microtubules in conjugate division in the dikaryon of Coprinus cinereus. Exp Mycology 18:338-348.

Taylor TN, Remy W and Hass H (1994). Allomyces in the Devonian. Nature 367:601-601.

Thara, VK, Fellers, JP, and Zhou, JM (2003). In planta induced genes of Puccinia triticina. Mol Plant Pathol 4:51-56.

Tian D, Traw MB, Chen JQ, Kreitman M and Bergelson J (2003). Fitness costs of R-gene-mediated resistance in Arabidopsis thaliana. Nature 423:74-77.

Tiburzy R, Martins EMF and Reisener HJ (1992). Isolation of haustoria of Puccinia graminis f. sp. tritici from wheat leaves. Exp Mycology 16:324-328.

Torreblanca J, Stumpferl S and Basse CW (2003). Histone deacetylase Hda1 acts as repressor of the Ustilago maydis biotrophic marker gene mig1. Fungal Genet Biol 38:22-32.

Trail F, Mahanti N and Linz J (1995). Molecular biology of aflatoxin biosynthesis. Microbiology 141 (Pt 4):755-765.

Trueheart J and Herskowitz I (1992). The a locus governs cytoduction in Ustilago maydis. J Bacteriol 174:7831-7833.

Uno I and Ishikawa T (1973). Metabolism of adenosine 3',5'-cyclic monophosphate and induction of fruiting bodies in Coprinus macrorhizus. J Bacteriol 113:1249-1255.

Uno I and Ishikawa T (1974). Effect of glucose on the fruiting body formation and adenosine 3',5'-cyclic monophosphate levels in Coprinus macrorhizus. J Bacteriol 120:96-100.

Uno I and Ishikawa Y (1971). Chemical and genetical control of induction of monokaryotic fruiting bodies in Coprinus macrorhizus. Mol Gen Genet 113:228-239.

Uno I, Yamaguchi M and Ishikawa T (1974). The effect of light on fruiting body formation and adenosine 3':5'-cyclic monophosphate metabolism in Coprinus macrorhizus. Proc Natl Acad Sci U S A 71:479-483.

Urban M, Kahmann R and Bolker M (1996 a). Identification of the pheromone response element in Ustilago maydis. Mol Gen Genet 251:31-37.

Urban M, Kahmann R and Bolker M (1996 b). The biallelic a mating type locus of Ustilago maydis: remnants of an additional pheromone gene indicate evolution from a multiallelic ancestor. Mol Gen Genet 250:414-420.

Valdivieso MH, Sugimoto K, Jahng KY, Fernandes PM and Wittenberg C (1993). FAR1 is required for posttranscriptional regulation of CLN2 gene expression in response to mating pheromone. Mol Cell Biol 13:1013-1022.

van der Valk P and Marchant R (1978). Hyphal ultrastructure in fruiting body primordia of the basidiomycetes Schizophyllum commune and Coprinus cinereus. Protoplasma 95:57-72.

van Wetter MA, Wosten HA and Wessels JG (2000). SC3 and SC4 hydrophobins have distinct roles in formation of aerial structures in dikaryons of Schizophyllum commune. Mol Microbiol 36:201-210.

Verrinder-Gibbins AM and Lu BC (1984). Induction of normal fruiting on originally monokaryotic cultures of Coprinus cinereus. Trans Br Mycol Soc 82:331-335.

Versele M, Lemaire K and Thevelein JM (2001). Sex and sugar in yeast: two distinct GPCR systems. EMBO Rep 2:574-579.

Voegele RT and Mendgen K (2003). Rust haustoria: Nutrient uptake and beyond. New Phytol 159:93-100.

Voegele RT, Struck C, Hahn M and Mendgen K (2001). The role of haustoria in sugar supply during infection of broad bean by the rust fungus Uromyces fabae. Proc Natl Acad Sci U S A 98:8133-8138.

Walser PJ, Velagapudi R, Aebi M and Kues U (2003). Extracellular matrix proteins in mushroom development. Recent Res Devel Microbiology 7:381-415.

Wang HX, Ng TB and Ooi VEC (1998). Lectins from mushrooms. Mycological Res 102:897-906.

Wanner R, Forster H, Mendgen K and Staples RC (1985). Synthesis of differentiation-specific proteins in germlings of the wheat stem rust fungus after heat shock. Exp Mycology 9:279-283.

Warren JM and Covert SF (2004). Differential expression of pine and Cronartium quercuum f. sp. fusiforme genes in fusiform rust galls. Appl Environ Microbiol 70:441-451.

Waugh MS, Nichols CB, DeCesare CM, Cox GM, Heitman J and Alspaugh JA (2002). *Ras1* and *Ras2* contribute shared and unique roles in physiology and virulence of *Cryptococcus neoformans*. Microbiology 148:191-201.

Weber I, Gruber C and Steinberg G (2003). A class-V myosin required for mating, hyphal growth, and pathogenicity in the dimorphic plant pathogen Ustilago maydis. Plant Cell 15:2826-2842.

Wedlich-Soldner R, Bolker M, Kahmann R and Steinberg G (2000). A putative endosomal t-SNARE links exo- and endocytosis in the phytopathogenic fungus Ustilago maydis. Embo J 19:1974-1986.

Wedlich-Soldner R, Schulz I, Straube A and Steinberg G (2002 a). Dynein supports motility of endoplasmic reticulum in the fungus Ustilago maydis. Mol Biol Cell 13:965-977.

Wedlich-Soldner R, Straube A, Friedrich MW and Steinberg G (2002 b). A balance of KIF1A-like kinesin and dynein organizes early endosomes in the fungus Ustilago maydis. Embo J 21:2946-2957.

Weinzierl G, Leveleki L, Hassel A, Kost G, Wanner G and Bolker M (2002). Regulation of cell separation in the dimorphic fungus Ustilago maydis. Mol Microbiol 45:219-231.

Wessels JGH (1992). Gene expression during fruiting in *Schizophyllum commune*. Mycol Res 96:609-620.

Wessels JGH (1996). Fungal hydrophobins: proteins that function at an interface. Trends Plant Sci 1:9-15.

Wessels JGH, Asgiersdottir SA, Birkenkamp KU, de Vries OMH, Lugones LG, Scheer JMJ, Schuren FHJ, Schuurs TA, van Wetter MA and Wosten HAB (1995). Genetic regulation of emergent growth in *Schizophyllum commune*. Can J Bot 73:S273-S281.

Wichmann G and Bergelson J (2004). Effector genes of Xanthamonas axonopodis pv. vesicatoria promote transmission and enhance other fitness traits in the field. Genetics 166:693-706.

Wietholter N, Horn S, Reisige K, Beike U and Moerschbacher BM (2003). In vitro differentiation of haustorial mother cells of the wheat stem rust fungus, Puccinia graminis f. sp. tritici, triggered by the synergistic action of chemical and physical signals. Fungal Genet Biol 38:320-326.

Williams PG (1984). Obligate parasitism and axenic culture. In The cereal rusts. Vol. I. Origins, specificity, structure and physiology., W. R. Bushnell, and A. P. Roelfs, eds. (Orlando, FL, Academic Press, Inc.), pp. 399-430.

Wirsel SG, Voegele RT and Mendgen KW (2001). Differential regulation of gene expression in the obligate biotrophic interaction of Uromyces fabae with its host Vicia faba. Mol Plant Microbe Interact 14:1319-1326.

Wosten HA (2001). Hydrophobins: multipurpose proteins. Annu Rev Microbiol 55:625-646.

Wosten HA, Bohlmann R, Eckerskorn C, Lottspeich F, Bolker M and Kahmann R (1996). A novel class of small amphipathic peptides affect aerial hyphal growth and surface hydrophobicity in Ustilago maydis. Embo J 15:4274-4281.

Wosten HA, van Wetter MA, Lugones LG, van der Mei HC, Busscher HJ and Wessels JG (1999). How a fungus escapes the water to grow into the air. Curr Biol 9:85-88.

Xiang X and Plamann M (2003). Cytoskeleton and motor proteins in filamentous fungi. Curr Opin Microbiol 6:628-633.

Xu H and Mendgen K (1997). Targeted cell wall degradation at the penetration site of cowpea rust basidiosporelings. Mol plant microb interact 10:87-94.

Xu JR (2000). MAP kinases in fungal pathogens. Fungal Genet Biol 31:137-152.

Xu JR and Hamer JE (1996). MAP kinase and cAMP signaling regulate infection structure formation and pathogenic growth in the rice blast fungus Magnaporthe grisea. Genes Dev 10:2696-2706.

Xu JR, Staiger CJ and Hamer JE (1998). Inactivation of the mitogen-activated protein kinase Mps1 from the rice blast fungus prevents penetration of host cells but allows activation of plant defense responses. Proceedings of the National Academy of Sciences of the United States of America 95:12713-12718.

Xuei X, Bhairi S, Staples RC and Yoder OC (1992). Characterization of INF56, a gene expressed during infection structure development of Uromyces appendiculatus. Gene 110:49-55.

Yamagishi K, Kimura T, Suzuki M and Shinmoto H (2002). Suppression of fruit-body formation by constitutively active G-protein alpha-subunits *ScGP-A* and *ScGP-C* in the homobasidiomycete *Schizophyllum commune*. Microbiology 148:2797-2809.

Yamagishi K, Kimura T, Suzuki M, Shinmoto H and Yamaki KJ (2004). Elevation of intracellular cAMP levels by dominant active heterotrimeric G protein alpha subunits *ScGP-A* and *ScGP-C* in homobasidiomycete, *Schizophyllum commune*. Biosci Biotechnol Biochem 68:1017-1026.

Yamazaki S and Katsuya Y (1988). Mating type of pine rust fungus, Cronartium-quercuum. Proceedings of the Japan Academy Series B-physical and biological sciences 64:197-200.

Yli-Mattila TMH, Ruiters J, Wessels JGH and Raudaskoski M (1989). Effect of inbreeding and light on monokaryotic and dikaryotic fruiting in the homobasidiomycete *Schizophyllum commune*. Mycological Res 93:535-542.

Yu X, Ekramoddoullah AK, Taylor DW and Piggott N (2002). Cloning and characterization of a cDNA of cro rI from the white pine blister rust fungus Cronartium ribicola. Fungal Genet Biol 35:53-66.

Yuki K, Akiyama M, Muraguchi H and Kamada T (2003). The *dst1* gene responsible for a photomorphogenetic mutation in *Coprinus cinereus* encodes a protein with high similarity to WC-1. Paper presented at: 22nd Fungal Biology Conference (Pacific Grove, CA, USA).

Zambino PJ, Kubelik AR and Szabo LJ (2000). Gene action and linkage of avirulence genes to DNA markers in the rust fungus Puccinia graminis. Phytopathology 90:819-826.

Zelter A, Bencina M, Bowman BJ, Yarden O and Read ND (2004). A comparative genomic analysis of the calcium signaling machinery in Neurospora crassa, Magnaporthe grisea, and Saccharomyces cerevisiae. Fungal Genetics and Biology 41:827-841.

Zhang L, Meakin H, and Dickinson M (2003). Isolation of genes expressed during compatible interactions between leaf rust (Puccinia triticina) and wheat using cDNA-AFLP. Mol Plant Pathol 4:469-477.

Zhang R, Dickinson MJ and Pryor A (1994). Double-stranded RNAs in the rust fungi. Annu rev phytopathol 32:115-133.

Zhao J and Kwan HS (1999). Characterization, molecular cloning, and differential expression analysis of laccase genes from the edible mushroom *Lentinus edodes*. Appl Environ Microbiol 65:4908-4913.

Keyword Index

Printed and bound by CPI Group (UK) Ltd, Croydon, CR0 4YY

08/05/2025

01865007-0005